Lecture Notes in Computer Science 6203

Commenced Publication in 1973
Founding and Former Series Editors:
Gerhard Goos, Juris Hartmanis, and Jan van Leeuwen

Shlomo Geva   Jaap Kamps
Andrew Trotman (Eds.)

Focused Retrieval and Evaluation

8th International Workshop of the Initiative
for the Evaluation of XML Retrieval, INEX 2009
Brisbane, Australia, December 7-9, 2009
Revised and Selected Papers

 Springer

Volume Editors

Shlomo Geva
Queensland University of Technology, Faculty of Science and Technology
GPO Box 2434, Brisbane Qld 4001, Australia
E-mail: s.geva@qut.edu.au

Jaap Kamps
University of Amsterdam, Archives and Information Studies/Humanities
Turfdraagsterpad 9, 1012 XT Amsterdam, The Netherlands
E-mail: kamps@uva.nl

Andrew Trotman
University of Otago, Department of Computer Science
P.O. Box 56, Dunedin 9054, New Zealand
E-mail: andrew@cs.otago.ac.nz

Library of Congress Control Number: 2010930655

CR Subject Classification (1998): H.3, H.3.3, H.3.4, H.2.8, H.2.3, H.2.4, E.1

LNCS Sublibrary: SL 3 – Information Systems and Application, incl. Internet/Web
and HCI

ISSN   0302-9743
ISBN-10   3-642-14555-8 Springer Berlin Heidelberg New York
ISBN-13   978-3-642-14555-1 Springer Berlin Heidelberg New York

springer.com

© Springer-Verlag Berlin Heidelberg 2010
Printed in Germany

Typesetting: Camera-ready by author, data conversion by Scientific Publishing Services, Chennai, India
Printed on acid-free paper   06/3180

Preface

Welcome to the proceedings of the 8th Workshop of the Initiative for the Evaluation of XML Retrieval (INEX)! Now in its eighth year, INEX is an established evaluation forum for XML information retrieval (IR), with over 100 organizations worldwide registered and over 50 groups participating actively in at least one of the tracks. INEX aims to provide an infrastructure, in the form of a large structured test collection and appropriate scoring methods, for the evaluation of focused retrieval systems.

XML IR plays an increasingly important role in many information access systems (e.g., digital libraries, Web, intranet) where content is a mixture of text, multimedia, and metadata, formatted according to the adopted W3C standard for information repositories, the so-called eXtensible Markup Language (XML). The ultimate goal of such systems is to provide the right content to their end-users. However, while many of today's information access systems still treat documents as single large (text) blocks, XML offers the opportunity to exploit the internal structure of documents in order to allow for more precise access, thus providing more specific answers to user requests. Providing effective access to XML-based content is therefore a key issue for the success of these systems.

INEX 2009 was an exciting year for INEX in which a new collection was introduced that is again based on Wikipedia but is more than four times larger, with longer articles and additional semantic annotation. In total, eight research tracks were included, which studied different aspects of focused information access:

Ad Hoc Track investigated the effectiveness of XML-IR and Passage Retrieval for four ad hoc retrieval tasks: Thorough, Focused, Relevant in Context, and Best in Context.

Book Track investigated information access to, and IR techniques for, searching full texts of digitized books.

Efficiency Track investigated both the effectiveness and efficiency of XML ranked retrieval approaches on real data and real queries.

Entity Ranking Track investigated entity retrieval rather than text retrieval: 1) Entity Ranking, 2) Entity List Completion.

Interactive Track investigated the behavior of users when interacting with XML documents, as well as developed retrieval approaches which are effective in user-based environments.

Question Answering Track investigated technology for accessing structured documents that can be used to address real-world focused information needs formulated as natural language questions.

Link the Wiki Track investigated link discovery between Wikipedia documents, both at the file level and at the element level.

XML Mining Track investigated structured document mining, especially the classification and clustering of structured documents.

The aim of the INEX 2009 workshop was to bring together researchers in the field of XML IR who participated in the INEX 2009 campaign. During the past year, participating organizations contributed to the building of large-scale XML test collections by creating topics, performing retrieval runs, and providing relevance assessments. The workshop concluded the results of this effort, summarized and addressed issues encountered, and devised a work plan for the future evaluation of XML retrieval systems. These proceedings report the final results of INEX 2009. We received 47 submissions, already being a selection of work at INEX, and accepted a total of 42 papers based on peer-reviewing, yielding a 89% acceptance rate.

All INEX tracks start from having available suitable text collections. We gratefully acknowledge the data made available by: Amazon (Interactive Track), New Zealand Ministry for Culture and Heritage (*Te Ara*, Link-the-Wiki Track), Microsoft (Book Track), Wikipedia, and Ralf Schenkel of the Max-Planck Institute for the conversion of the Wikipedia.

INEX has outgrown its previous home at *Schloss Dagstuhl* and was held in Brisbane, Australia. Thanks to Richi Nayak and the QUT team for preserving the unique atmosphere of INEX—a setting where informal interaction and discussion occur naturally and frequently—in the unique location of the Woodlands of Marburg. Thanks to HCSNet, the Australian Research Council's Research Network in Human Communication Science, for sponsoring the invited talks by Peter Bruza (QUT), Cécile Paris (CSIRO), and Ian Witten (Waikato). Finally, INEX is run for, but especially by, the participants. It was a result of tracks and tasks suggested by participants, topics created by particants, systems built by participants, and relevance judgments provided by participants. So the main thank you goes to each of these individuals!

April 2010   Shlomo Geva
                                                              Jaap Kamps
                                                          Andrew Trotman

Organization

Steering Committee

Charlie Clarke	University of Waterloo, Canada
Norbert Fuhr	University of Duisburg-Essen, Germany
Shlomo Geva	Queensland University of Technology, Australia
Jaap Kamps	University of Amsterdam, The Netherlands
Mounia Lalmas	Queen Mary, University of London, UK
Stephen Robertson	Microsoft Research Cambridge, UK
Andrew Trotman	University of Otago, New Zealand
Ellen Voorhees	NIST, USA

Chairs

Shlomo Geva	Queensland University of Technology, Australia
Jaap Kamps	University of Amsterdam, The Netherlands
Andrew Trotman	University of Otago, New Zealand

Track Organizers

Ad Hoc

Shlomo Geva	Queensland University of Technology, Australia
Jaap Kamps	University of Amsterdam, The Netherlands
Miro Lethonen	University of Helsinki, Finland
Ralf Schenkel	Max-Planck-Institut für Informatik, Germany
James A. Thom	RMIT, Australia
Andrew Trotman	University of Otago, New Zealand

Book

Antoine Doucet	University of Caen, France
Gabriella Kazai	Microsoft Research Cambridge, UK
Marijn Koolen	University of Amsterdam, The Netherlands
Monica Landoni	University of Lugano, Switzerland

Efficiency

Ralf Schenkel	Max-Planck-Institut für Informatik, Germany
Martin Theobald	Max-Planck-Institut für Informatik, Germany

Entity Ranking

Gianluca Demartini	L3S, Germany
Tereza Iofciu	L3S, Germany
Arjen de Vries	CWI, The Netherlands

Interactive

Thomas Beckers	University of Duisburg-Essen, Germany
Nisa Fachry	University of Amsterdam, The Netherlands
Norbert Fuhr	University of Duisburg-Essen, Germany
Ragnar Nordlie	Oslo University College, Norway
Nils Pharo	Oslo University College, Norway

Link the Wiki

Shlomo Geva	Queensland University of Technology, Australia
Darren Huang	Queensland University of Technology, Australia
Andrew Trotman	University of Otago, New Zealand

Question Answering

Patrice Bellot	University of Avignon, France
Veronique Moriceau	LIMSI-CNRS, France
Eric SanJuan	University of Avignon, France
Xavier Tannier	LIMSI-CNRS, France

XML Mining

Shlomo Geva	Queensland University of Technology, Australia
Ludovic Denoyer	University Paris 6, France
Chris De Vries	Queensland University of Technology, Australia
Patrick Gallinari	University Paris 6, France
Sangeetha Kutty	Queensland University of Technology, Australia
Richi Nayak	Queensland University of Technology, Australia

Table of Contents

Invited

Ad Hoc Track

Book Track

Efficiency Track

Question Answering Track

XML Mining Track

Erratum

Is There Something Quantum-Like about the Human Mental Lexicon?

Peter Bruza

Faculty of Science and Technology
Queensland University of Technology, Australia
p.bruza@qut.edu.au

Abstract. This talk proceeds from the premise that IR should engage in a more substantial dialogue with cognitive science. After all, how users decide relevance, or how they chose terms to modify a query are processes rooted in human cognition. Recently, there has been a growing literature applying quantum theory (QT) to model cognitive phenomena ranging from human memory to decision making. Two aspects will be highlighted. The first will show how concept combinations can be modelled in a way analogous to quantum entangled twin-state photons. Details will be presented of cognitive experiments to test for the presence of "entanglement" in cognition via an analysis of bi-ambiguous concept combinations. The second aspect of the talk will show how quantum inference effects currently being used to fit models of human decision making may be applied to model interference between different dimensions of relevance.

The underlying theme behind this talk is QT can potentially provide the theoretical basis of new genre of information processing models more aligned with human cognition.

Acknowledgments. This research is supported in part by the Australian Research Council Discovery grant DP0773341.

Supporting for Real-World Tasks: Producing Summaries of Scientific Articles Tailored to the Citation Context

Cécile Paris

Information and Communication Technology (ICT) Centre
CSIRO, Australia
Cecile.Paris@csiro.au

Abstract. The amount of scientific material available electronically is forever increasing. This makes reading the published literature, whether to stay up-to-date on a topic or to get up to speed on a new topic, a difficult task. Yet, this is an activity in which all researchers must be engaged on a regular basis. Based on a user requirements analysis, we developed a new research tool, called the Citation-Sensitive In-Browser Summariser (CSIBS), which supports researchers in this browsing task. CSIBS enables readers to obtain information about a citation at the point at which they encounter it. This information is aimed at enabling the reader to determine whether or not to invest the time in exploring the cited article further, thus alleviating information overload. CSIBS builds a summary of the cited document, bringing together meta-data about the document and a citation-sensitive preview that exploits the citation context to retrieve the sentences from the cited document that are relevant at this point. In this talk, I will briefly present our user requirements analysis, then describe the system and, finally, discuss the observations from an initial pilot study. We found that CSIBS facilitates the relevancy judgment task, by increasing the users' self-reported confidence in making such judgements.

S. Geva, J. Kamps, and A. Trotman (Eds.): INEX 2009, LNCS 6203, p. 2, 2010.
© Springer-Verlag Berlin Heidelberg 2010

Semantic Document Processing Using Wikipedia as a Knowledge Base

Ian H. Witten

Department of Computer Science
University of Waikato, New Zealand
ihw@cs.waikato.ac.nz

Abstract. Wikipedia is a goldmine of information; not just for its many readers, but also for the growing community of researchers who recognize it as a resource of exceptional scale and utility. It represents a vast investment of manual effort and judgment: a huge, constantly evolving tapestry of concepts and relations that is being applied to a host of tasks.

This talk will introduce the process of "wikification"; that is, automatically and judiciously augmenting a plain-text document with pertinent hyperlinks to Wikipedia articles – as though the document were itself a Wikipedia article. This amounts to a new semantic representation of text in terms of the salient concepts it mentions, where "concept" is equated to "Wikipedia article." Wikification is a useful process in itself, adding value to plain text documents. More importantly, it supports new methods of document processing.

I first describe how Wikipedia can be used to determine semantic relatedness, and then introduce a new, high-performance method of wikification that exploits Wikipedia's 60 M internal hyperlinks for relational information and their anchor texts as lexical information, using simple machine learning. I go on to discuss applications to knowledge-based information retrieval, topic indexing, document tagging, and document clustering. Some of these perform at human levels. For example, on CiteULike data, automatically extracted tags are competitive with tag sets assigned by the best human taggers, according to a measure of consistency with other human taggers.

Although this work is based on English it involves no syntactic parsing, and the techniques are largely language independent. The talk will include live demos.

S. Geva, J. Kamps, and A. Trotman (Eds.): INEX 2009, LNCS 6203, p. 3, 2010.

Overview of the INEX 2009 Ad Hoc Track

Shlomo Geva[1], Jaap Kamps[2], Miro Lethonen[3],
Ralf Schenkel[4], James A. Thom[5], and Andrew Trotman[6]

[1] Queensland University of Technology, Brisbane, Australia
s.geva@qut.edu.au
[2] University of Amsterdam, Amsterdam, The Netherlands
kamps@uva.nl
[3] University of Helsinki, Helsinki, Finland
miro.lehtonen@helsinki.fi
[4] Max-Planck-Institut für Informatik, Saarbrücken, Germany
schenkel@mpi-sb.mpg.de
[5] RMIT University, Melbourne, Australia
james.thom@rmit.edu.au
[6] University of Otago, Dunedin, New Zealand
andrew@cs.otago.ac.nz

Abstract. This paper gives an overview of the INEX 2009 Ad Hoc Track. The main goals of the Ad Hoc Track were three-fold. The first goal was to investigate the impact of the collection scale and markup, by using a new collection that is again based on a the Wikipedia but is over 4 times larger, with longer articles and additional semantic annotations. For this reason the Ad Hoc track tasks stayed unchanged, and the Thorough Task of INEX 2002–2006 returns. The second goal was to study the impact of more verbose queries on retrieval effectiveness, by using the available markup as structural constraints—now using both the Wikipedia's layout-based markup, as well as the enriched semantic markup—and by the use of phrases. The third goal was to compare different result granularities by allowing systems to retrieve XML elements, ranges of XML elements, or arbitrary passages of text. This investigates the value of the internal document structure (as provided by the XML mark-up) for retrieving relevant information. The INEX 2009 Ad Hoc Track featured four tasks: For the *Thorough Task* a ranked-list of results (elements or passages) by estimated relevance was needed. For the *Focused Task* a ranked-list of non-overlapping results (elements or passages) was needed. For the *Relevant in Context Task* non-overlapping results (elements or passages) were returned grouped by the article from which they came. For the *Best in Context Task* a single starting point (element start tag or passage start) for each article was needed. We discuss the setup of the track, and the results for the four tasks.

1 Introduction

This paper gives an overview of the INEX 2009 Ad Hoc Track. There are three main research questions underlying the Ad Hoc Track. The first main research

S. Geva, J. Kamps, and A. Trotman (Eds.): INEX 2009, LNCS 6203, pp. 4–25, 2010.

question is the impact of the new collection—four times the size, with longer articles, and additional semantic markup—on focused retrieval. That is, what is the impact of collection size? What is the impact of document length, and hence the complexity of the XML structure in the DOM tree? The second main research question is the impact of more verbose queries—using either the XML structure, or using multi-word phrases. That is, what is the impact of semantic annotation on both the submitted queries, and their retrieval effectiveness? What is the impact of explicitly annotated multi-word phrases? The third main research question is that of the value of the internal document structure (mark-up) for retrieving relevant information. That is, does the document structure help to identify where the relevant information is within a document?

To study the value of the document structure through direct comparison of element and passage retrieval approaches, the retrieval results were liberalized to arbitrary passages since INEX 2007. Every XML element is, of course, also a passage of text. At INEX 2008, a simple passage retrieval format was introduced using file-offset-length (FOL) triplets, that allow for standard passage retrieval systems to work on content-only versions of the collection. That is, the offset and length are calculated over the text of the article, ignoring all mark-up. The evaluation measures are based directly on the highlighted passages, or arbitrary best-entry points, as identified by the assessors. As a result it is possible to fairly compare systems retrieving elements, ranges of elements, or arbitrary passages. These changes address earlier requests to liberalize the retrieval format to ranges of elements [1] and to arbitrary passages of text [10].

The INEX 2009 Ad Hoc Track featured four tasks:

1. For the *Thorough Task* a ranked-list of results (elements or passages) by estimated relevance must be returned. It is evaluated by mean average interpolated precision relative to the highlighted (or believed relevant) text retrieved.

2. For the *Focused Task* a ranked-list of non-overlapping results (elements or passages) must be returned. It is evaluated at early precision relative to the highlighted (or believed relevant) text retrieved.

3. For the *Relevant in Context Task* non-overlapping results (elements or passages) must be returned, these are grouped by document. It is evaluated by mean average generalized precision where the generalized score per article is based on the retrieved highlighted text.

4. For the *Best in Context Task* a single starting point (element's starting tag or passage offset) per article must be returned. It is also evaluated by mean average generalized precision but with the generalized score (per article) based on the distance to the assessor's best-entry point.

We discuss the results for the four tasks, giving results for the top 10 participating groups and discussing their best scoring approaches in detail.

The rest of the paper is organized as follows. First, Section 2 describes the INEX 2009 ad hoc retrieval tasks and measures. Section 3 details the collection, topics, and assessments of the INEX 2009 Ad Hoc Track. In Section 4, we report the results for the Thorough Task (Section 4.2); the Focused Task (Section 4.3); the Relevant in Context Task (Section 4.4); and the Best in Context Task (Section 4.5). Finally, in Section 5, we discuss our findings and draw some conclusions.

2 Ad Hoc Retrieval Track

In this section, we briefly summarize the ad hoc retrieval tasks and the submission format (especially how elements and passages are identified). We also summarize the measures used for evaluation.

2.1 Tasks

Thorough Task. The core system's task underlying most XML retrieval strategies is the ability to estimate the relevance of potentially retrievable elements or passages in the collection. Hence, the Thorough Task simply asks systems to return elements or passages ranked by their relevance to the topic of request. Since the retrieved results are meant for further processing (either by a dedicated interface, or by other tools) there are no display-related assumptions nor user-related assumptions underlying the task.

Focused Task. The scenario underlying the Focused Task is the return, to the user, of a ranked list of elements or passages for their topic of request. The Focused Task requires systems to find the most focused results that satisfy an information need, without returning "overlapping" elements (shorter is preferred in the case of equally relevant elements). Since ancestors elements and longer passages are always relevant (to a greater or lesser extent) it is a challenge to chose the correct granularity.

The task has a number of assumptions:

Display the results are presented to the user as a ranked-list of results.
Users view the results top-down, one-by-one.

Relevant in Context Task. The scenario underlying the Relevant in Context Task is the return of a ranked list of articles and within those articles the relevant information (captured by a set of non-overlapping elements or passages). A relevant article will likely contain relevant information that could be spread across different elements. The task requires systems to find a set of results that corresponds well to all relevant information in each relevant article. The task has a number of assumptions:

Display results will be grouped per article, in their original document order, access will be provided through further navigational means, such as a document heat-map or table of contents.
Users consider the article to be the most natural retrieval unit, and prefer an overview of relevance within this context.

Best in Context Task. The scenario underlying the Best in Context Task is the return of a ranked list of articles and the identification of a best-entry-point from which a user should start reading each article in order to satisfy the information need. Even an article completely devoted to the topic of request will only have one best starting point from which to read (even if that is the beginning of the article). The task has a number of assumptions:

Display a single result per article.

Users consider articles to be natural unit of retrieval, but prefer to be guided to the best point from which to start reading the most relevant content.

2.2   Submission Format

Since XML retrieval approaches may return arbitrary results from within documents, a way to identify these nodes is needed. At INEX 2009, we allowed the submission of three types of results: XML elements, file-offset-length (FOL) text passages, and ranges of XML elements. The submission format for all tasks is a variant of the familiar TREC format extended with two additional fields.

```
topic Q0 file rank rsv run_id column_7 column_8
```

Here:

- The first column is the topic number.
- The second column (the query number within that topic) is currently unused and should always be Q0.
- The third column is the file name (without .xml) from which a result is retrieved, which is identical to the ¡id¿ of the Wikipedia
- The fourth column is the rank the document is retrieved.
- The fifth column shows the retrieval status value (RSV) or score that generated the ranking.
- The sixth column is called the "run tag" identifying the group and for the method used.

Element Results. XML element results are identified by means of a file name and an element (node) path specification. File names in the Wikipedia collection are unique, and (with the .xml extension removed) identical to the ⟨id⟩ of the Wikipedia document. That is, file `9996.xml` contains the article as the target document from the Wikipedia collection with ⟨id⟩ 9996.

Element paths are given in XPath, but only fully specified paths are allowed. The next example identifies the first "article" element, then within that, the first "body" element, then the first "section" element, and finally within that the first "p" element.

```
/article[1]/body[1]/section[1]/p[1]
```

Importantly, XPath counts elements from 1 and counts element types. For example if a section had a title and two paragraphs then their paths would be: `title[1]`, `p[1]` and `p[2]`.

A result element may then be identified unambiguously using the combination of its file name (or ⟨id⟩) in column 3 and the element path in column 7. Column 8 will not be used. Example:

```
1 Q0 9996 1 0.9999 I09UniXRun1 /article[1]/bdy[1]/sec[1]
1 Q0 9996 2 0.9998 I09UniXRun1 /article[1]/bdy[1]/sec[2]
1 Q0 9996 3 0.9997 I09UniXRun1 /article[1]/bdy[1]/sec[3]/p[1]
```

Here the results are from 9996 and select the first section, the second section, and the first paragraph of the third section.

FOL passages. Passage results can be given in File-Offset-Length (FOL) format, where offset and length are calculated in characters with respect to the textual content (ignoring all tags) of the XML file. A special text-only version of the collection is provided to facilitate the use of passage retrieval systems. File offsets start counting a 0 (zero).

A result element may then be identified unambiguously using the combination of its file name (or ⟨id⟩) in column 3 and an offset in column 7 and a length in column 8. The following example is effectively equivalent to the example element result above:

```
1 Q0 9996 1 0.9999 I09UniXRun1 465 3426
1 Q0 9996 2 0.9998 I09UniXRun1 3892 960
1 Q0 9996 3 0.9997 I09UniXRun1 4865 496
```

The results are from article 9996, and the first section starts at the 466th character (so 465 characters beyond the first character which has offset 0), and has a length of 3,426 characters.

Ranges of Elements. To support ranges of elements, elemental passages can be specified by their containing elements. We only allow elemental paths (ending in an element, not a text-node in the DOM tree) plus an optional offset.

A result element may then be identified unambiguously using the combination of its file name (or ⟨id⟩) in column 3, its start at the element path in column 7, and its end at the element path in column 8. Example:

```
1 Q0 9996 1 0.9999 I09UniRun1 /article[1]/bdy[1]/sec[1] /article[1]/bdy[1]/sec[1]
```

Here the result is again the first section from 9996. Note that the seventh column will refer to the beginning of an element (or its first content), and the eighth column will refer to the ending of an element (or its last content). Note that this format is very convenient for specifying ranges of elements, e.g., the first three sections:

```
1 Q0 9996 1 0.9999 I09UniXRun1 /article[1]/bdy[1]/sec[1] /article[1]/bdy[1]/sec[3]
```

2.3   Evaluation Measures

We briefly summarize the main measures used for the Ad Hoc Track. Since INEX 2007, we allow the retrieval of arbitrary passages of text matching the judges ability to regard any passage of text as relevant. Unfortunately this simple change has necessitated the deprecation of element-based metrics used in prior INEX campaigns because the "natural" retrieval unit is no longer an element, so elements cannot be used as the basis of measure. We note that properly evaluating the effectiveness in XML-IR remains an ongoing research question at INEX.

The INEX 2009 measures are solely based on the retrieval of highlighted text. We simplify all INEX tasks to highlighted text retrieval and assume that systems will try to return all, and only, highlighted text. We then compare the characters of text retrieved by a search engine to the number and location of characters of text identified as relevant by the assessor. For best in context we use the distance between the best entry point in the run to that identified by an assessor.

Thorough Task. Precision is measured as the fraction of retrieved text that was highlighted. Recall is measured as the fraction of all highlighted text that has been retrieved. Text seen before is automatically discounted. The notion of rank is relatively fluid for passages so we use an interpolated precision measure which calculates interpolated precision scores at selected recall levels. Since we are most interested in overall performance, the main measure is mean average interpolated precision (MAiP), calculated over over 101 standard recall points (0.00, 0.01, 0.02, ..., 1.00). We also present interpolated precision at early recall points (iP[0.00], iP[0.01], iP[0.05], and iP[0.10]),

Focused Task. As above, precision is measured as the fraction of retrieved text that was highlighted and recall is measured as the fraction of all highlighted text that has been retrieved. We use an interpolated precision measure which calculates interpolated precision scores at selected recall levels. Since we are most interested in what happens in the first retrieved results, the main measure is interpolated precision at 1% recall (iP[0.01]). We also present interpolated precision at other early recall points, and (mean average) interpolated precision over 101 standard recall points (0.00, 0.01, 0.02, ..., 1.00) as an overall measure.

Relevant in Context Task. The evaluation of the Relevant in Context Task is based on the measures of generalized precision and recall [6] over articles, where the per document score reflects how well the retrieved text matches the relevant text in the document. Specifically, the per document score is the harmonic mean of precision and recall in terms of the fractions of retrieved and highlighted text in the document. We use an F_β score with $\beta = 1/4$ making precision four times as important as recall:

$$F_\beta = \frac{(1 + \beta^2) \cdot Precision \cdot Recall}{(\beta^2 \cdot Precision) + Recall}.$$

We are most interested in overall performances, so the main measure is mean average generalized precision (MAgP). We also present the generalized precision scores at early ranks (5, 10, 25, 50).

Best in Context Task. The evaluation of the Best in Context Task is based on the measures of generalized precision and recall where the per document score reflects how well the retrieved entry point matches the best entry point in the document. Specifically, the per document score is a linear discounting function of the distance d (measured in characters)

$$\frac{n - d(x, b)}{n}$$

for $d < n$ and 0 otherwise. We use $n = 500$ which is roughly the number of characters corresponding to the visible part of the document on a screen. We are most interested in overall performance, and the main measure is mean average generalized precision (MAgP). We also show the generalized precision scores at early ranks (5, 10, 25, 50).

For further details on the INEX measures, we refer to [5]

3   Ad Hoc Test Collection

In this section, we discuss the corpus, topics, and relevance assessments used in the Ad Hoc Track.

3.1   Corpus

Starting in 2009, INEX uses a new document collection based on the Wikipedia. The original Wiki syntax has been converted into XML, using both general tags of the layout structure (like *article, section, paragraph, title, list* and *item*), typographical tags (like *bold, emphatic*), and frequently occurring link-tags. The annotation is enhanced with semantic markup of articles and outgoing links, based on the semantic knowledge base YAGO, explicitly labeling more than 5,800 classes of entities like persons, movies, cities, and many more. For a more technical description of a preliminary version of this collection, see [9].

The collection was created from the October 8, 2008 dump of the English Wikipedia articles and incorporates semantic annotations from the 2008-w40-2 version of YAGO. It contains 2,666,190 Wikipedia articles and has a total uncompressed size of 50.7 Gb. There are 101,917,424 XML elements of at least 50 characters (excluding white-space).

Figure 1 shows part of a document in the corpus. The whole article has been encapsulated with tags, such as the ⟨group⟩ tag added to the Queen page.

This allows us to find particular article types easily, e.g., instead of a query requesting articles about Freddie Mercury:

    //article[about(., Freddie Mercury)]

we can specifically ask about a group about Freddie Mercury:

    //group[about(., Freddie Mercury)]

which will return pages of (pop) groups mentioning Freddy Mercury. In fact, also all internal Wikipedia links have been annotated with the tags assigned to the

```
<article xmlns:xlink="http://www.w3.org/1999/xlink">
<holder confidence="0.9511911446218017" wordnetid="103525454">
<entity confidence="0.9511911446218017" wordnetid="100001740">
<musical_organization confidence="0.8" wordnetid="108246613">
<artist confidence="0.9511911446218017" wordnetid="109812338">
<group confidence="0.8" wordnetid="100031264">
<header>
<title>Queen (band)</title>
<id>42010</id>
...
</header>
<bdy>
...
<songwriter wordnetid="110624540" confidence="0.9173553029164789">
<person wordnetid="100007846" confidence="0.9508927676800064">
<manufacturer wordnetid="110292316" confidence="0.9173553029164789">
<musician wordnetid="110340312" confidence="0.9173553029164789">
<singer wordnetid="110599806" confidence="0.9173553029164789">
<artist wordnetid="109812338" confidence="0.9508927676800064">
<link xlink:type="simple" xlink:href="../068/42068.xml">
Freddie Mercury</link></artist>
</singer>
</musician>
</manufacturer>
</person>
</songwriter>
...
</bdy>
</group>
</artist>
</musical_organization>
</entity>
</holder>
</article>
```

Fig. 1. INEX 2009 Ad Hoc Track document `42010.xml` (in part)

page they link to, e.g., in the example about the link to Freddie Mercury gets the ⟨singer⟩ tag assigned. We can also use these tags to identify pages where certain types of links occur, and further refine the query as:

    //group[about(.//singer, Freddie Mercury)]

The exact NEXI query format used to express the structural hints will be explained below.

3.2 Topics

The ad hoc topics were created by participants following precise instructions. Candidate topics contained a short CO (keyword) query, an optional structured

```
<topic id="2009114" ct_no="310">
  <title>self-portrait</title>
  <castitle>//painter//figure[about(.//caption, self-portrait)]</castitle>
  <phrasetitle>"self portrait"</phrasetitle>
  <description>Find self-portraits of painters.</description>
  <narrative>
    I am studying how painters visually depict themselves in their
    work.  Relevant document components are images of works of art, in
    combination with sufficient explanation (i.e., a reference to the
    artist and the fact that the artist him/herself is depicted in the
    work of art).  Also textual descriptions of these works, if
    sufficiently detailed, can be relevant.  Document components
    discussing the portrayal of artists in general are not relevant, as
    are artists that figure in painters of other artists.
  </narrative>
</topic>
```

Fig. 2. INEX 2009 Ad Hoc Track topic 2009114

CAS query, a phrase title, a one line description of the search request, and narrative with a details of the topic of request and the task context in which the information need arose. For candidate topics without a ⟨castitle⟩ field, a default CAS-query was added based on the CO-query: //*[about(., "*CO-query*")]. Figure 2 presents an example of an ad hoc topic. Based on the submitted candidate topics, 115 topics were selected for use in the INEX 2009 Ad Hoc Track as topic numbers 2009001–2009115.

Each topic contains

Title. A short explanation of the information need using simple keywords, also known as the content only (CO) query. It serves as a summary of the content of the user's information need.

Castitle. A short explanation of the information need, specifying any structural requirements, also known as the content and structure (CAS) query. The castitle is optional but the majority of topics should include one.

Phrasetitle. A more verbose explanation of the information need given as a series of phrases, just as the ⟨title⟩ is given as a series of keywords.

Description. A brief description of the information need written in natural language, typically one or two sentences.

Narrative. A detailed explanation of the information need and the description of what makes an element relevant or not. The ⟨narrative⟩ should explain not only what information is being sought, but also the context and motivation of the information need, i.e., why the information is being sought and what work-task it might help to solve. Assessments will be made on compliance to the narrative alone; it is therefore important that this description is clear and precise.

The ⟨castitle⟩ contains the CAS query, an XPath expressions of the form: A[B] or A[B]C[D] where A and C are navigational XPath expressions using only the descendant axis. B and D are predicates using functions for text; the arithmetic operators $<$, $<=$, $>$, and $>=$ for numbers; or the connectives **and** and **or**. For text, the **about** function has (nearly) the same syntax as the XPath function **contains**. Usage is restricted to the form **about**(.*path*, *query*) where *path* is empty or contains only tag-names and descendant axis; and *query* is an IR query having the same syntax as the CO titles (i.e., query terms). The about function denotes that the content of the element located by the path is about the information need expressed in the query. As with the title, the castitle is only a hint to the search engine and does not have definite semantics.

The purpose of the phrasetitle field is to explicate the order and grouping of the query terms in the title. The absence of a phrasetitle implies the absence of a phrase, e.g., a query with independent words. The title and phrasetitle together make the "phrase query" for phrase-aware search. Some topics come with quotations marks in the title, in which case the phrasetitle is at least partially redundant. However, we have made sure that the phrasetitle does not introduce words other than those in the title and that the identified phrases are encapsulated in quotation marks. This setting helps us study whether systems can improve their performance when given explicit phrases as opposed to individual words as implicit phrases.

3.3   Judgments

Topics were assessed by participants following precise instructions. The assessors used the GPXrai assessment system that assists assessors in highlight relevant text. Topic assessors were asked to mark all, and only, relevant text in a pool of documents. After assessing an article with relevance, a separate best entry point decision was made by the assessor. The Thorough, Focused and Relevant in Context Tasks were evaluated against the text highlighted by the assessors, whereas the Best in Context Task was evaluated against the best-entry-points.

The relevance judgments were frozen on November 10, 2009. At this time 68 topics had been fully assessed. Moreover, some topics were judged by two separate assessors, each without the knowledge of the other. All results in this paper refer to the 68 topics with the judgments of the first assigned assessor, which is typically the topic author.

- The 68 assessed topics were numbered $2009n$ with n: 001–006, 010–015, 020, 022, 023, 026, 028, 029, 033, 035, 036, 039–043, 046, 047, 051, 053–055, 061–071, 073, 074, 076–079, 082, 085, 087–089, 091–093, 095, 096, 104, 105, 108–113, and 115

In total 50,725 articles were judged. Relevant passages were found in 4,858 articles. The mean number of relevant articles per topic is 71, and the mean number of passages per topic is was 117.

Assessors where requested to provide a separate best entry point (BEP) judgment, for every article where they highlighted relevant text.

3.4 Questionnaires

At INEX 2009, all candidate topic authors and assessors were asked to complete a questionnaire designed to capture the context of the topic author and the topic of request. The candidate topic questionnaire (shown in Table 1) featured 20

Table 1. Candidate Topic Questionnaire

B1	How familiar are you with the subject matter of the topic?
B2	Would you search for this topic in real-life?
B3	Does your query differ from what you would type in a web search engine?
B4	Are you looking for very specific information?
B5	Are you interested in reading a lot of relevant information on the topic?
B6	Could the topic be satisfied by combining the information in different (parts of) documents?
B7	Is the topic based on a seen relevant (part of a) document?
B8	Can information of equal relevance to the topic be found in several documents?
B9	Approximately how many articles in the whole collection do you expect to contain relevant information?
B10	Approximately how many relevant document parts do you expect in the whole collection?
B11	Could a relevant result be (check all that apply): a single sentence; a single paragraph; a single (sub)section; a whole article
B12	Can the topic be completely satisfied by a single relevant result?
B13	Is there additional value in reading several relevant results?
B14	Is there additional value in knowing all relevant results?
B15	Would you prefer seeing: only the best results; all relevant results; don't know
B16	Would you prefer seeing: isolated document parts; the article's context; don't know
B17	Do you assume perfect knowledge of the DTD?
B18	Do you assume that the structure of at least one relevant result is known?
B19	Do you assume that references to the document structure are vague and imprecise?
B20	Comments or suggestions on any of the above (optional)

questions capturing contextual data on the search request. The post-assessment questionnaire (shown in Table 2) featured 14 questions capturing further contextual data on the search request, and the way the topic has been judged (a few questions on GPXrai were added to the end).

The responses to the questionnaires show a considerable variation over topics and topic authors in terms of topic familiarity; the type of information requested; the expected results; the interpretation of structural information in the search request; the meaning of a highlighted passage; and the meaning of best entry points. There is a need for further analysis of the contextual data of the topics in relation to the results of the INEX 2009 Ad Hoc Track.

Table 2. Post Assessment Questionnaire

C1	Did you submit this topic to INEX?
C2	How familiar were you with the subject matter of the topic?
C3	How hard was it to decide whether information was relevant?
C4	Is Wikipedia an obvious source to look for information on the topic?
C5	Can a highlighted passage be (check all that apply): a single sentence; a single paragraph; a single (sub)section; a whole article
C6	Is a single highlighted passage enough to answer the topic?
C7	Are highlighted passages still informative when presented out of context?
C8	How often does relevant information occur in an article about something else?
C9	How well does the total length of highlighted text correspond to the usefulness of an article?
C10	Which of the following two strategies is closer to your actual highlighting: (I) I located useful articles and highlighted the best passages and nothing more, (II) I highlighted all text relevant according to narrative, even if this meant highlighting an entire article.
C11	Can a best entry point be (check all that apply): the start of a highlighted passage; the sectioning structure containing the highlighted text; the start of the article
C12	Does the best entry point correspond to the best passage?
C13	Does the best entry point correspond to the first passage?
C14	Comments or suggestions on any of the above (optional)

4 Ad Hoc Retrieval Results

In this section, we discuss, for the four ad hoc tasks, the participants and their results.

4.1 Participation

A total of 172 runs were submitted by 19 participating groups. Table 3 lists the participants and the number of runs they submitted, also broken down over the tasks (Thorough, Focused, Relevant in Context, or Best in Context); the used query (Content-Only or Content-And-Structure); whether it used the Phrase query or Reference run; and the used result type (Element, Range of elements, or FOL passage). Unfortunately, no less than 15 runs turned out to be invalid.

Participants were allowed to submit up to two element result-type runs per task and up to two passage result-type runs per task (for all four tasks). In addition, we allowed for an extra submission per task based on a reference run containing an article-level ranking using the BM25 model. This totaled to 20 runs per participant.[1] The submissions are spread well over the ad hoc retrieval tasks with 30 submissions for Thorough, 57 submissions for Focused, 33 submissions for Relevant in Context, and 37 submissions for Best in Context.

---

[1] As it turns out, one group submitted more runs than allowed: the *Queensland University of Technology* submitted 24 extra element runs (claiming that these runs in fact belong to the *University of Otago*). Some other groups submitted too many runs of a certain type or task. At this moment, we have not decided on any repercussions other than mentioning them in this footnote.

Table 3. Participants in the Ad Hoc Track

Id Participant	Thorough	Focused	Relevant in Context	Best in Context	CO query	CAS query	Phrase query	Reference run	Element results	Range of elements results	FOL results	# valid runs	# submitted runs
4 University of Otago	0	0	1	0	1	0	0	1	1	0	0	1	1
5 Queensland University of Technology	4	12	12	12	20	20	0	0	32	8	0	40	48
6 University of Amsterdam	4	2	2	2	7	3	0	0	10	0	0	10	10
10 Max-Planck-Institut Informatik	3	8	0	2	11	2	1	0	13	0	0	13	13
16 University of Frankfurt	0	2	0	0	0	2	0	0	2	0	0	2	2
22 ENSM-SE	0	4	0	0	4	0	4	0	4	0	0	4	4
25 Renmin University of China	1	3	3	2	7	2	0	0	9	0	0	9	9
29 INDIAN STATISTICAL INSTITUTE	0	2	0	0	2	0	0	0	2	0	0	2	2
36 University of Tampere	0	0	3	3	6	0	0	2	4	2	0	6	6
48 LIG	3	3	3	3	12	0	0	4	12	0	0	12	12
55 Doshisha University	0	1	0	0	0	1	0	0	1	0	0	1	1
60 Saint Etienne University	3	4	3	3	13	0	0	4	13	0	0	13	13
62 RMIT University	0	0	0	2	2	0	0	0	1	0	1	2	2
68 University Pierre et Marie Curie - LIP6	2	2	0	0	4	0	0	0	4	0	0	4	4
72 University of Minnesota Duluth	2	3	3	1	9	0	0	0	9	0	0	9	9
78 University of Waterloo	0	4	0	0	4	0	0	0	2	0	2	4	4
92 University of Lyon3	2	2	0	2	5	1	6	0	6	0	0	6	8
167 School of Electronic Engineering and Computer Science	3	3	1	3	10	0	0	4	10	0	0	10	12
346 University of Twente	3	2	2	2	0	9	0	4	9	0	0	9	12
Total runs	30	57	33	37	117	40	11	19	144	10	3	157	172

4.2   Thorough Task

We now discuss the results of the Thorough Task in which a ranked-list of non-overlapping results (elements or passages) was required. The official measure for the task was mean average interpolated precision (MAiP). Table 4 shows the best run of the top 10 participating groups. The first column gives the participant, see Table 3 for the full name of group. The second to fifth column give the interpolated precision at 0%, 1%, 5%, and 10% recall. The sixth column gives mean average interpolated precision over 101 standard recall levels (0%, 1%, . . . , 100%).

Here we briefly summarize what is currently known about the experiments conducted by the top five groups (based on official measure for the task, MAiP).

Table 4. Top 10 Participants in the Ad Hoc Track Thorough Task

Participant	iP[.00]	iP[.01]	iP[.05]	iP[.10]	MAiP
p48-LIG-2009-thorough-3T	0.5967	0.5841	0.5444	0.5019	0.2855
p6-UAmsIN09article	0.5938	0.5880	0.5385	0.4981	0.2818
p5-BM25thorough	0.6168	0.5983	0.5360	0.4917	0.2585
p92-Lyon3LIAmanlmnt*	0.5196	0.4956	0.4761	0.4226	0.2496
p60-UJM_15494	0.5986	0.5789	0.5293	0.4813	0.2435
p346-utCASartT09	0.5461	0.5343	0.4929	0.4415	0.2350
p10-MPII-CASThBM	0.5860	0.5537	0.4821	0.4225	0.2133
p167-09RefT	0.3205	0.3199	0.2779	0.2437	0.1390
p68-I09LIP6OWATh	0.3975	0.3569	0.2468	0.1945	0.0630
p25-ruc-base-coT	0.5440	0.4583	0.3020	0.1898	0.0577

LIG. Element retrieval run using the CO query. Description: Starting from 2K elements for each of the section types (sec, ss1, ss2, ss3, ss4) according to a multinomial language model with Dirichlet smoothing, we then interleave these five lists according to the score. We then group these results by the ranking of the reference run on articles, keeping within a document the element ranking. The run is based on the reference run.

University of Amsterdam. Element retrieval run using the CO query. Description: A standard run on an article index, using a language model with a standard linear length prior. The run is retrieving only articles.

Queensland University of Technology. Element retrieval run using the CO query. Description: Starting from a BM25 article retrieval run on an index of terms and tags-as-terms (produced by Otago), the top 50 retrieved articles are further processed by extracting the list of all (overlapping) elements which contained at least one of the search terms. The list is padded with the remaining articles, if needed.

University of Lyon3. A *manual* element retrieval run using the CO query. Description: Using Indri with Dirichlet smoothing and combining two language models: one of the full articles and one on the following tags: b, bdy, category, causal_agent, country, entry, group, image, it, list, location, p, person, physical_entity, sec, software, table, title. Special queries are created used NLP tools such as a summarizer and terminology extraction: the initial query based on the topic's phrase and CO title is expanded with related phrases extracted from the other topic fields and from an automatic summary of the top ranked documents by this initial query. In addition, standard query expansion are used, skip phrases are allowed, and occurrences in the title are extra weighted.

Saint Etienne University. Element retrieval run using the CO query. Description: Using BM25 on an element index with element frequency statistics. The b and k parameters were tuned on the INEX 2008 collection, leading to value different from standard document retrieval. The resulting run is filtered for elements from articles in the reference run, while retaining the original element ranking. The run is based on the reference run.

Table 5. Top 10 Participants in the Ad Hoc Track Focused Task

Participant	iP[.00]	iP[.01]	iP[.05]	iP[.10]	MAiP
p78-UWatFERBM25F	0.6797	0.6333	0.5006	0.4095	0.1854
p68-I09LIP6Okapi	0.6244	0.6141	0.5823	0.5290	0.3001
p10-MPII-COFoBM	0.6740	0.6134	0.5222	0.4474	0.1973
p60-UJM_15525	0.6241	0.6060	0.5742	0.4920	0.2890
p6-UamsFSsec2docbi100	0.6328	0.5997	0.5140	0.4647	0.1928
p5-BM25BOTrangeFOC	0.6049	0.5992	0.5619	0.5057	0.2912
p16-Spirix09R001	0.6081	0.5903	0.5342	0.4979	0.2865
p48-LIG-2009-focused-1F	0.5861	0.5853	0.5431	0.5055	0.2702
p22-emse2009-150*	0.6671	0.5844	0.4396	0.3699	0.1470
p25-ruc-term-coF	0.6128	0.4973	0.3307	0.2414	0.0741

Based on the information from these and other participants:

- All ten runs use retrieve element type results. Three out of ten runs retrieve only article elements: the second ranked *p6-UAmsIN09article*, sixth ranked *p346-utCASartT09*, and the eighth ranked *p167-09RefT*.
- Eight of the ten runs use the CO query, the runs ranked sixth, *p346-utCAS-artT09*, and seventh, *p10-MPII-CASThBM* use the structured CAS query.
- Three runs are based on the *reference run*: the first ranked *p48-LIG-2009-thorough-3T*, the fifth ranked *p60-UJM_15494*, and the eighth ranked *p167-09RefT*

4.3   Focused Task

We now discuss the results of the Focused Task in which a ranked-list of non-overlapping results (elements or passages) was required. The official measure for the task was (mean) interpolated precision at 1% recall (iP[0.01]). Table 5 shows the best run of the top 10 participating groups. The first column gives the participant, see Table 3 for the full name of group. The second to fifth column give the interpolated precision at 0%, 1%, 5%, and 10% recall. The sixth column gives mean average interpolated precision over 101 standard recall levels (0%, 1%, ..., 100%).

Here we briefly summarize what is currently known about the experiments conducted by the top five groups (based on official measure for the task, iP[0.01]).

University of Waterloo. FOL passage retrieval run using the CO query. Description: the run uses the Okapi BM25 model in Wumpus to score all content-bearing elements such as sections and paragraphs. It uses a fielded Okapi BM25F over two fields: a title composed of the concatenation of article and all ancestor's and current section titles, and a body field is the rest of the section. Training was done at element level and an average field length was used.

LIP6. Element retrieval run using the CO query. Description: A BM25 run with b=0.2 and k=2.0 and retrieving 1,500 articles for the CO queries, where negated words are removed from the query. For each document, the /article[1] element is retrieved. The run is retrieving only articles.

Max-Planck-Institut für Informatik. Element retrieval run using the CO query. Description: Using EBM25, an XML-specific extension of BM25 using element frequencies of individual tag-term pairs, i.e., for each distinct tag and term, we precompute an individual element frequency, capturing the amount of tags under which the term appears in the entire collection. A static decay factor for the TF component is used to make the scoring function favor smaller elements rather than entire articles.

Saint Etienne University. An element retrieval run using the CO query. Description: Using BM25 on an standard article index. The b and k parameters were tuned on the INEX 2008 collection. The run is retrieving only articles.

University of Amsterdam. Element retrieval run using the CAS query. Description: Language model run on a non-overlapping section index with top 100 reranked using a link degree prior. The link degree prior is the indegree+outdegree using local links from the retrieved sections. The link degree prior is applied to the article level, thus all sections from the same article have the same link prior.

Based on the information from these and other participants:

– Seven runs use the CO query. Three runs, the fifth ranked *p6-UamsFSsec2-docbi100*, the sixth ranked *p5-BM25BOTrangeFOC*, and the seventh ranked *p16-Spirix09R001* use the structured CAS query. The ninth run, *p22-emse-2009-150*, uses a manually expanded query using words from the description and narrative fields.

– Eight runs retrieve elements as results. The top ranked *p78-UWatFERBM25F* retrieves FOL passages, and the sixth ranked *p5-BM25BOTrangeFOC* retrieves ranges of elements.

– The systems at rank second, (*p68-I09LIP6Okapi*), fourth (*p60-UJM_15525*), and seventh (*p16-Spirix09R001*) are retrieving only full articles.

4.4   Relevant in Context Task

We now discuss the results of the Relevant in Context Task in which non-overlapping results (elements or passages) need to be returned grouped by the article they came from. The task was evaluated using generalized precision where the generalized score per article was based on the retrieved highlighted text. The official measure for the task was mean average generalized precision (MAgP).

Table 6 shows the top 10 participating groups (only the best run per group is shown) in the Relevant in Context Task. The first column lists the participant, see Table 3 for the full name of group. The second to fifth column list generalized precision at 5, 10, 25, 50 retrieved articles. The sixth column lists mean average generalized precision.

Here we briefly summarize the information available about the experiments conducted by the top five groups (based on MAgP).

Table 6. Top 10 Participants in the Ad Hoc Track Relevant in Context Task

Participant	gP[5]	gP[10]	gP[25]	gP[50]	MAgP
p5-BM25RangeRIC	0.3345	0.2980	0.2356	0.1786	0.1885
p4-Reference	0.3311	0.2936	0.2298	0.1716	0.1847
p6-UamsRSCMartCMdocbi100	0.3192	0.2794	0.2074	0.1660	0.1773
p48-LIG-2009-RIC-1R	0.3027	0.2604	0.2055	0.1548	0.1760
p36-utampere_given30_nolinks	0.3128	0.2802	0.2101	0.1592	0.1720
p346-utCASrefR09	0.2216	0.1904	0.1457	0.1095	0.1188
p60-UJM_15502	0.2003	0.1696	0.1311	0.0998	0.1075
p167-09RefR	0.1595	0.1454	0.1358	0.1205	0.1045
p25-ruc-base-casF	0.2113	0.1946	0.1566	0.1380	0.1028
p72-umd_ric_1	0.0943	0.0801	0.0574	0.0439	0.0424

Queensland University of Technology. Run retrieving ranges of elements using the CO query. Description: Starting from a BM25 article retrieval run on an index of terms and tags-as-terms (produced by Otago), the top 50 retrieved articles are further processed by identifying the first and last element in the article (in reading order) which contained any of the search terms. The focused result was then specified as a range of two elements (which could be one and the same). The list is padded with the remaining articles.

University of Otago. Element retrieval run using the CO query. Description: the run uses the Okapi BM25 model on an article index, with parameters trained on the INEX 2008 collection. The run is retrieving only articles and is based on the reference run—in fact, it is the original reference run.

University of Amsterdam. Element retrieval run using the CO query. Description: The results from section index are grouped and ranked based on the the the article ranking from the article index. The section run is reranked using the Wikipedia categories as background models before we cut-off the section run at 1,500 results per topic. The article run is similarly reranked using the Wikipedia categories as background models and link degree priors using the local incoming and outgoing links at article level.

LIG. Element retrieval run using the CO query. Description: First, separate lists of 2K elements are generated for the element types sec, ss1, ss2, ss3, and ss4, the five lists are merged according to score. Second, an article ranking is obtained using a mulinomial language model with Dirichlet smoothing. Third, the element results are group using the article ranking, by retaining with each article the reading order. Then we remove overlaps according to the reading order.

University of Tampere. Element retrieval run using the CO query. Description: For each document the only retrieved passage was between the first and the last link to the top 30 documents. If there were no such links, the whole article was returned. The run is based on the reference run.

Table 7. Top 10 Participants in the Ad Hoc Track Best in Context Task

Participant	gP[5]	gP[10]	gP[25]	gP[50]	MAgP
p5-BM25bepBIC	0.2941	0.2690	0.2119	0.1657	0.1711
p62-RMIT09titleO	0.3112	0.2757	0.2156	0.1673	0.1710
p10-MPII-COBIBM	0.2903	0.2567	0.2053	0.1598	0.1662
p48-LIG-2009-BIC-3B	0.2778	0.2564	0.1969	0.1469	0.1571
p6-UamsBAfbCMdocbi100	0.2604	0.2298	0.1676	0.1478	0.1544
p92-Lyon3LIAmanBEP*	0.2887	0.2366	0.1815	0.1482	0.1483
p36-utampere_given30_nolinks	0.2141	0.1798	0.1462	0.1234	0.1207
p346-utCASrefB09	0.1993	0.1737	0.1248	0.0941	0.1056
p25-ruc-term-coB	0.1603	0.1610	0.1274	0.0976	0.1013
p167-09LrnRefB	0.1369	0.1250	0.1181	0.1049	0.0953

Based on the information from these and other participants:

- The runs ranked sixth (*p346-utCASrefR09*) and ninth (*p25-ruc-base-casF*) are using the CAS query. All other runs use only the CO query in the topic's title field.
- The top scoring run retrieves ranges of elements, all other runs retrieve elements as results.
- Solid article ranking seems a prerequisite for good overall performance, with second best run, *p4-Reference* and the eighth best run, *p167-09RefR*, retrieving only full articles.

4.5   Best in Context Task

We now discuss the results of the Best in Context Task in which documents were ranked on topical relevance and a single best entry point into the document was identified. The Best in Context Task was evaluated using generalized precision but here the generalized score per article was based on the distance to the assessor's best-entry point. The official measure for the task was mean average generalized precision (MAgP).

Table 7 shows the top 10 participating groups (only the best run per group is shown) in the Best in Context Task. The first column lists the participant, see Table 3 for the full name of group. The second to fifth column list generalized precision at 5, 10, 25, 50 retrieved articles. The sixth column lists mean average generalized precision.

Here we briefly summarize the information available about the experiments conducted by the top five groups (based on MAgP).

Queensland University of Technology. Element retrieval run using the CO query. Description: Starting from a BM25 article retrieval run on an index of terms and tags-as-terms (produced by Otago), the top 50 retrieved articles are further processed by identifying the first element (in reading order) containing any of the search terms. The list is padded with the remaining articles.

RMIT University. Element retrieval run using the CO query. Description: Using Zettair with Okapi BM25 on an article-level index. The BEP is assumed to be at the start of the article. The run is retrieving only articles.

Max-Planck-Institut für Informatik. Element retrieval run using the CO query. Description: Using EBM25, an XML-specific extension of BM25 using element frequencies of individual tag-term pairs, i.e., for each distinct tag and term, we precompute an individual element frequency, capturing the amount of tags under which the term appears in the entire collection. A static decay factor for the TF component is used to make the scoring function favor smaller elements rather than entire articles, but the final run returns the start of the article as BEP. The run is retrieving only articles.

LIG. Element retrieval run using the CO query. Description: First, separate lists of 2K elements are generated for the element types sec, ss1, ss2, ss3, and ss4, the five lists are merged according to score. Second, an article ranking is obtained from the reference run. Third, for each article the best scoring element is used as the entry point. The run is based on the reference run.

University of Amsterdam. Element retrieval run using the CO query. Description: Article index run with standard pseudo-relevance feedback (using Indri), reranked with Wikipedia categories as background models and link degree priors using the local incoming and outgoing links at article level. The run is retrieving only articles.

Based on the information from these and other participants:

– The second best run (*p62-RMIT09titleO*) retrieves FOL passages, all other runs return elements as results. The FOL passage run is a degenerate case that always puts the BEP at the start of the article.
– As for the Relevant in Context Task, we see again that solid article ranking is very important. In fact, we see runs putting the BEP at the start of all the retrieved articles at rank two (*p62-RMIT09titleO*), rank three (*p10-MPII-COBIBM*), rank five (*p6-UamsBAfbCMdocbi100*), and rank ten (*p167-09LrnRefB*).
– With the exception of the run ranked eight (*p346-utCASrefB09*), which used the CAS query, all the other best runs per group use the CO query.

5   Discussion and Conclusions

In this paper we provided an overview of the INEX 2009 Ad Hoc Track that contained four tasks: For the *Thorough Task* a ranked-list of results (elements or passages) by estimated relevance was required. For the *Focused Task* a ranked-list of non-overlapping results (elements or passages) was required. For the *Relevant in Context Task* non-overlapping results (elements or passages) grouped by the article that they belong to were required. For the *Best in Context Task* a single starting point (element's starting tag or passage offset) per article was required. We discussed the results for the four tasks.

Given the efforts put into the fair comparison of element and passage retrieval approaches, the number submissions using FOL passages and range of elements was disappointing. Thirteen submissions used ranges of elements or FOL passage results, whereas 144 submissions used element results. In addition, several of the passage or FOL submissions used exclusively full articles as results. Still the non-element submissions were competitive with the top ranking runs for both the Focused and Relevant in Context Tasks, and the second ranking run for the Best in Context Task. There were too few submissions to draw any definite conclusions, but the outcome broadly confirms earlier results using passage-based element retrieval [3, 4].

There were also few submissions using the explicitly annotated phrases of the phrase query: ten in total. Phrase query runs were competitive with several of them in the overall top 10 results, but the impact of the phrases seemed marginal. Recall, that the exact same terms were present in the CO query, and the only difference was the phrase annotation. This is in line with earlier work. The use of phrases in queries has been studied extensively. In early publications, the usage of phrases and proximity operators showed improved retrieval results but rarely anything substantial [e.g., 2]. As retrieval models became more advanced, the usage of query operators was questioned. E.g., [7] conclude that when using a good ranking algorithm, phrases have no effect on high precision retrieval (and sometimes a negative effect due to topic drift). [8] combine term-proximity heuristics with an Okapi model, obtaining marginal improvements for early precision but with hardly observable impact on the MAP scores.

There were 19 submissions using the reference run providing a solid article ranking for further processing. These runs turned out to be competitive, with runs in the top 10 for all tasks. Hence the reference run was successful in helping participants to create high quality runs. However, runs based on the reference run were not directly comparable, since participants used the reference run in different ways leading to substantially different underlying article rankings.

Finally, the Ad Hoc Track had three main research questions. The first main research question was to study the effect of the new collection. We saw that the collection's size had little impact, with similar numbers of articles with relevance as for the INEX 2006–2008 collection. The second main research question was the impact of verbose queries using phrases or structural hints. The relatively few phase query submissions showed only marginal differences. The CAS query runs were in general less effective than the CO query runs, with one notable exception for the early precision measures of the Focused Task. The third main research question was the comparative analysis of element and passage retrieval approaches, hoping to shed light on the value of the document structure as provided by the XML mark-up. Despite the low number of non-element runs, we saw that some of the best performing system used FOL passages or ranges of elements. For all main research questions, we hope and expect that the resulting test collection will prove its value in future use. After all, the main aim of the INEX initiative is to create bench-mark test-collections for the evaluation of structured retrieval approaches.

Acknowledgments. Jaap Kamps was supported by the Netherlands Organization for Scientific Research (NWO, grants 612.066.513, 639.072.601, and 640.001.501).

References

[1] Clarke, C.L.A.: Range results in XML retrieval. In: Proceedings of the INEX 2005 Workshop on Element Retrieval Methodology, Glasgow, UK, pp. 4–5 (2005)

[2] Croft, W.B., Turtle, H.R., Lewis, D.D.: The use of phrases and structured queries in information retrieval. In: Proceedings of the 14th annual international ACM SIGIR conference on Research and development in information retrieval, pp. 32–45 (1991)

[3] Huang, W., Trotman, A., O'Keefe, R.A.: Element retrieval using a passage retrieval approach. In: Proceedings of the 11th Australasian Document Computing Symposium (ADCS 2006), pp. 80–83 (2006)

[4] Itakura, K.Y., Clarke, C.L.A.: From passages into elements in XML retrieval. In: Proceedings of the SIGIR 2007 Workshop on Focused Retrieval, pp. 17–22. University of Otago, Dunedin New Zealand (2007)

[5] Kamps, J., Pehcevski, J., Kazai, G., Lalmas, M., Robertson, S.: INEX 2007 evaluation measures. In: Fuhr, N., Kamps, J., Lalmas, M., Trotman, A. (eds.) INEX 2007. LNCS, vol. 4862, pp. 24–33. Springer, Heidelberg (2008)

[6] Kekäläinen, J., Järvelin, K.: Using graded relevance assessments in IR evaluation. Journal of the American Society for Information Science and Technology 53, 1120–1129 (2002)

[7] Mitra, M., Buckley, C., Singhal, A., Cardie, C.: An analysis of statistical and syntactic phrases. In: Proceedings of RIAO 1997 (1997)

[8] Rasolofo, Y., Savoy, J.: Term proximity scoring for keyword-based retrieval systems. In: Sebastiani, F. (ed.) ECIR 2003. LNCS, vol. 2633, pp. 207–218. Springer, Heidelberg (2003)

[9] Schenkel, R., Suchanek, F.M., Kasneci, G.: YAWN: A semantically annotated Wikipedia XML corpus. In: 12. GI-Fachtagung für Datenbanksysteme in Business, Technologie und Web (BTW 2007), pp. 277–291 (2007)

[10] Trotman, A., Geva, S.: Passage retrieval and other XML-retrieval tasks. In: Proceedings of the SIGIR 2006 Workshop on XML Element Retrieval Methodology, pp. 43–50. University of Otago, Dunedin New Zealand (2006)

A   Appendix: Full Run Names

Group	Run	Label	Task	Query	Results	Notes
4	617	Reference	RiC	CO	Ele	Reference run Article-only
5	757	BM25thorough	Tho	CO	Ele	
5	781	BM25BOTrangeFOC	Foc	CAS	Ran	Article-only
5	797	BM25RangeRIC	RiC	CO	Ran	Article-only
5	824	BM25bepBIC	BiC	CO	Ele	Article-only
6	634	UAmsIN09article	Tho	CO	Ele	Article-only
6	813	UamsFSsec2docbi100	Foc	CAS	Ele	
6	814	UamsRSCMartCMdocbi100	RiC	CO	Ele	
6	816	UamsBAfbCMdocbi100	BiC	CO	Ele	Article-only
10	619	MPII-COFoBM	Foc	CO	Ele	
10	620	MPII-CASThBM	Tho	CAS	Ele	
10	632	MPII-COBIBM	BiC	CO	Ele	Article-only
16	872	Spirix09R001	Foc	CAS	Ele	Article-only
22	672	emse2009-150	Foc	CO	Ele	Phrases Manual
25	727	ruc-base-coT	Tho	CO	Ele	
25	737	ruc-term-coB	BiC	CO	Ele	
25	739	ruc-term-coF	Foc	CO	Ele	
25	899	ruc-base-casF	RiC	CAS	Ele	
36	688	utampere_given30_nolinks	RiC	CO	Ele	Reference run
36	701	utampere_given30_nolinks	BiC	CO	Ele	Reference run
48	684	LIG-2009-thorough-3T	Tho	CO	Ele	Reference run
48	685	LIG-2009-focused-1F	Foc	CO	Ele	
48	714	LIG-2009-RIC-1R	RiC	CO	Ele	
48	719	LIG-2009-BIC-3B	BiC	CO	Ele	Reference run
60	822	UJM_15494	Tho	CO	Ele	Reference run
60	828	UJM_15502	RiC	CO	Ele	
60	868	UJM_15525	Foc	CO	Ele	Article-only
62	896	RMIT09titleO	BiC	CO	FOL	Article-only
68	679	I09LIP6Okapi	Foc	CO	Ele	Article-only
68	704	I09LIP6OWATh	Tho	CO	Ele	
72	666	umd_ric_1	RiC	CO	Ele	
78	707	UWatFERBM25F	Foc	CO	FOL	
92	695	Lyon3LIAmanBEP	BiC	CO	Ele	Phrases Manual Article-only
92	699	Lyon3LIAmanlmnt	Tho	CO	Ele	Phrases Manual
167	651	09RefT	Tho	CO	Ele	Reference run Article-only
167	657	09RefR	RiC	CO	Ele	Reference run Article-only
167	660	09LrnRefB	BiC	CO	Ele	Reference run Article-only
346	637	utCASartT09	Tho	CAS	Ele	Article-only
346	647	utCASrefR09	RiC	CAS	Ele	Reference run
346	648	utCASrefB09	BiC	CAS	Ele	Reference run

Analysis of the INEX 2009 Ad Hoc Track Results

Jaap Kamps[1], Shlomo Geva[2], and Andrew Trotman[3]

[1] University of Amsterdam, Amsterdam, The Netherlands
kamps@uva.nl
[2] Queensland University of Technology, Brisbane, Australia
s.geva@qut.edu.au
[3] University of Otago, Dunedin, New Zealand
andrew@cs.otago.ac.nz

Abstract. This paper analyzes the results of the INEX 2009 Ad Hoc Track, focusing on a variety of topics. First, we examine in detail the relevance judgments. Second, we study the resulting system rankings, for each of the four ad hoc tasks, and determine whether differences between the best scoring participants are statistically significant. Third, we restrict our attention to particular run types: element and passage runs, keyword and phrase query runs, and systems using a reference run with a solid article ranking. Fourth, we examine the relative effectiveness of content only (CO, or Keyword) search as well as content and structure (CAS, or structured) search. Fifth, we look at the ability of focused retrieval techniques to rank articles. Sixth, we study the length of retrieved results, and look at the impact of restricting result length.

1 Introduction

This paper provides analysis of the results of the INEX 2009 Ad Hoc Track, in addition to the overview of INEX 2009 Ad Hoc track's tasks and results in [1].

We focus on a variety of topics. First, we try to understand what constitutes a "highlighted" passage, and how the new and four times larger Wikipedia collection may affect the resulting test collection. For this purpose, we examine the relevance judgments in great detail. Second, we investigate the ability of the evaluation to distinguish between different retrieval approaches. We do this by studying the resulting system rankings, for each of the four ad hoc tasks, and determine whether differences between the best scoring participants are statistically significant. Third, we dig deeper in the effectiveness of particular focused retrieval approaches, by restricting our attention to particular run types: element and passage runs, keyword and phrase query runs, and systems using a reference run with a solid article ranking. Fourth, we try to grasp the impact of structural hints using either the original XML document structure, or automatically assigned YAGO tags. We examine the relative effectiveness of content only (CO, or Keyword) search as well as content and structure (CAS, or structured) search. Fifth, we relate the focused retrieval approaches to article retrieval, by looking at

S. Geva, J. Kamps, and A. Trotman (Eds.): INEX 2009, LNCS 6203, pp. 26–48, 2010.

the ability of focused retrieval techniques to rank articles. Sixth, we investigate the length of retrieved text per article, and the performance of focused retrieval systems under resource-limited conditions. In particular, we "cut off" the results after having retrieved the first 500 retrieved characters per article.

The rest of the paper is organized as follows. Section 2 analyzes the assessments of the INEX 2009 Ad Hoc Track. In Section 3, we report the results for the Thorough Task (Section 3.1); the Focused Task (Section 3.2); the Relevant in Context Task (Section 3.3); and the Best in Context Task (Section 3.4). Section 4 details particular types of runs (such as element versus passage, using phrases or using the reference run), and on particular subsets of the topics (such as topics with a non-trivial CAS query). Section 6 looks at the article retrieval aspects of the submissions, treating any article with highlighted text as relevant. We study the impact of result length in Section 7. Finally, in Section 8, we discuss our findings and draw some conclusions.

2   Analysis of the Relevance Judgments

In this section, we analyze the relevance assessments used in the Ad Hoc Track. The 2009 collection contains 2,666,190 Wikipedia articles (October 8, 2008 dump of the Wikipedia), which is four times larger than the earlier Wikipedia collection. What is the effect of this change in corpus size?

2.1   Topics

Topics were assessed by participants following precise instructions. The assessors used the GPXrai assessment system that assists assessors in highlighting relevant text. Topic assessors were asked to mark all, and only, relevant text in a pool of documents. After assessing an article with relevance, a separate best entry point decision was made by the assessor. The Thorough, Focused and Relevant in Context Tasks were evaluated against the text highlighted by the assessors, whereas the Best in Context Task was evaluated against the best-entry-points.

The relevance judgments were frozen on November 10, 2009. At this time 68 topics had been fully assessed. Moreover, some topics were judged by two separate assessors, each without the knowledge of the other. All results in this paper refer to the 68 topics with the judgments of the first assigned assessor, which is typically the topic author.

- The 68 assessed topics were numbered 2009n with n: 001–006, 010–015, 020, 022, 023, 026, 028, 029, 033, 035, 036, 039–043, 046, 047, 051, 053–055, 061–071, 073, 074, 076–079, 082, 085, 087–089, 091–093, 095, 096, 104, 105, 108–113, and 115

2.2   Highlighted Text

Table 1 presents statistics of the number of judged and relevant articles, and passages. In total 50,725 articles were judged. Relevant passages were found

Table 1. Statistics over judged and relevant articles per topic

| | total | | # per topic | | | | |
	topics	number	min	max	median	mean	st.dev
judged articles	68	50,725	380	766	754	746.0	49.0
articles with relevance	68	4,858	5	351	52	71.4	72.5
highlighted passages	68	7,957	5	594	75.5	117.0	121.5
highlighted characters	68	18,838,137	4,453	2,776,635	97,550.5	277,031.4	442,113.9

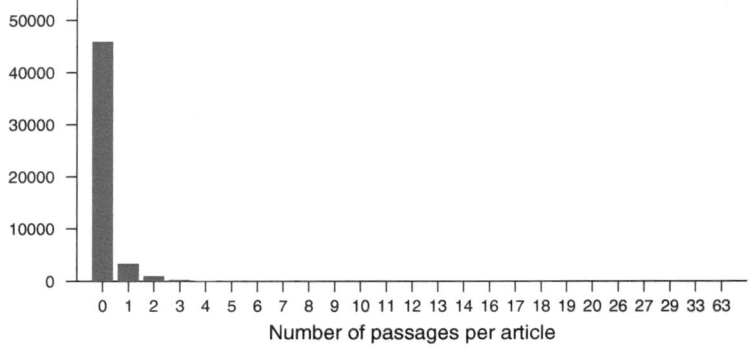

Fig. 1. Distribution of passages over articles

in 4,858 articles. The mean number of relevant articles per topic is 71, but the distribution is skewed with a median of 52. There were 7,957 highlighted passages. The mean was 117 passages and the median was 76 passages per topic.[1]

Figure 1 presents the number of articles with the given number of passages. The vast majority of relevant articles (3,339 out of 4,858) had only a single highlighted passage, and the number of passages quickly tapers off.

2.3 Best Entry Point

Assessors where requested to provide a separate best entry point (BEP) judgment, for every article where they highlighted relevant text. Table 2 presents statistics on the best entry point offset, on the first highlighted or relevant character, and on the fraction of highlighted text in relevant articles. We first look at the BEPs. The mean BEP is well within the article with 2,493 but the distribution is very skewed with a median BEP offset of only 311. Figure 2 shows the distribution of the character offsets of the 4,858 best entry points. It is clear that the overwhelming majority of BEPs is at the beginning of the article.

The statistics of the first highlighted or relevant character (FRC) in Table 2 give very similar numbers as the BEP offsets: the mean offset of the first relevant

---

[1] Note that for the Focused Task the main effectiveness measures is precision at 1% recall. Given that the average topic has 117 relevant passages in 52 articles, the 1% recall roughly corresponds to a relevant passage retrieved—for many systems this will be accomplished by the first or first few results.

Table 2. Statistics over relevant articles

	total		# per relevant article				
	topics	number	min	max	median	mean	st.dev
best entry point offset	68	4,858	2	86,545	311.5	2,493.2	6,481.8
first relevant character offset	68	4,858	2	86,545	295	2,463.0	6,375.6
length relevant documents	68	4,858	204	159,892	5,774.5	11,691.5	15,745.1
relevant characters	68	4,858	8	110,191	1,137	3,877.8	7,818.5
fraction highlighted text	68	4,858	0.00022	1.000	0.330	0.442	0.381

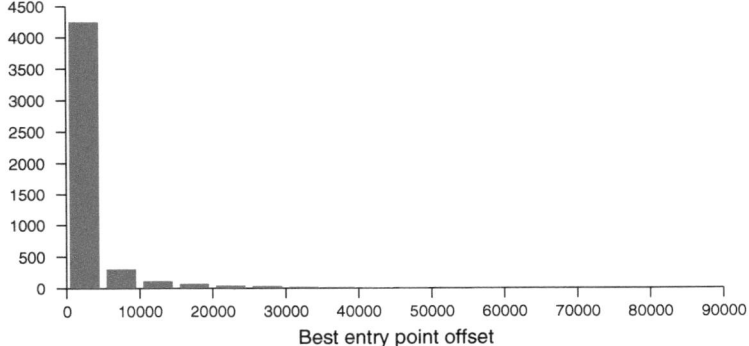

Fig. 2. Distribution of best entry point offsets

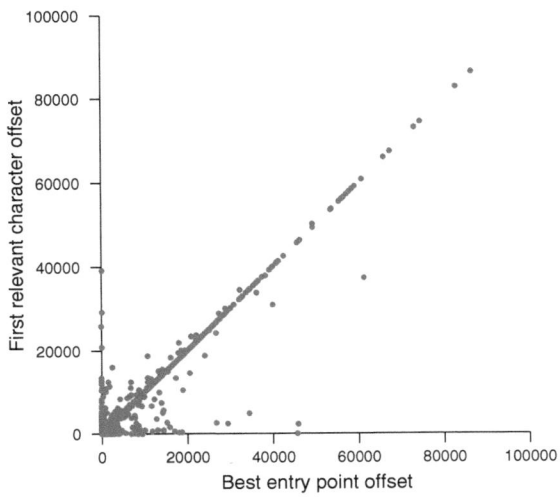

Fig. 3. Scatter plot of best entry point offsets versus the first relevant character

character is 2,463 but the median offset is only 295. This suggests a relation between the BEP offset and the FRC offset. Figure 3 shows a scatter plot the BEP and FRC offsets. Two observations present themselves. First, there is a clear

Table 3. Top 10 Participants in the Ad Hoc Track Thorough Task: Statistical significance (t-test, one-tailed, 95%)

Participant	MAiP	1 2 3 4 5 6 7 8 9 10
p48-LIG-2009-thorough-3T	0.2855	- - ★ - ★ - ★ ★ ★
p6-UAmsIN09article	0.2818	- ★ - ★ - ★ ★ ★
p5-BM25thorough	0.2585	★ - ★ - ★ ★ ★
p92-Lyon3LIAmanlmnt	0.2496	- - - ★ ★ -
p60-UJM_15494	0.2435	- - ★ ★ ★
p346-utCASartT09	0.2350	- ★ ★ ★
p10-MPII-CASThBM	0.2133	★ ★ ★
p167-09RefT	0.1390	- -
p68-I09LIP6OWATh	0.0630	-
p25-ruc-base-coT	0.0577	

diagonal where the BEP is positioned exactly at the first highlighted character in the article. Second, there is also a vertical line at BEP offset zero, indicating a tendency to put the BEP at the start of the article even when the relevant text appears later on.

Table 2 also shows statistics on the length of relevant articles. Many articles are relatively short with a median length of 5,775 characters, the mean length is 11,691 characters. This is considerably longer than the INEX 2008 collection, where the relevant articles had a median length of 3,030 and a mean length of 6,793. The length of highlighted text in characters is on average 3,876 (mean 1,137), in comparison to an average length of 2,338 (mean 838) in 2008. Table 2 also shows that the amount of relevant text varies from almost nothing to almost everything. The mean fraction is 0.44, and the median is 0.33, indicating that typically over one-third of the article is relevant. This is considerably less than the INEX 2008 collection, where over half of the text of articles was considered relevant. The observation that the majority of relevant articles contain such a large fraction of relevant text, plausibly explains that BEPs being frequently positioned on or near the start of the article.

3   Analysis of the Ad Hoc Tasks

In this section, we discuss, for the four ad hoc tasks, the participants and their results by looking at the significance of differences between participants.

3.1   Thorough Task

We tested whether higher ranked systems were significantly better than lower ranked systems, using a t-test (one-tailed) at 95%. Table 3 shows for the best runs of the 10 best scoring groups whether a run is significantly better (indicated by "★") than lower ranked runs.

For the Thorough Task, we see that the performance (measured by MAiP) of the top scoring run is significantly better than the runs at rank 4, 6, 8, 9, and

Table 4. Top 10 Participants in the Ad Hoc Track Focused Task: Statistical significance (t-test, one-tailed, 95%)

Participant	iP[0.01]	1	2	3	4	5	6	7	8	9	10
p78-UWatFERBM25F	0.6333		-	-	-	-	-	-	-	-	★
p68-I09LIP6Okapi	0.6141			-	-	-	-	-	★	-	★
p10-MPII-COFoBM	0.6134				-	-	-	-	-	-	★
p60-UJM_15525	0.6060					-	-	-	-	-	★
p6-UamsFSsec2docbi100	0.5997						-	-	-	-	★
p5-BM25BOTrangeFOC	0.5992							-	-	-	★
p16-Spirix09R001	0.5903								-	-	★
p48-LIG-2009-focused-1F	0.5853									-	★
p22-emse2009-150	0.5844										★
p25-ruc-term-coF	0.4973										

10. The same holds for the second and third best run. The fourth best run is significantly better than the runs at rank 8 and 9. The fifth, sixth, and seventh ranked runs are all significantly better than the runs at rank 8, 9, and 10. Of the 45 possible pairs of runs, there are 26 (or 58%) significant differences.

3.2   Focused Task

Table 4 shows for the best runs of the 10 best scoring groups whether a run is significantly better (indicated by "★") than lower ranked runs. For the Focused Task, we see that the early precision (at 1% recall) is a rather unstable measure. All runs are significantly better than the run at rank 10, the second best run also is significantly better than the run at rank 8. Of the 45 possible pairs of runs, there are only 10 (or 22%) significant differences. Hence we should be careful when drawing conclusions based on the Focused Task results.

The overall MAiP measure is more stable, see the analysis of the Thorough runs before.

3.3   Relevant in Context Task

Table 5 shows for the best runs of the 10 best scoring groups whether a run is significantly better (indicated by "★") than lower ranked runs. For the Relevant in Context Task, we see that the top run is significantly better than ranks 2 and 4 through 10. The second best run is significantly better than ranks 5 through 10. The third, fourth, and fifth ranked systems are significantly better than ranks 6 through 10. The sixth to ninth systems are significantly better than rank 10. Of the 45 possible pairs of runs, there are 33 (or 73%) significant differences, making MAgP a very discriminative measure.

3.4   Best in Context Task

Table 6 shows for the best runs of the 10 best scoring groups whether a run is significantly better (indicated by "★") than lower ranked runs. For the Best in Context Task, we see that the top run is significantly better than ranks 4 and

Table 5. Top 10 Participants in the Ad Hoc Track Relevant in Context Task: Statistical significance (t-test, one-tailed, 95%)

Participant	MAgP	1	2	3	4	5	6	7	8	9	10
p5-BM25RangeRIC	0.1885		★	-	★	★	★	★	★	★	★
p4-Reference	0.1847			-	-	★	★	★	★	★	★
p6-UamsRSCMartCMdocbi100	0.1773				-	-	★	★	★	★	★
p48-LIG-2009-RIC-1R	0.1760					-	★	★	★	★	★
p36-utampere_given30_nolinks	0.1720						★	★	★	★	★
p346-utCASrefR09	0.1188							-	-	-	★
p60-UJM_15502	0.1075								-	-	★
p167-09RefR	0.1045									-	★
p25-ruc-base-casF	0.1028										★
p72-umd_ric_1	0.0424										

Table 6. Top 10 Participants in the Ad Hoc Track Best in Context Task: Statistical significance (t-test, one-tailed, 95%)

Participant	MAgP	1	2	3	4	5	6	7	8	9	10
p5-BM25bepBIC	0.1711			-	-	★	★	-	★	★	★
p62-RMIT09titleO	0.1710			-	★	-	-	★	★	★	★
p10-MPII-COBIBM	0.1662				-	-	-	★	★	★	★
p48-LIG-2009-BIC-3B	0.1571					-	-	★	★	★	★
p6-UamsBAfbCMdocbi100	0.1544						-	★	★	★	★
p92-Lyon3LIAmanBEP	0.1483							-	★	★	★
p36-utampere_given30_nolinks	0.1207								-	-	★
p346-utCASrefB09	0.1056									-	-
p25-ruc-term-coB	0.1013										-
p167-09LrnRefB	0.0953										

5, and 7 through 10. The second best run is significantly better than than ranks 4 and 7 to 10. The third, fourth, and fifth ranked runs are significantly better than than ranks 7 to 10. The seventh ranked system is better than the systems ranked 8 to 10, and the eighth ranked system better than ranks 9 to 10. Of the 45 possible pairs of runs, there are 27 (or 60%) significant differences.

4   Analysis of Run Types

In this section, we will discuss the relative effectiveness of element and passage retrieval approaches, of phase and keyword queries, and of the reference run providing solid article ranking.

4.1   Elements Versus Passages

We received 13 submissions using ranges of elements of FOL-passage results, from in total 4 participating groups. We will look at the relative effectiveness of element and passage runs.

Table 7. Ad Hoc Track: Runs with ranges of elements or FOL passages

(a) Focused Task

Participant	iP[.00]	iP[.01]	iP[.05]	iP[.10]	MAiP
p78-UWatFERBM25F	0.6797	0.6333	0.5006	0.4095	0.1854
p5-BM25BOTrangeFOC	0.6049	0.5992	0.5619	0.5057	0.2912

(b) Relevant in Context Task

Participant	gP[5]	gP[10]	gP[25]	gP[50]	MAgP
p5-BM25RangeRIC	0.3345	0.2980	0.2356	0.1786	0.1885
p36-utampere_auth_40_top30	0.2717	0.2509	0.2006	0.1583	0.1185

(c) Best in Context Task

Participant	gP[5]	gP[10]	gP[25]	gP[50]	MAgP
p62-RMIT09titleO	0.3112	0.2757	0.2156	0.1673	0.1710

For three tasks there were high ranking runs using FOL passages or ranges of elements in the top 10 (discussed in Section 3). Table 7 shows the best runs using ranges of elements or FOL passages for three ad hoc tasks, there were no such submissions for the Thorough Task. As it turns out, the best focused run retrieving FOL passages was the top ranked run in Table 4; the best relevant in context retrieving ranges of elements was the top scoring run in Table 5; and the best best in context run retrieving FOL passages was the second best run in Table 6. Given the low number of submissions using passages or ranges of elements, this is an impressive result. However, looking at the runs in more detail, their character is often unlike what one would expect from a "passage" retrieval run. For Focused, *p5-BM25BOTrangeFOC* is an article retrieving run using ranges of elements, based on the CAS query. For Relevant in Context, *p5-BM25RangeRIC* is an article retrieving run using ranges of elements. For Best in Context, *p62-RMIT09titleO* is an article run using FOL passages. Hence, this is not sufficient evidence to warrant any conclusion on the effectiveness of passage level results. We hope and expect that the test collection and the passage runs will be used for further research into the relative effectiveness of element and passage retrieval approaches.

4.2 Phrase Queries

We received 10 submissions based on the phrase query. Table 8 shows the best runs using the phrase query for three of the ad hoc tasks, there were no valid submissions using the phrase title for Relevant in Context. The best phrase submission for the Thorough Task did rank 5th in the overall results. The best phrase submission for the Focused Task did rank 9th in the overall results. The best phrase submission for the Best in Context Task did rank 6th in the overall results.

Although few runs were submitted, the phrase title seems competitive, but not superior to the use of the CO query. The only participant submitting both types of runs, the *Max-Planck-Institute für Informatik* for the Focused Task, had marginally better performance for the CO query run over all 68 topics, and

Table 8. Ad Hoc Track: Runs using the phrase query

(a) Thorough Task

Participant	iP[.00]	iP[.01]	iP[.05]	iP[.10]	MAiP
p92-Lyon3LIAmanlmnt*	0.5196	0.4956	0.4761	0.4226	0.2496

(b) Focused Task

Participant	iP[.00]	iP[.01]	iP[.05]	iP[.10]	MAiP
p22-emse2009-150*	0.6671	0.5844	0.4396	0.3699	0.1470
p10-MPII-COArBPP	0.5563	0.5477	0.5283	0.4681	0.2566
p92-Lyon3LIAmanQE*	0.4955	0.4861	0.4668	0.4271	0.2522

(c) Best in Context Task

Participant	gP[5]	gP[10]	gP[25]	gP[50]	MAgP
p92-Lyon3LIAmanBEP*	0.2887	0.2366	0.1815	0.1482	0.1483

marginally better performance for the combined CO and Phrase title run over the 60 topics having a proper phrase in the Phrase title field. The differences between the query types are very small. A possible explanation for this is that all CO query have been expanded to contain the same terms as the more verbose phrase query. Hence the only difference is the explicit phrase markup, which requires special handling by the search engines. The available test collection with explicit phrases marked up in 60 topics is a valuable result of INEX 2009, and it can be studied in-depth in future experiments.

4.3   Reference Run

There were 19 submissions using the reference run. Table 9 shows the best runs using the reference runs for the four ad hoc tasks. For the Thorough Task, the best submission based on the reference run ranked first. For the Focused Task, the best submission based on the reference run would have ranked tenth. For the Relevant in Context Task, the best submission based on the reference run—in fact, the actual reference run itself—ranked second. For the Best in Context Task, the best submission based on the reference run ranked fourth. The results show that the reference run indeed provides competitive article ranking that forms a good basis for retrieval.

There are also considerable differences in performance of the runs based on the same reference run. This suggests that the runs do not retrieve the exact same set of articles. As explained later, in Section 6, we can look at the article rankings induced by the runs. Table 10 shows the best run of the top 10 participating groups, using the reference run. With the exception of *p36-utampere_given30_nolinks* the article rankings of the runs vary considerably.

5   Analysis of Structured Queries

In this section, we will discuss the relative effectiveness of systems using the keyword and structured queries.

Table 9. Ad Hoc Track: Runs using the reference run

(a) Thorough Task

Participant	iP[.00]	iP[.01]	iP[.05]	iP[.10]	MAiP
p48-LIG-2009-thorough-3T	0.5967	0.5841	0.5444	0.5019	0.2855
p60-UJM_15494	0.5986	0.5789	0.5293	0.4813	0.2435
p346-utCASrefF09	0.4834	0.4525	0.4150	0.3550	0.1982
p167-09RefT	0.3205	0.3199	0.2779	0.2437	0.1390

(b) Focused Task

Participant	iP[.00]	iP[.01]	iP[.05]	iP[.10]	MAiP
p48-LIG-2009-focused-3F	0.5946	0.5822	0.5344	0.5018	0.2732
p60-UJM_15518	0.5559	0.5136	0.4003	0.3104	0.1019
p346-utCASrefF09	0.4801	0.4508	0.4139	0.3547	0.1981
p167-09LrnRefF	0.3162	0.3072	0.2512	0.2223	0.1292

(c) Relevant in Context Task

Participant	gP[5]	gP[10]	gP[25]	gP[50]	MAgP
p4-Reference	0.3311	0.2936	0.2298	0.1716	0.1847
p48-LIG-2009-RIC-3R	0.3119	0.2790	0.2193	0.1629	0.1757
p36-utampere_given30_nolinks	0.3128	0.2802	0.2101	0.1592	0.1720
p346-utCASrefR09	0.2216	0.1904	0.1457	0.1095	0.1188
p167-09RefR	0.1595	0.1454	0.1358	0.1205	0.1045
p60-UJM_15503	0.1825	0.1548	0.1196	0.0953	0.1020

(d) Best in Context Task

Participant	gP[5]	gP[10]	gP[25]	gP[50]	MAgP
p48-LIG-2009-BIC-3B	0.2778	0.2564	0.1969	0.1469	0.1571
p36-utampere_given30_nolinks	0.2141	0.1798	0.1462	0.1234	0.1207
p346-utCASrefB09	0.1993	0.1737	0.1248	0.0941	0.1056
p167-09LrnRefB	0.1369	0.1250	0.1181	0.1049	0.0953
p60-UJM_15508	0.1274	0.1123	0.0878	0.0735	0.0795

Table 10. Top 10 Participants in the Ad Hoc Track: Article retrieval based on the reference run

Participant	P5	P10	1/rank	map	bpref
p4-Reference	0.6147	0.5294	0.8240	0.3477	0.3333
p36-utampere_given30_nolinks	0.6147	0.5294	0.8240	0.3477	0.3333
p48-LIG-2009-BIC-3B	0.6147	0.5294	0.8240	0.3463	0.3336
p60-UJM_15508	0.5324	0.4544	0.7020	0.2910	0.2925
p346-utCASrefB09	0.5441	0.4750	0.7494	0.2833	0.2768
p167-09RefT	0.3765	0.3603	0.5761	0.2443	0.2540

5.1   CO versus CAS

We now look at the relative effectiveness of the keyword (CO) and structured (CAS) queries. As we saw above, in Section 3, one of the best runs per group for the Relevant in Context Task, and two of the top 10 runs for the Best in Context Task used the CAS query.

Table 11. CAS query target elements over all 115 topics (YAGO tags slanted)

Target Element	Frequency
*	41
article	32
sec	9
group	5
p	4
music_genre	2
vehicles	1
theory	1
song	1
revolution	1
(p|sec|*person*)	1
(p|sec)	1
protest	1
(*person*|*chemist*|*alchemist*|*scientist*|*physicist*)	1
personality	1
museum	1
link	1
image	1
home	1
food	1
figure	1
facility	1
driver	1
dog	1
director	1
(*classical_music*|*opera*|*orchestra*|*performer*|*singer*)	1
bicycle	1
(article|sec|p)	1

All topics have a CAS query since artificial CAS queries of the form

```
//*[about(., keyword title)]
```

were added to topics without CAS title. Table 11 show the distribution of target elements, with YAGO tags in emphatic. In total 81 topics had a non-trivial CAS query.[2] These CAS topics are numbered $2009n$ with n: 001–009, 011–013, 015–017, 020–025, 028–032, 036, 037, 039–045, 048–053, 057, 058, 060, 061, 064–072, 074, 080, 085–096, 098, 099, 102, 105, 106, and 108–115. As it turned out, 50 of these CAS topics were assessed. The results presented here are restricted to only these 50 CAS topics.

Table 12 lists the top 10 participants measured using just the 50 CAS topics and for the Thorough Task (a and b) and the Focused Task (c and d). For the Thorough Task the best CAS run, *p5-BM25BOTthorough*, would have

---

[2] Note that some of the wild-card topics (using the "*" target) in Table 11 had non-trivial about-predicates and hence have not been regarded as trivial CAS queries.

Table 12. Ad Hoc Track CAS Topics: CO runs versus CAS runs

(a) Thorough Task: CO runs

Participant	iP[.00]	iP[.01]	iP[.05]	iP[.10]	MAiP
p48-LIG-2009-thorough-1T	0.5781	0.5706	0.5315	0.4834	0.2729
p6-UAmsIN09article	0.5900	0.5821	0.5149	0.4613	0.2629
p92-Lyon3LIAmanlmnt*	0.5365	0.5039	0.4794	0.4330	0.2450
p5-BM25thorough	0.6273	0.6023	0.5191	0.4620	0.2389
p60-UJM_15494	0.6034	0.5766	0.5131	0.4612	0.2280
p10-MPII-COThBM	0.6436	0.5916	0.5135	0.3783	0.1909
p167-09RefT	0.3245	0.3237	0.2682	0.2392	0.1291
p68-I09LIP6OWATh	0.4146	0.3651	0.2512	0.1963	0.0608
p25-ruc-base-coT	0.5328	0.4333	0.2538	0.1653	0.0505
p72-umd_thorough_3	0.4073	0.2893	0.1697	0.0999	0.0494

(b) Thorough Task: CAS runs

Participant	iP[.00]	iP[.01]	iP[.05]	iP[.10]	MAiP
p5-BM25BOTthorough	0.6460	0.6169	0.5359	0.4472	0.2279
p346-utCASartT09	0.5541	0.5381	0.4819	0.4136	0.2227
p10-MPII-CASThBM	0.5747	0.5308	0.4406	0.3627	0.1651

(c) Focused Task: CO runs

Participant	iP[.00]	iP[.01]	iP[.05]	iP[.10]	MAiP
p78-UWatFERBM25F	0.6742	0.6222	0.4905	0.3758	0.1737
p60-UJM_15525	0.6373	0.6127	0.5696	0.4585	0.2811
p10-MPII-COArBM	0.6201	0.6060	0.5387	0.4648	0.2684
p68-I09LIP6Okapi	0.6130	0.6005	0.5660	0.5064	0.2798
p5-ANTbigramsRangeFOC	0.6089	0.5936	0.5331	0.4531	0.2597
p48-LIG-2009-focused-3F	0.5971	0.5802	0.5205	0.4775	0.2583
p22-emse2009-150*	0.6453	0.5598	0.4211	0.3471	0.1371
p92-Lyon3LIAmanQE*	0.5185	0.5058	0.4815	0.4339	0.2472
p25-ruc-term-coF	0.6277	0.4955	0.2900	0.2065	0.0668
p167-09LrnRefF	0.3357	0.3234	0.2536	0.2211	0.1216

(d) Focused Task: CAS runs

Participant	iP[.00]	iP[.01]	iP[.05]	iP[.10]	MAiP
p6-UamsFSsec2docbi100	0.6151	0.5974	0.4851	0.4230	0.1718
p16-Spirix09R001	0.6201	0.5958	0.5386	0.4920	0.2794
p5-BM25BOTrangeFOC	0.6031	0.5954	0.5470	0.4789	0.2713
p10-MPII-CASFoBM	0.5643	0.5161	0.4454	0.3634	0.1644
p25-ruc-base-casF	0.5114	0.4775	0.4077	0.3214	0.1666
p346-utCASrefF09	0.4353	0.3955	0.3477	0.2781	0.1471
p55-doshisha09f	0.1273	0.0651	0.0307	0.0227	0.0060

ranked sixth amongst the CO runs on MAiP. The two participants submitting both CO and CAS runs had better MAiP scores for the CO runs. However, the best CAS run has higher scores on early precision, iP[0.00] through iP[0.05] than any of the CO submissions. For the Focused Task the best CAS run, *p6-UamsFSsec2docbi100*, would have ranked fifth amongst the CO runs. Two participants submitting both CO and CAS runs had better iP[0.01] scores for the CO

Table 13. Top 10 Participants in the Ad Hoc Track: Article retrieval

Participant	P5	P10	1/rank	map	bpref
p6-UamsTAbi100	0.6500	0.5397	0.8555	0.3578	0.3481
p48-LIG-2009-BIC-1B	0.6059	0.5338	0.8206	0.3573	0.3510
p62-RMIT09title	0.6029	0.5279	0.8237	0.3540	0.3488
p5-BM25ArticleRIC	0.6147	0.5294	0.8240	0.3477	0.3333
p4-Reference	0.6147	0.5294	0.8240	0.3477	0.3333
p36-utampere_given30_nolinks	0.6147	0.5294	0.8240	0.3477	0.3333
p68-I09LIP6OWA	0.6118	0.5147	0.8602	0.3420	0.3258
p10-MPII-COArBP	0.6353	0.5471	0.8272	0.3371	0.3458
p92-Lyon3LIAmanQE*	0.6265	0.5265	0.7413	0.3335	0.3416
p78-UWatFERBase	0.5765	0.5088	0.8093	0.3267	0.3205

runs, one participant had a better CAS run. For Relevant in Context Task (not shown), the best CAS run, *p5-BM25BOTrangeRIC*, would have ranked third among the CO runs. One participants submitting both CO and CAS runs had better MAgP scores for a CO run, another participant had a better CAS run. For the Best in Context Task (not shown), the best CAS run, *p5-BM25BOTbepBIC*, would rank seventh among the CO runs. All three participants submitting both CO and CAS runs had better MAgP scores for their CO runs. Overall, we see that teams submitting runs with both types of queries have higher scoring CO runs, with participant 5 as a notable exception for Focused.

6   Analysis of Article Retrieval

In this section, we will look in detail at the effectiveness of Ad Hoc Track submissions as article retrieval systems.

6.1   Article Retrieval: Relevance Judgments

We will first look at the topics judged during INEX 2009, but now using the judgments to derive standard document-level relevance by regarding an article as relevant if some part of it is highlighted by the assessor. We derive an article retrieval run from every submission using a first-come, first served mapping. That is, we simply keep every first occurrence of an article (retrieved indirectly through some element contained in it) and ignore further results from the same article.

We use `trec_eval` to evaluate the mapped runs and qrels, and use mean average precision (map) as the main measure. Since all runs are now article retrieval runs, the differences between the tasks disappear. Moreover, runs violating the task requirements are now also considered, and we work with all 172 runs submitted to the Ad Hoc Track.

Table 13 shows the best run of the top 10 participating groups. The first column gives the participant, see the companion article [1, Table 3] for the full name of group. The second and third column give the precision at ranks 5 and

10, respectively. The fourth column gives the mean reciprocal rank. The fifth column gives mean average precision. The sixth column gives binary preference measures (using the top R judged non-relevant documents). No less than seven of the top 10 runs retrieve exclusively full articles: only rank two (*p48-LIG-2009-BIC-1B*), rank six (*p36-utampere_given30_nolinks*) and rank ten (*p78-UWatFERBase*) retrieve elements proper. The relative effectiveness of these article retrieval runs in terms of their article ranking is no surprise. Furthermore, we see submissions from all four ad hoc tasks. A run from the Thorough task at rank 1; runs from the Best in Context task at ranks 2 and 3; runs from the Relevant in Context task at ranks 4, 5 and 6; and runs from the Focused task at ranks 7, 8, 9 and 10.

If we break-down all runs over the original tasks, shown in Table 14, we can compare the ranking to Section 3 above. We see some runs that are familiar from the earlier tables: five Thorough runs correspond to Table 3, four Focused runs correspond to Table 4, six Relevant in Context runs correspond to Table 5, and five Best in Context runs correspond to Table 6. More formally, we looked at how the two system rankings correlate using kendall's tau.

- Over all 30 Thorough Task submissions the system rank correlation is 0.646 between MAiP and map.
- Over all 57 Focused task submissions the system rank correlation is 0.420 between iP[0.01] and map, and 0.638 between MAiP and map.
- Over all 33 Relevant in Context submissions the system rank correlation between MAgP and map is 0.598.
- Over all 37 Best in Context submissions the system rank correlation between MAgP and map is 0.517.

Overall, we see a reasonable correspondence between the rankings for the ad hoc tasks in Section 3 and the rankings for the derived article retrieval measures. The correlation between article retrieval and the "in context" tasks was much higher (0.79) for the INEX 2008 collection. This is a likely effect of the increasing length of (relevant) Wikipedia articles in the INEX 2009 collection.

7   Analysis of Result Length

In this section, we will look in detail at the impact of result length on the effectiveness of Ad Hoc Track submissions.

7.1   Impact of Result Length

Focused retrieval and XML retrieval require all, but only, relevant text to be retrieval. This could be taken to suggest that a relatively short result length is optimal. In sharp contrast, researchers found that XML-IR require careful length normalization, effectively boosting the retrieval of longer elements [2, 3].

Let us look in detail at the length of results retrieved by top scoring runs. Table 15 shows for the best Thorough runs of the 10 best scoring groups statistics on number of articles, and characters retrieved (restricted to the 68

Table 14. Top 10 Participants in the Ad Hoc Track: Article retrieval per task

(a) Thorough Task

Participant	P5	P10	1/rank	map	bpref
p6-UamsTABi100	0.6500	0.5397	0.8555	0.3578	0.3481
p48-LIG-2009-thorough-1T	0.6118	0.5191	0.8042	0.3493	0.3392
p92-Lyon3LIAmanlmnt*	0.6382	0.5279	0.7706	0.3305	0.3374
p5-BM25thorough	0.6147	0.5294	0.8240	0.3188	0.3142
p10-MPII-COThBM	0.5853	0.5206	0.8084	0.3087	0.3138
p346-utCASartT09	0.5176	0.4588	0.7138	0.2913	0.2986
p60-UJM_15486	0.5647	0.4765	0.7149	0.2797	0.2884
p68-I09LIP6OWATh	0.4735	0.4353	0.7100	0.2665	0.2745
p72-umd_thorough_3	0.5382	0.4515	0.7406	0.2486	0.2674
p167-09RefT	0.3765	0.3603	0.5761	0.2443	0.2540

(b) Focused Task

Participant	P5	P10	1/rank	map	bpref
p48-LIG-2009-focused-1F	0.6059	0.5338	0.8206	0.3569	0.3506
p5-BM25ArticleFOC	0.6147	0.5294	0.8240	0.3477	0.3333
p68-I09LIP6OWA	0.6118	0.5147	0.8602	0.3420	0.3258
p10-MPII-COArBP	0.6353	0.5471	0.8272	0.3371	0.3458
p92-Lyon3LIAmanQE*	0.6265	0.5265	0.7413	0.3335	0.3416
p78-UWatFERBase	0.5765	0.5088	0.8093	0.3267	0.3205
p60-UJM_15525	0.5824	0.4926	0.8326	0.3256	0.3169
p16-Spirix09R002	0.5206	0.4588	0.7250	0.3133	0.3149
p6-UamsFSsec2docbi100	0.5941	0.4779	0.8958	0.2985	0.2994
p346-utCASartF09	0.5176	0.4588	0.7138	0.2913	0.2986

(c) Relevant in Context Task

Participant	P5	P10	1/rank	map	bpref
p48-LIG-2009-RIC-1R	0.6059	0.5338	0.8206	0.3569	0.3506
p6-UamsRSCMartCMdocbi100	0.6324	0.5309	0.9145	0.3523	0.3374
p5-BM25ArticleRIC	0.6147	0.5294	0.8240	0.3477	0.3333
p4-Reference	0.6147	0.5294	0.8240	0.3477	0.3333
p36-utampere_given30_nolinks	0.6147	0.5294	0.8240	0.3477	0.3333
p346-utCOartR09	0.5324	0.4882	0.7448	0.3120	0.3137
p72-umd_ric_2	0.5441	0.4544	0.7807	0.2708	0.2867
p167-09RefR	0.3765	0.3603	0.5761	0.2443	0.2540
p25-ruc-base-casF	0.4441	0.4176	0.6270	0.2243	0.2523
p60-UJM_15488	0.4382	0.3853	0.6043	0.2146	0.2343

(d) Best in Context Task

Participant	P5	P10	1/rank	map	bpref
p48-LIG-2009-BIC-1B	0.6059	0.5338	0.8206	0.3573	0.3510
p62-RMIT09title	0.6029	0.5279	0.8237	0.3540	0.3488
p5-BM25AncestorBIC	0.6147	0.5294	0.8240	0.3477	0.3333
p36-utampere_given30_nolinks	0.6147	0.5294	0.8240	0.3477	0.3333
p6-UamsBAfbCMdocbi100	0.6147	0.5118	0.8531	0.3361	0.3251
p10-MPII-COBIBM	0.5824	0.5191	0.8451	0.3325	0.3315
p92-Lyon3LIAmanBEP*	0.6382	0.5279	0.7706	0.3305	0.3374
p25-ruc-term-coB	0.5206	0.4779	0.7158	0.3197	0.3251
p346-utCOartB09	0.5324	0.4882	0.7448	0.3120	0.3137
p60-UJM_15508	0.5324	0.4544	0.7020	0.2910	0.2925

Table 15. Top 10 Participants in the Thorough Task: Result length

Participant	MAiP	# articles	# characters	# chars/art
p48-LIG-2009-thorough-3T	0.2855	588	5,621,997	9,554
p6-UAmsIN09article	0.2818	4,947	8,732,588	1,765
p5-BM25thorough	0.2585	632	5,195,586	8,213
p92-Lyon3LIAmanlmnt	0.2496	1,439	13,390,230	9,301
p60-UJM_15494	0.2435	551	1,461,857	2,648
p346-utCASartT09	0.2350	1,496	8,482,533	5,668
p10-MPII-CASThBM	0.2133	1,181	8,099,770	6,854
p167-09RefT	0.1390	1499	13,253,653	8,841
p68-I09LIP6OWATh	0.0630	976	4,400,118	4,508
p25-ruc-base-coT	0.0577	29	50,183	1,707

Table 16. Top 10 Participants in the Focused Task: Result length

Participant	iP[0.01]	# articles	# characters	# chars/art
p78-UWFERBM25F2	0.6333	1,130	1,613,095	1,426
p68-I09LIP6Okapi	0.6141	1,485	16,868,585	11,351
p10-MPII-COFoBM	0.6134	1,319	2,137,482	1,619
p60-UJM_15525	0.6060	1,485	10,420,397	7,016
p6-UamsFSsec2docbi100	0.5997	1,213	5,745,657	4,734
p5-BM25BOTrangeFOC	0.5992	1,498	13,236,136	8,835
p16-Spirix09R001	0.5903	1,496	8,355,434	5,584
p48-LIG-2009-focused-1F	0.5853	1,357	7,570,394	5,576
p22-emse2009-150	0.5844	1,410	6,306,031	4,470
p25-ruc-term-coF	0.4973	29	55,010	1,865

judged topics). There is an enormous spread in the average number of characters per article, which ranges from 1,707 to 9,554. The best run retrieves the highest number of characters per article. Recall from Section 2 that the length of a relevant article is 11,691 characters on average, and the number of relevant characters per article is 3,878 on average. Even runs that are relatively close in score seem to target radically different amounts of text per article.

Table 16 shows for the best Focused runs of the 10 best scoring groups statistics on number of articles, and characters retrieved. We see a similar spread in average number of characters per article, ranging from 1,426 to 11,351. The averages seem lower than for the Thorough Task. The best Focused run retrieves the lowest number of characters per article.

Table 17 shows for the best Relevant in Context runs of the 10 best scoring groups statistics on number of articles, and characters retrieved. We see again considerable spread in average number of characters per article, ranging from 677 to 8,841. The averages seem higher than for the Focused and Thorough Task. The second best Relevant in Context run retrieves the highest number of characters per article.

There is no analysis of result length for the Best in Context Task since for this task only a single best entry point is required.

Table 17. Top 10 Participants in the Relevant in Context Task: Result length

Participant	MAgP	# articles	# characters	# chars/art
p5-BM25RangeRIC	0.1885	1,498	13,215,573	8,821
p4-Reference	0.1847	1,499	13,253,653	8,841
p6-UamsRSCMartCMdocbi100	0.1773	1,230	10,157,349	8,254
p48-LIG-2009-RIC-1R	0.1760	1,357	7,570,394	5,576
p36-utampere_given30_nolinks	0.1720	1,498	10,555,338	7,046
p346-utCASrefR09	0.1188	1,055	8,186,120	7,758
p60-UJM_15502	0.1075	1,102	1,446,938	1,312
p167-09RefR	0.1045	1,499	13,253,653	8,841
p25-ruc-base-casF	0.1028	660	2,814,934	4,264
p72-umd_ric_1	0.0424	464	314,342	677

Table 18. Top 10 Participants in the Thorough Task: Restricted to 500 characters per result

Participant	iP[.00]	iP[.01]	iP[.05]	iP[.10]	MAiP
p5-BM25thorough	0.7032	0.6658	0.5511	0.4687	0.1625
p10-MPII-COThBM	0.6273	0.5009	0.3129	0.2187	0.0687
p6-UamsTSbi100	0.5908	0.4570	0.2900	0.1478	0.0445
p60-UJM_15500	0.6081	0.4896	0.2566	0.1143	0.0438
p48-LIG-2009-thorough-3T	0.6023	0.4792	0.2620	0.1283	0.0423
p25-ruc-base-coT	0.5334	0.4169	0.2387	0.1348	0.0414
p92-Lyon3LIAautolmnt	0.4651	0.3405	0.1878	0.0846	0.0280
p68-I09LIP6OkapiEl	0.3965	0.2839	0.1483	0.0692	0.0234
p72-umd_thorough_3	0.4235	0.2491	0.1103	0.0666	0.0216
p346-utCASartT09	0.5100	0.3574	0.1191	0.0288	0.0209

7.2   Limiting Result Length

In the previous section, we saw considerable spread in the numbers of characters per article retrieved. A partial explanation is the fact that making sure all relevant text is retrieved (avoiding false negatives) is easy, but making sure no non-relevant is retrieved (avoiding false positives) is very hard [4]. This leads to systems that prefer being "safe" (by retrieving whole articles or long elements) over being "sorry" (possibly missing relevant text by aiming for small elements). In many use-cases of focused retrieval there is a down-side to retrieving long excerpts or even entire documents. Think of mobile displays that can only show a certain number of characters, or think of query-biased summaries of documents that appear on the hit lists of modern search engines.

What if we limit the results to a maximum of 500 characters? For each run, we "cut off" each individual result after the first 500 retrieved characters. Table 18 shows the best Thorough run of the top 10 participating groups. This clearly hurts the overall performance, although run *p5-BM25thorough* is strikingly more effective than the other runs. Upon closer inspection, this run used a slice-and-dice approach to turn an article ranking into a list of all (overlapping) elements which contained at least one of the search terms. Recall from Table 15 that this

Table 19. Top 10 Participants in the Ad Hoc Track: Restricted to 500 characters per article

(a)Thorough Task

Participant	iP[.00]	iP[.01]	iP[.05]	iP[.10]	MAiP
p10-MPII-COThBM	0.6357	0.4649	0.1893	0.0805	0.0335
p5-ANTbigramsBOTthorough	0.6337	0.4911	0.2031	0.0643	0.0326
p60-UJM_15500	0.5934	0.4305	0.1486	0.0387	0.0245
p48-LIG-2009-thorough-3T	0.5728	0.4109	0.1387	0.0295	0.0231
p6-UAmsIN09article	0.5671	0.4080	0.1452	0.0265	0.0228
p92-Lyon3LIAmanlmnt*	0.5024	0.3352	0.1253	0.0429	0.0215
p346-utCASartT09	0.5100	0.3574	0.1191	0.0288	0.0209
p25-ruc-base-coT	0.5876	0.3704	0.0965	0.0236	0.0177
p68-I09LIP6OkapiEl	0.4137	0.2355	0.0728	0.0362	0.0149
p167-09RefT	0.3131	0.2143	0.0807	0.0245	0.0137

(b) Focused Task

Participant	iP[.00]	iP[.01]	iP[.05]	iP[.10]	MAiP
p78-UWatFERBM25F	0.6776	0.5304	0.2517	0.1179	0.0414
p10-MPII-COFoBM	0.6330	0.4684	0.2025	0.0828	0.0346
p5-BM25FOC	0.6159	0.4532	0.1746	0.0504	0.0278
p60-UJM_15525	0.5931	0.4337	0.1510	0.0388	0.0251
p16-Spirix09R002	0.5626	0.4226	0.1719	0.0499	0.0256
p68-I09LIP6Okapi	0.5885	0.4138	0.1212	0.0272	0.0220
p48-LIG-2009-focused-3F	0.5665	0.4044	0.1370	0.0341	0.0230
p25-ruc-term-coF	0.6311	0.4044	0.0879	0.0413	0.0206
p22-emse2009-150*	0.6223	0.3700	0.1218	0.0352	0.0225
p6-UamsFSsec2docbi100	0.6346	0.3672	0.1212	0.0220	0.0207

(c) Relevant in Context Task

Participant	gP[5]	gP[10]	gP[25]	gP[50]	MAgP
p5-ANTbigramsRIC	0.2308	0.2069	0.1743	0.1367	0.1291
p36-utampere_given30_nolinks	0.1952	0.1909	0.1444	0.1201	0.1215
p6-UamsRSCMartCMdocbi100	0.1835	0.1565	0.1314	0.1132	0.1182
p4-Reference	0.1755	0.1671	0.1317	0.1065	0.1116
p48-LIG-2009-RIC-3R	0.1704	0.1634	0.1288	0.1039	0.1082
p60-UJM_15502	0.1471	0.1325	0.1041	0.0806	0.0857
p25-ruc-base-casF	0.1894	0.1736	0.1396	0.1248	0.0828
p346-utCASrefR09	0.1206	0.1072	0.0831	0.0670	0.0720
p167-09RefR	0.0965	0.0873	0.0800	0.0729	0.0667
p72-umd_ric_1	0.0642	0.0571	0.0444	0.0342	0.0321

run still retrieved 8,213 characters per article, so a better and more realistic filter would be to limit the number of characters retrieved per article.

We change our analysis and for each run, we "cut off" the results after having retrieved the first 500 retrieved characters per article (so any further text from the same article is ignored, and the result is removed from the run). Table 19 shows the results of restricting the official submissions to maximally 500 characters per article. This naturally leads to a much lower score on the overall measures, and a somewhat lower score on early ranks. We see some familiar runs from Tables 3–5 before, but also some new runs. The best run for the Thorough Task is a variant of the seventh ranked run in Table 3; the best run for the Focused Task was also the best run in Table 4; and the best run for the Relevant in Context Task is a variant of the best run in Table 5.

The system rank correlation (Kendall's Tau) between the official ranking and the restricted run ranking is the following.

- Over all 30 Thorough Task submissions the system rank correlation is 0.545.
- Over all 57 Focused Task submissions the system rank correlation is 0.633.
- Over all 33 Relevant in Context Task submissions the system rank correlation is 0.655.

Overall, we see a reasonable correspondence between the rankings for the ad hoc tasks in Section 3 and the rankings for the restricted runs in this section. This comes as no surprise since both task share an important aspect: finding those articles that contain relevant information.

8   Discussion and Conclusions

In this paper we analyzed the results of the INEX 2009 Ad Hoc Track. For details of the tasks, measures, and outcomes, we refer to the INEX 2009 Ad Hoc track overview paper [1]. In this paper, we focused on six different aspects of the ad hoc track evaluation, which we will discuss in turn.

First, we examined in detail the relevance judgments. The 2009 collection contained 2,666,190 Wikipedia articles (October 8, 2008 dump of the Wikipedia), which is four times larger than the earlier Wikipedia collection. What was the effect of this change in corpus size? We saw that the collection's size had little impact, but that the relevant articles were much longer (a mean length 3,030 in 2008 and 5,775 in 2009, a 52% increase), leading to a lower fraction of highlighted text per article (a mean of 58% in 2008 and 33% in 2009). This also reduced the correlation between focused retrieval and article retrieval, e.g., from 79% for the "in context" tasks in 2008 to 51–58% in 2009.

Second, we studied the resulting system rankings, for each of the four ad hoc tasks, and determined whether differences between the best scoring participants are statistically significant. The early precision measure of the focused task, interpolated precision at 1% recall, is inherently unstable, and only very few of the differences between runs are statistically significant. The overall measures, the MAiP and MAgP variants of mean average precision, are able to distinguish the majority of pairs of runs. Almost 3/4 of system pairs are significantly different with the mean average generalized precision measure of the Relevant in Context task.

Third, we restricted our attention to particular run types: element and passage runs, keyword and phrase query runs, and systems using a reference run with a solid article ranking. Thirteen submissions used ranges of elements or FOL passage results, whereas 144 submissions used element results. Still the non-element submissions were competitive with the top ranking runs for both the

Focused and Relevant in Context Tasks, and the second ranking run for the Best in Context Task. Ten submissions used the explicitly annotated phrases of the phrase query. Phrase query runs were competitive with several of them in the overall top 10 results, but the impact of the phrases seemed marginal. Recall, that the exact same terms were present in the CO query, and the only difference was the phrase annotation. There were 19 submissions using the reference run providing a solid article ranking for further processing. These runs turned out to be competitive, with runs in the top 10 for all tasks. Hence the reference run was successful in helping participants to create high quality runs. However, runs based on the reference run were not directly comparable, since they had different underlying article rankings.

Fourth, we examined the relative effectiveness of content only (CO, or Keyword) search as well as content and structure (CAS, or structured) search. We found that for all tasks the best scoring runs used the CO query but some CAS runs were in the top 10 for all four tasks. Part of the explanation may be in the low number of CAS submissions (40) in comparison with the number of CO submissions (117). Only 50 of the 68 judged topics had a non-trivial CAS query, and the majority of those CAS queries made only reference to particular tags and not on their structural relations. The YAGO tags potentially expressing an information need naturally in terms of structural constraints, were popular: 36 CAS queries used them (21 of them judged). Over the 50 non-trivial CAS queries, most groups had a better performing run using the CO query. A notable exception was QUT who had better performance for CAS on the Focused Task.

Fifth, we looked at the ability of focused retrieval techniques to rank articles. As in earlier years, we saw that article retrieval is a reasonably effective at XML-IR: for each of the ad hoc tasks there were three article-only runs among the best runs of the top 10 groups. When looking at the article rankings inherent in all Ad Hoc Track submissions, we saw that again three of the best runs of the top 10 groups in terms of article ranking (across all three tasks) were in fact article-only runs. This also suggests that element-level or passage-level evidence is valuable for article retrieval. When comparing the system rankings in terms of article retrieval with the system rankings in terms of the ad hoc retrieval tasks, over the exact same topic set, we see a reasonable correlation. The systems with the best performance for the ad hoc tasks, also tend to have the best article rankings.

Sixth, we studied the length of retrieved results, and looked at the impact of restricting result length. We looked at the average number of characters per article that each run retrieved, and found that there is an enormous spread from less than 2,000 characters (less than the mean length of relevant text per article) to over 10,000 characters (longer than the mean length of relevant articles). Even runs scoring close on the Ad Hoc Track measures could apply radically different strategies. For many use-cases the result length is an issue, and we modified the official submission so that only the first retrieved 500 characters per article were retained. There resulting system rankings show agreement with the original scores, with a system rank correlation in the range 0.55–0.66, but also some new runs in the top 10 per task.

Acknowledgments. Jaap Kamps was supported by the Netherlands Organization for Scientific Research (NWO, grants 612.066.513, 639.072.601, and 640.001.501).

References

[1] Geva, S., Kamps, J., Lehtonen, M., Schenkel, R., Thom, J.A., Trotman, A.: Overview of the INEX 2009 ad hoc track. In: Geva, S., Kamps, J., Trotman, A. (eds.) Focused Retrieval and Evaluation. LNCS, pp. 4–25. Springer, Heidelberg (2010)

[2] Kamps, J., Marx, M., de Rijke, M., Sigurbjörnsson, B.: XML retrieval: What to retrieve? In: Clarke, C., Cormack, G., Callan, J., Hawking, D., Smeaton, A. (eds.) Proceedings of the 26th Annual International ACM SIGIR Conference on Research and Development in Information Retrieval, pp. 409–410. ACM Press, New York (2003)

[3] Kamps, J., de Rijke, M., Sigurbjörnsson, B.: Length normalization in XML retrieval (extended abstract). In: Verbrugge, R., Taatgen, N., Schomaker, L. (eds.) BNAIC-2004: Proceedings of the 16th Belgium-Netherlands Conference on Artificial Intelligence, pp. 369–370 (2004)

[4] Kamps, J., Koolen, M., Lalmas, M.: Locating relevant text within XML documents. In: Myaeng, S.-H., Oard, D.W., Sebastiani, F., Chua, T.-S., Leong, M.-K. (eds.) Proceedings of the 31st Annual International ACM SIGIR Conference on Research and Development in Information Retrieval, pp. 847–849. ACM Press, New York (2008)

A   Appendix: Full Run Names

Group	Run	Label	Task	Query	Results	Notes
4	617	Reference	RiC	CO	Ele	Reference run Article-only
5	744	BM25AncestorBIC	BiC	CO	Ele	Article-only
5	749	ANTbigramsRIC	RiC	CO	Ele	
5	757	BM25thorough	Tho	CO	Ele	
5	775	BM25ArticleFOC	Foc	CO	Ele	Article-only
5	776	BM25FOC	Foc	CO	Ele	
5	777	BM25RangeFOC	Foc	CO	Ran	Article-only
5	781	BM25BOTrangeFOC	Foc	CAS	Ran	Article-only
5	792	ANTbigramsRangeFOC	Foc	CO	Ran	Article-only
5	796	BM25ArticleRIC	RiC	CO	Ele	Article-only
5	797	BM25RangeRIC	RiC	CO	Ran	Article-only
5	804	BM25BOTrangeRIC	RiC	CAS	Ran	Article-only
5	807	ANTbigramsBOTthorough	Tho	CAS	Ele	
5	808	BM25BOTthorough	Tho	CAS	Ele	
5	824	BM25bepBIC	BiC	CO	Ele	Article-only
5	825	BM25BOTbepBIC	BiC	CAS	Ele	Article-only
6	634	UAmsIN09article	Tho	CO	Ele	Article-only
6	810	UamsTABi100	Tho	CO	Ele	Article-only
6	811	UamsTSbi100	Tho	CO	Ele	
6	813	UamsFSsec2docbi100	Foc	CAS	Ele	
6	814	UamsRSCMartCMdocbi100	RiC	CO	Ele	
6	816	UamsBAfbCMdocbi100	BiC	CO	Ele	Article-only
6	817	UamsBSfbCMsec2docbi100art1	BiC	CAS	Ele	Article-only
10	618	MPII-CASFoBM	Foc	CAS	Ele	
10	619	MPII-COFoBM	Foc	CO	Ele	
10	620	MPII-CASThBM	Tho	CAS	Ele	
10	621	MPII-COThBM	Tho	CO	Ele	
10	628	MPII-COArBM	Foc	CO	Ele	Article-only
10	632	MPII-COBIBM	BiC	CO	Ele	Article-only
10	700	MPII-COArBP	Foc	CO	Ele	Article-only
10	709	MPII-COArBPP	Foc	CO	Ele	Phrases Article-only
16	872	Spirix09R001	Foc	CAS	Ele	Article-only
16	873	Spirix09R002	Foc	CAS	Ele	Article-only
22	672	emse2009-150	Foc	CO	Ele	Phrases Manual
25	727	ruc-base-coT	Tho	CO	Ele	
25	737	ruc-term-coB	BiC	CO	Ele	
25	738	ruc-term-coF	RiC	CO	Ele	
25	739	ruc-term-coF	Foc	CO	Ele	
25	898	ruc-base-casF	Foc	CAS	Ele	
25	899	ruc-base-casF	RiC	CAS	Ele	
36	688	utampere_given30_nolinks	RiC	CO	Ele	Reference run
36	701	utampere_given30_nolinks	BiC	CO	Ele	Reference run
36	708	utampere_auth_40_top30	RiC	CO	Ran	
48	682	LIG-2009-thorough-1T	Tho	CO	Ele	
48	684	LIG-2009-thorough-3T	Tho	CO	Ele	Reference run
48	685	LIG-2009-focused-1F	Foc	CO	Ele	
48	686	LIG-2009-focused-3F	Foc	CO	Ele	Reference run

Continued on Next Page...

Group	Run	Label	Task	Query	Results	Notes
48	714	LIG-2009-RIC-1R	RiC	CO	Ele	
48	716	LIG-2009-RIC-3R	RiC	CO	Ele	Reference run
48	717	LIG-2009-BIC-1B	BiC	CO	Ele	
48	719	LIG-2009-BIC-3B	BiC	CO	Ele	Reference run
55	836	doshisha09f	Foc	CAS	Ele	
60	819	UJM_15518	Foc	CO	Ele	Reference run
60	820	UJM_15486	Tho	CO	Ele	
60	821	UJM_15500	Tho	CO	Ele	
60	822	UJM_15494	Tho	CO	Ele	Reference run
60	827	UJM_15488	RiC	CO	Ele	
60	828	UJM_15502	RiC	CO	Ele	
60	829	UJM_15503	RiC	CO	Ele	Reference run
60	830	UJM_15490	BiC	CO	Ele	
60	832	UJM_15508	BiC	CO	Ele	Reference run
60	868	UJM_15525	Foc	CO	Ele	Article-only
62	895	RMIT09title	BiC	CO	Ele	Article-only
62	896	RMIT09titleO	BiC	CO	FOL	Article-only
68	679	I09LIP6Okapi	Foc	CO	Ele	Article-only
68	681	I09LIP6OWA	Foc	CO	Ele	Article-only
68	703	I09LIP6OkapiEl	Tho	CO	Ele	
68	704	I09LIP6OWATh	Tho	CO	Ele	
72	666	umd_ric_1	RiC	CO	Ele	
72	667	umd_ric_2	RiC	CO	Ele	
72	870	umd_thorough_3	Tho	CO	Ele	
78	706	UWatFERBase	Foc	CO	FOL	
78	707	UWatFERBM25F	Foc	CO	FOL	
92	694	Lyon3LIAautoBEP	BiC	CAS	Ele	Phrases
92	695	Lyon3LIAmanBEP	BiC	CO	Ele	Phrases Manual Article-only
92	697	Lyon3LIAmanQE	Foc	CO	Ele	Phrases Manual Article-only
92	698	Lyon3LIAautolmnt	Tho	CO	Ele	Phrases
92	699	Lyon3LIAmanlmnt	Tho	CO	Ele	Phrases Manual
167	651	09RefT	Tho	CO	Ele	Reference run Article-only
167	654	09LrnRefF	Foc	CO	Ele	Reference run Article-only
167	657	09RefR	RiC	CO	Ele	Reference run Article-only
167	660	09LrnRefB	BiC	CO	Ele	Reference run Article-only
346	637	utCASartT09	Tho	CAS	Ele	Article-only
346	638	utCASartF09	Foc	CAS	Ele	Article-only Invalid
346	639	utCOartR09	RiC	CO	Ele	Article-only Invalid
346	640	utCOartB09	BiC	CO	Ele	Article-only Invalid
346	645	utCASrefF09	Tho	CAS	Ele	Reference run
346	646	utCASrefF09	Foc	CAS	Ele	Reference run
346	647	utCASrefR09	RiC	CAS	Ele	Reference run
346	648	utCASrefB09	BiC	CAS	Ele	Reference run

ENSM-SE at INEX 2009 : Scoring with Proximity and Semantic Tag Information

Michel Beigbeder[1], Amélie Imafouo[1], and Annabelle Mercier[2]

[1] École Nationale Supérieure des Mines de Saint-Étienne
158, cours Fauriel
F-42023 Saint Etienne Cedex 2, France
[2] LCIS Lab - Grenoble University
50, rue Barthelemy de Laffemas
BP 54 - F-26902 Valence Cedex 9, France

Abstract. We present in this paper some experiments on the Wikipedia collection used in the INEX 2009 evaluation campaign with an information retrieval method based on proximity. The idea of the method is to assign to each position in the document a fuzzy proximity value depending on its closeness to the surrounding keywords. These proximity values can then be summed on any range of text – including any passage or any element – and after normalization this sum is used as the relevance score for the extent. To take into account the semantic tags, we define a contextual operator which allow to consider at query time only the occurrences of terms that appear in a given semantic context.

1 Introduction

Within the context of structured documents, the idea of adhoc retrieval is a generalization of what it is in the flat document context: searching a static set of documents using a new set of topics, and returning arbitrary XML elements or passages instead of whole documents.

For the full document retrieval task, with many of the popular models the documents and the queries are represented by bags of words. The ranking scores are computed by adding the contributions of the different query terms to the score of one document.

With the bag of words representation the positions of words in the documents are not used for scoring. While the absolute position of terms does not seem to be significant for scoring, we can have some intuition that the relative positions of the query terms could be relevant. For instance with a query with the two words **roman** and **architecture**, one occurrence of each of these terms in a document gives the same contribution to the score either these two occurrences are close or not in a document. Taking into account the relative positions at least involves that two query terms does appear in a document so that some proximity of the occurrences of these two terms can be considered. Thus some kind of conjunctive interpretation of the query has to be done if the query itself is a simple list of words.

S. Geva, J. Kamps, and A. Trotman (Eds.): INEX 2009, LNCS 6203, pp. 49–58, 2010.
© Springer-Verlag Berlin Heidelberg 2010

Section 2 is devoted to a brief state of the art of techniques and models used in flat information retrieval to introduce the proximity of query term occurrences in the score of documents. In the next section we successively present our scoring model for flat documents, its extension for hierarchically structured documents and how we used semantic tags. Our experiments with this model at the INEX 2009 campaign are detailed in the section 5.

2   A Brief State of the Art of Proximity Usage

One of the first use of term position appeared within implementations of the boolean model with the NEAR operator. But there are two main problems with this operator, the first one is that the semantics of this operator is not clean and it leads to some inconsistency problems as it was noticed by [1]. The second one is that it is stuck to boolan retrieval: documents verify or not the query and there is no ranking.

As suggested by the sample query with the two terms **roman** and **architecture**, there are relationships between the use of proximity and indexing/ranking with phrases. Thus one highly tested idea is to discover phrases in the corpus and then to index the documents with both these phrases and the usual single terms. The conclusion [2] is that this method does not improve retrieval effectiveness except for low quality retrieval methods.

In the phrase discovery works, phrases are only looked for with adjacent words and only some of them are then retained for both indexation and scoring. To relax this constraint about phrases a proposition is to compute the score of a document as the sum of two scores [3], the first one is the usual Okapi BM25 score, and the second one is the proximity score. Later investigated [4], the conclusion was that proximity use is more useful as documents are longer and as the collection is larger.

Another idea to take into account the term proximity with the Okapi BM25 framework circumvents the problem of coherence between word scoring and relaxed phrase scoring [5,6].

As a synthesis of all these experiments we can derive a conclusion that it is better to consider several terms of the query in the proximity consideration. A second intuition is that their co-occurrence must be considered in a relaxed way by accepting that a quite large number of other words could intermix with the query terms. This second conclusion was also formulated by [7]. She experimented a quite simple method with a usual vectorial ranking but the results were then filtered through a conjunctive filter which impose that all the query terms occur in a returned document and in a more strict version that they occur in a passage shorter than 100 to 300 words.

To take into account proximity as phrase relaxation with a continuous paradigm two ideas were experimented. The first one was developed for the TREC 1995 campaign by two independant teams [8,9]. Intervals that contain all the query terms are searched. Each selected interval receives a score, higher as the interval is shorter. [10] later experimented their method and concluded that it is quite beneficial for short queries and for the first recall levels.

The last idea is based on the influence that each query term occurrence exercises on its neighbouring [11]. Each position in the text receives an influence from the occurrences of the query terms, and all these influences are added. Then either the maximum or the summation of this function is the score of the document. This influence idea was reused in a boolean framework, mostly used in a conjunctive way[12]. This last method is developped in the next section as it is the basis for the model we will describe for structured documents.

3   Fuzzy Proximity Model

3.1   Fuzzy Proximity Model for Flat Documents

As in any other information retrieval model, the textual documents are represented with the terms that occur in them. In the sequel, we will call T the set of terms appearing in the collection.

We want to represent each document with the positions of the term occurrences. This can easily be achieved by representing a document d with a function over $\mathbb{N}$, the set of positive integers. We will also call d this function:

$$d : \mathbb{N} \to T$$
$$x \mapsto d(x)$$

and $d(x)$ is the term that appears at position x in the document d. With this notation $d^{-1}(t)$ is the set of positions where the term t occurs in the document d — in a positional inverted file this set is sorted and represented by the list of postings for the term t in the document d.

A collection is a set of documents. Figure 1 displays an example of a collection of four documents (d_0 to d_3) where only two different elements of T, A and B, are showed. In this example, we have for instance: $d_3(2) = A$, $d_3(3) = A$, and $d_3^{-1}(A) = \{2, 3\}$.

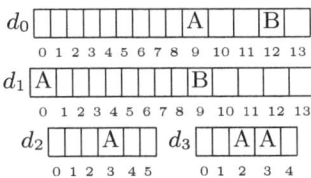

Fig. 1. Example of a collection C, A and B are two distinct elements of T

Fuzzy proximity to a term. Instead of trying to define a proximity measure between the query terms by taking into account the position of the term occurrences as it is done for instance by the length of intervals containing all the query terms, our approach defines a proximity to the query at each position in the document.

Thus the first step is to define a (fuzzy) proximity between a position in the text and one term. Formally, given some $t \in T$, we define $\mu_t^d : \mathbb{Z} \to [0, 1]$ with

$$\mu_t^d(x) = \max_{i \in d^{-1}(t)} \left(\max \left(\frac{k - |x - i|}{k}, 0 \right) \right),$$

where k is some integral parameter which controls to which extent one term occurrence spreads its influence. The function μ_t^d reaches its maximum (the value 1) where the term t appears and it decreases with a constant slope down to zero on each sides of this maximum. In other terms, this function has a triangular shape at each occurrence of the term t. Fig. 2.a shows $(\mu_A^d)_{d \in C}$ and $(\mu_B^d)_{d \in C}$ for the collection C shown in Fig. 1, with k set to 4. This function can be interpreted as the membership degree of a text position x in the document d to a fuzzy set $P(d, t)$.

a) $(\mu_A^d)_{d \in C}$ (plain lines) and $(\mu_B^d)_{d \in C}$ (dotted lines)

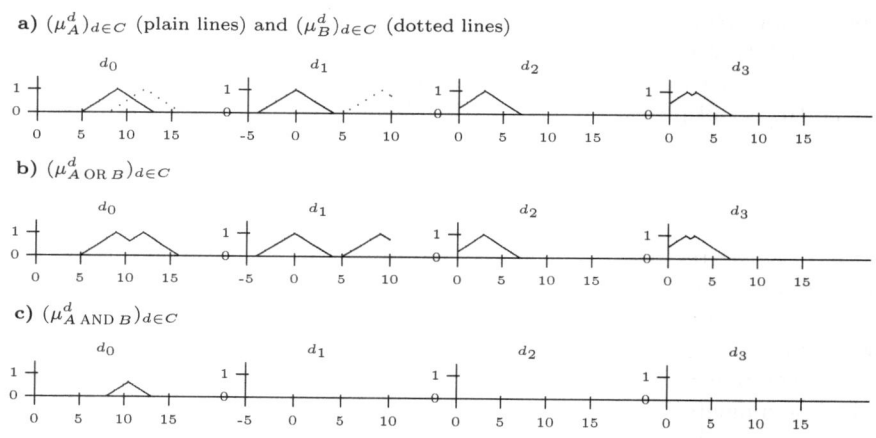

b) $(\mu_{A \text{ OR } B}^d)_{d \in C}$

c) $(\mu_{A \text{ AND } B}^d)_{d \in C}$

Fig. 2. Different $(\mu_q^d)_{d \in C}$ for the collection C of Fig. 1 with $k = 4$

Fuzzy proximity to a query. We will now generalize the fuzzy proximity to a term to a fuzzy proximity to a query. Again it is a local proximity as it is defined for each position in the document. We use boolean queries, which thus can be represented as trees. The leaves of these trees are the terms that appear in the query, and the internal nodes are boolean operators: AND, OR and NOT.

The functions μ_t^d defined in the previous section are associated to the leaves of the query tree. Considering a conjunctive node with two terms A and B, we want to measure for each position how close it is from both terms. This can easily be achieved by using a formula for computing the degree of membership of an intersection in the fuzzy set theory. They are known as *t-norms*. In our experiments we used the minimum t-norm: $\top_{\min}(a, b) = \min(a, b)$. For a disjunctive node, we used the complementary co-norm $\mu_{q_1 \text{OR} q_2}^d = \max(\mu_{q_1}^d, \mu_{q_2}^d)$ Finally for a complement node, we define $\mu_{\text{NOT} q}^d = 1 - \mu_q^d$.

Thus with these definitions μ_q^d is defined for every boolean query q. In the implementation, these formulas are recursively applied with a post-order tree traversal. The result is the function at the root of the tree, which we call *local fuzzy proximity to the query*. This function can be interpreted as the membership degree of the text positions in the document d to the fuzzy set $P(d,q)$. With a purely conjunctive query, this function has higher values as all the query terms are close to a given position.

Fig. 2.b (resp. Fig. 2.c) plots $\mu_{A\,OR\,B}^d$ (resp. $\mu_{A\,AND\,B}^d$) for the documents of the collection C of Fig. 1. Note that the function $\mu_{A\,AND\,B}^{d_1}$ is uniformly zero, though the document d_1 contains both the terms A and B because their occurrences are not close enough.

Score of documents and passages. With the local proximity μ_q^d defined in the previous section it is easy to define a global proximity of any range of positions, either for a full document or for any passage between the positions x_1 and x_2 with

$$\sum_{x_1 \leq x \leq x_2} \mu_q^d(x)$$

To take into account the specificity of the range to the query we then normalize this score by the length of the passage, and finally the score of a passage p between the positions x_1 and x_2 is

$$s(q,p) = \frac{\sum_{x_1 \leq x \leq x_2} \mu_q^d(x)}{x_2 - x_1 + 1}. \tag{1}$$

3.2   Fuzzy Proximity Model for Structured Documents

Given the proximity model presented in the previous section, we will now deal with its extension to the structured case. We just deal with the hierarchical aspect of structure. In this hierarchy, the components are the nested sections and their title. So, we have to model the influence of an occurrence of a query term by taking into account in which type of elements it appears in, either in a section-like element or in a title-like element.

For a term occurrence which appears in a section-like element, the basis is the same as in flat text: A decreasing value in regards to the distance to the occurrence. We add another constraint, the influence is limited to the section in which the occurrence appears.

For a term occurrence which appears in a title-like element its influence is extended to the full content of the surrounding (ancestor) section-like element and recursively to the section-like elements contained in it. Here the assumption is that the title is descriptive of the content of the section it entitles. In the proximity paradigm, any title term in a title-like element should be close to any term occurrence in the corresponding section. So our choice is that the influence of a title term is set to the maximum value (value 1) over the whole section.

For a single term some term influence areas (triangles) can overlap and it results in the "mountain" aspect in the document representation. Truncated triangles can also be viewed, which indicates that the influence of an occurrence of the term has been limited to the boundaries of the section-like element it belongs to. Finally, rectangles can be seen where an occurrence appears in a title-like element and the influence was uniformly extended to the whole bounding section-like element.

Once the proximity function is computed for a document at the root of the query tree, to answer to the focused task we have to select some elements and compute their relevance value. For a given document, the scores of every section-like elements are computed according to formula 1. Those elements that are relevant (i.e. their relevance score is not zero) are inserted in a list. This list is sorted according to the scores in decreasing order. Then an iterative algorithm is applied to this list of elements. The top element is removed from the list and inserted in the output list. All its descendants and all the elements that belongs to its path to the document tree root are removed from the list in order to avoid overlapping elements in the output list. Then the process is repeated until the list is empty. When all the documents have been processed, the output list is sorted by decreasing relevance score.

3.3   Fuzzy Proximity Model with Semantic Tags

In the INEX 2009 Wikipedia collection, the documents are annotated with the 2008-w40-2 version of YAGO. Thus the collection includes semantic annotations for articles and outgoing links, based on the WordNet concepts that YAGO assigns to Wikipedia articles. Our previously presented proximity model for structured documents was extended to take into account these semantic tags. The terms contained in the name of a semantic tag are added to the content of the article within a new tag `<t>`. Thus it will be possible at query time time to know if a given occurrence of a term appeared in the original text or if it was added as a member of a semantic tag. For instance, the tag `<literary_composition>`... is replaced by `<literary_composition><t>literary composition</t>`...

A new binary operator `<>` is introduced in the boolean query model to help using the semantic tags for the disambiguation of queries. This operator connects two boolean expressions: $expr_1$ `<>` $expr_2$. For ease of input, this operator could also be written with the following syntax: `<`$expr_1$`>` $expr_2$. The idea is to dictate conditions defined by $expr_1$ on the occurrences of terms found in $expr_2$ for them to be taken into consideration. For instance if an ambiguous term appears in a query, e.g. *queen*, the user could be interested either in a sovereign, or the actual name of someone, or the rock group. Figure 3 displays three short and simplified extracts from three documents where the term `queen` appears with these three senses.

The query `queen` considers the three occurrences displayed in Fig. 3. But `<person>` `queen` will eliminate the third one, on the contrary `<group>` `queen` will only consider the third one. Finally, the context itself is evaluated as an expression, so that the query `<anarchist & artist>` `queen` will consider the occurrence of Doc. 4291000 in our example.

Doc. 194000   Doc. 4291000   Doc. 5662000

```
<sovereign>       <person>              <music>
  <person>          <anarchist>           <musical_organization>
   Queen Mary         <artist>              <group>
  </person>            Oliver Queen          Queen
</sovereign>         </artist>             </group>
```

Fig. 3. Three simplified extracts from three documents

Expression $expr_1$ is represented by a tree whose leaves are terms, as any boolean expression. This expression is evaluated as any expression in regards to the boolean operators, but the proximity to an occurrence of a leaf term is different. If an occurrence of a term appears outside of any <t> </t> couple, it simply is ignored. If an occurrence of a term appears in some <t> </t> couple the proximity to this occurrence is a rectangle that spans the father of this element which thus is an element labelled with a semantic tag.

For instance, remembering that the tag <literary_composition>... is replaced by <literary_composition><t>literary composition</t>... The first expression (literary) of the query <literary> hercule will produce a proximity equal to one over any element whose tag is literary_composition, or literary_journal or literary_movement, etc. And thus only the occurrences of hercule that appear in the content of semantic tags that contain the word literary will be considered for scoring the documents.

4   Experiments

4.1   Semantic Tag Name Transformation

As stated in section 3.3, a semantic tag <ST>CONTENT</ST> is expanded in the form <ST><t>text(ST)</t>CONTENT</ST> where text(ST) denotes a text extracted from the tag name ST.

To extract a text from a tag name, the first applied transformation is a character conversion: é to e, ô to o, etc. Some tags contain some strings (e.g.: http, www) that show that they were generated from an URL. In this case no text is generated from the tag name and text(ST) is empty. Non alphabetical characters like _ and numerical digits are replaced by the space character, consecutive space characters are replaced by a single space and all the space characters at the beginning or at the end of a tag name are removed. Here are two examples of initial tags and their transformations to text: fictional_character becomes fictional character, __Columbus_Jets-_____International_League_____1955-1970____seating_capacity becomes Columbus Jets International League seating capacity.

4.2   Query Construction

The queries for our experiments were manually built. Our queries are based on the `<title>` field of the topics. In fact, this field could be used as is for the queries because spaces between words or phrases are analyzed by our query parser as AND operators, for instance the title field `yoga exercise` of topic 2009003 would be interpreted as `yoga & exercise`. Though, as the conjunctive interpretation is very restrictive, we often added some variations of terms with disjunctive operators. These variations were either flexional variations or found in the narrative field or derived by common knowledge. Sometimes some clues about the information needs do not appear in the title field, and we were able to add some hints in the query. For instance, again for topic 2009003, the title field doesn't mention the term `history` though the narrative field precises that yoga history is not required. The final query for this topic was `yoga (exercise | exercises | courses | lessons | postures | poses | positions) ! history` (The `!` operator denotes the unary NOT operator of the boolean model).

In some cases, the castitle field contains clues about the semantic context. For instance the castitlle field of topic 2009006 is `//(classical_music | opera | orchestra | performer | singer)[about(., italian spanish opera singer -soprano)]` and our query was `<singer | performer | orchestra | classical music> ((italian | spanish) opera ! soprano)`.

5   Experimental Results and Conclusion

For the INEX 2009 campaign, we submit four focused runs. Their effectiveness with the official measure are given in figure 4. Experimental runs are based on variations of three parameters:

- *Influence extent, k.* The parameter k controls the influence extent of an occurrence of a term (cf. section 3.1). The measure unit of this parameter is in number of terms. We chose $k = 200$ to match the length of a paragraph and $k = 1000$ to match the length of a section.
- *Stemming or not.* Stemming was applied or not to the documents and the query terms.
- *Propagation or not.* The propagation technique explained in section 3.2 was used or not. If not, the text is considered as a flat text with no structure.

Rank	$iP[0.01]$	Run name	Features
21	0.5844	emse2009-150	$k = 200$, no stemming, propagation
25	0.5733	emse2009-153	$k = 1000$, no stemming, propagation
35	0.5246	emse2009-151	$k = 200$, stemming, propagation
47	0.3360	emse2009-152	$k = 1000$, no stemming, no propagation

Fig. 4. Runs submitted for the focused task with our proximity based method

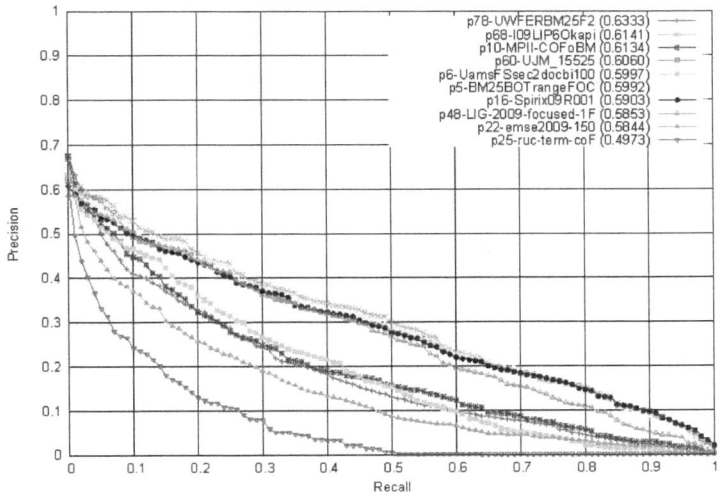

Fig. 5. Results for focused task taken on the INEX 2009 Web site

Our best official run uses title propagation, no stemming and k=200. We can notice a small effectiveness degradation with the enlargement of k to one thousand. Though the effectiveness difference is small and further experiments are needed to conclude about the value of this parameter.

Stemming does not seem to be a useful feature. As our queries contain by themselves some variations of terms (flexional variations among other ones), it is likely that stemming adds some noise.

The features that lead to our worst result within our experiments were: no stemming, no propagation and k=1000. The sole difference with our second ranked run is the absence of propagation, and results are significantly worse without propagation. This is an important clue that structure can be helpful for information retrieval effectiveness.

Another particularity of our method is that precision quickly get worse as recall increases and quicker than most other methods as it can be seen on Fig. 5. This figure displays the results obtained by the ten best runs per institute, our best run (emse2009-150) was one of them.

Our conclusion is that taking into account the structure by giving more influence to title terms is beneficial to effectiveness. A second conclusion is that scoring with proximity give good results in terms of precision but recall still has to be improved.

Acknowledgements

This work is supported by the *Web Intelligence Project* of the "Informatique, Signal, Logiciel Embarqué" cluster of the Rhône-Alpes region.

References

1. Mitchell, P.C.: A note about the proximity operators in information retrieval. In: Proceedings of the 1973 meeting on Programming languages and information retrieval, pp. 177–180. ACM Press, New York (1974)
2. Mitra, M., Buckley, C., Singhal, A., Cardie, C.: An analysis of statistical and syntactic phrases. In: Proceedings of RIAO 1997, 5th International Conference "Recherche d'Information Assistee par Ordinateur", pp. 200–214 (1997)
3. Rasolofo, Y., Savoy, J.: Term proximity scoring for keyword-based retrieval systems. In: Sebastiani, F. (ed.) ECIR 2003. LNCS, vol. 2633, pp. 207–218. Springer, Heidelberg (2003)
4. Büttcher, S., Clarke, C.L.A., Lushman, B.: Term proximity scoring for ad-hoc retrieval on very large text collections. In: ACM SIGIR '06, pp. 621–622. ACM, New York (2006)
5. Song, R., Taylor, M.J., Wen, J.R., Hon, H.W., Yu, Y.: Viewing term proximity from a different perspective. In: Macdonald, C., Ounis, I., Plachouras, V., Ruthven, I., White, R.W. (eds.) ECIR 2008. LNCS, vol. 4956, pp. 346–357. Springer, Heidelberg (2008)
6. Vechtomova, O., Karamuftuoglu, M.: Lexical cohesion and term proximity in document ranking. Information Processing and Management 44(4), 1485–1502 (2008)
7. Hearst, M.A.: Improving full-text precision on short queries using simple constraints. In: Proceedings of the 5th Annual Symposium on Document Analysis and Information Retrieval (SDAIR), pp. 217–232 (1996)
8. Clarke, C.L.A., Cormack, G.V., Burkowski, F.J.: Shortest substring ranking (multitext experiments for TREC-4). [13]
9. Hawking, D., Thistlewaite, P.: Proximity operators - so near and yet so far. [13]
10. Clarke, C.L.A., Cormack, G.V.: Shortest-substring retrieval and ranking. ACM Transactions on Information Systems 18(1), 44–78 (2000)
11. de Kretser, O., Moffat, A.: Effective document presentation with a locality-based similarity heuristic. In: ACM SIGIR '99, pp. 113–120. ACM, New York (1999)
12. Beigbeder, M., Mercier, A.: An information retrieval model using the fuzzy proximity degree of term occurences. In: Liebrock, L.M. (ed.) SAC 2005: Proceedings of the 2005 ACM symposium on Applied computing. ACM Press, New York (2005)
13. Harman, D.K. (ed.): The Fourth Text REtrieval Conference (TREC-4). Number 500-236, Department of Commerce, National Institute of Standards and Technology (1995)

LIP6 at INEX'09: OWPC for Ad Hoc Track

David Buffoni, Nicolas Usunier, and Patrick Gallinari

UPMC Paris 6 - LIP6
4, place Jussieu, 75005 Paris, France
{buffoni,usunier,gallinari}@poleia.lip6.fr

Abstract. We present a Retrieval Information system for XML documents using a Machine Learning Ranking approach. We then propose a way to annotate and build a training set for our learning to rank algorithm OWPC [1]. Finally, we apply our algorithm to the INEX'09 collection and present the results we obtained.

1 Introduction

Learning to rank algorithms have been used in the Machine Learning field for a while now. In the field of IR, they have first been used to combine features or preferences relations in the meta search [2], [3]. Learning ranking functions has also lead to improved performances in a series of tasks such as passage classification or automatic summarization [4]. More recently, they have been used for learning the rank function of search engines [5], [6], and [1].

Ranking algorithms work by combining features which characterize the data elements to be ranked. In our case, these features will depend on the document or the element itself, its structural context and its internal structure. Ranking algorithms will learn to combine these different features in an optimal way, according to a specific loss function related to IR criteria, using a set of examples. This set of examples is in fact a set of queries where for each one, a list of documents is given. In ranking, starting from this list, we make a set of pairs of documents where one is relevant to the query and the other is irrelevant.

The main problem in ranking is that the loss associated to a predicted ranked list is the mean of the pairwise classification losses. This loss is inadequate for IR tasks where we prefer high precision on the top of the predicted list. We propose, here to use a ranking algorithm, named OWPC [1] which optimizes loss functions focused on the top of the list.

In this paper, we describe the way we choose to annotate a set of examples and the selection of features which represent a document or an element according a query (section 2). We then present the two ranking models our baseline BM25 and a learning to rank model, OWPC, in section 3. Finally, in section 4 we discuss the results obtained by our models for the adhoc track.

2 Data Preparation

This year, INEX uses a new document collection based on the Wikipedia where the original Wiki syntax has been converted into XML [7]. This collection is

S. Geva, J. Kamps, and A. Trotman (Eds.): INEX 2009, LNCS 6203, pp. 59–69, 2010.
© Springer-Verlag Berlin Heidelberg 2010

now composed of 2,666,190 documents, four times more than was in previous years, and for each document we can separate semantic annotation elements and content elements. In our case, we indexed only content elements without doing any preprocessing step along the corpus as stemming or using a stop word list. To use a machine learning algorithm on this new collection we have to build manually a set of examples (i.e of queries) as a learning base. For each query, we must create a pool of retrieved elements, assess them and represente them as a vector of extracted features. We present in this section the approach we chose to build this learning dataset.

Assessment step. We assessed manually 20 randomly chosen queries from previous INEX competitions (from 2006 to 2008). We used only the title part of the queries without negated words and without a stemming process. Frow now on, we assume that each element can be judged with binary relevance as a value of relevant, irrelevant.

The assessment protocol was made by a pooling technique where we took the top t results of n models for each query. We hope by this technique to grow the diversity of the result list. So, in pratice, for each query we ran three models (BM25 [8] with $b = 0.5$ and $k = 1.2$, LogTF [9] and a Language Model with an Absolute Discount smoothing function [10] with $\delta = 0.88$). Then for even result list returned by each model, we assessed relevant documents in the top 50 of the list. By opposition, all non-judged documents are considered as irrelevant according to the query.

Finally, for each relevant document we select all the relevant elements to the query. We chose to consider only elements that could be members of the following XML tags : $\{article, title, bdy, section, p, item\}$. An element is described as relevant when it contains any relevant text in relation to the query. For 20 queries a total of 214 documents were found relevant, which gave us 1285 relevant elements and a lot of irrelevant ones.

Selection strategy of elements. Now, after annotating a set of queries, we can select the elements we will propose to our learning algorithm. Thus, we have to select documents or elements according to a query. The strategy used to select objects according to a query is important because it can achieve different performances for the same learning to rank algorithm. An experimental study to understand this behaviour can be found in [11] where the authors suggest selecting elements by pooling.

Therefore we decided to build our learning base as a pool of the top k of three information retrieval models as BM25, LogTF and a Language Model with Absolute Discount smoothing function as in the assessment step with a different value of t. Fixing the parameter t leads influences performances and will be further discussed in section 4. With this approach we aim to obtain a better diversity of results than a selection made by a single model.

Extracted features for learning. Once, we have obtained a set of elements according a query, we have to represent each element by a feature representation.

In our case each feature is a similarity function between a query and an element. We can separate features in 3 families :

- **content feature:** it's a similarity function based on the content of an element and the content of the query. For example, models such as BM25, TF-IDF or Language Models can be put in this class.
- **corpus structure feature:** it gives us degree of importance of an element in the corpus. In web search, we can cite Pagerank[12], Hits[13]... as corpus structure feature.
- **document structure feature:** it provides us information on the internal structure of an element or a document. For example, the layout information such as the number of format tags and the number of links... can be discriminative on the usefulness of the article.

We sum up in Table 1 all the features used in our experiments for the Focused and the Thorough tasks[1]. We denote $tf(t, x)$ as the term frequency of the term t of a query q in the element x, $[x]$ as the length of x, Col as the collection and $df(t)$ as the number of elements in the corpus containing t.

3 Models

We employed two models, one used as a baseline is a BM25 similarity function and a second, a learning to rank algorithm formulated as a SVM for interdependent output spaces [14].

3.1 BM25 Model

The BM25 model computes a score for an element e according a query of multiple terms $q = \{t_1, ..., t_n\}$. We define the query independent term weights in the formulas as inverse element frequencies $ief(t) = log \frac{[Col] - df(t) + 0.5}{df(t) + 1}$ where $[Col]$ is the number of elements in the collection. Similarly average (avl) and actual length $([e])$ are computed for elements. So the BM25 score of an element e and for a query q is :

$$BM25(e, q) = \sum_{t \in q} ief(t) * \frac{(k_1 + 1) * tf(t, e)}{k_1 * ((1 - b) + b * \frac{[e]}{avl(e)}) + tf(t, e)}$$

In our experiments, b and k parameters are fixed on the annotated set of documents as described in section 2.

3.2 OWPC Model

We outline here the learning to rank model described more in detail in [1]. We consider a standard setting for learning to rank. The ranking system receives

---

[1] These features are commonly employed in IR but we leave the investigation of the effects of the features for future works.

Table 1. Extracted features for the joint representation of a document according to a query. For the Focused track x represents the whole article although for the Thorough track, x is used for the document (globality information), and for the element (specificity information). We additionnaly use as information, the ratio $\frac{score_{element}}{score_{document\ it\ belongs\ to}}$. The 36-42th features give a boolean attribute for the family tag of the current element $x \in \{sec, ss, ss1, ss2, ss3, ss4, p\}$, used only for the Thorough track.

ID	Feature Description
1	$[x]$
2	# of unique words in x
3	$\sum_{t \in q \cap x} tf(t,x)$ in x
4	$\sum_{t \in q \cap x} log(1 + tf(t,x))$ in x
5	$\sum_{t \in q \cap x} \frac{tf(t,x)}{[x]}$ in x
6	$\sum_{t \in q \cap x} log(1 + \frac{tf(t,x)}{[x]})$ in x
7	$\sum_{t \in q \cap x} log(\frac{[Col]+1}{df(t)+0.5})$ in x
8	$\sum_{t \in q \cap x} log(1 + log(\frac{[Col]+1}{df(t)+0.5}))$ in x
9	$\sum_{t \in q \cap x} log(1 + \frac{[Col]+1}{tf(t,x)})$ in x
10	$\sum_{t \in q \cap x} log(1 + \frac{tf(t,x)}{[x]} * \frac{[Col]+1}{df(t)+0.5})$ in x
11	$\sum_{t \in q \cap x} log(1 + \frac{tf(t,x)}{[x]} * \frac{[Col]}{tf(t,Col)})$ in x
12	$\sum_{t \in q \cap x} log(1 + tf(t,x)) * log(\frac{[Col]+1}{0.5+df(t)})$ in x
13	$\sum_{t \in q \cap x} tf(t,x) * log(\frac{[Col]+1}{0.5+df(t)})$ in x
14	$BM25^{k=2}_{b=0.2}$ in x
15	$log(1 + BM25^{k=2}_{b=0.2})$ in x
16	$LM^{Jelinek-Mercer}_{\lambda=0.5}$ in x
17	$LM^{Dirichlet}_{\mu=2500}$ in x
18	$LM^{AbsoluteDiscount}_{\delta=0.88}$ in x
19-20	$\# <section>$ and $log(1 + \# <section>)$ in the document
21-22	$\# <p>$ and $log(1 + \# <p>)$ in the document
23-24	$\#categories$ and $log(1 + \#categories)$ in the document
25-26	$\#format_{tags}$ and $log(1 + \#format_{tags})$ in the document
27-28	$\# <weblink>$ and $log(1 + \# <weblink>)$ in the document
29-30	$\# <template>$ and $log(1 + \# <template>)$ in the document
31-32	$\#in <link>$ and $log(1 + \#in <link>)$ in the document
33-34	$\#out <link>$ and $log(1 + \#out <link>)$ in the document
35	pagerank of the document
36-37-38-39-40-41-42	*family tag of the element*

as input a query q, which defines a sequence of candidates (XML elements) $\mathbf{X}(q) \stackrel{def}{=} (\mathbf{X}_1(q), ..., \mathbf{X}_{[q]}(q))$ (where $[q]$ is used as a shortcut for $[\mathbf{X}(q)]$). In our case, the sequence is a subset of the collection filtered by the pooling technique as described in section 2. $\mathbf{X}_j(q)$ corresponds to the similarity representation, i.e the vector of extracted similarities, of the j-th object. A score (i.e. real-valued) function f takes as input the similarity representation of an element, thus $f(X_j(q))$, denoted $f_j(z)$ for simplicity, is the score of the j-th element. The output of the ranking system is the list of the candidates sorted by decreasing scores.

Learning step. For clarity in the discussion, we assume a binary relevance of the elements: a sequence $\mathbf{y}$ contains the indexes of the *relevant objects* labeled by a human expert ($\bar{\mathbf{y}}$ contains the indexes of the *irrelevant* ones).

Given a training set $S = (q_i, \mathbf{y}_i)_{i=1}^m$ of m examples, learning to rank consists in choosing a score function f that will minimize a given *ranking error function* $\hat{R}^\Phi(f, S)$:

$$\hat{R}^\Phi(f, S) \overset{def}{=} \underset{(q,\mathbf{y})\sim S}{\hat{\mathbb{E}}} \, \mathbf{err}\big(\mathbf{rank}(f, q, \mathbf{y})\big)$$

$\hat{\mathbb{E}}_{(q,\mathbf{y})\sim S}$ is the mean on S of ranking errors and $\mathbf{err}$ is the number of misranked relevant documents in the predicted list.

Rank function. We define the *rank* of a relevant document for a given query q, its relevant candidates $\mathbf{y}$, and a score function f, as follows:

$$\forall y \in \mathbf{y}, rank_y(f, q, \mathbf{y}) \overset{def}{=} \sum_{\bar{y} \in \bar{\mathbf{y}}} \mathrm{I}\,(f_y(q) \le f_{\bar{y}}(q)) \tag{1}$$

where $\mathrm{I}\,(f_y(q) \le f_{\bar{y}}(q))$ indicates whether the score of a relevant document is lower than the score of an irrelevant one. However, directly minimizing the ranking error function $\hat{R}^\Phi(f, S)$ is difficult due to the non-differentiable and non-convex properties of function $\mathrm{I}\,(f_y(q) \le f_{\bar{y}}(q))$ of $rank_y(f, q, \mathbf{y})$. To solve this problem we generally take a convex upper bound of the indicator function which is differentiable and admits only one minimum. In [1], this bound is denoted $\ell\big(f_y(q) - f_{\bar{y}}(q)\big)$ and is set to the hinge loss function $\ell : t \mapsto [1 - t]_+$ (where $[1 - t]_+$ stands for $max(0, 1 - t)$ and $t = f_y(q) - f_{\bar{y}}(q)$).

Error function. With equation 1, we can define a general form of the ranking error functions $\mathbf{err}$ of a real valued function f on $(q, \mathbf{y})$ as:

$$\mathbf{err}(f, q, \mathbf{y}) \overset{def}{=} \frac{1}{[\mathbf{y}]} \sum_{y \in \mathbf{y}} \Phi_{[\bar{\mathbf{y}}]}\,(rank_y(f, q, \mathbf{y}))$$

where $\Phi_{[\bar{\mathbf{y}}]}$ is an aggregation operator over the position of each relevant document in the predicted list. Traditionnaly, this aggregation operator $\Phi_{[\bar{\mathbf{y}}]}$ was set to the mean in learning to rank algorithms as in [15, 16]. Yet, optimizing the mean of the rank of relevant document does not constitute a related ranking error function to classical Information Retrieval measures. In fact, we obtain the same ranking error for a relevant element ranked on the top or on the bottom of the list. This behaviour is not shared by IR metrics where more consideration is given to the rank of the relevant documents on the top of the list.

To overcome this problem, the authors of [1] showed that fixing $\Phi_{[\bar{\mathbf{y}}]}$ by the convex Ordered Weighted Aggregation (OWA) operators [17] we can affect the degree to which the ranking loss function focuses on the top of the list. The definition of the OWA operator is given as follows :

Definition 1 (OWA operator [17]). *Let $\boldsymbol{\alpha} = (\alpha_1, ..., \alpha_n)$ be a sequence of n non-negative numbers with $\sum_{j=1}^{n} \alpha_j = 1$. The* Ordered Weighted Averaging *(OWA) Operator associated to $\boldsymbol{\alpha}$, is the function* owa$^{\boldsymbol{\alpha}} : \mathbb{R}^n \to \mathbb{R}$ *defined as follows:*

$$\forall \mathbf{t} = (t_1, ..., t_n) \in \mathbb{R}^n, \text{owa}^{\boldsymbol{\alpha}}(\mathbf{t}) = \sum_{j=1}^{n} \alpha_j t_{\sigma(j)}$$

where $\sigma \in \mathfrak{S}_n$ (set of permutations) such that $\forall j, t_{\sigma(j)} \geq t_{\sigma(j+1)}$.

It is then used by the authors to rewrite the ranking error function as follows:

$$\text{err}(f, q, \mathbf{y}) \stackrel{def}{=} \frac{1}{[\mathbf{y}]} \sum_{y \in \mathbf{y}} \underset{\bar{y} \in \bar{\mathbf{y}}}{\text{owa}} (rank_y(f, q, \mathbf{y})) \tag{2}$$

SVM formulation. Thus, according to the authors, this provides a regularized version of the empirical risk of equation (2) and can be solved using existing algorithms as Support Vector Machines for structured output spaces [14].

$$\min_{w} \frac{1}{2} ||w||^2 + C \sum_{(q,\mathbf{y}) \in S} \frac{1}{[\mathbf{y}]} \sum_{y \in \mathbf{y}} \underset{\bar{y} \in \bar{\mathbf{y}}}{\text{owa}} [1 - \langle w, X_y(q) - X_{\bar{y}}(q) \rangle]_+ \tag{3}$$

where the learning algorithm according to the training set S and the ranking error function $\text{err}(f, q, \mathbf{y})$ will determine the parameter vector w. This weight vector will be used in the prediction step to compute the score of a document (f function). C is a trade-off parameter fixed by the user, to balance the learning model complexity $||w||^2$ and the upper bounded ranking loss function $\text{err}(f, q, \mathbf{y})$.

To sum up, this algorithm learns a score function by minimizing a ranking error function focused on the top of the list. The user has to fix the non-increasing weights α of the OWA operators to vary the consideration on the errors incurred on the top of the list. It's the typical behaviour of a IR evaluation measure.

Ranking prediction. Given an unlabeled query q_u with a candidate $X_j(q_u)$ of the sequence of elements $\mathbf{X}(q_u)$, the corresponding predicted score based on the learned weight vector w is:

$$f_j(z) = \langle w, X_j(q_u) \rangle$$

where $\langle ., . \rangle$ is the scalar product between the weight vector and the similarity representation of $X_j(q_u)$. This allows us to sort all elements of $\mathbf{X}(q_u)$ by decreasing scores.

4   Experiments - Results

In this section, we present our experiments and results for the Focused and the Thorough tasks in Adhoc track for INEX'09. We concentrated on these two tasks to validate the selection of elements which are provided to the learning to rank algorithm. The learning set built in section 2 is used for fixing the parameters of our BM25 model and our learning to rank algorithm OWPC. To select them we use a standard evaluation metrics in IR: the Mean Average Precision (MAP) [9].

For a set of 20 annotated queries, we manually fixed the BM25's parameters. We computed for each run, its MAP score. The one achieving the best MAP score on this set was chosen. Finally, we chose the values of the parameters: $b = 0.2$ and $k = 2.0$.

For the OWPC, we set the weights of the OWA operator to be linearly decreasing as suggested by [1]. We fixed the C parameter of the equation (3) among $\{1, 10, 100\}$ using the MAP score on the training set. In our experiments, $C = 100$ gave us the best performances according to the MAP score on the training set.

4.1   Focused Task

We present in this section the performances of our models for the Focused track. For this track, the two models retrieve only the article elements to point out the efficiency of this strategy. In previous INEX competitions, to seek articles gave better results than element/passage retrieval. We would like to figure out if this strategy is still valid on the new version of Wikipedia.

In addition, we would like to compare a baseline run, a BM25, and a run enhanced with a learning to rank technique. To be able to compare the two approaches we limit some potential side effects due to the removal of overlap step and we aim to return only articles for each query.

Moreover, in additionnal runs, we would like to evaluate the importance of the size of the pool for the training and the testing set. For our official run, we took the top 800 articles of three models but we additionnaly tried with $t = 1500$ and $t = 3000$.

We report the performances of our models in the Focused track in Table 2 in terms of IP[0.01] and MAP measures. BM25-Art is a BM25 model which retrieves only articles, and OWPC-Art_t $with$ $t \in \{800, 1500, 3000\}$ is our learning to rank algorithm returning only articles and with different size of pool. We can see that in terms of IP[0.01] the BM25 outperforms the learning to rank algorithm but the inverse is true for the MAP metric. In addition, the size of the pool affects the performances of OWPC with a better MAP and IP[0.01] score.

We can assume that the poor performance of the OWPC on the IP[0.01] metric is due to the annotation step which is not related to the element retrieval evaluation measure. As told before, an annotation of an element is a binary

Table 2. Test performances of OWPC-Art_t and BM25-Art in the Focused track in terms of IP[0.01] and MAP. Runs with a star are official submissions for INEX 09.

	IP[0.01]	MAP
BM25-Art^*	**0.6141**	0.3417
OWPC-$Art_{800}{}^*$	0.5357	0.3420
OWPC-Art_{1500}	0.5544	0.3560
OWPC-Art_{3000}	0.5516	**0.3567**

label although the element retrieval measure depends of the quantity of the relevant text in an element. In case of learning, relevant articles have the same label but in reality they don't have the same quantity of relevant text. This way, the learning algorithm doesn't differentiate between the different degree of relevance for example between perfectly relevant and fairly relevant. In the case of document retrieval, when all degrees of relevance are the same, we notice that our learning model obtains better results than the baseline.

4.2   Thorough Task

We present here only the additionnal runs experimented in the following of our Focused track runs. We want to experimentaly prove the asumption made previously in regards to the annotation of elements.

For our models, we use a fetch and browse strategy. We retrieve the top 1500 articles for the fetch step and we extract the list of all overlapping elements which contained at least one of the search terms for the browse step. We strive to collect only small elements and we limit the domain to types $\in \{sec, ss, ss1, ss2, ss3, ss4, p\}$. Thus, the side effects due to waste labeling are reduced.

Our BM25 run on the elements, BM25-$Elems$, returns the top 1500 elements of the browse step. For the learning algorithm, OWPC-$Elems_t$, after a pooling on the document for the fetch step, as in the Focused track, we additionnaly pool the elements by taking the top $t \in \{1500, 3000, All\}$. We want to investigate the influence of the size of the pool.

The performances of our models are summarized in Table 3 and in Figure 1 for the Thorough track. We can compare our runs on elements such as BM25-$Elems$ and OWPC-$Elems_t$ $with$ $t \in \{1500, 3000, All\}$ with the runs used in the Focused track.

As expected, runs which return articles have better performances in MAP and MAiP than runs returning only small elements. This shows that the strategy of retrieving articles is most informative both with respect to precision and recall. A simple explanation for this, is the effect of the limitation of the results list. In fact, only the top 1500 elements for each query are evaluated and in the case of the Thorough track, where the overlap is permitted, this penalizes runs returning a lot of small elements rather than one article.

Table 3. Test performances of BM25 and OWPC models in the Thorough task in terms of IP[0.01], MAiP and MAP

	IP[0.01]	MAiP	MAP
BM25-Art	0.6141	**0.3000**	0.3417
OWPC-Art_{3000}	0.5516	0.2753	**0.3567**
BM25-$Elems$	0.5515	0.1347	0.2431
OWPC-$Elems_{1500}$	0.5692	0.1744	0.3082
OWPC-$Elems_{3000}$	0.5854	0.1881	0.3073
OWPC-$Elems_{All}$	**0.6220**	0.2024	0.3064

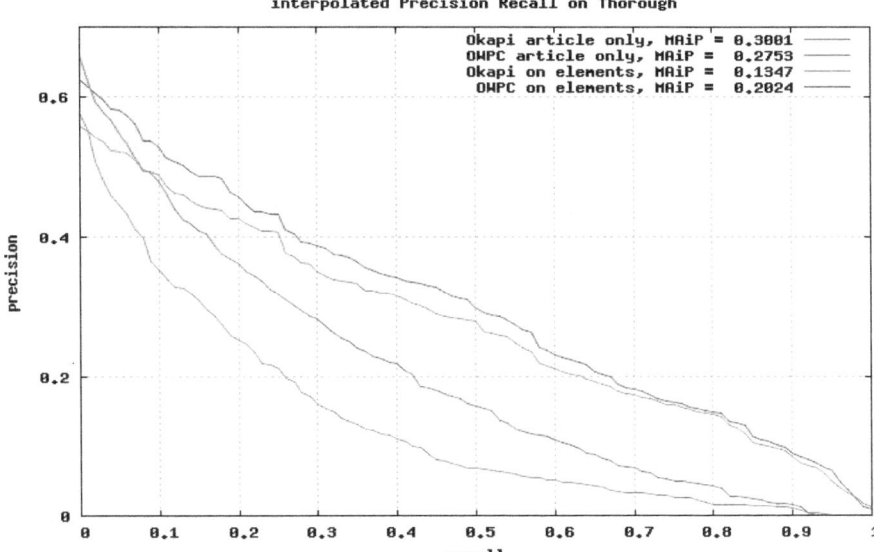

Fig. 1. Interpolated Precision Recall for the Thorough task of our models BM25 and OWPC (on elements and on articles)

For models retrieving elements, OWPC-*Elems*'s performances improve according the size of the pool of elements for the browse step. Providing to our learning algorithm an exhautive list of elements ($t = All$), gives the best results in IP[0.01], more precise than models retrieving articles.

In end, as expected OWPC-*Elems$_t$* outperforms BM25-*Elems* in terms of MAP but also on MAiP and IP[0.01]. This experimentaly proves our asumption that the binary relevance labeling must be only done on small elements. Thus, a further direction of this work would be to define learning to rank loss functions able to take into consideration different degrees of relevance for articles. We hope by adding this granularity to tie in better with the INEX measure as IP[0.01] and MAiP.

5 Conclusion

In conclusion, this year the collection changed and we had to build a training set for our learning model named OWPC. In this paper we proposed a way to annotate elements, a strategy to select elements and a feature representation of them.

We figured out the importance of these three steps which can easily lead several unwanted behaviours of learning algorithms. We concentrated this year on the first two steps. In terms of selecting elements, a pooling technique seems to be a judicious approach but the size of this pool is a important factor in the performances of the OWPC. In addition, as in previous INEX competitions, retrieving articles rather than smaller elements gives better results. However, OWPC on articles performed well on the MAP metric but it didn't perform as expected on IP[0.01] and MAiP where these measures depend on the ratio between the quantity of the relevant text of an element and its total size. We experimentaly proved our binary labeling is not related to INEX measures for articles and works well for very small elements where this ratio is less deterministic.

In future works we propose to define a new loss function of our learning to rank algorithm which takes into account different degrees of relevance of elements. We hope to prove that a learning to rank algorithm can outperform baseline as BM25 on articles.

Acknowledgment

This work was supported in part by the IST Programme of the European Community, under the PASCAL2 Network of Excellence, IST-2007-216886 and by Agence Nationale de la Recherche, ANR-08-CORD-009-Fragrances . This publication only reflects the authors' views.

References

[1] Usunier, N., Buffoni, D., Gallinari, P.: Ranking with ordered weighted pairwise classification. In: Proceedings of the 26th Annual International Conference on Machine Learning, ICML 2009, Montreal (2009)

[2] Cohen, W.W., Schapire, R.E., Singer, Y.: Learning to order things. In: Advances in Neural Information Processing Systems, NIPS '98, vol. 10 (1998)

[3] Freund, Y., Iyer, R., Schapire, R.E., Singer, Y.: An efficient boosting algorithm for combining preferences. In: Proceedings of the 15th Annual International Conference on Machine Learning, ICML '98 (1998)

[4] Amini, M.R., Usunier, N., Gallinari, P.: Automatic text summarization based on word-clusters and ranking algorithms. In: Losada, D.E., Fernández-Luna, J.M. (eds.) ECIR 2005. LNCS, vol. 3408, pp. 142–156. Springer, Heidelberg (2005)

[5] Burges, C.J.C., Shaked, T., Renshaw, E., Lazier, A., Deeds, M., Hamilton, N., Hullender, G.N.: Learning to Rank with Nonsmooth Cost Functions. In: Advances in Neural Information Processing Systems, NIPS '06, vol. 19 (2006)

[6] Yue, Y., Finley, T., Radlinski, F., Joachims, T.: A support vector method for optimizing average precision. In: Proceedings of the 30th Annual International Conference on Research and Development in Information Retrieval, SIGIR '07 (2007)

[7] Schenkel, R., Suchanek, F.M., Kasneci, G.: YAWN: A Semantically Annotated Wikipedia XML Corpus. In: Datenbanksysteme in Business, Technologie und Web, BTW '07 (2007)

[8] Robertson, S.E., Walker, S., Hancock-Beaulieu, M., Gull, A., Lau, M.: Okapi at TREC. In: Text REtrieval Conference, pp. 21–30 (1992)

[9] Baeza-Yates, R., Ribeiro-Neto, B.: Modern Information Retrieval (1999)

[10] Zhai, C., Lafferty, J.: A study of smoothing methods for language models applied to information retrieval. ACM Trans. Inf. Syst. (2004)

[11] Aslam, J.A., Kanoulas, E., Pavlu, V., Savev, S., Yilmaz, E.: Document selection methodologies for efficient and effective learning-to-rank. In: Proceedings of the 32nd Annual International Conference on Research and Development in Information Retrieval, SIGIR '09 (2009)

[12] Page, L., Brin, S., Motwani, R., Winograd, T.: The PageRank Citation Ranking: Bringing Order to the Web. Technical Report (1999)

[13] Kleinberg, J.M.: Authoritative sources in a hyperlinked environment. Journal of ACM (1999)

[14] Tsochantaridis, I., Joachims, T., Hofmann, T., Altun, Y.: Large Margin Methods for Structured and Interdependent Output Variables. Journal of Machine Learning Research (2005)

[15] Joachims, T.: Optimizing search engines using clickthrough data. In: Proceedings of the Eighth ACM SIGKDD International Conference on Knowledge Discovery and Data Mining (2002)

[16] Cao, Y., Xu, J., Liu, T.-Y., Hang, L., Huang, Y., Hon, H.-W.: Adapting ranking SVM to document retrieval. In: Proceedings of the 29th Annual International Conference on Research and Development in Information Retrieval, SIGIR '06 (2006)

[17] Yager, R.R.: On ordered weighted averaging aggregation operators in multi-criteria decision making. IEEE Transactions on Systems, Man and Cybernetics (1988)

A Methodology for Producing Improved Focused Elements

Carolyn J. Crouch, Donald B. Crouch, Dinesh Bhirud,
Pavan Poluri, Chaitanya Polumetla, and Varun Sudhakar

Department of Computer Science
University of Minnesota Duluth
Duluth, MN 55812
(218) 726-7607
ccrouch@d.umn.edu

Abstract. This paper reports the results of our experiments to consistently produce highly ranked focused elements in response to the Focused Task of the INEX Ad Hoc Track. The results of these experiments, performed using the 2008 INEX collection, confirm that our current methodology (described herein) produces such elements for this collection. Our goal for 2009 is to apply this methodology to the new, extended 2009 INEX collection to determine its viability in this environment. (These experiments are currently underway.) Our system uses our method for dynamic element retrieval [4], working with the semi-structured text of Wikipedia [5], to produce a rank-ordered list of elements in the context of focused retrieval. It is based on the Vector Space Model [15]; basic functions are performed using the Smart experimental retrieval system [14]. Experimental results are reported for the Focused Task of both the 2008 and 2009 INEX Ad Hoc Tracks.

1 Introduction

In 2008, our INEX investigations centered on integrating our methodology for the dynamic retrieval of XML elements [4] with traditional article retrieval to facilitate in particular the Focused Task of the Ad Hoc Track. Our goal was to produce what we refer to as *good focused elements*—i.e., elements which when evaluated were competitive with others in the upper ranges of the 2008 rankings. Our best results for these experiments, as reported in [6], accomplished this goal but also produced indications that further, significant overall improvements were possible with more investigation. Thus, our efforts in 2009 are directed at producing not merely good but rather exceptionally good focused elements (which consistently rank near or above those listed at the top of the official rankings). This paper reports the results of these experiments, as performed using the INEX 2008 collection. Our goal for 2009 is to produce equivalent results—i.e., superior focused elements--using the new, larger INEX 2009 collection. But as our system requires tuning, this work is in progress at the present time.

Dynamic element retrieval—i.e., the dynamic retrieval of elements at the desired degree of granularity—has been the focus of our INEX investigations for some

S. Geva, J. Kamps, and A. Trotman (Eds.): INEX 2009, LNCS 6203, pp. 70–80, 2010.

time [4, 5]. We have shown that our method works well for both structured and semi-structured text and that it produces a result identical to that produced by the search of the same query against the corresponding all-element index [11]. In [5], we show that dynamic element retrieval (with terminal node expansion) produces a result considerably higher than that reported by the top-ranked participant for the INEX 2006 Thorough task. However, when the task changed to focused retrieval, our results fell into the mid-range of participant scores [6]. This paper (1) describes our current methodology for producing focused elements and (2) reports the results of experiments performed on the INEX 2008 collection which clearly establish that it produces superior (i.e., highly ranked) focused elements in this environment. The remainder of the paper is directed towards establishing whether the same methodology can be successfully applied in the environment of INEX 2009—i.e., whether it is also able to produce superior focused elements for this much larger, scaled up version of Wikipedia.

2 Producing Superior Focused Elements

In this section, we describe our methodology for producing high-ranking focused elements. The experiments performed using this approach are detailed and their results reported in Section 3. Basic functions are performed using Smart. Negative terms are removed from the queries.

2.1 Focused Task Methodology

Two important issues arise with respect to retrieving good focused elements in response to a query: (1) how the documents of interest are identified, and (2) the method by which the focused elements are selected from those documents. The approach described herein incorporates article retrieval (to identify the articles of interest) with dynamic element retrieval (to produce the elements). We then apply one of three focusing strategies to the list of elements that results. A slightly lower-level view of the process follows.

For each query, we retrieve n articles or documents. We then use dynamic element retrieval to produce the elements themselves. Dynamic element retrieval, described in detail in [4, 9], allows us to build the document tree at execution time, based on a stored schema of the document and a paragraph (or terminal-node) index of the collection. *Lnu-ltu* term weighting [16], designed to deal with differences in the lengths of vectors, is utilized to produce a rank-ordered list of elements from each document. The lists are merged, and one of the three focusing strategies described below is then used to remove overlap. A set of focused elements, $\leq m$ in size, is then reported. (Here m represents either the number of focused elements reported for each query or an upper bound on that number. Differences in determining m resulted in many variations of this basic experiment. See [2, 13] for more detail.)

Three focusing or overlap removal strategies were investigated in these experiments. The *section strategy* chooses the highest correlating non-body element along a path as the focused element. (Most of these elements turn out in fact to be sections. A body element appears in this list of focused elements if and only if none of its child elements are ranked within the top m.) The *correlation strategy* chooses the highest correlating element along a path as the focused element, without restriction on element type. And the *child strategy* chooses the terminal element along a path as the focused element (i.e., ignores correlation and always gives preference to the child rather than the parent).

2.2 Early Focused Results

The best results produced by our approach for the INEX 2007 Focused Task, as reported in [6], exceeded the iP[0.01] value produced by the entry at rank 1. When we applied the same methodology to the INEX 2008 Focused Task (reported in [1, 6]) our best results were competitive with those in the upper ranges of the official ranking. These results led us to believe that our basic methods could be applied to produce more highly ranked focused elements. (Most of our effort during 2009 centered on validating this belief.) Final results, presented in Section 3, clearly show that such focused elements are produced by the current methodology.

3 Experiments in Focused Retrieval

The description of our experiments in focused retrieval and the results produced are described here in Sections 3.1 and 3.2, respectively. All of these experiments were performed using the INEX 2008 collection. Observations and analysis follow in Section 3.3. Focused results performed to date on the new INEX 2009 collection are reported in Section 3.4.

3.1 Overview

Recall the basic approach described above. Given a query, the top-ranked n articles are identified based on Smart retrieval. Dynamic element retrieval is then used to build the document trees. As the trees are built, bottom up, each *Lnu*-weighted element vector is correlated with the *ltu*-weighted query using inner product. For each tree, a rank-ordered list of elements is produced. This list includes elements representing untagged text in the document, which must be present in order to generate the trees properly but which do not physically exist as elements per se. These elements must be removed from the element list along with overlapping elements to produce the set of focused elements for the tree. The focused elements from each tree are then reported, in article order, for evaluation.

Two basic sets of experiments in focused retrieval were performed in 2009. Both used the 2008 INEX collection. One set, based on what we call the Upper Bound method, uses m as the bound on the number of rank-ordered elements (of any type) retrieved by a query. From these m elements, those representing untagged text are removed, focusing is applied, and the resultant set of focused elements for each tree is reported until the window is filled. The Exact method, on the other hand, guarantees that if m focused elements are available (i.e., retrieved by the query), the top-ranked m

are reported until the window is filled. Given the ranked-ordered list of all m focused elements associated with the query, this approach outputs its results in two ways: (1) by reporting the top m elements in article order (RAC reporting) and (2) by grouping the elements by article and reporting the top m (RBC reporting). Note that RAC and RBC result in slightly different sets of focused elements being reported.

The results of these experiments are reported below. In all cases, n, the number of articles retrieved, ranges from 25 to 500, and m, representing focused elements, ranges from 50 to 4000. For each approach determining which focused elements to report (i.e., Upper Bound and Exact), three focusing strategies (Section, Child, and Correlation) were applied.

3.2 Results

Tables 1, 2, and 3 show the results of applying the Upper Bound method in conjunction with the Section, Child, and Correlation focusing strategies, respectively. Tables 4-9 show the results of applying the Exact method in conjunction with the Section, Child, and Correlation focusing strategies, respectively. Tables 4-6 use RAC reporting whereas Tables 7-9 utilize RBC reporting.

3.3 Observations and Analysis

As can be seen from the tables, each set of experiments produces fairly consistent results. The Upper Bound data, reported in Tables 1-3, may be more difficult to interpret since m in this case represents an upper bound on the total number of elements, so we have little insight as to how the window is being filled with focused elements (if these are available). If we assume that each query must retrieve in excess of 1500 elements so as potentially to produce 1500 focused elements (which is well justified by the evidence), then reasonable results can be seen at values of m exceeding 1500. All of these values, across Tables 1-3, show iP[0.01] in excess of 0.72.

Tables 4-6 show results produced by the Exact method using RAC reporting. Thus, from a ranked list of focused elements, the top m elements are selected, grouped by article, and reported. RAC reporting guarantees that the highest ranked focused elements from the top ranked articles are reported (but not that all focused elements from each such article are reported). For values of $m \geq 1500$, iP[0.01] exceeds 0.72 in each case. The results obtained by applying the Exact method with RBC reporting are seen in Tables 7-9. Given a ranked list all of focused elements associated with the query, this method guarantees, for each article, that all of its focused elements are reported. At $m = 1500$, the window is filled (if possible). Again, all values of iP[0.01] for $m \geq 500$ exceed 0.72. Meaningful differences in the results produced by the three focusing strategies are difficult to observe.

The tables show that the basic methodology consistently produces highly ranked focused elements (as compared to the best value at 0.6896 of iP[0.01] in the 2008 ranking for the Focused Task). This being the case, we prefer the Section strategy, which favors the higher, non-body element, in conjunction with the Exact method using RAC reporting, which reports all focused elements for each document. This represents a clear and straight forward approach to focused retrieval.

Table 1. 2008 Upper Bound Method Section Focusing Strategy

NUMBER OF DOCUMENTS	NUMBER OF ELEMENTS										
	50	100	150	200	250	500	1000	1500	2000	3000	4000
25	0.6971	0.7147	0.7041	0.7076	0.7191	0.7198	0.7211	0.7211	0.7211	0.7211	0.7211
50	0.6844	0.7131	0.7197	0.7039	0.7065	0.7200	0.7211	0.7222	0.7222	0.7222	0.7222
100	0.6923	0.7096	0.7171	0.7191	0.7190	0.7078	0.7210	0.7217	0.7229	0.7229	0.7229
150	0.6977	0.7092	0.7162	0.7192	0.7171	0.7111	0.7199	0.7228	0.7215	0.7233	0.7233
200	0.6918	0.7057	0.7084	0.7145	0.7185	0.7190	0.7086	0.7210	0.7223	0.7224	**0.7235**
250	0.6896	0.7029	0.7103	0.7153	0.7177	0.7207	0.7080	0.7197	0.7218	0.7221	0.7229
500	0.6817	0.7034	0.7108	0.7108	0.7128	0.7185	0.7100	0.7088	0.7221	0.7203	0.7233

Table 2. 2008 Upper Bound Method Child Focusing Strategy

NUMBER OF DOCUMENTS	NUMBER OF ELEMENTS										
	50	100	150	200	250	500	1000	1500	2000	3000	4000
25	0.7063	0.7100	0.7038	0.7068	0.7254	0.7258	0.7205	0.7205	0.7205	0.7205	0.7205
50	0.6925	0.7203	0.7188	0.6984	0.7044	0.7283	0.7212	0.7218	0.7218	0.7218	0.7218
100	0.7039	0.7169	0.7213	0.7169	0.7206	0.7074	0.7268	0.7210	0.7224	0.7224	0.7224
150	0.7052	0.7190	0.7224	0.7159	0.7165	0.7090	0.7218	0.7290	0.7274	0.7227	0.7227
200	0.7022	0.7128	0.7159	0.7173	0.7184	0.7133	0.7090	0.7274	0.7293	0.7222	0.7230
250	0.6982	0.7112	0.7180	0.7170	0.7154	0.7139	0.7082	0.7221	0.7278	0.7214	0.7227
500	0.6897	0.7104	0.7194	0.7136	0.7139	0.7179	0.7077	0.7087	0.7236	0.7268	0.7307

Table 3. 2008 Upper Bound Method Correlation Focusing Strategy

NUMBER OF DOCUMENTS	NUMBER OF ELEMENTS										
	50	100	150	200	250	500	1000	1500	2000	3000	4000
25	0.7064	0.7217	0.7197	0.7210	0.7195	0.7191	0.7203	0.7203	0.7203	0.7203	0.7203
50	0.6949	0.7169	0.7229	0.7214	0.7227	0.7199	0.7202	0.7213	0.7213	0.7213	0.7213
100	0.6997	0.7141	0.7205	0.7225	0.7228	0.7223	0.7211	0.7208	0.7219	0.7219	0.7219
150	0.7043	0.7111	0.7198	0.7236	0.7209	**0.7265**	0.7207	0.7223	0.7205	0.7222	0.7222
200	0.6987	0.7083	0.7149	0.7180	0.7222	0.7235	0.7214	0.7213	0.7215	0.7213	0.7223
250	0.6966	0.7085	0.7164	0.7188	0.7219	0.7254	0.7217	0.7206	0.7217	0.7211	0.7217
500	0.6895	0.7087	0.7151	0.7169	0.7168	0.7241	0.7246	0.7226	0.7228	0.7207	0.7226

Table 4. 2008 Exact Method Section Focusing Strategy RAC Reporting

NUMBER OF DOCUMENTS	NUMBER OF ELEMENTS										
	50	100	150	200	250	500	1000	1500	2000	3000	4000
25	0.7060	0.7183	0.7172	0.7207	0.7206	0.7211	0.7211	0.7211	0.7211	0.7211	0.7211
50	0.6956	0.7071	0.7068	0.7193	0.7199	0.7214	0.7222	0.7222	0.7222	0.7222	0.7222
100	0.6853	0.7104	0.7079	0.7086	0.7064	0.7216	0.7229	0.7229	0.7229	0.7229	0.7229
150	0.6847	0.7078	0.7055	0.7126	0.7078	0.7187	0.7219	0.7233	0.7233	0.7233	0.7233
200	0.6819	0.7045	0.7062	0.7056	0.7131	0.7202	0.7225	0.7227	0.7235	0.7235	0.7235
250	0.6770	0.7032	0.7076	0.7064	0.7103	0.7075	0.7215	0.7222	0.7229	0.7236	0.7236
500	0.6791	0.7004	0.7088	0.7077	0.7065	0.7100	0.7198	0.7207	0.7227	0.7229	0.7236

Table 5. 2008 Exact Method Child Focusing Strategy RAC Reporting

NUMBER OF DOCUMENTS	NUMBER OF ELEMENTS										
	50	100	150	200	250	500	1000	1500	2000	3000	4000
25	0.6823	0.6945	0.7195	0.7188	0.7179	0.7205	0.7205	0.7205	0.7205	0.7205	0.7205
50	0.6803	0.6956	0.6972	0.7132	0.7215	0.7206	0.7218	0.7218	0.7218	0.7218	0.7218
100	0.6761	0.6906	0.6972	0.7008	0.6988	0.7223	0.7223	0.7224	0.7224	0.7224	0.7224
150	0.6791	0.6852	0.6911	0.6977	0.6992	0.7147	0.7211	0.7223	0.7227	0.7227	0.7227
200	0.6651	0.6813	0.6938	0.7011	0.6972	0.7135	0.7227	0.7223	0.7225	0.7230	0.7230
250	0.6593	0.6825	0.6937	0.6927	0.6978	0.6993	0.7221	0.7213	0.7225	0.7230	0.7230
500	0.6618	0.6815	0.6855	0.6903	0.6882	0.7000	0.7156	0.7176	0.7245	0.7220	0.7226

Table 6. 2008 Exact Method Correlation Focusing Strategy RAC Reporting

NUMBER OF DOCUMENTS	NUMBER OF ELEMENTS										
	50	100	150	200	250	500	1000	1500	2000	3000	4000
25	0.7212	0.7194	0.7176	0.7190	0.7197	0.7203	0.7203	0.7203	0.7203	0.7203	0.7203
50	0.7156	0.7249	0.7220	0.7194	0.7200	0.7209	0.7213	0.7213	0.7213	0.7213	0.7213
100	0.7087	0.7218	0.7218	0.7237	0.7213	0.7209	0.7219	0.7219	0.7219	0.7219	0.7219
150	0.7097	0.7178	0.7240	0.7244	0.7232	0.7207	0.7215	0.7222	0.7222	0.7222	0.7222
200	0.7098	0.7198	0.7201	0.7238	0.7252	0.7197	0.7208	0.7217	0.7223	0.7223	0.7223
250	0.7113	0.7141	0.7206	0.7242	0.7254	0.7219	0.7219	0.7216	0.7223	0.7223	0.7223
500	0.7112	0.7167	0.7168	0.7206	0.7246	0.7261	0.7204	0.7231	0.7214	0.7217	0.7225

Table 7. 2008 Exact Method Section Focusing Strategy RBC Reporting

NUMBER OF DOCUMENTS	NUMBER OF ELEMENTS										
	50	100	150	200	250	500	1000	1500	2000	3000	4000
25	0.6899	0.7110	0.7186	0.7211	0.7211	0.7211	0.7211	0.7211	0.7211	0.7211	0.7211
50	0.6900	0.7111	0.7187	0.7212	0.7220	0.7222	0.7222	0.7222	0.7222	0.7222	0.7222
100	0.6899	0.7111	0.7187	0.7212	0.7219	0.7224	0.7229	0.7229	0.7229	0.7229	0.7229
150	0.6900	0.7111	0.7187	0.7213	0.7220	0.7224	0.7233	0.7233	0.7233	0.7233	0.7233
200	0.6900	0.7111	0.7187	0.7212	0.7219	0.7224	0.7235	0.7235	0.7235	0.7235	0.7235
250	0.6900	0.7111	0.7187	0.7212	0.7220	0.7224	0.7235	0.7236	0.7236	0.7236	0.7236
500	0.6899	0.7110	0.7186	0.7211	0.7219	0.7220	0.7234	0.7236	0.7236	0.7236	0.7236

Table 8. 2008 Exact Method Child Focusing Strategy RBC Reporting

NUMBER OF DOCUMENTS	NUMBER OF ELEMENTS										
	50	100	150	200	250	500	1000	1500	2000	3000	4000
25	0.6725	0.7036	0.7133	0.7183	0.7194	0.7205	0.7205	0.7205	0.7205	0.7205	0.7205
50	0.6725	0.7036	0.7133	0.7183	0.7194	0.7218	0.7218	0.7218	0.7218	0.7218	0.7218
100	0.6725	0.7036	0.7133	0.7183	0.7194	0.7221	0.7224	0.7224	0.7224	0.7224	0.7224
150	0.6725	0.7036	0.7133	0.7183	0.7194	0.7221	0.7227	0.7227	0.7227	0.7227	0.7227
200	0.6725	0.7036	0.7133	0.7183	0.7194	0.7221	0.7228	0.7230	0.7230	0.7230	0.7230
250	0.6725	0.7036	0.7133	0.7183	0.7194	0.7221	0.7228	0.7230	0.7230	0.7230	0.7230
500	0.6725	0.7036	0.7133	0.7183	0.7194	0.7221	0.7228	0.7230	0.7230	0.7230	0.7230

Table 9. 2008 Exact Method Correlation Focusing Strategy RBC Reporting

NUMBER OF DOCUMENTS	NUMBER OF ELEMENTS										
	50	100	150	200	250	500	1000	1500	2000	3000	4000
25	0.6936	0.7125	0.7183	0.7203	0.7203	0.7203	0.7203	0.7203	0.7203	0.7203	0.7203
50	0.6937	0.7126	0.7184	0.7213	0.7213	0.7213	0.7213	0.7213	0.7213	0.7213	0.7213
100	0.6937	0.7126	0.7184	0.7213	0.7213	0.7219	0.7219	0.7219	0.7219	0.7219	0.7219
150	0.6938	0.7127	0.7185	0.7214	0.7214	0.7221	0.7222	0.7222	0.7222	0.7222	0.7222
200	0.6937	0.7126	0.7184	0.7213	0.7213	0.7220	0.7223	0.7223	0.7223	0.7223	0.7223
250	0.6937	0.7126	0.7184	0.7213	0.7213	0.7220	0.7223	0.7223	0.7223	0.7223	0.7223
500	0.6936	0.7125	0.7183	0.7212	0.7212	0.7220	0.7223	0.7225	0.7225	0.7225	0.7225

Table 10. 2009 Exact Method Section Focusing Strategy RBC Reporting

NUMBER OF DOCUMENTS	NUMBER OF ELEMENTS										
	50	100	150	200	250	500	1000	1500	2000	3000	4000
25	0.5462	0.5489	0.5539	0.5539	0.5539	0.5539	0.5539	0.5539	0.5539	0.5539	0.5539
50	0.5465	0.5493	0.5539	0.5546	0.5557	0.5557	0.5557	0.5557	0.5557	0.5557	0.5557
100	0.5465	0.5492	0.5537	0.5544	0.5556	0.5573	0.5573	0.5573	0.5573	0.5573	0.5573
150	0.5463	0.5494	0.5539	0.5545	0.5558	0.5585	0.5588	0.5588	0.5588	0.5588	0.5588
200	0.5459	0.5493	0.5538	0.5546	0.5556	0.5583	0.5589	0.5593	0.5593	0.5593	0.5593
250	0.5464	0.5493	0.5538	0.5546	0.5556	0.5582	0.5587	0.5594	0.5594	0.5594	0.5594
500	0.5465	0.5493	0.5539	0.5546	0.5557	0.5586	0.5592	0.5602	0.5602	0.5602	0.5602

3.4 Experiments with the 2009 INEX Collection

Working with a new collection is always a challenge, and the change from the 2008 to the 2009 INEX collection is no exception. Our approach requires tuning, which usually takes more time than is available before results are due. Thus our 2009 Focused Task results are based on parameter settings found to produce good results for the corresponding task in 2008.

Our official results for 2009 place us at the bottom of the ranking. This poor showing was due primarily to errors in reporting the xpaths. Many of the 2009 articles contain tags preceding the body tag, whereas the 2008 collection did not. Correcting this simple error made the xpaths recognizable. We were left with one problem. Our method generates each document tree according to its schema, created during parsing. In this parsing, subsections were not retained as a part of the document structure. (The content is present but is not recognizable as a subsection.) This means that the xpath of any element occurring at or below subsection level is not recognizable and hence cannot be evaluated. Removing such elements markedly decreased the number of unrecognizable elements in the set presented for evaluation. Once these modifications were made to our official result set, we resubmitted this data for evaluation. This produced an iP[0.01] value of 0.5286 (up from the original value of 0.3296). Thus our original run, with corrected xpaths, produces a result that would appear at rank 10 in the official 2009 Focused task results.

The improved evaluation resulting from the correction of xpaths renews our confidence in the methodology. This approach worked well, consistently producing highly ranked focused elements, in 2008. Without tuning, it produces acceptable results (as demonstrated by our corrected results) for the new INEX 2009 collection. As further evidence, we offer Table 10, which duplicates the methods of the experiment reported in Table 7 on the INEX 2009 collection. (That is, it uses the Exact Method with Section focusing strategy and RBC reporting.) Best results occur at 500 documents and 1500 or more elements (as expected); this produces an iP[0.01] value of 0.5602 which would rank again at 10 in the official results. We have yet to clear up the lower level xpath problem or do significant tuning, but we believe the results can only improve from this point.

We found in 2008 [6] that element retrieval itself was insufficient as a basis for focused retrieval but when combined with article retrieval was able to produce good focused elements. Although a number of the top reporting groups at INEX 2008 used BM25-based approaches for their focused runs [3, 7, 8, 10], we find that Singhal's *Lnu-ltu* term-weighting [16] produces competitive results when compared to what may be viewed as this state-of-the-art approach. Salton's classic Vector Space Model [15], adapted to meet the particulars of XML representation, has served well in this environment.

4 Conclusions

Our goal for this year centered on establishing that our basic methodology could be strengthened to consistently produce highly ranked focused elements. We have met this goal, as the results reported here for the 2008 collection demonstrate. (Details of these experiments may be found in [2, 12, 13].) With current work in progress, we hope to report similar results for the 2009 collection next year.

References

[1] Bapat, S.: Improving the results for focused and relevant-in-context tasks. M.S. Thesis, University of Minnesota Duluth (2008),
http://www.d.umn.edu/cs/thesis/bapat.pdf

[2] Bhirud, D.: Focused retrieval using upper bound methodology. M.S. Thesis, University of Minnesota Duluth (2009), http://www.d.umn.edu/cs/thesis/bhirud.pdf

[3] Broschart, A., Schenkel, R., Theobald, M.: Experiments with proximity-aware scoring for XML retrieval at INEX 2008. In: Geva, S., Kamps, J., Trotman, A. (eds.) INEX 2008. LNCS, vol. 5631, pp. 29–32. Springer, Heidelberg (2009)

[4] Crouch, C.: Dynamic element retrieval in a structured environment. ACM TOIS 24(4), 437–454 (2006)

[5] Crouch, C., Crouch, D., Kamat, N., Malik, V., Mone, A.: Dynamic element retrieval in the Wikipedia collection. In: Fuhr, N., Kamps, J., Lalmas, M., Trotman, A. (eds.) INEX 2007. LNCS, vol. 4862, pp. 70–79. Springer, Heidelberg (2008)

[6] Crouch, C., Crouch, D., Bapat, S., Mehta, S., Paranjape, D.: Finding good elements for focused retrieval. In: Geva, S., Kamps, J., Trotman, A. (eds.) INEX 2008. LNCS, vol. 5631, pp. 33–38. Springer, Heidelberg (2009)

[7] Gery, M., Largeron, C., Thollard, F.: UMJ at INEX 2008: pre-impacting of tags weights. In: Geva, S., Kamps, J., Trotman, A. (eds.) INEX 2008. LNCS, vol. 5631, pp. 46–53. Springer, Heidelberg (2009)

[8] Itakura, K., Clarke, C.: University of Waterloo at INEX 2008: adhoc, book, and link-the Wiki tracks. In: Geva, S., Kamps, J., Trotman, A. (eds.) INEX 2008. LNCS, vol. 5631, pp. 132–139. Springer, Heidelberg (2009)

[9] Khanna, S.: Design and implementation of a flexible retrieval system. M.S. Thesis, University of Minnesota Duluth (2005),
http://www.d.umn.edu/cs/thesis/khanna.pdf

[10] Lehtonen, M., Doucet, A.: Enhancing keyword search with a keyphrase index. In: Geva, S., Kamps, J., Trotman, A. (eds.) INEX 2008. LNCS, vol. 5631, pp. 65–70. Springer, Heidelberg (2009)

[11] Mone, A.: Dynamic element retrieval for semi-structured documents. M.S. Thesis, University of Minnesota Duluth (2007),
http://www.d.umn.edu/cs/thesis/mone.pdf

[12] Polumetla, C.: Improving results for the relevant-in-context task. M.S. Thesis, University of Minnesota Duluth (2009),
http://www.d.umn.edu/cs/thesis/polumetla_c.pdf

[13] Poluri, P.: Focused retrieval using exact methodology. M.S. Thesis, University of Minnesota Duluth (2009), http://www.d.umn.edu/cs/thesis/poluri.pdf

[14] Salton, G. (ed.): The Smart Retrieval System—Experiments in Automatic Document Processing. Prentice-Hall, Englewood Cliffs (1971)

[15] Salton, G., Wong, A., Yang, C.S.: A vector space model for automatic indexing. ACM Comm. 18(11), 613–620 (1975)

[16] Singhal, A., Buckley, C., Mitra, M.: Pivoted document length normalization. In: Proc. of the 19th Annual International ACM SIGIR Conference, pp. 21–29 (1996)

ListBM: A Learning-to-Rank Method for XML Keyword Search

Ning Gao, Zhi-Hong Deng, Yong-Qing Xiang, and Yu Hang

Key Laboratory of Machine Perception (Ministry of Education)
School of Electronic Engineering and Computer Science, Peking University
{Nanacream,pkucthh}@gmail.com,
{xiangyq,zhdeng}@cis.pku.edu.cn

Abstract. This paper describes Peking University's approach to the Ad Hoc Track. In our first participation, results for all four tasks were submitted: the Best In Context, the Focused, the Relevance In Context and the Thorough. Based on retrieval method Okapi BM25, we implement two different ranking methods NormalBM25 and LearningBM25 according to different parameter settings. Specially, the parameters used in LearningBM25 are learnt by a new learning method called ListBM. The evaluation result shows that LearningBM25 is able to beat NormalBM25 in most tasks.

Keywords: XML keyword learn-to-rank.

1 Introduction

INEX Ad Hoc Track [1] aims to evaluate performance in retrieving relevant results (e.g. XML elements or documents) to a certain query. Based on lots of research and comparative experiments, Okapi BM25 [2] is confirmed to be an effective ranking method. Plus, evaluation results of Ad Hoc Track show that Okapi BM25 performs better than some other frequently cited ranking models, such as TF*IDF [3] and so on. Motivated by BM25's excellent performance, many participants prefer BM25 as their basic retrieval model. In INEX 2008 Ad Hoc Track, University of Waterloo [4] outperforms in all three tasks of Measured as Focused Retrieval, known as Best in Context, Focused and Relevance in Context. The ranking system Waterloo used is "a biased BM25 and language modeling, in addition to Okapi BM25" [4].

However, in Okapi BM25 formula, there are several parameters used to adjust the proportion of element length and term frequency (tf) in the final score and they are frequently set manually by participants. Here we note that different parameter settings might lead to totally different evaluation results. Thus, in order to get a more rigorous, evidence-based and data-based parameter setting, a listwise machine learning method to learn the parameter settings is proposed. We call it listBM.

In detail, figure 1 shows the architecture of our ranking system. Firstly, when user submits a query, a results recognizer will calculate the result elements by matching the *Keywords* and the *Inverted Index*. An element is defined as a result element only if it contains all the keywords. The output of results recognizer is a Results Set, in which result elements are disordered. Therefore, a ranking method BM25 is introduced to

S. Geva, J. Kamps, and A. Trotman (Eds.): INEX 2009, LNCS 6203, pp. 81–87, 2010.

sort these result elements according to their relevance to the query. However, according to different parameter settings used in BM25, we implement two different ranking models NormalBM25 and LearningBM25. In NormalBM25, the parameter setting we used is same as what Waterloo used in INEX 2008. We call this parameter setting the Origin Parameter Setting. The result list ranked by using this parameter setting is called *NormalBM25 Results*. While in LearningBM25 model, the parameters are learned by a machine learning method called ListBM, to be introduced in section 3. The result list ranked by BM25 using this *Learnt Parameter Setting* is defined as *LearningBM25 Results*. We submit both the *NormalBM25 results* and the *LearningBM25 results* in four tasks of INEX 2009 Ad Hoc Track. The evaluation results show that *LearningBM25 results* perform better than *NormalBM25 results*, indicating that our learning method ListBM indeed help to improve the performance.

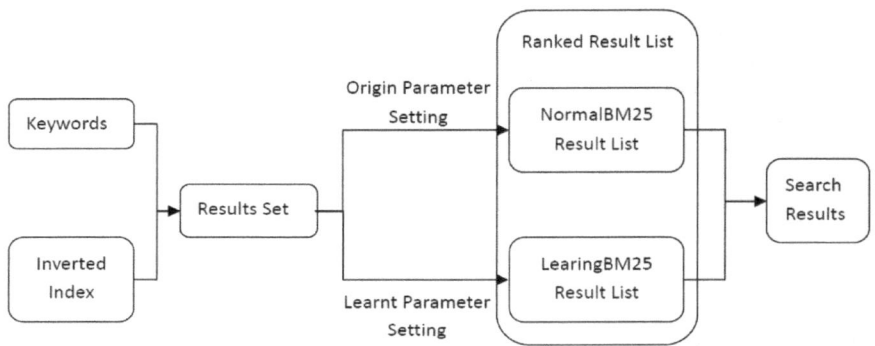

Fig. 1. Architecture of Ranking System

In section 2, we introduce the concepts of BM25 and background of machine learning method we used. Section 3 describes our learning method ListBM. In section 4, we show the evaluation results. Section 5 is the conclusion and future work.

2   Related Work

2.1   Okapi BM25[2]

BM25 is a widely quoted ranking method. It shows excellent performance referring to the evaluation results of INEX in the past years. For our method, to score an element according to its relevance to a certain query, we chose BM25 as our basic ranking model. In detail, the score is defined as follows.

$$score(e,Q) = \sum_{t \in Q} W_t \cdot \frac{(k_1+1) \cdot tf(t,e)}{k_1 \cdot (1-b+b \cdot \frac{len(e)}{avel}) + tf(t,e)} \qquad (1)$$

Score(e,Q) measures the relevance of element e to a certain query Q; W_t is the weight of term t indicating the inverse term frequency(IDF) of term t in collection. tf(t,e) is the frequency of term t appearing in element e; len(e) denotes the length of element e and avel denotes the average length of elements in whole collection. Two parameters, k_1 and b, are used to balance the weight of term frequency (tf) and element length(len) in final score.

2.2 Learning-to-Rank Methods

Learning-to-rank methods focus on using machine learning algorithms for better ranking. Many learning-to-rank algorithms have been proposed. According to different "instance" they use, learning-to-rank methods can be classified into three categories [5]: pointwise, pairwise and listwise.

In pointwise methods, documents are used as learning instance. The relevant score of a document is calculated by its features such as term frequency (tf), tag name and document links. This kind of algorithm attempts to find classification engine that can mark document as relevant or irrelevant correctly.

Pairwise methods, such as RankBoost [6] and RankSVM [7], take document pair as learning instance. Consider two documents d_1 and d_2, if d_1 is more relevant than d_2 to a certain query Q, then the document pair (d_1, d_2) is set to 1, otherwise it is set to -1. Pairwise methods target at training a learning engine to find the best document pair preferences.

In listwise methods, document list is taken as learning instance to train ranking engines. To find the best ranked list is the final goal. There are several well-known listwise methods such as Listnet [5], ListMLE [8], SVM-MAP[9] and so on.

Comparative tests [10] have shown that listwise methods perform best in these three categories.

3 Learning-to-Rank Method ListBM

3.1 ListBM

In NormalBM model, the parameter setting used in BM25 is same as Waterloo set in INEX 2008. In LearningBM model, the parameters are learnt by a learning method called ListBM. We use INEX 2008 data collection as the training data base.

In training, there is a set of query $Q=\{q^1, q^2, ..., q^m\}$. Each query q^i is associated with a ranked list of relevant documents $D^i = (d^i_1, d^i_2, ... , d^i_n)$, where d^i_j denotes the j-th relevant document to query q^i. These relevant documents lists, downloaded from the website of INEX, are used as standard results in our training. What's more, for each query q^i, we use basic ranking method BM25 mentioned in 2.1 to get a list of search results RL^i. Results returned by BM25 are all in form of elements. The first n result elements of RL^i are recorded in $R^i = (r^i_1\ r^i_2, ..., r^i_n)$. Then each documents list D^i and elements list R^i form a "instance".

The loss function is defined as the "distance" between standard results lists D^i and search results lists R^i. Therefore, the objective of learning is formalized as minimization of the total losses with respect to the training data.

$$\sum_{i=1}^{m} L(D^i, R^i) \tag{2}$$

Suppose there are two search results R, R' and a standard result D, the definition of loss function should meet the following two criterions:

- The loss value should be inversely proportional to the recall. If R contains more relevant results appeared in D than R' does, then the loss value of R should be smaller than the loss value of R'.
- The loss value should be inversely proportional to the precision. If the relevant content contained by R has the higher relevance degree (they appear in the top-ranking documents in D), then the loss value of R should be lower.

According to these two criterions, we define the loss function for a single query q^i as:

$$L\left(D^i, R^i\right) = \frac{mn^2}{\sum_{j=1}^n rank_j^i} \tag{3}$$

$$rank_j^i = \begin{cases} 0, & \text{(if the j-th result in } R^i \text{ doesn' appear in } D^i) \\ k, & \text{(if the j-th result in } R^i \text{ is contained by the k-th result in } D^i) \end{cases} \tag{4}$$

Where m is the number of queries in Q and n is the number of relevant documents in D^i corresponding to a certain query q^i. $rank_j^i$ is divided into two parts: (1) if the j-th result element in R^i isn't contained by any relevant documents in D^i, then $rank_j^i$ is set to 0; (2) if the j-th result element in R^i is contained by the k-th relevant document d_k^i in D^i, then $rank_j^i$ is set to k. Apparently, an element can only be contain by one document. Since the documents in D are different, it is impossible that there are more than one documents in D contains the j-th result in R.

Using the pair of standard results D and searching results R as "instance", L defined above as loss function, we implement a learning method ListBM to learn the two parameters k_1 and b in BM25 separately. Algorithm 1 describes the procedure of learning parameter k_1.

Algorithm 1. ListBM: The process of learning k_1

Input: query Q, relevant documents D
Parameter: k_1, number of iterations T,
Initialize: b=0.8, $\sum_{i=1}^m L(D^i, R^i) = +\infty$
for t=1 **to** T **do**
    for i=1 **to** m **do**
        search the q^i using BM25, get R^i;
        compute $L(D^i, R^i)$
        update $k_1 += \frac{1}{L(D^i, R^i)}$
    end for
    if $\sum_{i=1}^m L(D^i, R^i)$ increases, then break;
end for
output the pair of $\{k_{min}, \min \sum_{i=1}^m L(D^i, R^i); \}$

When learning k_1, parameter b will be initialized to 0.8. While randomly entering an original value of k_1, the loop won't end until the new updated k_1 begin to increase the loss value. The output is a pair of $\{k_{min}$, $\min \sum_{i=1}^{m} L(D^i, R^i)$; $\}$, in which $\min \sum_{i=1}^{m} L(D^i, R^i)$ is the minimum loss value got in the process and k_{min} is the corresponded k_1. While learning parameter b, k_1 is a content value 4 and b is updated according to the loss function. What's more, the output will be the pair of $\{b_{min}$, $\min \sum_{i=1}^{m} L(D^i, R^i)$; $\}$, in which b_{min} is the better b leading to minimum loss value.

3.2 Experiments

We implemented our ranking system in C++. The data collection we use is the English files of wiki provided by INEX 2008 Ad Hoc Track. The total size with these 659,388 files is 4.6G. In the process of reading and analyzing these XML files, we remove all the stop word from a standard stop word list before stemming. We use 8 queries from INEX 2008 topic pool in the training process. Totally 4800 documents are signed as relevant results. Figure 2 and figure 3 show the learning results of k_1 and b.

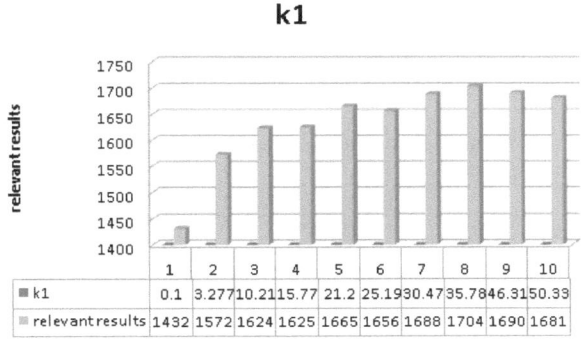

Fig. 2. Learning result of k_1

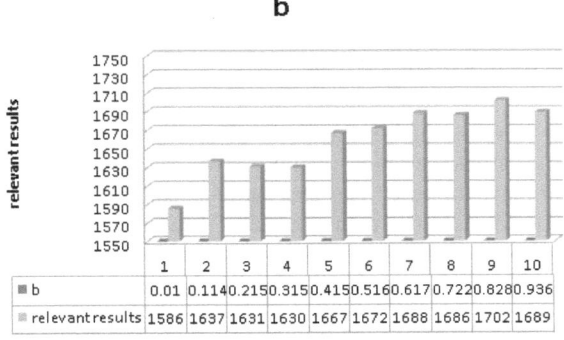

Fig. 3. Learning result of b

Figure 2 illustrates the learning results of k_1. In NormalBM model, the k_1 is set to 4. As is shown, in total 4800 search results, only 1572 results are relevant according to the standard results when k_1 is set to 3.277. The best set of k_1 is 35 so that the relevant results can reach up to 1704.

Figure 3 shows the learning result of b. The origin set of b is 0.8 according to the parameter setting of Waterloo. Result shows that b=0.8 is indeed the best set leading to a better searching performance. Hence, the parameter setting of {k_1, b} is set to {35, 0.8} in LearningBM model and {4, 0.8} in NormalBM model.

4 Evaluation Results

We submit both NormalBM results and LearningBM results for all four tasks to compare the performance. The original results are taken as the results of Thorough task. For Best in Context task, we return the most relevant element with the highest score as the best entry of a document. Moreover, based on the original result, for Focused task, we remove the elements which are contained by other element. We get evaluation results for three of four tasks. Table 1 shows the evaluation results of Measured as Focused Retrieval in which the search results are elements. Table 2 describes the search performance of Measured as Document Retrieval tasks. We can say that, in most conditions, the retrieval effectiveness of LearningBM is better than that of NormalBM.

Table 1. Evaluation results of element retrieval

	Best in Context	Focused	Thorough
LearningBM25	0.0953	0.3072	0.0577
NormalBM25	0.0671	0.1779	0.0521

Table 2. Evaluation results of document retrieval

	Best in Context	Focused	Thorough
LearningBM25	0.2382	0.2382	0.1797
NormalBM25	0.1849	0.1847	0.1689

5 Conclusion and Future Work

We propose a new learning method ListBM to learn the parameters in ranking method BM25. ListBM is a listwise learning method, using the data source of INEX 2008 Ad Hoc Track as training data base. The evaluation results show that parameter setting learnt by ListBM performs better than parameter setting set manually.

For the future work, we will continue to work on the following problems:

- The training data we used is the collection of INEX 2008 Ad Hoc Track. However, data collection has changed a lot for INEX 2009. We will study the learning to rank method on the new collection in the future.

- There are only 8 queries used in training. For the further study, more queries from INEX 2009 topics pool will be searched in the learning process.
- We will propose new definition of "distance" between the search result list and the standard result list. Furthermore, a more reasonable loss function and a new updating method will be introduced.

Acknowledgments

This work was supported in part by the National High Technology Research and Development Program of China (863 Program) under Grant No. 2009AA01Z136 and the National Natural Science Foundation of China under Grant No. 90812001.

References

[1] http://www.inex.otago.ac.nz/
[2] Theobald, M., Schenkel, R., Wiekum, G.: An Efficient and Versatile Query Engine for TopX Search. In: VLDB, pp. 625–636 (2005)
[3] Carmel, D., Maarek, Y.S., Mandelbrod, M., et al.: Searching XML documents via XML fragments. In: SIGIR, pp. 151–158 (2003)
[4] Itakura, K.Y., Clarke, C.L.A.: University of Waterloo at INEX 2008: Adhoc, Book, and Link-the-Wiki Tracks. In: Geva, S., Kamps, J., Trotman, A. (eds.) INEX 2008. LNCS, vol. 5631, pp. 116–122. Springer, Heidelberg (2009)
[5] Cao, Z., Qin, T., Liu, T.-Y., Tsai, M.-F., Li, H.: Learning to Rank: From Pairwise Approach to Listwise Approach. Microsoft technique report
[6] Freund, Y., Iyer, R., Schapire, R.E., Singer, Y.: An efficient boosting algorithm for combining preferences. J. Mach. Learn. Res., 933–969 (2003)
[7] Herbrich, R., Graepel, T., Obermayer, K.: Support vector learning for ordinal regression. In: Proceedings of ICANN, pp. 97–102 (1999)
[8] Xia, F., Liu, T.-Y., Wang, J., Zhang, W., Li, H.: Listwise approach to learning to rank: theory and algorithm. In: ICML 2008: Proceedings of the 25th international conference on Machine learning, pp. 1192–1199 (2008)
[9] Yue, Y., Finley, T., Radlinski, F., Joachims, T.: A Support Vector Method for Optimizing Average Precision. In: Proceedings of the ACM Conference on Research and Development in Information Retrieval, SIGIR (2007)
[10] Zhang, M., Kuang, D., Hua, G., Liu, Y., Ma, S.: Is learning to rank effective for Web search? In: SIGIR 2009 workshop (2009)

UJM at INEX 2009 Ad Hoc Track

Mathias Géry and Christine Largeron

Université de Lyon, F-42023, Saint-Étienne, France
CNRS UMR 5516, Laboratoire Hubert Curien
Université de Saint-Étienne Jean Monnet, F-42023, France
{mathias.gery,christine.largeron}@univ-st-etienne.fr

Abstract. This paper[1] presents our participation to the INEX 2009 Ad-Hoc track. We have experimented the tuning of various parameters using a "training" collection (i.e. INEX 2008) quite different than the "testing" collection used for 2009 INEX Ad-Hoc track. Several parameters have been studied for article retrieval as well as for element retrieval, especially the two main BM25 weighting function parameters: b and k_1.

1 Introduction

The focused information retrieval (IR) aims at exploiting the documents structure in order to retrieve the relevant elements (parts of documents) matching the user information need. The structure can be used to emphasize some words or some parts of the document: the importance of a term depends on its formatting (*e.g.* bold font, italic, etc.), and also on its position in the document (*e.g.*, title versus text body). During our previous INEX participations, we have developed a probabilistic model that learns a weight for each XML tag, representing its capability to emphasize relevant text fragments [3] [2]. One interesting result was that article retrieval based on BM25 weighting gives good results against element retrieval, even when considering a precision oriented measure ($iP[0.01]$): 3 article retrieval runs appear in the top-10 of the focused task (2nd, 4th and 8th, cf. [6]), and the 3 best $MAiP$ runs are 3 article retrieval runs! Thus a question comes: "Is BM25 suitable for element retrieval"? Indeed, we can imagine that, BM25 being developed for article retrieval, its adaptation to element retrieval is challenging. This problem has been addressed *e.g.* with BM25e [8].

Our objective during INEX 2009 was to answer to two questions, using the 2008 INEX collection as a training collection:

- is it possible to reuse the parameters tuned with INEX 2008 collection?
- is it still possible to obtain good results with article retrieval against element retrieval, regarding MAiP as well as $iP[0.01]$?

We present the experimental protocol in section 2, then our system overview in section 3, our tuning experiments using INEX 2008 in section 4, and finally our INEX 2009 results in section 5.

---

[1] This work has been partly funded by the Web Intelligence project (région Rhône-Alpes, cf. http://www.web-intelligence-rhone-alpes.org)

S. Geva, J. Kamps, and A. Trotman (Eds.): INEX 2009, LNCS 6203, pp. 88–94, 2010.
© Springer-Verlag Berlin Heidelberg 2010

2 Experimental Protocol

We have used the INEX Ad-Hoc 2008 collection as a training collection, and the INEX 2009 collection as a test collection. INEX 2008 collection contains 70 queries and 659,388 XML articles extracted from the English Wikipedia in early 2006 [1], while INEX 2009 collection contains 115 queries and 2,666,190 XML articles extracted from Wikipedia in 2008 [10]. We used the main INEX measures: $iP[x]$ the precision value at recall x, AiP the *interpolated average precision*, $MAiP$ the *mean AiP* and MAgP the *generalized mean average precision* [7]. The main INEX ranking is based on $iP[0.01]$ instead of the overall measure $MAiP$, allowing to emphasize the precision at low recall levels.

Given that every experiment is submitted to INEX in the form of a ranked list of 1,500 XML elements for each query, such measures favor, in terms of recall, the experiments for which whole articles are found (thereby providing a greater quantity of information for 1,500 documents). This is an issue, because focused answers may be penalized even if it is the very purpose of Focused IR to be able to return better granulated answers (i.e. relevant elements extracted from a whole article). Thus, we also calculated $R[1500]$, the recall rate for 1,500 documents, and $S[1500]$, the size (in Mb) of the 1,500 documents.

3 System Overview

Our system is based on the BM25 weighting function [9], that processes articles a_j a well as elements e_j:

$$w_{ji} = \frac{tf_{ji} \times (k_1 + 1)}{k_1 \times ((1 - b) + (b * ndl)) + tf_{ji}} \times \log \frac{N - df_i + 0.5}{df_i + 0.5} \tag{1}$$

with:

- tf_{ji}: the frequency of t_i in article a_j (resp. element e_j).
- N: the number of articles (resp. elements) in the collection.
- df_i: the number of articles (resp. elements) containing the term t_i.
- ndl: the ratio between the length of articles a_j (resp. elements e_j) and the average article (resp. element) length (*i.e.* its number of terms occurrences).
- k_1 and b: the classical BM25 parameters.

Parameter k_1 allows to control the term frequency saturation. Parameter b allows to set the importance of ndl, *i.e.* the importance of document length normalization. This is particularly important in focused IR as the length variation for elements is greater than that of articles, as each article is fragmented into elements (the largest article contains about 35,000 words).

Our system also considers some other parameters, *e.g.*:

- *logical_tags*: list of XML tags which the system will consider either at indexing and querying step (the system will therefore not be able to return an element that does not belong to this list);
- *minimum_size*: minimum size of documents (articles/elements) (# of terms);

- $level_{max}$: maximum depth of documents (depth of XML tree);
- df: df_i value for each term, computed on articles (INEX 2008: $max(df_i) = 659,388$); or on elements (INEX 2008: $max(df_i)$ between 1 and 52 millions);
- stop words: using a stop words list;
- parameters concerning queries handling: mandatory or banned query terms (+/- operators), etc.

4   Parameters Tuning (INEX 2008)

4.1   System Settings

All our runs have been obtained automatically, and using only the query terms (*i.e* the *title* field of INEX topics). We thus do not use the fields *description*, *narrative* nor *castitle*. Several parameters have been studied for article retrieval as well as for element retrieval. Some parameters were set after a few preliminary experiments, *e.g.*:

- *logical_tags* for article retrieval: *article*;
- *logical_tags* for element retrieval: *article, li, row, template, cadre, normallist, section, title, indentation1, numberlist, table, item, p, td, tr*;
- *minimum_size_{terms}*: 10 terms. Some analysis on the INEX 2008 assessments (not presented here) have shown that it is not useful to consider elements smaller than 10 terms, because these small elements are either non-relevant or their father is 100% relevant, and in this case it is better to return the father. Note that [5] has shown, using former INEX 2002 collection, that an optimal value for this parameter is to be set around 40;
- $level_{max}$: 1 for article retrieval, 23 for element retrieval;
- *df*: computed on articles (resp. elements) while indexing articles (resp. elements), instead of computing an overall *df* (*e.g.* at article level) used while indexing articles as well as elements. Note that [11] compute an overall *df*.
- stop words: 319 words from Glasgow Information Retrieval Group[2],

Two important parameters were studied more thoroughly: b and k_1, using a 2D grid: b varying from 0.1 to 1, with 0.1 steps, and k_1 varying from 0.2 to 3.8 with 0.2 graduations), thus a total of 380 runs (article and element retrieval).

4.2   INEX 2008 Tuning Results

The results presented in this section were computed after INEX 2008 using the official evaluation program *inex-eval* (version 1.0).

Figure 1 presents the behavior of article retrieval, showing the MAiP and $iP[0.01]$ changes according to b (resp. k_1). For a given b (resp. k_1), the $iP[0.01]$ and $MAiP$ measures drawn are obtained using the optimal k_1 (resp. b).

The best (b, k_1) values for article retrieval are slightly higher for MAiP ($(b, k_1) = (0.6, 2.2)$) than for iP[0.01] ($(b, k_1) = (0.4, 1.6)$). These values are not far from the classical values in the literature (*e.g.* (0.7, 1.2)). The results obtained with these optimal parameters are presented in table 1.

---

[2] http://www.dcs.gla.ac.uk/idom/ir_resources/linguistic_utils/stop_words

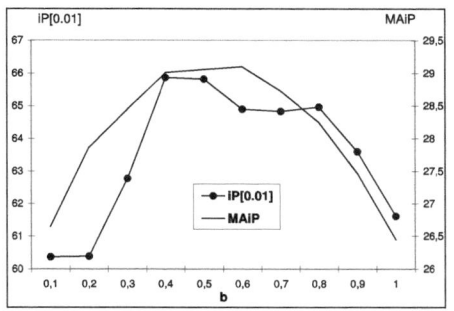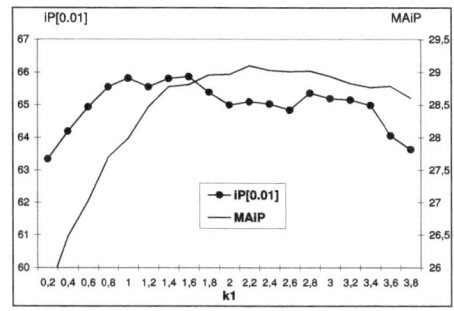

Fig. 1. Article retrieval in function of b and k_1

Table 1. Optimal b and k_1 parameters for article retrieval ($iP[0.01]$ and $MAiP$)

Run	Granularity	b	k_1	Optimized results	#doc	#art	R[1500]	S[1500]
R1	Articles	0.4	1.6	$iP[0.01] = 0.6587$	1,457	1,457	**0.8422**	8.22
R2	Articles	0.6	2.2	$MAiP = \mathbf{0.2910}$	1,457	1,457	0.8216	6.15

Figure 2 presents the behavior of the BM25 model in focused IR.

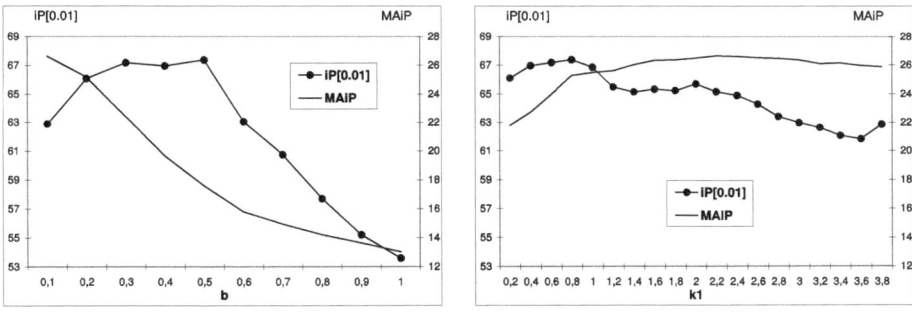

Fig. 2. Focused IR in function of b and k_1

The best (b, k_1) values are different for MAiP ($(b, k_1) = (0.1, 2.2)$) than for iP[0.01] ($(b, k_1) = (0.5, 0.8)$). The best MAiP is reached with the minimum value $b = 0.1$. The length normalization of BM25 (through b) seems to be counter-productive when optimizing recall in focused IR. But on the other hand, it is still useful in order to optimize precision (best value: $b = 0.5$). The tf saturation (through k_1) seems to be less important for focused IR: both $iP[0.01]$ and $MAiP$ slightly fluctuate with k_1. The results obtained with these optimal parameters are presented in table 2.

Table 2. Optimal b and k_1 parameters for focused IR ($iP[0.01]$ and $MAiP$)

Run	Granularity	b	k_1	Optimized results	#doc	#art	R[1500]	S[1500]
R3	Elements	0.5	0.8	$iP[0.01] = \mathbf{0.6738}$	1,463	1,257	0.4134	**1.65**
R4	Elements	0.1	2.2	$MAiP = 0.2664$	1,459	1,408	0.7476	5.24

5   INEX 2009 Results

We present in this section the official results obtained during INEX 2009. We submitted 17 runs: 5 runs to the Focused task and 4 runs to the Best In Context, the Relevant In Context and the Thorough tasks. One run per task is based on the BM25 reference run (article-level ranking) given by the INEX organizers in order to facilitate cross-system comparisons [4].

5.1   System Settings

All our runs have been obtained automatically, and using only the *title* field of INEX topics. Most of the settings given in section 4.1 have been reused for our INEX 2009 runs, except:

- *logical_tags* for element retrieval: article, list, p, reflist, sec, ss1, ss2, ss3, ss4, table, template (manually chosen);
- b and k_1: 0.6 and 2.2 for article retrieval (in order to maximize $MAiP$);
- b and k_1: 0.5 and 0.8 for element retrieval (in order to maximize $iP[0.01]$);
- $level_{max}$: 1 for article retrieval; 100 for element retrieval;
- df: computed on articles while indexing articles ($max(df_i) = 2,666,190$) and computed on elements while indexing elements ($max(df_i) = 444,540,453$).

5.2   Results: Focused Task

Table 3 presents the official results of our runs, compared to UWFERBM25F2 (Waterloo University) which was the winning run of the Focused task.

Table 3. Official "Focused" task results (57 runs)

Run	Granularity	Reference run	b	k_1	$iP[0.01]$	Rank
UWFERBM25F2	Element	-	-	-	**0.6333**	1
UJM_15525	Article	-	0.6	2.2	0.6060	6
UJM_15479	Article	-	0.6	2.2	0.6054	7
UJM_15518	Element	INEX organizers	0.5	0.8	0.5136	36
UJM_15484	Element	-	0.5	0.8	0.4296	45

Our system gives very interesting results compared to the best INEX systems. Article retrieval, i.e. the BM25 model applied on full articles, achieves the best results in terms of precision: $iP[0.01] = 0.6060$ by UJM_15525 (differences between UJM_15525 and UJM_15479 settings are not significant). The article retrieval runs outperform our focused IR run: $iP[0.01] = 0.4296$ (UJM_15484), despite the fact that BM25 parameter `ndl` is designed to take into account different documents lengths and thus documents granularities. This confirms the results obtained during INEX 2008. Note that our focused IR is improved when an article-level run (the reference run) is used as a pre-filter: $iP[0.01] = 0.5136$ by UJM_15518.

5.3   Relevant In Context (RIC), Best In Context (BIC), Thorough

Our BIC, RIC and Thorough runs have not been computed specifically. In order to respect the order and coverage rules of the RIC, BIC and Thorough tasks, our "focused" runs were reranked and filtered, using the same parameters settings than for the run UJM_15518. These results are presented in tables 4, 5 and 6.

Table 4. Official "Best In Context" task results (37 runs)

Run	Granularity	Reference run	b	k_1	$MAgP$	Rank
BM25bepBIC	Element	-	-	-	**0.1711**	1
UJM_15490	Element	UJM_15479	0.5	0.8	0.0917	28
UJM_15506	Element	UJM_15479	0.5	0.8	0.0904	30
UJM_15508	Element	INEX organizers	0.5	0.8	0.0795	34

Table 5. Official "Relevant In Context" results (33 runs)

Run	Granularity	Reference run	b	k_1	$MAgP$	Rank
BM25RangeRIC	Element	-	-	-	**0.1885**	1
UJM_15502	Element	UJM_15479	0.5	0.8	0.1075	21
UJM_15503	Element	INEX organizers	0.5	0.8	0.1020	26
UJM_15488	Element	UJM_15479	0.5	0.8	0.0985	27

Table 6. Official "Thorough" task results (30 runs)

Run	Granularity	Reference run	b	k_1	$MAiP$	Rank
LIG-2009-thorough-3T	Element	-	-	-	**0.2855**	1
UJM_15494	Element	INEX organizers	0.5	0.8	0.2435	9
UJM_15500	Element	UJM_15479	0.5	0.8	0.2362	12
UJM_15486	Element	-	0.5	0.8	0.1994	17

Runs UJM_15488 and UJM_15490 have been **filtered** with our best article run (UJM_15479), while UJM_15500, UJM_15502 and UJM_15506 have been **filtered and re-ranked** with the same article run (UJM_15479).

6 Conclusion

Our run UJM_15525 is ranked sixth of the competition according to the $iP[0.01]$ ranking. That means that a basic BM25 article retrieval run (article retrieval) gives better "precision" results ($iP[0.01]$) than BM25 element retrieval (focused IR), and should also give better "recall" results ($MAiP$).

These results confirm that article retrieval gives very good results against focus retrieval (as in INEX 2008 [3] [2]), even considering precision (that was not the case in 2008). However, we don't know if it comes from BM25, which is perhaps not suitable for elements indexing, or if it comes from a non optimal parameters settings. It is perhaps not so easy to reuse settings of parameters tuned on a different collection. We have to experiment more deeply on 2009 collection, using the same 2D grid for b and k_1, but also varying other parameters in order to better understand these results.

References

1. Denoyer, L., Gallinari, P.: The Wikipedia XML corpus. SIGIR Forum 40(1), 64–69 (2006)
2. Géry, M., Largeron, C., Thollard, F.: Integrating structure in the probabilistic model for information retrieval. In: Web Intelligence, Sydney, Australia, December 2008, pp. 763–769 (2008)
3. Géry, M., Largeron, C., Thollard, F.: UJM at INEX 2008: Pre-impacting of tags weights. In: Geva, S., Kamps, J., Trotman, A. (eds.) INEX 2008. LNCS, vol. 5631, pp. 46–53. Springer, Heidelberg (2009)
4. Geva, S., Kamps, J., Lethonen, M., Schenkel, R., Thom, J.A., Trotman, A.: Overview of the INEX 2009 Ad Hoc track. In: Geva, S., Kamps, J., Trotman, A. (eds.) INEX 2009, LNCS. LNCS, vol. 6203, pp. 88–94. Springer, Heidelberg (2010)
5. Kamps, J., de Rijke, M., Sigurbjörnsson, B.: The importance of length normalization for XML retrieval. Inf. Retr. 8(4), 631–654 (2005)
6. Kamps, J., Geva, S., Trotman, A., Woodley, A., Koolen, M.: Overview of the INEX 2008 Ad Hoc track. In: Geva, S., Kamps, J., Trotman, A. (eds.) INEX 2008. LNCS, vol. 5631, pp. 1–28. Springer, Heidelberg (2009)
7. Kamps, J., Pehcevski, J., Kazai, G., Lalmas, M., Robertson, S.: INEX 2007 evaluation measures. In: Fuhr, N., Kamps, J., Lalmas, M., Trotman, A. (eds.) INEX 2007. LNCS, vol. 4862, pp. 24–33. Springer, Heidelberg (2008)
8. Lu, W., Robertson, S., MacFarlane, A.: Field-weighted XML retrieval based on BM25. In: Fuhr, N., Lalmas, M., Malik, S., Kazai, G. (eds.) INEX 2005. LNCS, vol. 3977, pp. 161–171. Springer, Heidelberg (2006)
9. Robertson, S., Sparck Jones, K.: Relevance weighting of search terms. JASIST 27(3), 129–146 (1976)
10. Schenkel, R., Suchanek, F., Kasneci, G.: Yawn: A semantically annotated wikipedia xml corpus. In: GI-Fachtagung fur Datenbanksysteme in Business, Technologie und Web. LNI, vol. 103, pp. 277–291. GI (2007)
11. Taylor, M., Zaragoza, H., Craswell, N., Robertson, S., Burges, C.: Optimisation methods for ranking functions with multiple parameters. In: 15th ACM conference on Information and knowledge management (CIKM '06), New York, NY, USA, pp. 585–593 (2006)

Language Models for XML Element Retrieval

Rongmei Li[1] and Theo van der Weide[2]

[1] University of Twente, Enschede, The Netherlands
[2] Radboud University, Nijmegen, The Netherlands

Abstract. In this paper we describe our participation in the INEX 2009 ad-hoc track. We participated in all four retrieval tasks (thorough, focused, relevant-in-context, best-in-context) and report initial findings based on a single set of measure for all tasks. In this first participation, we test two ideas: (1) evaluate the performance of standard IR engines used in full document retrieval and XML element retrieval; (2) investigate if document structure can lead to more accurate and focused retrieval result. We find: 1) the full document retrieval outperforms the XML element retrieval using language model based on Dirichlet priors; 2) the element relevance score itself can be used to remove overlapping element results effectively.

1 Introduction

INEX offers a framework for cross comparison among content-oriented XML retrieval approaches given the same test collections and evaluation measures. The INEX ad-hoc track is to evaluate system performance in retrieving relevant document components (e.g. XML elements or passages) for a given topic of request. The relevant results should discuss the topic exhaustively and have as little non-relevant information as possible (specific for the topic). The ad-hoc track includes four retrieval tasks: the Thorough task, the Focused task, the Relevant in Context task, and the Best in Context task.

The 2009 collection is the English Wikipedia with XML format. The ad-hoc topics are created by the participants to represent real life information need. Each topic consists of five fields. The `<title>` field (CO query) is the same as the standard keyword query. The `<castitle>` field (CAS query) adds structural constraints to the CO query by explicitly specifying where to look and what to return. The `<phrasetitle>` field (Phrase query) presents explicitly a marked up query phrase. The `<description>` and `<narrative>` fields provide more information about topical context. Especially the `<narrative>` field is used for relevance assessment.

The paper documents our first participation in the INEX 2009 ad-hoc track. Our aims are to: 1) evaluate the performance of standard IR engines (Indri search engine) used in full document retrieval and XML element retrieval; 2) investigate if document structure can lead to a more accurate and focused retrieval result. We adopt the language modeling approach [2] and tailor the estimate of query term generation from a document to an XML element according to the user

S. Geva, J. Kamps, and A. Trotman (Eds.): INEX 2009, LNCS 6203, pp. 95–102, 2010.
© Springer-Verlag Berlin Heidelberg 2010

request. The retrieval results are evaluated as: 1) XML element retrieval; 2) full document retrieval.

The rest of the paper describes our experiments in the ad-hoc track. The pre-processing and indexing steps are given in section 2. Section 3 explains how to convert a user query to an Indri structured query. The retrieval model and strategies are summarized in section 4. We present our results in section 5 and conclude this paper with a discussion in section 6.

2   Pre-processing and Indexing

The original English XML Wikipedia is not stopped or stemmed before indexing. The 2009 collection has 2,666,190 documents taken on 8 October 2008. It is annotated with the 2008-w40-2 version of YAGO ([3]).

We index mainly the queried XML fields as follows: `category`, `actor`, `actress`, `adversity`, `aircraft`, `alchemist`, `article`, `artifact`, `bdy`, `bicycle`, `caption`, `catastrophe`, `categories`, `chemist`, `classical_music`, `conflict`, `director`, `dog`, `driver`, `group`, `facility`, `figure`, `film_festival`, `food`, `home`, `image`, `information`, `language`, `link`, `misfortune`, `mission`, `missions`, `movie`, `museum`, `music_genre`, `occupation`, `opera`, `orchestra`, `p`, `performer`, `person`, `personality`, `physicist`, `politics`, `political_party`, `protest`, `revolution`, `scientist`, `sec`, `section`, `series`, `singer`, `site`, `song`, `st`, `theory`, `title`, `vehicles`, `village`.

3   Query Formulation

We handle CO queries by full article retrieval while ignoring boolean operators (e.g. "-"or "+") in the `<title>` field. For CAS queries, we adopt two different strategies to formulate our Indri structured query ([4]) for retrieving full articles or XML elements respectively. The belief operator #combine is used in both cases. Neither CO queries nor CAS queries are stemmed or stopped. Similar to the document, we assume the query terms form a bag-of-words [1] when phrase constraints (noted as "") are not added in `<title>` or `<castitle>`.

- CAS queries for full article retrieval: we extract all `<castitle>` terms within "about" and the XML tags that have semantic meaning as our query terms. Boolean operators (e.g. "-"or "+") in `<castitle>` are ignored. For instance, for INEX query (id=2009009) we have:
  ```
  <castitle>
  //(p|sec)[about(.//(political_party|politics),
  election +victory australian labor party state council -federal)]
  </castitle>
  ```
  Extraction leads to the following query terms: `election`, `victory`, `australian`, `labor`, `party`, `state`, `council`, `federal`, `political`, `party`, `politics`.

- CAS queries for XML element retrieval: we extract all `<castitle>` terms within "about" and use the Indri belief operator #not to exclude ("-") certain terms. At the same time, we add XML element constraints (e.g. `<song>`, `<p>`) and phrase constraints to our new Indri queries. For instance, for INEX query (id=2009021) we have:

    `<castitle>//article[about(., ''wonder girls'')]</castitle>`

  The formulated Indri query looks like:

    `#combine[article](#1(wonder girls))`

  For CAS queries for any XML element type (noted as *), we retrieve either `aritle` element only or additional elements such as `bdy`, `link`, `p`, `sec`, `section`, `st`, and `title`. An example CAS query is `<castitle>//*[about(., Dwyane Wade)]</castitle>` (id=2009018).

4   Retrieval Model and Strategies

We use the run of full article retrieval as the baseline for both CO and CAS queries. The retrieval model for the baseline runs is based on cross-entropy scores for the query model and the document model that is smoothed using Dirichlet priors. It is defined as follows:

$$score(D|Q) = \sum_{i=1}^{l} P_{ml}(t_i|\theta_Q) \cdot \log \left(\frac{tf(t_i, D) + \mu P_{ml}(t_i|\theta_C)}{|D| + \mu} \right) \qquad (1)$$

where l is the length of the query, $P_{ml}(t_i|\theta_Q)$ and $P_{ml}(t_i|\theta_C)$ are the Maximum Likelihood (ML) estimates of respectively the query model θ_Q and the collection model θ_C. $tf(t_i, D)$ is the frequency of query term t_i in a document D. $|D|$ is the document length. μ is the smoothing parameter.

For XML element retrieval, we compute the relevance score ($score(E|Q)$) of the queried XML field (E) in regard to the given CAS query (Q). The smoothed document model (expressed in the log function) is tailored to compute the ML estimate of the XML element model $P_{ml}(t_i|\theta_E)$.

We set up our language model and model parameters based on the experimental results of similar tasks for INEX 2008. Here μ is considered to be 500.

4.1   Baselines

Baseline runs retrieve full articles for CO or CAS queries. Only the #combine operator is used. We submitted the results of CAS query for the Thorough and the Focused tasks and the results of CO query for the Relevant in Context and the Best in Context tasks. This baseline performance indicates the performance of the Indri search engine in the setting of XML element retrieval.

4.2   Strategies for Overlapping Removal

Within the ad-hoc XML retrieval track, there are four sub-tasks:

- The `Thorough task` requires the system to estimate the relevance of elements from the collection. It returns elements or passages in order of relevance (where specificity is rewarded). Overlap is permitted.
- The `Focused task` requires the system to return a ranked list of elements or passages. Overlap is removed. When equally relevant, users prefer shorter results over longer ones.
- The `Relevant in Context task` requires the system to return relevant elements or passages clustered per article. For each article, it returns an unranked set of results, covering the relevant material in the article. Overlap is not permitted.
- The `Best in Context task` asks the system to return articles with one best entry point. Overlap is not allowed.

Because of the hierarchical structure of an XML document, sometimes a parent element is also considered relevant if its child element is highly relevant to a given topic. As a result, we obtain a number of overlapping elements. To fulfill the overlap-free requirement for the Focused task, the Relevant in Context task and the Best in Context task, we adopt the following strategies to remove overlapping element paths based on the result of the Thorough task:

- `Relevance Score`: The result of the Thorough task is scanned from most to less relevant. When an overlapped element path is found within a document, then the element path with lower relevance score is removed. (see Figure 1). In case that overlapped elements in the same document have the same relevance score, we choose the element with the higher rank.

docID	relevance score	element path
7575091	5.33165	/article[1]/physicist[1]/scientist[1]/chemist[1]
63140	**5.30598**	**/article[1]/scientist[1]/chemist[1]**
~~63140~~	~~5.10164~~	~~/article[1]/scientist[1]/chemist[1]/chemist[1]/~~
14353374	5.00247	/article[1]/physicist[1]/scientist[1]/chemist[1]
63140	**3.39020**	**/article[1]/person[1]/alchemist[1]**

Fig. 1. Example result of the Focused task (qid=2009005)

Next the overlap-free result is grouped by article. For each query, the articles are ranked based on their highest relevancy score. For each article, the retrieved element paths keep the rank order of relevance (see Figure 2).

For the Best in Context task, we choose the most relevant XML element path of each article as our result.

docID	relevance score	element path
7575091	5.33165	/article[1]/physicist[1]/scientist[1]/chemist[1]
63140	**5.30598**	**/article[1]/scientist[1]/chemist[1]**
63140	**3.39020**	**/article[1]/person[1]/alchemist[1]**
14353374	5.00247	/article[1]/physicist[1]/scientist[1]/chemist[1]

Fig. 2. Example result of the Relevant in Context task (qid=2009005)

- **Relevance Score and Full Article Run:** In addition to the Relevance Score strategy, we combine our overlap-free result with the result of the full article run (baseline run for CO query). We remove XML element paths whose article does not appear in the result of the full article runs (see Figure 3). The filtered result follows the rank order of our baseline run. We adopt the same strategy for the Reference task as well.

docID	relevance score	element path
7575091	5.33165	/article[1]/physicist[1]/scientist[1]/chemist[1]
14353374	5.00247	/article[1]/physicist[1]/scientist[1]/chemist[1]

Fig. 3. Example result of the Reference task (qid=2009005)

5 Results

For each of the four sub-tasks, we submitted two XML element results and one extra result for the reference task. On the whole, we had 12 submissions to the ad-hoc track. Among them, 9 runs are qualified runs. Additionally, we report more results in this paper.

5.1 Full Document Retrieval

One of our goals for the ad-hoc track is to compare the performance of Indri search engine used in full document retrieval and XML element retrieval. The observation will be used to analyze our element language models and improve our overlapping removal strategies. For official runs, we submitted full document retrieval using both CO and CAS queries for four sub-tasks. Except the Thorough task, one run of the rest tasks was disqualified because of overlapped results. The qualified run for the Thorough task is an automatic run for CAS query (see Table 1). In the same table, we have an extra run for CO query. For full document retrieval, the result of CAS query is slightly worse than that of the CO query. This performance gap indicates the difference between two expressions for the same topic of interest.

Table 1. Results of full document retrieval

tasks	performance metrics				
	iP[.00]	iP[.01]	iP[.05]	iP[.10]	MAiP
thorough (article.CO additional)	0.5525	0.5274	0.4927	0.4510	0.2432
thorough (article.CAS official)	0.5461	0.5343	0.4929	0.4415	0.2350

5.2   XML Element Retrieval

For official submission, we presented our results using the strategy of **Relevance Score and Full Article Run**. All qualified runs use CAS query. The result of the Thorough task is in Table 2. The additional runs are the original results without the help of full article runs. The run of `element1.CAS` returns `article` element type only for any element type (noted as * in `<castitle>`) requests while the run of `element2.CAS` returns all considered element types.

Table 2. Results of XML element retrieval

tasks	performance metrics				
	iP[.00]	iP[.01]	iP[.05]	iP[.10]	MAiP
thorough (element.CAS.ref official)	0.4834	0.4525	0.4150	0.3550	0.1982
thorough (element1.CAS additional)	0.4419	0.4182	0.3692	0.3090	0.1623
thorough (element.CAS.baseline official)	0.4364	0.4127	0.3574	0.2972	0.1599
thorough (element2.CAS additional)	0.4214	0.3978	0.3468	0.2876	0.1519

The full document retrieval outperforms the element retrieval in locating all relevant information. The same finding shows in the performance difference between `element1.CAS` and `element2.CAS`. Our results again agree with the observation in previous INEX results. System wise, the given reference result (element.CAS.baseline) has better performance over Indri search engine (element.CAS.ref).

Focused Task. The official and additional results of the Focused task are in Table 3. Our run (`element.CAS.baseline`) successfully preserves the retrieval result of the Thorough task and brings moderate improvement. The full document runs still have the highest rank in the Focused task.

Relevant in Context Task. As explained earlier, we rank documents by the highest element score in the collection and rank element paths by their relevance score within the document. Overlapping elements are removed as required. The retrieval result is in Table 4. The full document runs still dominate the performance and reference runs continue boosting retrieval results.

Best in Context Task. This task is to identify the best entry point for accessing the relevant information in a document. Our strategy is to return the element path with highest relevance score in a document. The retrieval result is in Table 5. Using the relevance score as the only criteria has lead to a promising

Table 3. Results of XML element retrieval

tasks	performance metrics				
	iP[.00]	iP[.01]	iP[.05]	iP[.10]	MAiP
focus (article.CO additional)	0.5525	0.5274	0.4927	0.4510	0.2432
focus (article.CAS additional)	0.5461	0.5343	0.4929	0.4415	0.2350
focus (element.CAS.ref official)	0.4801	0.4508	0.4139	0.3547	0.1981
focus (element.CAS.baseline official)	0.4451	0.4239	0.3824	0.3278	0.1695
focus (element1.CAS additional)	0.4408	0.4179	0.3687	0.3092	0.1622
focus (element2.CAS additional)	0.4228	0.3999	0.3495	0.2909	0.1527

Table 4. Results of XML element retrieval

tasks	performance metrics				
	gP[5]	gP[10]	gP[25]	gP[50]	MAgP
relevant-in-context (article.CO additional)	0.2934	0.2588	0.2098	0.1633	0.1596
relevant-in-context (article.CAS additional)	0.2853	0.2497	0.2132	0.1621	0.1520
relevant-in-context (element.CAS.ref official)	0.2216	0.1904	0.1457	0.1095	0.1188
relevant-in-context (element.CAS.baseline official)	0.1966	0.1695	0.1391	0.1054	0.1064
relevant-in-context (element1.CAS additional)	0.1954	0.1632	0.2150	0.1057	0.0980
relevant-in-context (element2.CAS additional)	0.1735	0.1453	0.1257	0.1003	0.0875

result when we compare the original runs (`element1.CAS` and `element2.CAS`) with the run boosted by the baseline (`element.CAS.baseline`). However, the run (`element1.CAS`) containing more `article` returns is still better than the run with other element returns.

Table 5. Results of XML element retrieval

tasks	performance metrics				
	gP[5]	gP[10]	gP[25]	gP[50]	MAgP
best-in-context (article.CO additional)	0.2663	0.2480	0.1944	0.1533	0.1464
best-in-context (article.CAS additional)	0.2507	0.2305	0.1959	0.1499	0.1372
best-in-context (element.CAS.ref official)	0.1993	0.1737	0.1248	0.0941	0.1056
best-in-context (element1.CAS additional)	0.2257	0.1867	0.1426	0.1125	0.1015
best-in-context (element2.CAS additional)	0.2089	0.1713	0.1343	0.1084	0.0924
best-in-context (element.CAS.baseline official)	0.1795	0.1449	0.1143	0.0875	0.0852

6   Conclusion

In our official runs, we present our baseline results and results filtered by our document retrieval and the given reference run. In this paper, we provide additional results of the original document and XML element retrieval. The Indri search engine can provide reasonable result of XML element retrieval compared to results of full article retrieval and results of other participating groups. We can also use relevance score as the main criteria to deal with the overlapping problem effectively. However, the full document runs are still superior than XML element runs. When the results of reference and baseline run are used, the result of the Thorough task is improved. This may imply that the search engine is able to locate relevant elements within documents effectively. Besides the accuracy of relevance estimation, retrieval performance also depends on the effective formulation of Indri structured query. For example, the two ways of wildcard interpretation in `element1.CAS` and `element2.CAS` present different results in our experiments. The step of overlapping removal is the other key factor that may harm the retrieval performance. In our case, our result (`element1.CAS`) ranks high in the Thorough task but low in the Focus and Relevant in Context tasks.

Except using the relevance score for removing the overlapping element paths, we may try other criteria such as the location of the element within a document. This is especially important for the Best in Context task as users tend to read a document from top-down.

Acknowledgments

This work is sponsored by the Netherlands Organization for Scientific Research (NWO), under project number 612-066-513.

References

1. Blei, D.M., Ng, A.Y., Jordan, M.I.: Latent dirichlet allocation. J. Mach. Learn. Res. 3, 993–1022 (2003)
2. Zhai, C.X., Lafferty, J.: A Study of Smoothing Methods for Language Models Applied to Information Retrieval. ACM Trans. on Information Systems 22(2), 179–214 (2004)
3. Schenkel, R., Suchanek, F.M., Kasneci, G.: YAWN: A Semantically Annotated Wikipedia XML Corpus. In: 12. GI-Fachtagung fr Datenbanksysteme in Business, Technologie und Web, Aachen, Germany (March 2007)
4. Strohman, T., Metzler, D., Turtle, H., Croft, W.B.: Indri: A Language-model Based Search Engine for Complex Queries. In: Proceedings of ICIA (2005)

Use of Language Model, Phrases and Wikipedia Forward Links for INEX 2009

Philippe Mulhem[1] and Jean-Pierre Chevallet[2]

[1] LIG - CNRS, Grenoble, France
Philippe.Mulhem@imag.fr
[2] LIG - Université Pierre Mendès France, Grenoble, France
Jean-Pierre.Chevallet@imag.fr

Abstract. We present in this paper the work of the Information Retrieval Modeling Group (MRIM) of the Computer Science Laboratory of Grenoble (LIG) at the INEX 2009 Ad Hoc Track. Our aim this year was to twofold: first study the impact of extracted noun phrases taken in addition to words as terms, and second using forward links present in Wikipedia to expand queries. For the retrieval, we use a language model with Dirichlet smoothing on documents and/or doxels, and using an Fetch and Browse approach we select rank the results. Our best runs according to doxel evaluation get the first rank on the Thorough task, and according to the document evaluation we get the first rank for the Focused, Relevance in Context and Best in Context tasks.

1 Introduction

This paper describes the approach used by the MRIM/LIG research team for the Ad Hoc Track of the INEX 2009 competition. Our goal here is to experiment enrichment of documents and queries in two directions: first, using the annotation provided by Ralph Schenker and his colleagues [5] we expand the vocabulary to annotated noun phrases, and second, using forward pages extracted from wikipedia (dump processed in July 2009), we expand the user's queries. So, the vocabulary extension comes from internal Wikipedia data (categories for instance) and external data (Wordnet), and the forward information comes from only internal Wikipedia data.

Our work integrates structured documents during retrieval according to a Fetch and Browse framework [1], as retrieval is achieved in two steps: the first one focuses on whole articles, and the second one process integrates the non-article doxels in a way to provide *focused* retrieval according to the retrieved documents parts.

First of all, we define one term: a *doxel* is any part of an XML document between its opening and closing tag. We do not make any kind of difference between a doxel describing the logical structure of the document (like a title or a paragraph) or not, like anchors of links or words that are emphasized), a relation between doxels may come from the structural composition of the doxels, or from any other source. Assume that an xml document is "<A>This

S. Geva, J. Kamps, and A. Trotman (Eds.): INEX 2009, LNCS 6203, pp. 103–111, 2010.

is an <B>example</B> of <C>XML</C> document</A>". This document contains 3 doxels: the first is delimited by the tag A, the second is delimited by the tag B, and the third is delimited by the tag C. We also consider that a compositional link relates A to B, and A to C. We will also depict B and C as direct structural components of A.

The remaining of this paper is organized as follows: after commenting shortly in section 2 related works, we describe the vocabulary expansion based on noun phrases in part 3. Then the query expansion according to the forward links extracted from Wikipedia are presented in section 4. Section 5 introduces our *matching* process. Results of the INEX 2009 Ad Hoc track are presented in Section 6, where we present the twelve (three for each of the four tasks) officially submitted runs by the LIG this year, and we discuss in more detail some aspects of our proposal. We conclude in part 7.

2   Related Works

The language modeling approach to information retrieval exists from the end of the 90s [4]. In this framework, the relevance status value of a document for a given query is estimated by the probability of generating the query from the document. In the case of structured documents indexing, two problems arise. First, doxels have very different sizes (from few words to whole articles). The smoothing may then depend on the considered doxel. Second, a smoothing according to the whole corpus may generate uniform models, especially for small doxels. To cope with the first point, we propose to integrate a dirichlet smoothing, as described in [6]. With such smoothing the probability of a query term to be generated by a document model is:

$$p_\mu(w|\theta_D) = (1 - \frac{\mu}{|D| + \mu})\frac{c(w, D)}{|D|} + \frac{\mu}{|D| + \mu}p(w|C) \tag{1}$$

$$= \frac{c(w, D) + \mu p(w|C)}{|D| + \mu} \tag{2}$$

where θ_D is the model of document D, $c(w, D)$ the number of occurrences of term w in D, $|D|$ the size of document D, $p(w|C)$ the probability of the corpus C to generate w and μ a constant that weights the importance of the corpus. The value μ represents the sum of (non integer) pseudo occurrences of a term w in any document.

In a way to avoid a unique smoothing for all doxels, we choose to generate a different dirichlet smoothing for each type of doxels t considered. Here, a type of doxel corresponds a tag t in an XML document. We have then:

$$p_{\mu_t}(w|\theta_{D_t}) = \frac{c(w, D_t) + \mu_t p(w|C_t)}{|D_t| + \mu_t} \tag{3}$$

associated to the type t of a doxel D_t.

Using noun phrases for information retrieval has also proved to be effective in some contexts like medical documents [3] and [2]. In the medical context, trustworthy data like UMLS[1] for instance are available. In the case of Inex 2009 we choose to experiment extraction of noun phrases according to results provided by YAWN [5].

3   Extraction and Use of Noun Phrases from YAWN

Here, we use the result YAWN [5] provided with the INEX 2009 collection. However, we do not use the tags that characterize the phrases, but the words themselves as additional terms of the vocabulary. For instance, in the document 12000.xml, the following text appears:

```
<music>
<composer wordnetid="109947232" confidence="0.9173553029164789">
<artist wordnetid="109812338" confidence="0.9508927676800064">
<link xlink:type="simple" xlink:href="../434/419434.xml">
Koichi Sugiyama</link></artist>
</composer>
</music>
```

The noun phrase surrounded by YAWN tags is the *Koichi Sugiyama*. In our case, the noun phrase *Koichi Sugiyama* is then used as a term that will be considered in the indexing vocabulary. In fact, the terms *Koichi* and *Sugiyama* are also indexing terms. We keep only the noun phrases that are described by YAWN because we consider that they are trustworthy to be good indexing terms. An inconsistency may appear, though: we noticed that all the noun phrases surrounded by YAWN tags were not always tagged (for instance the phrase "Koichi Sugiyama" may occur in other documents without YAWN tags). That is why we ran one pass that generates the noun phrases terms for each occurrence of the YAWN tagged noun phrases. In the following, we refer to the initial vocabulary (keywords only) as *KBV*, the phrase-based only vocabulary as *Phrase*, and the union of both vocabulary as *KBV+Phrase*.

4   Extraction and Use of Wikipedia Forward Links

The original wikipedia documents have also a very nice feature worth studying, which is related to forwarding links. To reuse a case taken from the query list of INEX 2009, if on Wikipedia we look for the page "VOIP", the page is almost empty and redirects to the page titled "Voice over Internet Protocol". In this case, we clearly see that some very strong semantic relationship exists between the initial term and the meaning of the acronym. We use this point in to consideration by storing all these forward links, and by expending the query expressions by the terms that occur in the redirected page (i.e. the title of the page pointed to). In this case, a query "VOIP" is then translated into "VOIP Voice over Internet Protocol".

So, this expansion does not impact the IR model used, but only the query expression.

---

[1] http://www.nlm.nih.gov/research/umls/

5   Indexing and Matching of Doxels

As it has been done in several approaches in the previous years, we choose to select doxels according to their type. In our case, we chose to consider only doxel that contain one title to be potentially relevant for queries. The list called L_d is then: article, ss, ss1, ss2, ss3, ss4.

We generated several indexing of the doxels using a language model that uses a dirichlet smoothing, using the Zettair system. One of the indexing is achieved according to *KBV*, and another one according to *KBV+Phrase*.

In our experiments, we worked on two sets of potential retrieved doxels: the article only doxels , and the doxels of type in L_d.

We integrated also the reference run provided by the INEX Ad-Hoc track organizers, by considering that the vocabulary is *KBV*, that the target doxels are articles only and that the model used is BM25.

We have defined a Fetch and Browse framework that can be processed in four steps:

- The first step generates a ranked list L according to a set of doxel types, namely S,
- The second step generates a ranked list L' according to another set of doxel type S' (where no intersection exists between S and S').
- The third step fuses L and L' into Lr . We transform L into Lr by inserting some doxels of L' into L. For each element l of L, we append all the elements of L' that have a relationship with l. In the work reported here the relationship is the structural composition. For instance, elements of L are articles, and elements of L' are subsections. This inclusion is called a Rerank.
- The last step postprocesses the results. It removes overlaps in Lr when needed. In this case we give priority on the first results: we remove any of the doxels that overlap an already seen doxel in Lr. This postprocessing may select the best entry point of a document.

The lists L and L' may be generated by different information retrieval models IRM and IRM', based on different vocabularies Voc and Voc', using different query preprocessing QP and QP', leading to one Fetch and Browse processing noted as:

Remove_ovelap(Rerank((QP, IRM, Voc, S),(QP', IRM', Voc , S')))

6   Experiments and Results

The INEX 2009 Adhoc track consists of four retrieval tasks: the Thorough task, Focused Task, Relevant In Context Task, and Best In Context Task. We submitted 3 runs for each of these tasks. Two ranking functions were tested: $Rerank_{rsv}$ that preserves the rsv orser of the L', and $Rerank_{reading}$ that ranks L' according to tho original reading ordre of the documents. We have used language models and BM25 as Information Retrieval Models IRM and IRM'. We tested two different vocabularies KBV and KBV+Phrase, and no query preprocessing (None) and query expansion using forward links (Forward_links_expansion).

6.1 Thorough Task

For the Thorough task, we considered

- $1T$: No fetch an browse here is processed, we just have a one step processing, (none, LM, KBV, L_d), which ranks all the potential doxels according to their rsv. This result can be considered as a baseline for our LM based approach ;
- $2T$: In this run, we applied a Fetch and Browse approach: No_remove_overlap(Rerank$_{rsv}$((none, LM, KBV, {article}), (none, LM, KBV, $L_d \setminus$ {article}))). Here the idea is to put priority on the whole article matching, and then to group all the documents doxels according to their rsv.
- $3T$: For our third Thorough run, the reference run is used as the Fetch step: No_remove_overlap(Rerank$_{rsv}$((none, BM25, KBV, {article}),(none, LM, KBV, $L_d \setminus$ {article})))

Table 1. Thorough Task for LIG at INEX2008 Ad Hoc Track

Run	precision at 0.0 recall	precision at 0.01 recall	precision at 0.05 recall	precision at 0.10 recall	MAiP (rank / 30)	MAiP Doc. (rank / 30)
$1T$	0.575	0.570	0.544	0.496	0.281 (1)	0.344 (3)
$2T$	0.572	0.567	0.542	0.492	0.273 (4)	0.342 (4)
$3T$	0.582	0.569	0.535	0.493	0.281 (2)	0.333 (5)

From the table 1, we see that the language model and the BM25 with fetch and browse outperforms the other runs at recall 0.00. We notice also that the Fetch and Browse approach with language model underperforms the simple process of run 1T, which is not the case with the BM25 fetching.

6.2 Focused Task

The INEX 2009 Focused Task is dedicated to find the most focused results that satisfy an information need, without returning "overlapped" elements. In our focused task, we experiment with two different rankings.

We submitted three runs :

- $1F$: This run is similar to the run 2T, but we apply the Remove_overlap operation, so the run is described by: remove_overlap(Rerank$_{rsv}$((none, LM, KBV, {article}),(none, LM, KBV, $L_d \setminus$ {article})));
- $2F$: In this second Focused run, we study the impact of using the query expansion using the forward links of Wikipedia and the extented vocabulary in the first step of the matching: remove_overlap(Rerank$_{rsv}$((Forward_links_expansion, LM, KBV+Phrase, {article}), (none, LM, KBV, $L_d \setminus$ {article})));
- $3F$: This run based on the reference run is achieved by removing the overlaps on the run 3T: Remove_overlap(Rerank$_{rsv}$((none, BM25, KBV, {article}), (none, LM, KBV, $L_d \setminus$ {article}))).

Table 2. Focused Task for LIG at INEX2009 Ad Hoc Track

Run	ip[0.00]	ip[0.01] (rank / 57)	ip[0.05]	ip[0.10]	MAiP	MAiP Doc. (rank / 62)
1F	0.574	0.573 (22)	0.534	0.497	0.267	0.351 (1)
2F	0.551	0.532 (31)	0.490	0.453	0.236	0.300 (33)
3F	0.581	0.568 (24)	0.525	0.493	0.269	0.341 (3)

The results obtained by using *doxel* based evaluation show that the use of query expansion and noun phrases underperforms the other approaches, which seems to indicate that the matching of documents (for the fetching) is less accurate using these two expansions. The reason should come form the fact that the query expansion is quite noisy, but we will check this hypothesis in the future. Here again, using the language model outperforms the BM25 baseline provided, except for the ip[0.00] measure.

6.3   Relevant In Context Task

For the Relevant In Context task, we take "default" focused results and reordered the first 1500 doxels such that results from the same document are clustered together. It considers the article as the most natural unit and scores the article with the score of its doxel having the highest RSV. The runs submitted are similar to our Focused runs, but we apply a reranking on the document parts according to the reading order (the $\text{Rerank}_{reading}$ function), before taking care of removing overlaping parts. We submitted three runs :

- 1R: This run is similar to the run 1F, but we apply the $\text{Rerank}_{reading}$ instead of the Rerank_{rsv} function. So the 1R run is described by: remove_overlap($\text{Rerank}_{reading}$((none, LM, KBV, {article}), (none, LM, KWV, $L_d \smallsetminus$ {article}))));
- 2R: Compared to the run 2F, we apply similar modifications than for the 1R run, leading to: remove_overlap($\text{Rerank}_{reading}$((Forward_links_expansion, LM, KBV+Phrase, {article}),(none, LM, KBV, $L_d \smallsetminus$ {article}))));
- 3R: Compared to the run 3F, we apply similar modifications than for the 1R run, leading to: Remove_overlap($\text{Rerank}_{reading}$((none, BM25, KBV, {article}), (none, LM, KWV, $L_d \smallsetminus$ {article}))).

The results are presented in table 3, and the measures for the *doxel* based evaluations are using generalized precision where the results for the document based evaluation is the MAiP.

In the case of retrieval in context, event of the best MAgP value is for 1R, the BM25 based run performs better for all the gP values presented in table 3. Here the run 2R that uses query expansion and noun phrases obtains much lower results.

Table 3. Relevant In Context Task for INEX2009 Ad Hoc

Run	gP[5]	gP[10]	gP[25]	gP[50]	MAgP (rank / 33)	MAiP Doc. (rank / 42)
1R	0.295	0.256	0.202	0.152	0.173 (12)	0.351 (1)
2R	0.189	0.171	0.135	0.105	0.091 (28)	0.168 (41)
3R	0.305	0.273	0.216	0.160	0.173 (13)	0.341 (9)

6.4 Best In Context Task

For this task, we take the focused results, from which we filter the best entry point by selecting the doxel with the higher rsv per document (the keep_best function).

We submitted three runs :

- 1B: This run is similar to the run 1F, but we apply the keep_best instead of the remove_overlap function. So the 1B run is described by: keep_best(Rerank$_{rsv}$((none, LM, KBV, {article}), (none, LM, KBV, $L_d \diagdown${article}))));
- 2B: Compared to the run 2F, we apply similar modifications than for the 1B run, leading to: keep_best(Rerank$_{reading}$((Forward_links_expansion, LM, KBV+Phrase, {article}),(none, LM, KBV, $L_d \diagdown${article}))));
- 3B: Compared to the run 3F, we apply similar modifications than for the 1B run, leading to: keep_best(Rerank$_{reading}$((none, BM25, KBV, {article}), (none, LM, KBV, $L_d \diagdown${article}))).

The results are presented in table 4.

Table 4. Best In Context Task for INEX2009 Ad Hoc

Run	gP[5]	gP[10]	gP[25]	gP[50]	MAgP (rank / 37)	MAiP Doc. (rank / 38)
1B	0.244	0.226	0.182	0.139	0.154 (17)	0.351 (1)
2B	0.169	0.159	0.128	0.101	0.087 (31)	0.168 (38)
3B	0.271	0.251	0.193	0.144	0.154 (16)	0.341 (7)

Here, for the first time with our runs, the BM25 run outperforms (very slightly: 0.1544 versus 0.1540) the language model one for the official evaluation measure on doxels. Which is not the case for the document based evaluation. Here also, our proposed expansions do not perform well.

6.5 Discussion

In this section, we concentrate on some runs with noun phrases and query expansion in a way to find out why our proposal did not work as planned, and if there is room for improvements.

First of all, on major problem that occur with our proposal is that the query expansion generates too many additional words to be really effective. For

Table 5. Relevant In Context Task for INEX2009 Ad Hoc, query-based comparisons

Query	gP[10]			AgP		
	R1	R2	R3	R1	R2	R3
2009_001	0.279	0.379	0.279	0.268	0.280	0.314
2009_015	0.686	0.687	0.686	0.395	0.295	0.421
2009_026	0.110	0.170	0.110	0.284	0.273	0.307
2009_029	0.736	0.926	0.636	0.369	0.439	0.287
2009_069	0.299	0.199	0.500	0.289	0.332	0.285
2009_088	0.427	0.391	0.462	0.429	0.450	0.486
2009_091	0.099	0.653	0.182	0.189	0.511	0.221
Average (relative difference)	0.377 (-22.6%)	0.487	0.408 (-16.2%)	0.318 (-13.8%)	0.369	0.332 (-10.0%)

instance, the query 2009_005, "chemists physicists scientists alchemists periodic table elements", is in fact translated into "chemists chemist periodic_table periodic table elements periodic table chemical elements periodic table elements periodic system periodic system elements periodic table mendeleev periodic table table elements periodic properties natural elements element symbol list groups periodic table elements representative element mendeleev periodic chart peroidic table elements mendeleev table periodic table elements periodicity elements organization periodic table periodic table chemical elements fourth period periodic table elements periodic patterns group 2a nuclear symbol", which is a very long query. In this expanded query *periodic_table* is a noun phrase term. We see in this expansion that we find the word "mendeleev", which is a very good term for the query, but on the other side we see many words that are unrelated to the initial query, like "representative" "organization" "system", "properties", "fourth", and so on. So the main problem that we face here is that the expanded queries are not narrowed enough to be really useful, leading to poor results according to the evaluation measure during the INEX 2009 evaluation campaign.

Things are not hopeless, though: consider the seven (i.e., 10% of the 69 INEX 2009 queries) best results obtained for our expansion based *Relevance in Context* run 2R according to MAgP, we get the results presented in table 5. There we see that potential improvement may be achieved through the use of our proposal. The query 091, "Himalaya trekking peak", has been translated into "trekking_peak himalaya himalayas": in this case the query result has been clearly improved by using the noun phrase "trekking_peak" instead of using the two words 'trekking" and "peak". On these queries (table 5), the increase of generalized precision at 10 results is around 23% shows the potential of our proposal.

7   Conclusion

In the INEX 2009 Ad Hoc track, we proposed several baseline runs limiting the results to specific types of doxels. From the official INEX 2009 measures wether

MRIM/LIG reached the first place on 4 measures, mostly for the evaluations based on documents. The new directions studied that we proposed are based on vocabulary expansion using noun phrases and query expansions using forward links extracted from Wikipedia. The results obtained using these extensions underperformed the more crude approaches we proposed, but we shown here that there is great room for improvements using extracted date from Wikipedia on a subset of the official INEX 2009 queries.

References

1. Chiaramella, Y.: Information retrieval and structured documents. In: Agosti, M., Crestani, F., Pasi, G. (eds.) ESSIR 2000. LNCS, vol. 1980, pp. 286–309. Springer, Heidelberg (2001)
2. Chevallet, J.-P., Lim., J. H., Le, T.H.D.: Domain knowledge conceptual inter-media indexing, application to multilingual multimedia medical reports. In: ACM Sixteenth Conference on Information and Knowledge Management, CIKM 2007 (2007)
3. Lacoste, C., Chevallet, J.-P., Lim, J.-H., Hoang, D.L.T., Wei, X., Racoceanu, D., Teodorescu, R., Vuillenemot, N.: Inter-media concept-based medical image indexing and retrieval with umls at ipal. In: Peters, C., Clough, P., Gey, F.C., Karlgren, J., Magnini, B., Oard, D.W., de Rijke, M., Stempfhuber, M. (eds.) CLEF 2006. LNCS, vol. 4730, pp. 694–701. Springer, Heidelberg (2007)
4. Ponte, J.M., Croft, W.B.: A language modeling approach to information retrieval. In: Research and Development in Information Retrieval (1998)
5. Schenkel, R., Suchanek, F.M., Kasneci, G.: Yawn: A semantically annotated wikipedia xml corpus. In: 2. GI-Fachtagung fur Datenbanksysteme in Business, Technologie und Web (BTW 2007), pp. 277–291 (2007)
6. Zhai, C.: Statistical Language Models for Information Retrieval. Morgan and Claypool (2008)

Parameter Tuning in Pivoted Normalization for XML Retrieval: ISI@INEX09 Adhoc Focused Task

Sukomal Pal[1], Mandar Mitra[1], and Debasis Ganguly[2]

[1] Information Retrieval Lab, CVPR Unit
Indian Statistical Institute, Kolkata, India
{sukomal_r,mandar}@isical.ac.in
[2] Synopsys,
Bangalore, India
debforit@gmail.com

Abstract. This paper describes the work that we did at Indian Statistical Institute towards XML retrieval for INEX 2009. Since there has been an abrupt quantum jump in the INEX corpus size (from 4.6 GB with 659,388 articles to 50.7 GB with 2,666,190 articles), retrieval algorithms and systems were put to a 'stress test' in the INEX 2009 campaign. We tuned our text retrieval system (SMART) based on the Vector Space Model (VSM) that we have been using since INEX 2006. We submitted two runs for the adhoc focused task. Both the runs used VSM-based document-level retrieval with blind feedback: an initial run (*indsta_VSMpart*) used only a small fraction of INEX 2009 corpus; the other used the full corpus (*indsta_VSMfb*). We considered *Content-Only* (CO) retrieval, using the Title and Description fields of the INEX 2009 adhoc queries (2009001-2009115). Our official runs, however, used incorrect topic numbers. This led to very dismal performance. Post-submission, the corrected version of both baseline and with-feedback document-level runs achieved competitive scores. We performed a set of experiments to tune our pivoted normalization-based term-weighting scheme for XML retrieval. The scores of our best document-level runs, both with and without blind feedback, seemed to substantially improve after tuning of normalization parameters. We also ran element-level retrieval on a subset of the document-level runs; the new parameter settings seemed to yield competitive results in this case as well. On the evaluation front, we observed an anomaly in the implementation of the evaluation-scripts while interpolated precision is being calculated. We raise the issue since a XML retrievable unit (passage/element) can be partially relevant containing a portion of non-relevant text, unlike document retrieval paradigm where a document is considered either completely relevant or completely non-relevant.

1 Introduction

Traditional Information Retrieval systems return whole documents in response to queries, but the challenge in XML retrieval is to return the most relevant

S. Geva, J. Kamps, and A. Trotman (Eds.): INEX 2009, LNCS 6203, pp. 112–121, 2010.

parts of XML documents which meet the given information need. Since INEX 2007 [1], arbitrary passages are permitted as retrievable units, besides the usual XML elements. A retrieved passage consists of textual content either from within an element or spanning a range of elements. Since INEX 2007, the adhoc retrieval task has also been classified into three sub-tasks: a) the FOCUSED task which asks systems to return a ranked list of non-overlapping elements or passages to the user; b) the RELEVANT in CONTEXT task which asks systems to return relevant elements or passages grouped by article; and c) the BEST in CONTEXT task which expects systems to return articles along with one best entry point to the user. Along with these, INEX 2009 saw the return of the d) THOROUGH task, where all the relevant items (either passages or elements) from a document are retrieved. Here, overlap among the elements are permitted but the elements are ranked according to relevance order.

Each of the four subtasks can be again sub-classified based on different retrieval approaches:

- element retrieval versus passage retrieval.
- Standard keyword query or Content-Only(CO) retrieval versus structured query or Content-And-Structure (CAS) retrieval.
- Standard keyword query (CO) retrieval versus phrase query retrieval.

In the CO task, a user poses a query in free text and the retrieval system is supposed to return the most relevant elements/passages. A CAS query can provide explicit or implicit indications about what kind of element the user requires along with a textual query. Thus, a CAS query contains structural hints expressed in XPath [2] along with an *about()* predicate. Phrase queries were introduced in INEX 2009. These queries use explicit multi-word phrases. The aim was to make the query more verbose and to see what impact verbose queries have on retrieval effectiveness.

This year we submitted two adhoc focused runs, both using a Vector Space Model (VSM) based approach with blind feedback. VSM sees both the document and the query as bags of words, and uses their *tf-idf*- based weight-vectors to measure the inner product *similarity* as a measure of closeness between the document and the query. The documents are retrieved and ranked in decreasing order of the similarity-value.

We used a modified version of the SMART system for the experiments at INEX 2009. Since the corpus used for the adhoc track this year is huge compared to the earlier INEXes ([3], [4]), the system was really put to a 'stress test' both in terms of robustness and time-efficiency. To make sure that at least one retrieval run was completed within the stipulated time, one of our submissions (*indsta_VSMpart*) was run on partial data (top-10060 documents from INEX 2009 corpus in the alphabetical order). The other run (*indsta_VSMfb*) used, however, the complete corpus. For both the runs, retrieval was at the whole-document level, using query expansion based on blind feedback after the initial document retrieval. We considered CO queries only using *title* and *description* fields. Post-submission we found that our official runs suffered from a serious flaw: during query processing queries were assigned incorrect QIDs by the system. Correcting this error improved the performance noticeably.

In our earlier endeavours at INEX, we did not explore the issue of proper length normalization for XML retrieval in the VSM approoach. Instead, we used parameter-settings of the pivoted normalization approach that were tuned for document retrieval with TREC test collections. This time we tuned the parameters for the INEX XML collection (both pre-2009 and 2009 version) to suit both document-retrieval and element-retrieval.

We also observed an anomaly in the evaluation-scripts shared by the INEX organizers. For the adhoc focused task, definitions of *precision* and *recall* are customized by replacing documents with characters (amount of text in terms of #characters) within documents. Like document retrieval, interpolated precision (iP) is here estimated from the natural precision-values once they are calculated at all natural recall-points, i.e. after each relevant text is encountered. But unlike document retrieval, each retrieved and relevant text unit in XML retrieval may not be entirely relevant; often some portion of the retrieved text is relevant and the retrieved text does contain some portion which is not relevant. The question therefore arises whether estimation of interpolated precision should be done only at the end of the entire retrieved text which contain some relevant text? Or should it be done just after the very character where interpolated recall-level points to (no matter whether it lies within the retrieved text)? We demonstrate the issue in more detail with some example in section 3.

In the following section, we describe our general approach for the runs, present results in section 4 and finally conclude.

2   Approach

2.1   Indexing

We first shortlisted 74 tags from previous INEX Wikipedia corpus [3]. Each of these tags occurs at least twice in the pre-2009 Wikipedia collection, contain at least 10 characters of text and are listed in the INEX 2007 qrels. Examples of such tags are: *<article>*, *<body>*, *<caption>*, *<center>*, *<collectionlink>*, *<definitionitem>*, *<defintionlist>*, *<div>*, *<em >*, *<figure>*, *<gallery>*, *<item>*, *<outsidelink>*, *<p>*, *<section>*, *<wikipedialink>*, etc. Documents were parsed using the libxml2 parser, and only the textual portions included within the selected tags were used for indexing. Similarly, for the topics, we considered only the *title* and *description* fields for indexing, and discarded the *inex-topic*, *castitle* and *narrative* tags. No structural information from either the queries or the documents was used.

The extracted portions of the documents and queries were indexed using single terms and a controlled vocabulary (or pre-defined set) of statistical phrases following Salton's blueprint for automatic indexing [5].

2.2   Document-Level Retrieval

Stopwords that occur in the standard stop-word list included within SMART were removed from both documents and queries. Words were stemmed using a variation of the Lovins' stemmer implemented within SMART. Frequently occurring word bi-grams (loosely referred to as phrases) were also used as indexing units. We used the N-gram Statistics Package (NSP)[1] on the English Wikipedia text corpus from INEX 2006 and selected the 100,000 most frequent word bi-grams as the list of candidate phrases. Documents and queries were weighted using the *Lnu.ltn* [6] term-weighting formula. For the initial run, we used *slope* = 0.2 and *pivot* = 120 and retrieved 1500 top-ranked XML documents for each of 115 adhoc queries (2009001 - 2009115).

Next we used blind feedback to retrieve whole documents. We applied automatic query expansion following the steps given below for each query (for more details, please see [7]).

1. For each query, collect statistics about the co-occurrence of query terms within the set $\mathcal{S}$ of 1500 documents retrieved for the query by the baseline run. Let $df_{\mathcal{S}}(t)$ be the number of documents in $\mathcal{S}$ that contain term t.
2. Consider the 50 top-ranked documents retrieved by the baseline run. Break each document into overlapping 100-word windows.
3. Let $\{t_l, \ldots, t_m\}$ be the set of query terms (ordered by increasing $df_{\mathcal{S}}(t_i)$) present in a particular window. Calculate a similarity score *Sim* for the window using the following formula:

$$Sim = idf(t_1) + \sum_{i=2}^{m} idf(t_i) \times \min_{j=1}^{i-1}(1 - P(t_i|t_j))$$

   where $P(t_i|t_j)$ is estimated based on the statistics collected in Step 1 and is given by

$$\frac{\#\ documents\ in\ \mathcal{S}\ containing\ words\ t_i\ and\ t_j}{\#\ documents\ in\ \mathcal{S}\ containing\ word\ t_j}$$

   This formula is intended to reward windows that contain multiple matching query words. Also, while the first or "most rare" matching term contributes its full idf (inverse document frequency) to *Sim*, the contribution of any subsequent match is deprecated depending on how strongly this match was predicted by a previous match — if a matching term is highly correlated to a previous match, then the contribution of the new match is correspondingly down-weighted.
4. Calculate the maximum *Sim* value over all windows generated from a document. Assign to the document a new similarity equal to this maximum.
5. Rerank the top 50 documents based on the new similarity values.

---

[1] http://www.d.umn.edu/~tpederse/nsp.html

6. Assuming the new set of top 20 documents to be relevant and all other documents to be non-relevant, use Rocchio relevance feedback to expand the query. The expansion parameters are given below:

$$number\ of\ words = 20$$
$$number\ of\ phrases = 5$$
$$Rocchio\ \alpha = 4$$
$$Rocchio\ \beta = 4$$
$$Rocchio\ \gamma = 2.$$

For each topic, 1500 documents were retrieved using the expanded query.

Post-submission, we revisited the issue of pivoted length normalization in our term-weighting scheme. In the VSM approach [8] that we use, the length normalization factor is given by

$$(1 - slope) * pivot + slope * length.$$

At earlier INEXes [9] we blindly set the slope and pivot parameters to 0.20 and 80 respectively for our runs following [10]. On an adhoc basis we took two more *slope* values, viz. 0.3 and 0.4, for the same pivot value. But these parameters were actually tuned for the TREC adhoc document collection. The INEX Wikipedia XML collection is both syntactically and semantically different (not from the news genre). More importantly the average length of a Wiki-page is much smaller than a general English text document from the TREC collection. Thus the issue of length normalization needs a revisit. In *Lnu.ltn*, document length normalization factor is rewritten as

$$normalization = 1 + \frac{slope}{(1 - slope)} * \frac{\#unique\ terms}{pivot}$$

where *length* is given by #unique terms in the document. Instead of varying both pivot and slope separately we considered the combined term $\frac{slope}{(1-slope)*pivot}$. With $0 \leq slope < 1$ and $pivot \geq 1$, the factor $\in [0,1]$. For the sake of simplicity, we chose $pivot = 1$ which reduces the factor to $\frac{slope}{(1-slope)}$ (we call it *pivot-slope factor*) and varied the factor from 0 to 1.

2.3   Element-Level Run

For element-level retrieval, we adopted a 2-pass strategy. In the first pass, we retrieved 1500 documents for each query using the method described in 2.1.

In the second pass, these documents were parsed using the libxml2 parser. All elements in these 1500 documents that contain text were identified, indexed and compared to the query. The elements were then ranked in decreasing order of similarity to the query. In order to avoid any overlap in the final list of retrieved elements, the nodes for a document are sorted in decreasing order of similarity, and all nodes that have an overlap with a higher-ranked node are eliminated.

The issue of length normalization is similarly explored in element-retrieval as well. The findings are described in section 4.

3   Evaluation

During our post-hoc analysis, another issue related to evaluation of XML retrieval popped up. For the adhoc focused task, definitions of *precision* and *recall* at rank r (resp. $P[r]$ and $R[r]$) are adapted from the traditional definitions by replacing documents with characters (amount of text in terms of #characters). These definitions thus implicitly regards a character as the unit of retrieval (as opposed to complete documents).

As in the case of document retrieval, interpolated precision (iP) for the focused task is calculated using the precision-values obtained at each natural recall-point. Since the definition effectively regards a character as the unit of retrieval, it would be natural to define recall-points at the character-level. In practice, however, when computing interpolated precision (iP), recall levels are calculated only after each passage / element is retrieved, rather than after each *character* is retrieved. This may give rise to inconsistencies because, unlike in document retrieval, each retrieved and relevant text unit in XML retrieval may not be entirely relevant: often, only a part of the retrieved unit is relevant, with the remainder being non-relevant. For example, consider a system (say A) that retrieves some passage/element X from a document D in response to a topic t. Assume that

- X starts from the 200-th character of D and runs through the 499-th character (thus the length of X is 300 characters),
- X contains 50 characters of relevant text, running from the 230-th character through the 279-th character, and
- the total amount of relevant text in the collection for t is 1000 characters.

Suppose further that A has already retrieved 450 characters of text prior to X, of which 80 are relevant. Then, before X is retrieved, $R = 80/1000 = 0.08$; after X is retrieved, $R = (80 + 50)/1000 = 0.13$. Note that the true 10% recall point is achieved at a point within X (when 100 relevant characters have been retrieved). The precision at this point is $P_{R=10\%} = 100/450 + (230 + 20 - 200) = 100/500 = 0.2$. Thus, $iP[0.10]$ should be at least 0.2. However, if precision is only measured at the end of X (as done by the official evaluation scripts), we get $iP[0.10] = (80 + 50)/(450 + 300) = 130/750 = 0.173$ (assuming that the precision value decreases further at all subsequent ranks).

Since we are not correctly estimating interpolated precision if we compute precision values only at the ends of relevant passages, there is a possibility that we are underestimating / overestimating system performance, and therefore ranking systems incorrectly. If all the systems are similarly affected by this anomaly, there would not be much change in relative system ranks, but this may not be the case. The point we would like to make is that, in the document retrieval scenario, a document is regarded as an atomic unit both in the definition and in the calculation of evaluation metrics. In contrast, while the metrics for XML retrieval are defined at the character level, their computation is done at the element / passage level.

In line with the JAVA implementation of the evaluation-program, we implemented a C-version making the necessary changes to calculate interpolated precision by including additional recall points that lie within a retrieved text. We evaluated each of the submissions to the adhoc focused task using the C-version and compared their score with the reported scores. The findings are reported in the next section.

4   Results

4.1   Retrieval

Though there were 115 topics for the INEX 2009 adhoc task, relevance judgments were available for 68 topics. After results were released and qrels and evaluation-scripts were made available, we conducted some post-hoc analysis. We started by tuning the pivot and slope parameters using INEX 2008 data for element-retrieval over an initial retrieved document-set. We first varied pivot-slope ($\frac{slope}{1-slope}$) from 0 to 1 in steps of 0.1, and observed the change in retrieval scores (mainly based on $MAiP$) and thus narrowed down the range of interest. We found that setting the value of pivot-slope factor to 0.1 also seemed to cause over-normalization. Searching in the $[0.0, 0.1]$ range, we got a magic-value 0.00073 for the factor that gave best $MAiP$ score.

The experiment was repeated for INEX 2007 data where we observed a similar pattern. This bolstered our confidence to repeat it for the INEX 2009 element-level and document retrieval runs.

A brief summary of varying pivot-slope during element retrieval with INEX 2009 data is given by Table 1.

Table 1. Element-retrieval over result of *indsta_VSMfb*, pivot $= 1$

slope	0	0.01	0.05	0.10	0.20	0.30	0.40	0.50	1.0	0.00073	0.005
iP[0.00]	0.4146	**0.4908**	0.3221	0.1934	0.0698	0.0460	0.0265	0.0094	0.0044	0.4432	0.4881
iP[0.01]	0.4139	0.4495	0.2149	0.1049	0.0244	0.0170	0.0036	0.0029	0.0000	0.4425	**0.4703**
iP[0.05]	0.3815	0.3519	0.0873	0.0222	0.0122	0.0124	0.0000	0.0000	0.0000	**0.4030**	0.3836
iP[0.10]	0.3294	0.2634	0.0604	0.0115	0.0000	0.0000	0.0000	0.0000	0.0000	**0.3468**	0.3061
MAiP	0.1542	0.0992	0.0168	0.0054	0.0015	0.0011	0.0003	0.0001	0.0000	**0.1596**	0.1231

Similarly our experiment on pivot-slope for baseline document retrieval with INEX 2009 data is summarized in Table 2.

With feedback as described above, the best score improved by about 5% (MAP $= 0.2743$).

Table 2. Variation of pivot-slope for document-retrieval, pivot = 1

slope	0	0.00073	0.005	0.10	0.20	0.30	0.40	0.50
MAP	0.2572	**0.2612**	0.1463	0.0250	0.0177	0.0149	0.0130	0.0121

Table 3. Subdocument-level (element/passage) evaluation for the FOCUSED, CO task

Retrieval	Run Id	iP@0.01
doc-level	indsta_VSMfb(official)	0.0078
	indsta_VSMfb (corrected)	0.4531
	indsta_VSMpart (official)	0.0003
	indsta_VSM-base	0.4607
element-level	indsta_VSMfbEltsPvt1.0slope0.005(our best)	0.4703
	UWFERBM25F2 (best)	0.6333

Table 4. Document-level evaluation for the FOCUSED, CO task

Run Id	MAP
indsta_VSMfb (official)	0.0055
indsta_VSMpart (official)	0.0000
indsta_VSMfb (corrected)	0.2149
VSMbase-pvt1.0slope0.00073	0.2612
VSMfb-pvt1.0slope0.00073 (our best)	0.2743
LIG-2009-focused-1F (best)	0.3569

Our official performance as reported in the INEX09 website along with our results of our post-submission runs are shown in Table 3 and Table 4.

Following figure 1 demonstrates our performance vis-a-vis the best element-level retrieval.

Because of erroneous query processing, performance of our official submissions was really dismal, but on correction performance improved considerably. Though our runs are way behind the best performer (*UWFERBM25F2*), so far early precisions are concerned, on an average our runs are competitive, specifically our corrected document-level submission (*indsta_VSMfb*). For element-retrieval (*indsta_VSMfbEltsPvt1.0slope0.005* & *indsta_VSMfbElts-slope0.005*), our parameter tuning experiments paid a promising results, but still there is enough room for improvement.

4.2   Evaluation

Once submission files were made available, we were in a position to a make a comparative study. We evaluated all runs which were submitted for the adhoc focused task using both the official JAVA version of the evaluation program and its C-clone with our modified method for calculating interpolated precision. The system-scores obtained using the JAVA and C versions are compared for all the five metrics ($iP[0.00]$, $iP[0.01]$, $iP[0.05]$, $iP[0.10]$, $MAiP$).

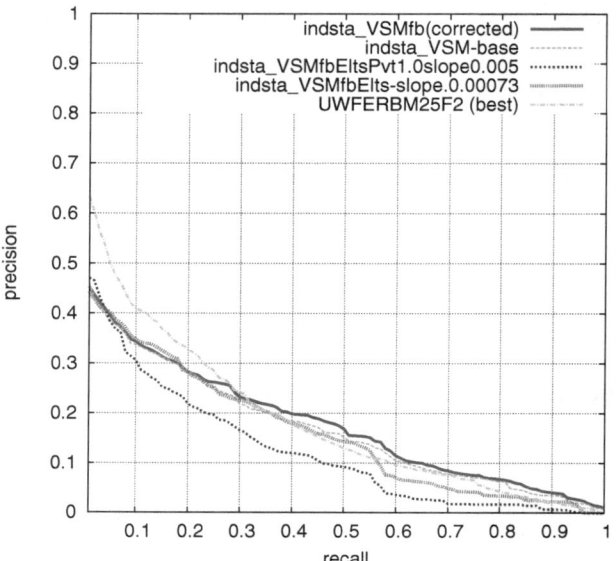

Fig. 1. Interpolated P - R graph for ISI runs

Table 5. Kendall's τ between system rankings, adhoc focused task, 56 systems

metric	τ
iP[0.00]	0.9244
iP[0.01]	0.9580
iP[0.05]	0.9827
iP[0.10]	0.9958
MAiP	0.9954

Pairwise t-test between system-scores using any of the metrics came statistically significant even at 1% level of significance(both-sided). However, Table 5 shows that Kendall's τ between system rankings for none of the metrics using these two version seemed to fall below 0.9.

In summary, though absolute changes are significantly different, one can rest assured that relative ranking is not appreciably affected, so this may not be a major concern.

5   Conclusion

The test collection used in INEX 2009 was really resource-hungry and put the models and their implementations into a 'stress test'. We approached the job very late this year and found ourselves inadequately prepared. We therefore submitted two simple runs: one using a minimal portion of corpus and another using the whole corpus. Both the runs were at the document-level retrieval using VSM-based approach. The results of our official submission were dismal because we made mistake in the query processing. However post-submission, we rectified the error and found our performance competitive. We also made a study on pivoted length normalization for XML retrieval and tuned our parameter for INEX Wikipedia collection. This parameters yielded good results for pre-2009 and INEX 2009 test collections for both document retrieval as well as element-retrieval. But we need to enhance performance for element-retrieval. On the evaluation front, we found an issue as to how interpolated precision should be calculated. We demonstrated that the issue can potentially jeoparadise evaluation results. However INEX 2009 results seemed not seriously affected by the issue. Even though individual system performances were significantly affected in terms of their abolute scores, their relative rankings were effectively unchanged.

References

1. INEX: Initiative for the Evaluation of XML Retrieval (2009),
   http://www.inex.otago.ac.nz
2. W3C: XPath-XML Path Language(XPath) Version 1.0,
   http://www.w3.org/TR/xpath
3. Denoyer, L., Gallinari, P.: The Wikipedia XML corpus. In: Fuhr, N., Lalmas, M., Trotman, A. (eds.) INEX 2006. LNCS, vol. 4518, pp. 12–19. Springer, Heidelberg (2006)
4. Schenkel, R., Suchanek, F.M., Kasneci, G.: Yawn: A semantically annotated wikipedia xml corpus. In: BTW, pp. 277–291 (2007)
5. Salton, G.: A Blueprint for Automatic Indexing. ACM SIGIR Forum 16(2), 22–38 (Fall 1981)
6. Buckley, C., Singhal, A., Mitra, M.: Using Query Zoning and Correlation within SMART: TREC5. In: Voorhees, E., Harman, D. (eds.) Proc. Fifth Text Retrieval Conference (TREC-5), NIST Special Publication 500-238 (1997)
7. Mitra, M., Singhal, A., Buckley, C.: Improving automatic query expansion. In: SIGIR '98, Melbourne, Australia, pp. 206–214. ACM, New York (1998)
8. Singhal, A., Buckley, C., Mitra, M.: Pivoted document length normalization. In: SIGIR '96: Proceedings of the 19th annual international ACM SIGIR conference on Research and development in information retrieval, pp. 21–29. ACM, New York (1996)
9. Pal, S., Mitra, M.: Indian statistical institute at inex 2007 adhoc track: Vsm approach. In: Fuhr, N., Kamps, J., Lalmas, M., Trotman, A. (eds.) INEX 2007. LNCS, vol. 4862, pp. 122–128. Springer, Heidelberg (2008)
10. Singhal, A.: Term Weighting Revisited. PhD thesis, Cornell University (1996)

Combining Language Models with NLP and Interactive Query Expansion

Eric SanJuan[1] and Fidelia Ibekwe-SanJuan[2]

[1] LIA & IUT STID, Université d'Avignon
339, chemin des Meinajaries, Agroparc BP 1228,
84911 Avignon Cedex 9, France
eric.sanjuan@univ-avignon.fr
[2] ELICO, Université de Lyon 3
4, Cours Albert Thomas, 69008 Lyon, France
fidelia.ibekwe-sanjuan@univ-lyon3.fr

Abstract. Following our previous participation in INEX 2008 Ad-hoc track, we continue to address both standard and focused retrieval tasks based on comprehensible language models and interactive query expansion (IQE). Query topics are expanded using an initial set of Multiword Terms (MWTs) selected from top n ranked documents. In this experiment, we extract MWTs from article titles, narrative field and automatically generated summaries. We combined the initial set of MWTs obtained in an IQE process with automatic query expansion (AQE) using language models and smoothing mechanism. We chose as baseline the Indri IR engine based on the language model using Dirichlet smoothing. We also compare the performance of bag of word approaches (TFIDF and BM25) to search strategies elaborated using language model and query expansion (QE). The experiment is carried out on all INEX 2009 Ad-hoc tasks.

1 Introduction

This year (2009) represents our second participation in the INEX Ad-hoc track. The three tasks defined in the previous years were maintained: focused retrieval (element, passage), Relevant-in-Context (RiC), Best-in-Context (BiC). A fourth task called "thorough task" was added to this year's edition. The thorough task can be viewed as the generic form of the focused task in that systems are allowed to retrieve overlapping elements whereas this is not allowed in the focused task. In the 2008 edition, we explored the effectiveness of NLP, in particular that of multiword terms (MWTs) combined with query expansion mechanisms - Automatic Query Expansion (AQE) and Interactive Query Expansion (IQE). Previous experiments in IR have sought to determine the effectiveness of NLP in IR. A study by [1] concluded that the issue of whether NLP and longer phrases would improve retrieval effectiveness depended more on query representation rather than on document representation within IR models because no matter how rich and elaborate the document representation, a poor representation of

S. Geva, J. Kamps, and A. Trotman (Eds.): INEX 2009, LNCS 6203, pp. 122–132, 2010.

the information need (short queries of 1-2 words) will ultimately lead to poor retrieval performance.

A few years later, [2] applied NLP in order to extract noun phrases (NPs) used in an IQE process. The IQE approach described in her study shares similar points with that of [1] except that instead of using the abstracts of the top n-ranked documents to expand the queries, [2] extracted NPs from query topics using a part-of-speech tagger and a chunker. She tested different term weighting functions for selecting the NPs: idf, C-value and log-likelihood. We refer the reader to [3] for a detailed description and comparison of these measures. The ranked lists of NPs were displayed to the users who selected the ones that best described the information need expressed in the topics. Documents were then ranked based on the expanded query and on the BM25 probabilistic model [4]. By setting optimal parameters, the IQE experiment in [2] showed significant precision gains but surprisingly only from high recall levels.

Based on these earlier findings and on our own performance in 2008's Ad-Hoc track evaluation, we pursue our investigation of the effectiveness of representing queries with MultiWord Terms (MWTs). MWTs is understood here in the sense defined in computational terminology [5] as textual denominations of concepts and objects in a specialized field. Terms are linguistic units (words or phrases) which taken out of context, refer to existing concepts or objects of a given field. As such, they come from a specialized terminology or vocabulary [6]. MWTs, alongside noun phrases, have the potential of disambiguating the meaning of the query terms out of context better than single words or statistically-derived n-grams and text spans. In this sense, MWTs cannot be reduced to words or word sequences that are not linguistically and terminologically grounded. Our approach was successfully tested on two corpora: the TREC Enterprise track 2007 and 2008 collections, and INEX 2008 Ad-hoc track [7] but only at the document level.

We ran search strategies implementing IQE based on terms from different fields of the topic (title, phrase, narrative). We tested many new features in the 2009 edition including:

- XML element retrieval. In 2008 edition, we only did full article retrieval;
- more advanced NLP approaches including automatic multi-document summarization as additional source of expansion terms;
- expansion of terms based on related title documents from Wikipedia;
- comparison of other IR models, namely bag of word models (TFIDF, BM25) without query expansion (QE);
- a combination of different query expansion mechanisms (IQE+AQE, IQE alone, AQE alone) with the language model implemented in Indri.

Our query expansion process runs as follows. First a seed query consisting of the title field is sent to the Indri search engine which served as our baseline. The system returns a ranked list of documents. Our system automatically extracts MWTs from the top n-ranked documents and from the topic fields (title, phrase, narrative). The expanded query resulting from the IQE process is further expanded using the automatic query expansion process (AQE) implemented in

Indri. Indri is based on standard IR Language Models for document ranking. Our system also generates automatic summaries from these top ranked documents and resulting MWTs. Thus the user has multiple sources - topic fields or top ranked documents, from which to select MWTs with which to expand the initial seed query in an Interactive Query Expansion (IQE) process. Our aim was to set up a comprehensive experimental framework in which competing models and techniques could be compared. Our IR system thus offers a rich framework in which language models are compared against bag of word models in combination with different query expansion techniques (IQE, AQE) as well as advanced NLP techniques (MWT extraction, automatic document summarization). A novelty in INEX 2009's Ad-Hoc track is that a new phrase (ph) field has been added to the topic description fields. These phrases are quite similar to the MWTs we automatically extract from the other topic fields. This may have an impact on the performance of some of our IR strategies. We will discuss this issue further in the §4.

The rest of the paper is structured as follows: section §2 presents the language model and its application to the IR tasks; section §3 presents the results on the Wikipedia collection in the INEX 2009 Ad-hoc track; finally, section §4 discusses the lessons learned from these experiments.

2   Probabilistic IR Model

2.1   Language Model

Language models are widely used in NLP and IR applications [8,4]. In the case of IR, smoothing methods play a fundamental role [9]. We first describe the probability model that we use.

Document Representation: probabilistic space and smoothing. Let us consider a finite collection $\mathcal{D}$ of documents, each document D being considered as a sequence $(D_1, ..., D_{|D|})$ of $|D|$ terms D_i from a language $\mathcal{L}$, i.e. $\mathcal{D}$ is an element of $\mathcal{L}^*$, the set of all finite sequences of elements in $\mathcal{L}$. Our formal framework is the following probabilistic space $(\Omega, \wp(\Omega), P)$ where Ω is the set of all occurrences of terms from $\mathcal{L}$ in some document $D \in \mathcal{D}$ and P is the uniform distribution over Ω. Language Models (LMs) for IR rely on the estimation of the a priori probability $P_D(q)$ of finding a term $q \in \mathcal{L}$ in a document $D \in \mathcal{D}$. We chose the Dirichlet smoothing method because it can be viewed as a maximum *a priori* document probability distribution. Given an integer μ, it is defined as:

$$P_D(q) = \frac{f_{q,D} + \mu \times P(q)}{|D| + \mu} \tag{1}$$

In the present experiment, documents can be full wikipedia articles, sections or paragraphs. Each of them define a different probabilistic space that we combine in our runs.

Query Representation and ranking functions. Our purpose is to test the efficiency of MWTs in standard and focused retrieval compared to a bag-of-word model or statistically-derived phrases. For that, we consider phrases (instead of single terms) and a simple way of combining them. Given a phrase $s = (s_0, ..., s_n)$ and an integer k, we formally define the probability of finding the sequence s in the corpus with at most k insertions of terms in the following way. For any document D and integer k, we denote by $[s]_{D,k}$ the subset of $D_i \in D$ such that: $D_i = s_1$ and there exists n integers $i < x_1, ..., x_n \leq i + n + k$ such that for each $1 \leq j \leq n$ we have $s_j = D_{x_j}$.

We can now easily extend the definition of probabilities P and P_D to phrases s by setting $P(s) = P([s]_{.,k})$ and $P_D(s) = P_D([s]_{D,k})$. Now, to consider queries that are set of phrases, we simply combine them using a weighted geometric mean as in [10] for some sequence $w = (w_1, ..., w_n)$ of positive reals. Unless stated otherwise, we suppose that $w = (1, ..., 1)$, i.e. the normal geometric mean. Therefore, given a sequence of weighted phrases $Q = \{(s_1, w_1), ..., (s_n, w_n)\}$ as query, we rank documents according to the following scoring function $\Delta_Q(D)$ defined by:

$$\Delta_Q(D) = \prod_{i=1}^{n} (P_D(s_i))^{\frac{w_i}{\sum_{j=1}^{n} w_j}} \tag{2}$$

$$\stackrel{\text{rank}}{=} \sum_{i=1}^{n} \left(\frac{w_i}{\sum_{j=1}^{n} w_j} \times \log(P_D(s_i)) \right) \tag{3}$$

This plain document ranking can easily be computed using any passage information retrieval engine. We chose for this purpose the Indri engine [11] since it combines a language model (LM) [8] with a bayesian network approach which can handle complex queries [10]. However, in our experiments, we use only a very small subset of the weighting and ranking functionalities available in Indri.

2.2 Query Expansion

We propose a simple QE process starting with an approximate short query $Q_{T,S}$ of the form (T, S) where $T = (t_1, ..., t_k)$ is an approximate document title consisting of a sequence of k words, followed by a possibly empty set of phrases: $S = \{S_1, ..., S_i\}$ for some $i \geq 0$. In our case, each S_i will be a MWT.

Baseline document ranking function. By default, we rank documents according to :

$$\Delta_{T,S} = \Delta_T \times \prod_{i=1}^{|S|} \Delta_{S_i} \tag{4}$$

Therefore, the larger S is, the less the title part T is taken into account. Indeed, S consists of a coherent set of MWTs found in a phrase query field or chosen by the user. If the query can be expanded by coherent clusters of terms, then we are no more in the situation of a vague information need and documents should be ranked according to precise MWTs. For our baseline, we generally consider S to be made of the phrases given in the query.

Interactive Query Expansion Process. The IQE process is implemented on a html interface available at *http://master.termwatch.es/*. Given an INEX topic identifier, this interface uses the title field as the seed query. The interface is divided into two sections:

Topic section: a column displays terms automatically extracted from the topic description fields (title, phrase, narrative). A second column in this section displays titles of documents related to the query. These are titles of Wikipedia documents found to be related to the seed query terms by the language model implemented in Indri. The user can then select terms either from topic fields (title, phrase, narrative) and/or from related titles.

Document summary section: this section displays short summaries from the top twenty ranked documents of Δ_Q ranking together with the document title and the MWTs extracted from the summary. The summaries are automatically generated using a variant of TextRank algorithm. The user can select MWTs in context (inside summaries) or directly from a list without looking at the sentence from which they were extracted.

MWTs are extracted from the summaries based on shallow parsing and proposed as possible query expansions. The user selects all or a subset S' of them. This leads to acquiring sets of synonyms, abbreviations, hypernyms, hyponyms and associated terms with which to expand the original query terms. The selected multiword terms S'_i are added to the initial set S to form a new query $Q' = Q_{T,S \cup S'}$ leading to a new ranking $\Delta_{Q'}$ computed as in §2.2.

Automatic Query Expansion. we also consider Automatic Query Expansion (AQE) to be used with or without IQE. In our model, it consists in the following: let $D_1, ..., D_K$ be the top ranked documents by the initial query Q. Let $C = \cup_{i=1}^{K} D_i$ be the concatenation of these K top ranked documents. Terms c occurring in D can be ranked according to $P_C(c)$ as defined by equation (1). We consider the set E of the N terms $\{c_1, ..., c_N\}$ with the highest probability $P_C(c_i)$. We then consider the new ranking function Δ'_Q defined by $\Delta'_Q = \Delta_Q^{\lambda} \times \Delta_E^{1-\lambda}$ where $\lambda \in [0, 1]$.

Unless stated otherwise, we take $K = 4$, $N = 50$ and $\lambda = 0.1$ since these were the parameters that gave the best results on previous INEX 2008 ad-hoc track.

We now explore in which context IQE based on MWTs is effective. Our baseline is an automatic document retrieval based on equation 2 in §2.1.

3   Results

We submitted eight runs to the official INEX evaluation: one automatic and one manual for each of the four tasks (focused, thorough, BiC, RiC). Our Relevant-in-Context (RiC) runs were disqualified because they had overlapping elements.

Here we focus on analysing results from focused and thorough tasks. Focused task is measured based on interpolated precision at 1% of recall (iP[0.01]) while thorough is measured based on Mean Average interpolated Precision (MAiP), so the two are complementary. Moreover, on document retrieval, computing MAiP on focused results or on thorough's is equivalent.

We compare our officially submitted runs to additional ones we generated after the official evaluation. Among them, are two baselines runs. It appears that these baseline runs outperform our submitted runs based on the qrels released by the organizers. Our runs combine features from the following:

Xml: these runs retrieve XML elements, not full articles. Each element is evaluated in the probabilistic space of all elements sharing the same tag. Elements are then ranked by decreasing probability. The following elements were considered: b, bdy, category, causal_agent, country, entry, group, image, it, list, location, p, person, physical_entity, sec, software, table, title.

Doc: only full articles are retrieved.

AQE: Automatic Query Expansion is performed.

ph: the query is expanded based on the phrases furnished this year in the topic fields. These phrases are similar to MWTs.

IQE: Interactive Query Expansion (IQE) is performed based on the interface described previously (see §2.2).

Ti: elements in documents whose title overlaps the initial query or its expansion terms are favoured.

All our submitted runs were using a default language model. After the official evaluation, we generated runs based on other ranking methods, namely TFIDF and BM25 that are also implemented in the Indri system. We also test the impact of stemming on these different methods.

3.1   Search Strategies

We consider the following runs. They are all based on LM and they all use the topic Title and PhraseTitle fields.

Lyon3LIAautolmnt combines **Xml** element extraction, **Ti** heuristic, and **AQE**. It was submitted to the thorough task.

Lyon3LIAmanlmnt adds IQE on the top of the previous one and was also submitted to the thorough task.

Lyon3LIAautoQE is similar to Lyon3LIAmanlmnt but retrieves documents **Doc** instead of XML elements. It was submitted to the focused task.

Lyon3LIAmanQE adds IQE on the top of the previous one and was also submitted to the focused task.

LMDoc baseline run was not submitted. It retrieves full documents without using any of **Ti**, **AQE**, nor **IQE**.

LMDocIQE the same baseline run with IQE, also not submitted.

Table 1 summarizes these runs and gives their IP[0.01] and MAiP scores.

Table 1. Results of submitted runs and two non submitted baselines. All of them use the Language Model

Name of the run	XML	Doc	ph	Ti	AQE	IQE	Submitted	IP[0.01]	MAiP
Lyon3LIAmanlmnt	×	-	×	×	×	×	thorough	0.4956	0.2496
Lyon3LIAmanQE	-	×	×	×	×	×	thorough	0.4861	0.2522
Lyon3LIAautolmnt	×	-	×	×	×	-	focus	0.4646	0.2329
Lyon3LIAautoQE	-	×	×	×	×	-	focus	0.4645	0.2400
LMDocIQE	-	×	×	-	-	×	-	0.5840	0.2946
LMDoc	-	×	×	-	-	-	-	0.5527	0.2826

It appears that Ti and AQE degraded the results since the non submitted baselines LMDocIQE and LMDoc outperformed all submitted runs. Retrieving Xml elements lightly improves iP[0.01] score but degrades MAiP. However these differences are not statistically significant (paired t-test, p-value=0.1). Following our observation in 2008's edition, adding IQE improved scores. However, on this year's corpus, the improvement is not significant. None of these differences are statistically significant but we can check on precision/recall curves if there is a general tendency. Figure 1 shows the Interpolated generalized precision curves based on thorough evaluation measures for all these runs. The curves from the two baselines are clearly on the top but it appears that the baseline with IQE only outperforms the automatic baseline at a recall level lower than 0.2. After this level, the two curves are almost identical. It also appears that runs retrieving XML elements only outperform their full document counterpart at very low recall levels.

As in the INEX 2008 corpus, it appears that IQE based on MWTs can improve document or XML element retrieval whereas AQE does not. This is a surprising difference with our 2008 results [7]. Contrary to the runs we submitted in 2008, this year (2009), we used AQE in all our submitted runs because it had improved performance previously.

Moreover, it appears on these baseline runs that the difference between our automatic baseline run and the one with IQE is not statistically significant. In fact with an Ip[0.01] of 0.56 and a MAiP of 0.28, our baseline run performs much better than in 2008 meanwhile the score of our MWT runs is unchanged. The reason could be the availability this year in the topics of a new topic field with phrases that is used by all of our runs (**ph** feature), including the baselines. This makes the input to the baseline runs somewhat similar to the runs using MWTs. We need to investigate this issue further in order to confirm our intuitions. More experiments are performed hereafter.

3.2   The Impact of Language Element Models and AQE

First, we explored the impact of two wikipedia XML elements on LM results: document title and paragraphs.

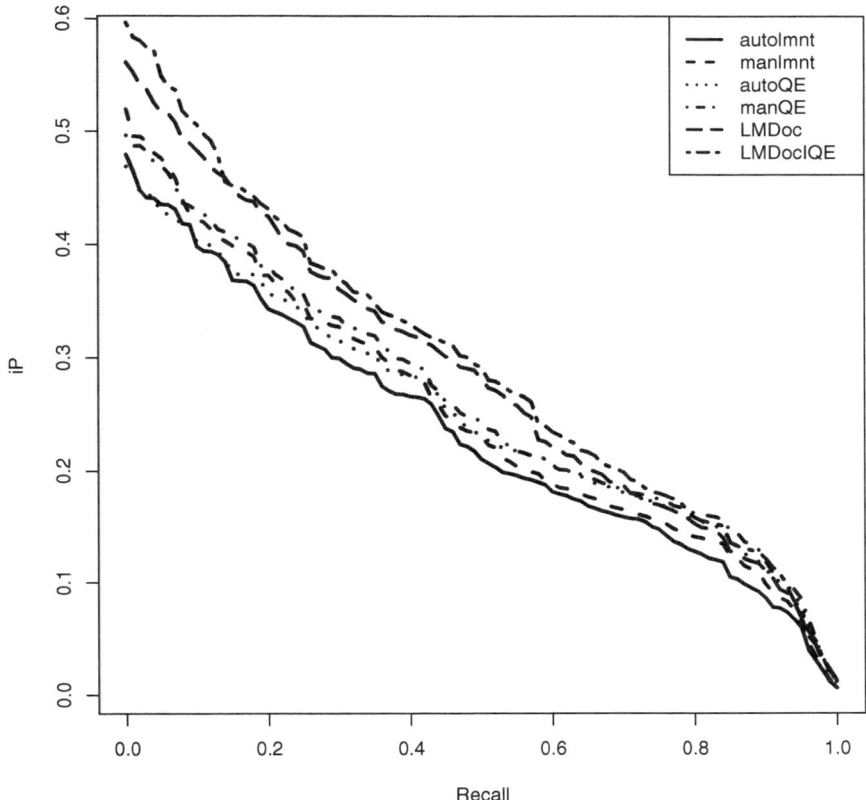

Fig. 1. Interpolated generalized precision curves at INEX 2009 on thorough task

Our idea was to favour documents in which the title is related to the query (shared common terms). We then consider the product of two LM ranking functions, one on the whole document, the other on document titles. It appears that this combination had a clear negative impact on the overall performance as it can be observed on Table 1 and in Figure 1 where the two runs LMDoc and LMDocIQE outperform the two submitted runs autoQE and manQE that only differ on the use of AQE and document titles.

The same observation can be made about AQE. The resulting ranking function is the product of the initial ranking with the expansion. As we already mentioned, on 2008 AQE significantly improved overall performance. In our 2009 runs, it turned out to be a drawback.

For paragraphs we apply the same query on different LM models: one on paragraphs, the other on documents. We then merge the two rankings based on their scores. Since this score is a probability, we wanted to check if it was possible to use it to merge different rankings, despite the fact that these probabilities are estimated on different spaces. This experiment was possible on the thorough task since it allowed overlapping passages.

It appears that considering XML elements does not significantly improve LM baseline. However, on thorough task it appears that iP[0.01] score of full document runs Lyon3LIAmanQE and Lyon3LIAautoQE can be improved by simply merging them with XML element runs as done for runs Lyon3LIAmanlmnt and Lyon3LIAautolmnt.

3.3   Extended Baseline Runs

As many parameters are combined in our runs, it is necessary to try to isolate the impact of each of them. To do this, we generated more baselines. We then tested successively the effects of stemming and query expansion with MWTs on both the LM and the bag of word models.

Three supplementary baselines based only on the title field and retrieving full documents were generated. No preprocessing was performed on this field. The first baseline **LM** uses language model based on equation 1. The difference between this LM baseline and the previous LMDoc is that we do not use the topic's phrase field. The two other baselines are based on the bag of word models - TFIDF and BM25 using their default parameters in Indri[1] ($k1 = 1.2$, $b = 0.75$, $k3 = 7$). We observed that LM performed better on a non stemmed index whereas TFIDF and BM25 performed better on a stemmed index. Therefore in the remainder of the analysis, we only consider LM runs on non stemmed index and bag of word model runs on stemmed corpus. In Table 2, the "No ph" columns give the scores of the baselines without using phrases from the phrase field whereas "With ph" does the opposite.

Table 2. Supplementary baselines based on bag of words model and Language Model

Measure	No ph		With ph	
	IP[0.01]	MAiP	IP[0.01]	MAiP
TFIDF	0.6114	0.3211	0.5631	0.3110
BM25	0.5989	0.3094	0.5891	0.3059
LM	0.5389	0.2758	0.5527	0.2826

From these results, it appears that the strongest baseline is TFIDF followed by BM25. What is surprising is that with a MAiP of 0.32, this baseline outperforms all official runs submitted to the thorough task and with an $iP[0.01]$ of 0.61138, it would have been ranked fifth (and third by team) in the focused task. A t-test shows that the difference between TFIDF and BM25 is not statistically significant, but they are between these bag of word models and LM for both IP[0.001] and MAiP (p-value < 0.01 for TFIDF and < 0.05 for BM25).

We now investigate the impact of MWTs to expand the seed query on the performance of the language model (LM) and the bag of word models, without

---

[1] http://www.lemurproject.org/doxygen/lemur/html/IndriParameters.html

any tuning to account for the nature of MWTs. First, we only consider MWTs from the topic's phrase field. The resulting runs are still considered as automatic because there is no manual intervention. We shall refer to them as **ph** runs. It appears that this incremental process handicaps both TFIDF and BM25 as it can be observed in the columns "with ph" of Table 2. By adding phrases, the scores for TFIDF drop even more than BM25's although the latter's scores also drop. On the other hand, the LM model naturally takes advantage of this new information and each addition of MWTs improves its performance. LM with ph corresponds to our initial automatic baseline LMDoc. It appears that difference between LMDoc and the best TFIDF scores is not more significant after adding MWTs in topic phrase field. Adding more MWTs following our IQE process only generates more noise in TFIDF and BM25 runs, but improves LM performance, even though the improvement is not statistically significant.

4   Discussion

We used Indri with Dirichlet smoothing and we combined two language models, one on the documents and one on elements. The results from both models are then merged together and ranked by decreasing probability.

For query representation, we used NLP tools (summarizer and terminology extraction). We started from the topic phrase and title, then we added related Multiword Terms (MWT) extracted from the other topic fields and from an automatic summary of the top ranked documents by this initial query. We also used standard Automatic Query Expansion when applied to the document model.

Other features tested are the Indri operators to allow insertions of words (up to 4) into the MWTs and favoring documents in which the MWTs appear in the title.

As observed on the previous INEX corpus (2008), IQE based on MWTs still improves retrieval effectiveness but only for search strategies based on the language model. On the contrary, automatic query expansion (AQE) using the same parameters has the reverse effect on the 2009 corpus. At the baseline level, we observe that TFIDF performs significantly better than LM, but LM naturally allows an incremental and interactive process. This suggests that users can more easily interact with an LM in an IR system. For two consecutive years, we have observed that stemming does not have any significant impact on the performance of the different strategies we have tested. Another interesting finding in this year's experiments is that baselines generated using the bag of word models with their default parameters as implemented in Indri and without recourse to any NLP (MWTs) nor to query expansion, outperformed language models that combined interactive query expansion based on MWTs. We need to investigate these findings further. We also need to ascertain the exact impact of the phrases furnished this year to represent topic's contents with regard to MWTs that we exctracted from other sources.

References

1. Perez-Carballo, J., Strzalkowski, T.: Natural language information retrieval: progress report. Information Processing and Management 36(1), 155–178 (2000)
2. Vechtomova, O.: The role of multi-word units in interactive information retrieval. In: Losada, D.E., Fernández-Luna, J.M. (eds.) ECIR 2005. LNCS, vol. 3408, pp. 403–420. Springer, Heidelberg (2005)
3. Knoth, P., Schmidt, M., Smrz, P., Zdrahal, Z.: Towards a framework for comparing automatic term recognition methods. In: ZNALOSTI 2009, Proceedings of the 8th annual conference, p. 12. Vydavatelstvo STU (2009)
4. Jones, K.S., Walker, S., Robertson, S.E.: A probabilistic model of information retrieval: development and comparative experiments. Inf. Process. Manage. 36(6), 779–840 (2000)
5. Kageura, K.: The dynamics of Terminology: A descriptive theory of term formation and terminological growth. John Benjamins, Amsterdam (2002)
6. Ibekwe-SanJuan, F.: Constructing and maintaining knowledge organization tools: a symbolic approach. Journal of Documentation 62, 229–250 (2006)
7. Ibekwe, F., SanJuan, E.: Use of multiword terms and query expansion for interactive information retrieval. In: Geva, S., Kamps, J., Trotman, A. (eds.) INEX 2008. LNCS, vol. 5631, pp. 54–64. Springer, Heidelberg (2009)
8. Ponte, J.M., Croft, W.B.: A language modeling approach to information retrieval. In: SIGIR '98: Proceedings of the 21st annual international ACM SIGIR conference on Research and development in information retrieval, pp. 275–281. ACM, New York (1998)
9. Zhai, C., Lafferty, J.: A study of smoothing methods for language models applied to information retrieval. ACM Trans. Inf. Syst. 22(2), 179–214 (2004)
10. Metzler, D., Croft, W.B.: Combining the language model and inference network approaches to retrieval. Information Processing and Management 40(5), 735–750 (2003)
11. Strohman, T., Metzler, D., Turtle, H., Croft, W.B.: Indri: A language-model based search engine for complex queries (extended version). IR 407, University of Massachusetts (2005)

Exploiting Semantic Tags in XML Retrieval

Qiuyue Wang, Qiushi Li, Shan Wang, and Xiaoyong Du

School of Information, Renmin University of China
and
Key Laboratory of Data Engineering and Knowledge Engineering, MOE,
Beijing 100872, P.R. China
{qiuyuew,qiushili,swang,duyong}@ruc.edu.cn

Abstract. With the new semantically annotated Wikipedia XML corpus, we attempt to investigate the following two research questions. Do the structural constraints in CAS queries help in retrieving an XML document collection containing semantically rich tags? How to exploit the semantic tag information to improve the CO queries as most users prefer to express the simplest forms of queries? In this paper, we describe and analyze the work done on comparing CO and CAS queries over the document collection at INEX 2009 ad hoc track, and we propose a method to improve the effectiveness of CO queries by enriching the element content representations with semantic tags. Our results show that the approaches of enriching XML element representations with semantic tags are effective in improving the early precision, while on average precisions, strict interpretation of CAS queries are generally superior.

1 Introduction

With the growth of XML, there has been increasing interest in studying structured document retrieval. A key characteristic that distinguishes the XML retrieval task from a traditional retrieval task is the existence of structural information in the former one. The structural information not only enables the system to retrieve the document fragments rather than the whole documents relevant to users' queries, but also provides new dimensions to be exploited to improve the retrieval performance.

For example, a common approach to exploit the hierarchical structure in XML documents is to score the leaf elements that directly contain terms and propagate the scores up to their ancestors. Thus the scores of elements up in the tree are calculated as weighted combinations of their descendants' scores. Such a score propagation strategy can reflect the hierarchical level of the elements (the lower elements are considered as more specific than the upper elements), and also the weights can be set to reflect the importance of different element types. Another well-accepted idea of utilizing structural information to improve retrieval performance is to formulate more precise queries by specifying structural conditions in the query. For example, by specifying that the return element type should be *movie*, or *John* should be a *director*, the retrieval precision can be greatly improved. Structural conditions add more semantics into the query as discussed in [1] by *specifying target information type, disambiguating keywords, specifying search term context,* and/or *relating*

S. Geva, J. Kamps, and A. Trotman (Eds.): INEX 2009, LNCS 6203, pp. 133–144, 2010.
© Springer-Verlag Berlin Heidelberg 2010

search query terms. They can be explicitly specified by the user or automatically inferred by the system.

There are various structured query languages, e.g. XML fragments [2], NEXI [3], InQuery that is used in Lemur/Indri [4], XQuery Full-Text [5] and etc. They differ in their expressive power on specifying structural constraints, from the simplest one (with only the return element type) to the most complex one (with the support of full-fledged XQuery). For most users, however, complex structured queries are hard to construct. Moreover, once such queries were incorrectly formulated or inferred, possibly due to the imprecise knowledge about the document structure or the semantic gap between the query and data, strict matching of these queries would greatly harm the retrieval precision instead of improving it. To overcome this problem, we can make the hard constraints soft by treating the structural conditions as "hints" rather than "constraints". The structural hints however won't help much in XML retrieval as analyzed from the previous INEX Ad hoc tracks [6][21]. The reasons may partially lie in the document collections used in previous INEX tracks. Both the collection of IEEE journal articles used from INEX 2002 to 2005 and the Wikipedia document collection used from INEX 2006 to 2008 contain very few semantic tags, such as *movie, director*, but mostly structural tags, like *article, sec, p* and etc. When expressing queries with only structural tags, the user intends to constrain the size of the results rather than make the query semantically clearer. For example, when the user specifies the return element type to be a section or a paragraph, it has nothing to do with the topic relevance of the query. Users are in general bad at giving such structural hints however [6]. Thus the performance improvement is not significant.

In INEX 2009, document collection is changed to a semantically annotated Wikipedia XML corpus [7]. In this corpus, the Wikipedia pages, as well as the links and templates in the pages, are annotated with semantically rich tags using the concepts from WordNet and etc. For Example, Fig. 1 shows an excerpt of an example XML document (4966980.xml) in the collection. With this new semantics-annotated document collection, we attempt to investigate the following research questions:

1. Do the structural constraints in CAS queries help in retrieving such an XML document collection containing semantically rich tags?
2. How to exploit the semantic tag information to improve the CO queries as most users prefer to express the simplest forms of queries?

In this paper, we describe and analyze the work done on comparing CO and CAS queries on such a semantically annotated XML corpus, and propose to improve the effectiveness of CO queries by enriching element content representations with semantic tags. Our experimental results show that the approaches of enriching XML element representations with semantic tags are effective in improving the early precision, while on average precisions, strict interpretation of CAS queries are generally superior.

The paper is organized as follows. Section 2 describes the baseline retrieval models for CO and CAS queries in XML retrieval used in our comparisons. In Section 3, we discuss the methodology for exploiting semantic tags in evaluating CO queries for XML retrieval. The experiment results are presented in Section 4. Finally, Section 5 concludes the paper and describes future work.

```
<article xmlns:xlink="http://www.w3.org/1999/xlink">
  <physical_entity confidence="0.8" wordnetid="100001930">
    <communicator confidence="0.8" wordnetid="109610660">
      <person confidence="0.8" wordnetid="100007846">
        <causal_agent confidence="0.8" wordnetid="100007347">
          <writer confidence="0.8" wordnetid="110794014">
            <dramatist confidence="0.8" wordnetid="110030277">
              <header>
                <title>Sam Ukala</title>
                <id>4966980</id>
                <revision>
                  <timestamp>2007-08-12T14:57:18Z</timestamp>
                  <contributor><username>Cydebot</username></contributor>
                </revision>
                <categories>
                  <category>Nigerian dramatists and playwrights</category>
                </categories>
              </header>
              <bdy>
                <b>Sam Ukala</b> is a
                <link xlink:type="simple" xlink:href="../383/21383.xml">Nigerian</link>
playwright, poet, short story writer, actor, theatre director and academic. He has been
Professor of Drama and Theatre Arts at a number of Nigerian universities, including
                <region wordnetid="108630985" confidence="0.8">
                  <administrative_district wordnetid="108491826" confidence="0.8">
                    <location wordnetid="100027167" confidence="0.8">
                      <district wordnetid="108552138" confidence="0.8">
                        <country wordnetid="108544813" confidence="0.8">
                          <link xlink:type="simple" xlink:href="../627/2227627.xml">Edo State</link>
                        </country>
                      </district>
                    </location>
                  </administrative_district>
                </region> University and ......
```

Fig. 1. An excerpt of document 4966980.xml in the semantically annotated Wikipedia collection

2   Baseline Approaches

Language modeling is a newly developed and promising approach to information retrieval. It has a solid statistical foundation, and can be easily adapted to model various kinds of complex and special retrieval problems, such as structured document retrieval. In particular, mixture models [8] and hierarchical language models [9][10][11] are proposed to be applied in XML retrieval. We base our work on language modeling approaches for XML retrieval. In this section, we describe the baseline XML retrieval models for both CO and CAS queries compared in our experiments.

We model an XML document as a node-labeled tree, where each node in the tree corresponds to an element in the document and the node label corresponds to the tag name of the element. The hierarchical structure represents the nesting relationship between the elements. The content of each element can be modeled using its *full content* or *weighted combination of leaf contents*. The full content of an element consists of all the text contained in the subtree rooted at the element, while the leaf content of an element consists of all the text directly contained in the element. The weighted combination of leaf contents of an element refers to the weighted combination of the leaf contents of all the elements in the subtree rooted at the element. The full content is in fact a special case of weighted combination of leaf contents, where all the weights are equal to 1. In this paper, we represent the element content by its full content. How to set optimal weights for combining leaf contents to improve the retrieval performance is orthogonal to the techniques addressed in this paper, and beyond the scope of the paper.

The basic idea of language modeling approaches in information retrieval is to estimate a language model for each document (θ_D) and the query (θ_Q), and then rank the document in one of the two ways: by estimating the probability of generating the query string with the document language model, i.e. $P(Q/\theta_D)$, as in Equation 1, or by computing the Kullback-Leibler divergence of the query language model from the document language model, i.e. $D(\theta_Q \| \theta_D)$, as in Equation 2.

$$P(Q \mid \theta_D) = \sum_{w \in Q} P(w \mid \theta_D).$$

(1)

$$-D(\theta_Q \| \theta_D) = -\sum_{w \in V} P(w \mid \theta_Q) \log \frac{P(w \mid \theta_Q)}{P(w \mid \theta_D)}.$$
$$\propto \sum_{w \in V} P(w \mid \theta_Q) \log P(w \mid \theta_D)$$

(2)

On the surface, the KL-divergence model appears to be quite different from the query likelihood method. However, it turns out that the KL-divergence model covers the query likelihood method as a special case when we use the empirical distribution to estimate the query language model, i.e. maximum-likelihood estimate. By introducing the concept of query language model, the KL-divergence model offers opportunities of leveraging feedback information to improve retrieval accuracy. This can be done by re-estimating the query language model with the feedback information [12]. In this paper, we do not consider the effect of feedback information. So we adopt the query likelihood method in our experiments.

Thus, the entire retrieval problem is reduced to the problem of estimating document language models. The most direct way to estimate a language model given some observed text is to use the maximum likelihood estimate, assuming an underlying multinomial model. However, the maximum likelihood estimate assigns zero probability to the unseen words. This is clearly undesirable. Smoothing plays a critical role in language modeling approaches to avoid assigning zero probability to unseen words and also to improve the accuracy of estimated language models in general. Traditionally, most smoothing methods mainly use the global collection information to smooth a document language model [13][14]. Recently, corpus graph structures, e.g. the similarity between documents, have been exploited to provide more accurate smoothing of document language models [15]. Such a local smoothing strategy has been shown to be effective.

In XML retrieval, the element language model is usually smoothed by interpolating it with the global information such as the whole collection model or the language model specific to the element type, and further more, possibly with other related element language models in the tree structure, e.g. its parent, children, or even descendants and ancestors if the smoothing is done recursively on the tree [8][9][10][11]. However, according to the previous study [10][16], the recursive smoothing strategy exploiting the hierarchical structure of the XML tree only improve the retrieval effectiveness slightly. Thorough experimental study on effective smoothing methods for XML retrieval is needed, and we leave it to our future work. As for the baseline approaches in this paper, we adapt the two-stage smoothing method proposed in [14] to XML retrieval as it was shown to be effective in our previous experiments [16]. In the first stage, the element language model is smoothed using a Dirichlet prior with the document language model as the reference model. In the second stage, the smoothed element language model is further interpolated with a query background model. With no sufficient data to estimate the query background model, the collection language model is assumed to be a reasonable approximation of the query background model. Thus, we get the estimation of each element language model as shown in Equation 3, where $tf(w,e)$ is the term frequency of w in the element e, $len(e)$ is length of e, μ is the scale parameter for Dirichlet smoothing and λ is the interpolation parameter for Jelinek-Mercer smoothing.

$$P(w \mid \theta_e) = (1-\lambda)\frac{tf(w,e) + \mu \cdot P(w \mid \theta_D)}{len(e) + \mu} + \lambda P(w \mid \theta_C). \qquad (3)$$

2.1 CO Queries

CO queries at INEX ad hoc track are given in the title fields of topics. We remove all the signs, i.e. +, -, and quotes, i.e. "" in the title field. That is, a CO query is simply a bag of keywords in our system, $Q = \{w_1, w_2, \ldots, w_m\}$. We estimate a language model for each element in the collection using its full content and two-stage smoothing method. Each element is scored independently based on the query likelihood as stated above, and a ranked list of elements is returned.

We submitted results to the four tasks of the ad hoc track, i.e. focused, best in context, relevant in context and thorough. For the first three tasks, overlap in the result list has to be removed. We adopt the simplest strategy of removing overlap, i.e. keeping only the highest ranked element on each path. For the in-context tasks, all the elements from the same document are clustered together, and the clusters (corresponding to documents) are ordered by their maximum element scores. For the best in context task, all the elements except the max-scored one in each document are removed from the result list. That is, a ranked list of documents is returned for best in context task. The best entry point for each document is set to be the max-scored element in the document. For the thorough task, no overlapping is removed from the result list.

2.2 CAS Queries

In INEX ad hoc track, CAS queries are given in the castile fields of topics. The queries are expressed in the NEXI query language [3]. For example, consider the CAS query of topic 85,

```
//article[about(., operating system)]//sec[about(.//p,
mutual exclusion)]
```

which requests section components in which some paragraph is about "mutual exclusion" and such sections are in an article about "operating system". There can be strict and vague interpretations for the structural constraints expressed in the query. As how to vaguely interpret the structural constraints properly remains a challenge problem, in this paper for the comparisons, we adopt the strict interpretation strategy as implemented in the Lemur/Indri system [4].

When evaluating the above example query, for each section that appears in an article, a score depending on its relevance to the query condition *about(.//p, mutual exclusion)]* is first calculated; to calculate this score, a list of relevance scores of all paragraphs in the section to the keyword query "*mutual exclusion*" are computed using the query likelihood method, and then the section's score is set to be its best matching paragraph's score. Finally, this score is combined (probabilistic AND) with the relevance score of the article containing this section to the keyword query "*operating system*" to form the final score of the section. A ranked list of sections is returned based on the final scores.

As in CO queries, we ignore all the phrases and +, - signs in CAS queries. Removing overlaps and presenting the results as required in context tasks are handled in the same way as that for CO queries as described in Section 2.2.

3   Exploiting Semantic Tags in Evaluating CO Queries

With the semantically rich tags present in the XML document, we assume that providing more semantic information expressed as structural conditions in queries could be more effective, as discussed in previous studies [1]. However, most users are not willing or not good at providing such structural conditions in queries. Many users in general are only willing to submit the simplest forms of queries, i.e. keyword queries.

To assists its users, an XML retrieval system can shift the burden of specifying effective structured queries from the user to the system. That is, the system automatically infers or generates CAS queries from the CO queries submitted by users [17]. This process is typically divided into three steps: firstly, generating all possible structured queries from the input unstructured query by incorporating the knowledge of schemas, data statistics, heuristics, user/pseudo relevance feedback and etc.; secondly, ranking the structured queries according to their likelihood of matching user's intent; thirdly, selecting the top-k structured queries and evaluating them. However, if the inferred structured queries are not intended by the user, we can not expect to get the right result. Another line of research work done in this direction is to infer some structural hints, not necessarily complete structured queries, from the input keyword query, such as in [18][19].

In this paper, we investigate the problem of how to exploit the semantic tag information to improve the performance of CO queries not from the point of view of modifying queries but from the point of view of enriching element representations with semantic tags. The idea is similar to that of enriching web document representations with aggregated anchor text [20], even though we are in different settings, web documents versus XML elements and aggregated anchor text versus semantic tags. Although enriching XML element representations with the semantic tags of all its related elements or even the ones from other linked documents would be an interesting research issue, we leave it to our future work. In this paper we investigated two ways of enriching element content representations with its own semantic tags. One is text representation and the other is new field representation.

3.1 Text Representation

When a user issues a keyword query, it often contains keywords matching the semantic tags of the relevant elements, e.g. to specify the target information type or to specify the context of search terms. For example, among many others, topic 5 "chemists physicists scientists alchemists periodic table elements" is looking for *chemists, physicists, scientists, alchemists* who studied elements and the periodic table, and topic 36 "notting hill film actors" requests all the actors starring in the film "Nottting Hill", where *actor* gives the target information type and *film* specifies the search context of "Notting Hill".

Thus, the terms in a keyword query has to match not only the text content of an element but also the semantic tags. The simplest way of enriching the element representation to match the query with both the content and semantic tags of the element is to propagate all the semantic tags of the element to its raw text content. Note that there could be more than one semantic tag added to the raw text content of an element. For example, in Fig. 1, the text content of the *article* element will be augmented with additional terms, *article, physical_entity, communicator, person, causal_agent, writer, dramatist*; the text content of the second *link* element will be augmented to be "*region, administrative-district, location, district, country, link, Edo State*".

There are altogether 32311 distinct tags in the Wikipedia collection of INEX 2009. Only a small portion of them (less than 0.1%) are for structural or formatting uses, e.g. *article, sec, p, bdy, b, it*, while all others are of semantic use. Among all the semantic tags, only about 5245 of them are from the WordNet concepts, i.e. with the "*wordnetid*" attribute in the start tag. Since most tags have semantic meanings, we did not differentiate them, but chose to propagate all of them to their respective elements.

3.2 New Field Representation

When evaluating a query, it may be helpful if we assign different weights to the case when a search term matches a semantic tag and to the case when it matches the raw text of an element. This can be achieved by the new field representation, in which all the semantic tags of an element are added to a new subelement of this element. The new subelement is given special tag name "*semantics*", which is not among the existing 32311 distinct tag names.

When evaluating a CO query over the new field representation, the generative model for each element is first estimated using the Equation 3. Next, the model for each element is further smoothed by interpolating it with all its children's smoothed language models as shown in Equation 4. This can be done non-recursively or recursively from the bottom up to the top of the tree as discussed in the hierarchical language modeling approaches [10][11]. If by non-recursive smoothing, $P(w/\ \theta_c')$ in Equation 4 should be changed to $P(w/\ \theta_c)$.

$$P(w \mid \theta_e') = \frac{1}{\lambda_e + \sum\limits_{c \in children(e)} \lambda_c} [\lambda_e P(w \mid \theta_e) + \sum\limits_{c \in children(e)} \lambda_c P(w \mid \theta_c')] \tag{4}$$

In such a model, we can set different weights for different fields. By default, all the fields have equal weights that are equal to 1. To stress the "*semantics*" field, we can set a higher weight on it, e.g. in our experiments, we set $\lambda_{semantics}$ to be 2. In this representation, the matching of terms with semantic tags and with raw text can be differently weighted, which may be useful.

4 Experiments

To compare CO and CAS queries and evaluate the methodology we proposed to exploit semantic tags in executing CO queries, we planed to submit four runs for each task at the ad hoc track of INEX 2009. These four runs, named as *base_co*, *base_cas*, *text_co* and *semantics_co*, correspond to the baseline CO, baseline CAS, text representation, and new field representation approaches respectively. Due to the limit of time, not all runs were successfully submitted before the deadline. We ignore the Thorough task as it may have similar results as the Focused task and present the results on the other three tasks in this section.

4.1 Setup

We implemented the four retrieval strategies inside the Lemur/Indri IR system [5], which is based on language modeling approaches. Since in this paper we intend to investigate whether the semantic tags in XML documents could be useful, and many other retrieval strategies are orthogonal to our approaches, e.g. by incorporating the element length priors, the proximity of keywords, contextualization and etc., we did not employ them in this set of experiments. We expect the retrieval performance be further improved by incorporating the above mentioned techniques.

At the ad hoc track of INEX 2009, the data collection consists of 2,666,190 semantically annotated Wikipedia XML documents. There are more than 762M elements and 32311 distinct tags in the collection. We indexed all the elements in the XML documents, and the index was built using the Krovetz stemmer and the shortest list of stop words that contains only three words {"a", "an", "the"}. To make Lemur/Indri possible to index 32311 different fields, we modified its internal data structures. Among the list of 32311 tags, there are many cases that two tags only differ on the letter cases, e.g. "Mission" and "mission", or a list of tags share the same prefix while their different suffixes form a sequence of numbers, like "country1", "country2", "country3", "country4", and etc. We specify some rules to conflate these tags when indexing the documents. For example, all the tags with capital letters are conflated to its small cases version, and multiple tags with same word prefix but different number suffixes are conflated to their common prefix word. Thus, "country1", "country2", and etc. are all conflated to "country". This can improve the retrieval performance of CAS queries when matching the tag names exactly. So this has similar effect as the tag equivalence strategy, but is done at the indexing time.

The parameters in the retrieval models are set as its default values in Lemur/Indri system, e.g. μ and λ in Equation 3 are set to be 2500 and 0.4 respectively, and we set the $\lambda_{semantics}$ in Equation 4 to be 2 while all other λs in Equation 4 are set to be 1. For each run, our system returns the top 1500 elements.

The measures used in INEX 2009 are the same as that in INEX 2007. For Focused task, interpolated precisions at 101 recall levels, i.e. $iP(i)$, $i=0.0, 0.01, 0.02, ..., 1.0$, and mean average interpolated precision $(MAiP)$ are computed. For In-Context tasks, generalized precisions and recalls at different ranks, i.e. $gP[r]$ and $gR[r]$, $r=1, 2, 3, 5, 10, 25, 50$, and mean average generalized precision $(MAgP)$ are computed.

4.2 Results

The results of the four approaches in different tasks are shown in Table 1. The results annotated with "*" are not official runs submitted to the INEX 2009 but were evaluated with the INEX 2009 assessments afterwards. Due to some implementation issues, we did not finish all the *semantics_co* runs. However, from the results of best in context task, we can draw similar conclusions for this approach in other tasks. It does not perform better than other approaches, especially the text representation approach. This may be due to that we did not tune the parameter $\lambda_{semantics}$.

Table 1. Experimental results of different retrieval strategies at INEX 2009

Tasks Runs	Focused $(iP[0.01])$	Focused $(MAiP)$	Relevant in context $(gP[10])$	Relevant in context $(MAgP)$	Best in context $(gP[10])$	Best in context $(MAgP)$
base_co	0.4322	0.0565	0.1934	0.0645	0.1465*	0.0958*
base_cas	0.4876[1]	**0.1472**[13]	0.1946	**0.1028**[13]	0.1444*	0.0971*
text_co	**0.4973**[1]	0.0741[1]	**0.2381**[12]	0.0807[1]	**0.1610**[1]	**0.1013**[1]
semantics_co	-	-	-	-	0.1484	0.0923

We carried out Wilcoxon signed-rank tests on the results. Let number *1*, *2*, and *3* denote the three approaches, *base_co*, *base_cas*, *text_co*, respectively. In the tables, the superscript of a figure indicates that this approach is significantly (at $p < 0.01$) better than the approaches denoted by the numbers in the superscript for the same task.

From Table 1, we made the following observations. The text representation approach is useful in retrieving the relevant results earlier, while the baseline CAS retrieval strategy, i.e. strict interpretation of the structural constraints, performs better on average precisions.

To analyze these observations more deeply, we classify the 68 assessed CAS queries in INEX 2009 into three categories:

1. CAS queries with no structural constraints, i.e. identical to CO queries. For example, *//*[about(., Bermuda Triangle)]*. There are 15 of them.
2. CAS queries with only structural tags, such as *article*, *sec*, *p*, etc. For example, *//article[about(., Nobel prize)]*, *//article[about(.,IBM)]//sec[about (., computer)]*, and *//article[about(., operating system)]//sec[about(.//p, mutual exclusion)]*. There are 25 of them.
3. CAS queries with semantic tags, such as *vehicle*, *music_genre*, *language*, etc. For example, *//vehicles[about(., fastest speed)]*, *//article[about(., musician)] //music_genre[about(., Jazz)]*, and *//article[about(.//language, java) OR about(., sun)]//sec[about(.//language, java)]*. There are 28 of them.

For each class of CAS queries, we compared their results as shown in Table 2, Table 3 and Table 4 respectively.

Table 2. Results over the CAS queries with no structural constraints

Tasks \ Runs	Focused (iP[0.01])	Focused (MAiP)	Relevant in context (gP[10])	Relevant in context (MAgP)	Best in context (gP[10])	Best in context (MAgP)
base_co	0.4864	0.0718	0.2263	0.0814	0.1194*	0.0923*
base_cas	0.4865	0.0718	0.2263	0.0814	0.1194*	**0.0933***
text_co	**0.5302**	**0.0974**[12]	**0.2818**[12]	**0.1024**[12]	**0.1219**	0.0855

For the first class of CAS queries, CAS queries are identical to their CO versions. From Table 2, we can observe that text representation performs better than the baseline CO/CAS queries.

Table 3. Results over the CAS queries with only structural tags

Tasks \ Runs	Focused (iP[0.01])	Focused (MAiP)	Relevant in context (gP[10])	Relevant in context (MAgP)	Best in context (gP[10])	Best in context (MAgP)
base_co	0.4170	0.0580	0.1807	0.0669	0.1629*	0.0996*
base_cas	**0.5582**[13]	**0.2296**[13]	**0.2054**	**0.1242**[13]	0.1627*	**0.1099***
text_co	0.4545	0.0771[1]	0.2032[1]	0.0814[1]	**0.1778**	**0.1099**

For the second class of CAS queries, baseline CAS approach performs better than other approaches both in terms of early precisions and in terms of average precisions. This may be because that for querying INEX Wikipedia collection, most of the time the whole article is relevant. CAS queries constraining the results to be articles or sections would avoid returning many small sub-elements, thus returning more relevant components in the top 1500 returned elements.

Table 4. Results over the CAS queries with semantic tags

Tasks / Runs	Focused (iP[0.01])	Focused (MAiP)	Relevant in context (gP[10])	Relevant in context (MAgP)	Best in context (gP[10])	Best in context (MAgP)
base_co	0.4166	0.0648	0.1872	0.0534	0.1465*	0.0942*
base_cas	0.4155	**0.1141**[13]	0.1679	**0.0951**[13]	0.1415*	0.0876*
text_co	**0.5179**[1]	0.0589[1]	**0.2459**[12]	0.0684[1]	**0.1670**	**0.1020**[1]

For the third class of queries, we can observe the same trend as that on the whole query set. The reason why CAS queries are not always beneficial may be that if the CAS query is badly formed, strict matching of the query would hurt the performance greatly. For example, //*[about(., notting hill actors) AND about(.//category, film)] does not clarify that "notting hill" should occur in the "film" context and "actor" be the desired result element type, so the baseline CAS performance for this query is much worse than other approaches. Another example is //article[about(., rally car)]//driver[about(., female) OR about(., woman)], the user submit this query intends to retrieve "cars", however the badly formulated CAS query will return "drivers". To avoid such problems, it is better for the system to approximately match the CAS queries and to infer some hints from user submitted CO queries instead of asking the user to specify these hints. It is hard for them to formulate good queries.

5   Conclusions and Future Work

In this paper, we did experiments over the semantically annotated Wikipedia XML corpus at INEX 2009 ad hoc track, attempting to investigate the following two research questions:

1. Do the structural constraints in CAS queries help in retrieving an XML document collection containing semantically rich tags?
2. How to exploit the semantic tag information to improve the CO queries as most users prefer to express the simplest forms of queries?

The results show that CAS queries are helpful in retrieving more relevant elements. When the query was badly formed, however, the performance could be hurt greatly. The simplest way of enriching element representations with semantic tags can improve the performance slightly, especially in terms of the early precisions.

As our future work, we are going to study how to infer structural hints from CO queries and match them with data approximately. We are also interested in how to evaluate top-k queries with complex scoring models efficiently.

Acknowledgements

The research work is supported by the 863 High Tech. Project of China under Grant No. 2009AA01Z149.

References

1. Chu-Carroll, J., Prager, J., Czuba, K., Ferrucci, D., Duboue, P.: Semantic Search via XML Fragments: A High-Precision Approach to IR. In: SIGIR 2006 (2006)
2. Carmel, D., Maarek, Y.S., Mandelbrod, M., et al.: Searching XML documents via XML fragments. In: SIGIR 2003 (2003)
3. Trotman, A., Sigurbjörnsson, B.: Narrowed extended xPath I (NEXI). In: Fuhr, N., Lalmas, M., Malik, S., Szlávik, Z. (eds.) INEX 2004. LNCS, vol. 3493, pp. 16–40. Springer, Heidelberg (2005)
4. Lemur/Indri, http://www.lemurproject.org
5. XQuery Full-Text, http://www.w3.org/TR/xpath-full-text-10/
6. Trotman, A., Lalmas, M.: Why Structural Hints in Queries do not Help XML-Retrieval? In: SIGIR 2006 (2006)
7. Schenkel, R., Suchanek, F., Kasneci, G.: YAWN: A Semantically Annotated Wikipedia XML Corpus. In: BTW 2007 (2007)
8. Hiemstra, D.: Statistical Language Models for Intelligent XML Retrieval. In: Blanken, H., et al. (eds.) Intelligent Search on XML Data. LNCS, vol. 2818, pp. 107–118. Springer, Heidelberg (2003)
9. Ogilvie, P., Callan, J.: Language Models and Structured Document Retrieval. In: INEX 2003 (2003)
10. Ogilvie, P., Callan, J.: Hierarchical Language Models for XML Component Retrieval. In: Fuhr, N., Lalmas, M., Malik, S., Szlávik, Z. (eds.) INEX 2004. LNCS, vol. 3493, pp. 224–237. Springer, Heidelberg (2005)
11. Ogilvie, P., Callan, J.: Parameter Estimation for a Simple Hierarchical Generative Model for XML Retrieval. In: Fuhr, N., Lalmas, M., Malik, S., Kazai, G. (eds.) INEX 2005. LNCS, vol. 3977, pp. 211–224. Springer, Heidelberg (2006)
12. Zhai, C.: Statistical Language Models for Information Retrieval: A Critical Review. Foundations and Trends in Information Retrieval 2(3) (2008)
13. Zhai, C., Lafferty, J.: A Study of Smoothing Methods for Language Models Applied to Ad Hoc Information Retrieval. In: SIGIR 2001 (2001)
14. Zhai, C., Lafferty, J.: Two-Stage Language Models for Information Retrieval. In: SIGIR 2002 (2002)
15. Mei, Q., Zhang, D., Zhai, C.: A General Optimization Framework for Smoothing Language Models on Graph Structures. In: SIGIR 2008 (2008)
16. Wang, Q., Li, Q., Wang, S.: Preliminary Work on XML Retrieval. In: Pre-Proceedings of INEX 2007 (2007)
17. Pektova, D., Croft, W.B., Diao, Y.: Refining Keyword Queries for XML Retrieval by Combining Content and Structure. In: ECIR 2009 (2009)
18. Kim, J., Xue, X., Croft, W.B.: A Probabilistic Retrieval Model for Semistructured Data. In: ECIR 2009 (2009)
19. Bo, Z., Ling, T.W., Chen, B., Lu, J.: Effective XML Keyword Search with Relevance Oriented Ranking. In: ICDE 2009 (2009)
20. Metzler, D., Novak, J., Cui, H., Reddy, S.: Building Enriched Document Representations using Aggregated Anchor Text. In: SIGIR 2009 (2009)
21. Kamps, J., Marx, M., de Rijke, M., Sigurbjörnsson, B.: Structured Queries in XML Retrieval. In: CIKM 2005 (2005)

Overview of the INEX 2009 Book Track

Gabriella Kazai[1], Antoine Doucet[2], Marijn Koolen[3], and Monica Landoni[4]

[1] Microsoft Research, United Kingdom
v-gabkaz@microsoft.com
[2] University of Caen, France
doucet@info.unicaen.fr
[3] University of Amsterdam, Netherlands
m.h.a.koolen@uva.nl
[4] University of Lugano
monica.landoni@unisi.ch

Abstract. The goal of the INEX 2009 Book Track is to evaluate approaches for supporting users in reading, searching, and navigating the full texts of digitized books. The investigation is focused around four tasks: 1) the Book Retrieval task aims at comparing traditional and book-specific retrieval approaches, 2) the Focused Book Search task evaluates focused retrieval approaches for searching books, 3) the Structure Extraction task tests automatic techniques for deriving structure from OCR and layout information, and 4) the Active Reading task aims to explore suitable user interfaces for eBooks enabling reading, annotation, review, and summary across multiple books. We report on the setup and the results of the track.

1 Introduction

The INEX Book Track was launched in 2007, prompted by the availability of large collections of digitized books resulting from various mass-digitization projects [1], such as the Million Book project[1] and the Google Books Library project[2]. The unprecedented scale of these efforts, the unique characteristics of the digitized material, as well as the unexplored possibilities of user interactions present exciting research challenges and opportunities, see e.g. [3].

The overall goal of the INEX Book Track is to promote inter-disciplinary research investigating techniques for supporting users in reading, searching, and navigating the full texts of digitized books, and to provide a forum for the exchange of research ideas and contributions. Toward this goal, the track aims to provide opportunities for exploring research questions around three broad topics:

- Information retrieval techniques for searching collections of digitized books,
- Mechanisms to increase accessibility to the contents of digitized books, and
- Users' interactions with eBooks and collections of digitized books.

---

[1] http://www.ulib.org/
[2] http://books.google.com/

S. Geva, J. Kamps, and A. Trotman (Eds.): INEX 2009, LNCS 6203, pp. 145–159, 2010.
© Springer-Verlag Berlin Heidelberg 2010

Based around these main themes, the following four tasks were defined:

1. The Book Retrieval (BR) task, framed within the user task of building a reading list for a given topic of interest, aims at comparing traditional document retrieval methods with domain-specific techniques, exploiting book-specific features, e.g., back-of-book index, or associated metadata, e.g., library catalogue information,
2. The Focused Book Search (FBS) task aims to test the value of applying focused retrieval approaches to books, where users expect to be pointed directly to relevant book parts,
3. The Structure Extraction (SE) task aims at evaluating automatic techniques for deriving structure from OCR and layout information for building hyperlinked table of contents, and
4. The Active Reading task (ART) aims to explore suitable user interfaces enabling reading, annotation, review, and summary across multiple books.

In this paper, we report on the setup and the results of each of these tasks at INEX 2009. First, in Section 2, we give a brief summary of the participating organisations. In Section 3, we describe the corpus of books that forms the basis of the test collection. The following three sections detail the four tasks: Section 4 summarises the two search tasks (BR and FBS), Section 5 reviews the SE task, and Section 6 discusses ART. We close in Section 7 with a summary and plans for INEX 2010.

2   Participating Organisations

A total of 84 organisations registered for the track (compared with 54 in 2008, and 27 in 2007), of which 16 took part actively throughout the year (compared with 15 in 2008, and 9 in 2007); these groups are listed in Table 1.

In total, 7 groups contributed 16 search topics comprising a total of 37 topic aspects (sub-topics), 4 groups submitted runs to the SE task, 3 to the BR task, and 3 groups submitted runs to the FBS task. Two groups participated in ART, but did not submit results. 9 groups contributed relevance judgements.

3   The Book Corpus

The track builds on a collection of 50,239 out-of-copyright books[3], digitized by Microsoft. The corpus is made up of books of different genre, including history books, biographies, literary studies, religious texts and teachings, reference works, encyclopedias, essays, proceedings, novels, and poetry. 50,099 of the books also come with an associated MAchine-Readable Cataloging (MARC) record, which contains publication (author, title, etc.) and classification information. Each book in the corpus is identified by a 16 character long bookID – the name of the directory that contains the book's OCR file, e.g., A1CD363253B0F403.

---

[3] Also available from the Internet Archive (although in a different XML format).

Table 1. Active participants of the INEX 2009 Book Track, contributing topics, runs, and/or relevance assessments (BR = Book Retrieval, FBS = Focused Book Search, SE = Structure Extraction, ART = Active Reading Task)

ID	Institute	Topics	Runs	Judged topics (book/page level)
6	University of Amsterdam	8, 11	2 BR, 4 FBS	Book: 3, 5, 7, 8, 11, 14, 15; Page: 8, 11, 14
7	Oslo University College	1, 2	10 BR, 10 FBS	Book 1, 2; Page: 1, 2
12	University of Granada			Book: 1, 16; Page: 1
14	Uni. of California, Berkeley		9 BR, ART	
29	Indian Statistical Institute			Book: 16
41	University of Caen	7, 9	3 SE	SE
43	Xerox Research Centre Europe		3 SE	SE
52	Kyungpook National Uni.	3, 4	ART	
54	Microsoft Research Cambridge	10, 16		Book: 3, 5, 7, 9, 10, 16; Page: 3, 5, 7, 9, 10, 16
78	University of Waterloo	5, 6	4 FBS	Book: 5, 6; Page: 5, 6
86	University of Lugano	12, 13, 14, 15		
125	Microsoft Dev. Center Serbia		1 SE	
335	Fraunhofer IAIS			SE
339	Universita degli Studi di Firenze			SE
343	Noopsis Inc.		1 SE	
471	Peking University, ICST			SE
	Unkown			Book: 13, 16

The OCR text of the books has been converted from the original DjVu format to an XML format referred to as BookML, developed by Microsoft Development Center Serbia. BookML provides additional structure information, including markup for table of contents entries. The basic XML structure of a typical book in BookML is a sequence of pages containing nested structures of regions, sections, lines, and words, most of them with associated coordinate information, defining the position of a bounding rectangle ([coords]):

```
<document>
  <page pageNumber="1" label="PT_CHAPTER" [coords] key="0" id="0">
    <region regionType="Text" [coords] key="0" id="0">
      <section label="SEC_BODY" key="408" id="0">
        <line [coords] key="0" id="0">
          <word [coords] key="0" id="0" val="Moby"/>
          <word [coords] key="1" id="1" val="Dick"/>
        </line>
        <line [...]><word [...] val="Melville"/>[...]</line>[...]
      </section>    [...]
    </region>     [...]
  </page>       [...]
</document>
```

BookML provides a set of labels (as attributes) indicating structure information in the full text of a book and additional marker elements for more complex structures, such as a table of contents. For example, the first label attribute in the XML extract above signals the start of a new chapter on page 1 (label="PT_CHAPTER"). Other semantic units include headers (SEC_HEADER), footers (SEC_FOOTER), back-of-book index (SEC_INDEX), table of contents (SEC_TOC). Marker elements provide detailed markup, e.g., for table of contents, indicating entry titles (TOC_TITLE), and page numbers (TOC_CH_PN), etc.

The full corpus, totaling around 400GB, was made available on USB HDDs. In addition, a reduced version (50GB, or 13GB compressed) was made available for download. The reduced version was generated by removing the word tags and propagating the values of the `val` attributes as text content into the parent (i.e., line) elements.

4   Information Retrieval Tasks

Focusing on IR challenges, two search tasks were investigated: 1) Book Retrieval (BR), and 2) Focused Book Search (FBS). Both these tasks used the corpus described in Section 3, and shared the same set of topics (see Section 4.3).

4.1   The Book Retrieval (BR) Task

This task was set up with the goal to compare book-specific IR techniques with standard IR methods for the retrieval of books, where (whole) books are returned to the user. The user scenario underlying this task is that of a user searching for books on a given topic with the intent to build a reading or reference list, similar to those appended to an academic publication or a Wikipedia article. The reading list may be for research purposes, or in preparation of lecture materials, or for entertainment, etc.

Participants of this task were invited to submit either single runs or pairs of runs. A total of 10 runs could be submitted, each run containing the results for all the 16 topics (see Section 4.3). A single run could be the result of either a generic (non-specific) or a book-specific IR approach. A pair of runs had to contain both types, where the non-specific run served as a baseline, which the book-specific run extended upon by exploiting book-specific features (e.g., back-of-book index, citation statistics, book reviews, etc.) or specifically tuned methods. One automatic run (i.e., using only the topic title part of a topic for searching and without any human intervention) was compulsory. A run could contain, for each topic, a maximum of 1,000 books (identified by their bookID), ranked in order of estimated relevance.

A total of 21 runs were submitted by 3 groups (2 runs by University of Amsterdam (ID=6); 9 runs by University of California, Berkeley (ID=14); and 10 runs by Oslo University College (ID=7)), see Table 1. The 21 runs contained a total of 316,000 books, 1,000 books per topic (4 runs from Oslo University College only contained results for 11 of the 16 topics).

4.2   The Focused Book Search (FBS) Task

The goal of this task was to investigate the application of focused retrieval approaches to a collection of digitized books. The task was thus similar to the INEX ad hoc track's Relevant in Context task, but using a significantly different collection while also allowing for the ranking of book parts within a book. The user scenario underlying this task was that of a user searching for information in a library of books on a given subject, where the information sought may be 'hidden' in some books (i.e., it forms only a minor theme) while it may be the main focus of some other books. In either case, the user expects to be pointed directly to the relevant book parts. Following the focused retrieval paradigm, the task of a focused book search system is then to identify and rank (non-overlapping) book parts that contain relevant information and return these to the user, grouped by the books they occur in.

Participants could submit up to 10 runs, where one automatic and one manual run was compulsory. Each run could contain, for each of the 37 topic aspects (see Section 4.3), a maximum of 1,000 books estimated relevant to the given aspect, ordered by decreasing value of relevance. For each book, a ranked list of non-overlapping book parts, i.e., XML elements or passages, estimated relevant were to be listed in decreasing order of relevance. A minimum of one book part had to be returned for each book in the ranking. A submission could only contain one type of result, i.e., only XML elements or only passages.

A total of 18 runs were submitted by 3 groups (4 runs by the University of Amsterdam (ID=6); 10 runs by Oslo University College (ID=7); and 4 runs by the University of Waterloo (ID=78)), see Table 1. The 18 runs contained a total of 444,098 books and 2,638,783 pages; 5.94 pages per book. All runs contained XML elements, and in particular page level elements, with the exception of two runs by the University of Waterloo, which also contained title elements.

4.3   Topics

Topics are representations of users' information needs that may be more or less generic or specific. Reflecting this, a topic may be of varying complexity and may comprise one or multiple aspects (sub-topics). We encouraged participants to create multiple aspects for their topics, where aspects should be focused (narrow) with only a few expected relevant book parts (e.g., pages).

Participants were recommended to use Wikipedia when preparing their topics. The intuition behind the introduction of Wikipedia is twofold. First, Wikipedia can be seen as a real world application for both the BR and FBS tasks: articles often contain a reading list of books relevant to the overall topic of the article, while they also often cite related books in relation to a specific statement in the article. Thus, we anticipated that browsing through Wikipedia entries could provide participants with suggestions about topics and their specific aspects of interest. Second, Wikipedia, can also provide participants with insights and relevant terminology to be used for better searches and refinements that should lead to a better mapping between topics and collection.

```
<topic id=''10'' cn_no=''60''>
<task>Find relevant books and pages to cite from the Wikipedia article on
      Cleopatra's needle</task>
<title>Cleopatra needle obelisk london paris new york</title>
<description>I am looking for reference material on the obelisks known as
             Cleopatra's needle, three of which have been erected: in London,
             Paris, and New York.</description>
<narrative>I am interested in the obelisks' history in Egypt, their transportation,
           their physical descriptions, and current locations. I am, however, not
           interested in the language of the hieroglyphics.</narrative>
<wikipedia-title>Cleopatra's needle</wikipedia-title>
<wikipedia-url>http://en.wikipedia.org/wiki/Cleopatra's_Needle</wikipedia-url>
<wikipedia-text>Cleopatra's Needle is the popular name for each of three Ancient
               Egyptian obelisks [...] </wikipedia-text>
<aspect aspect_id=''10.1''>
<aspect-title>Description of the London and New York pair</aspect-title>
<aspect-narrative>I am looking for detailed physical descriptions of the London and
                  New York obelisks as well as their history in Egypt. When and
                  where they were originally erected and what happened to them when
                  they were moved to Alexandria.</aspect-narrative>
<aspect-wikipedia-text>The pair are made of red granite, stand about 21 meters
                       (68 ft) high, weigh [...] </aspect-wikipedia-text>
</aspect>
<aspect aspect_id=''10.2''>
<aspect-title>London needle</aspect-title>
<aspect-narrative>I am interested in details about the obelisk that was moved to
                  London. When and where was it moved, the story of its
                  transportation. Information and images of the needle and the two
                  sphinxes are also relevant.</aspect-narrative>
<aspect-wikipedia-text>The London needle is in the City of Westminster, on the
                       Victoria Embankment [...] </aspect-wikipedia-text>
</aspect>
<aspect aspect_id=''10.3''>
<aspect-title>New York needle</aspect-title>
<aspect-narrative>I am looking for information and images on the obelisk that was
                  moved to New York. Its history, its transportation and
                  description of its current location.</aspect-narrative>
<aspect-wikipedia-text>The New York needle is in Central Park. In 1869, after the
                       opening of the Suez Canal, [...] </aspect-wikipedia-text>
</aspect>
<aspect aspect_id=''10.4''>
<aspect-title>Paris needle</aspect-title>
<aspect-narrative>Information and images on the Paris needle are sought. Detailed
                  description of the obelisk, its history, how it is different from
                  the London and New York pair, its transportation and current
                  location are all relevant.</aspect-narrative>
<aspect-wikipedia-text>The Paris Needle (L'aiguille de Cleopatre) is in the Place
                       de la Concorde. The center [...] </aspect-wikipedia-text>
</aspect>
</topic>
```

Fig. 1. Example topic from the INEX 2009 Book Track test set

Fig. 2. Screenshot of the relevance assessment module of the Book Search System, showing the list of books in the assessment pool for a selected topic in game 1. For each book, its metadata, its table of contents (if any) and a snippet from a recommended page is shown.

An example topic is shown in Figure 1. In this example, the overall topic includes all three Egyptian obelisks known as Cleopatra's needle, which were erected in London, Paris, and New York. The topic aspects focus on the history of the individual obelisks or on their physical descriptions. Paragraphs in the associated Wikipedia page (¡wikipedia-url¿) relate to the individual topic aspects, while the whole article relates to the overall topic.

Participants were asked to create and submit 2 topics, ideally with at least 2 aspects each, for which relevant books could be found in the corpus. To aid participants with this task, an online Book Search System (see Section 4.4) was developed, which allowed them to search, browse and read the books in the collection.

A total of 16 new topics (ID: 1-16), containing 37 aspects (median 2 per topic), were contributed by 7 participating groups (see Table 1). The collected topics were used for retrieval in the BR task, while the topic aspects were used in the FSB task.

4.4   Relevance Assessment System

The Book Search System (`http://www.booksearch.org.uk`), developed at Microsoft Research Cambridge, is an online tool that allows participants to search, browse, read, and annotate the books of the test corpus. Annotation includes

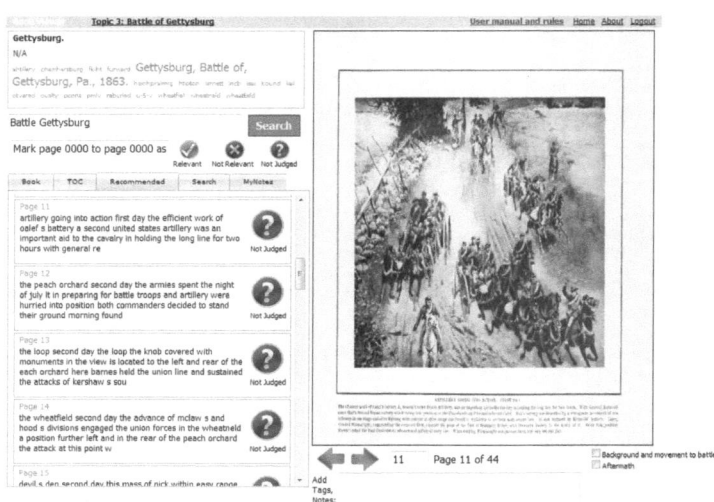

Fig. 3. Screenshot of the relevance assessment module of the Book Search System, showing the Book Viewer window with Recommended tab listing the pooled pages to judge with respect to topic aspects in game 2. The topic aspects are shown below the page images.

the assignment of book and page level relevance labels and recording book and page level notes or comments. The system supports the creation of topics for the test collection and the collection of relevance assessments. Screenshots of the relevance assessment module are shown in Figures 2 and 3.

In 2008, a game called the Book Explorers' Competition was developed to collect relevance assessments, where assessors competed for prizes [4]. The competition involved reading books and marking relevant content inside the books for which assessors were rewarded points. The game was based on two competing roles: *explorers*, who discovered relevant content inside books and *reviewers*, who checked the quality of the explorers' assessments.

Based on what we learnt in 2008, we modified the game this year to consist of three separate, but interconnected 'Read and Play' games: In game 1, participants had the task of finding books relevant to a given topic and then ranking the top 10 most relevant books. In game 2, their task was to explore the books selected in game 1 and find pages inside them that are relevant to a given topic aspect. Finally, in game 3, their task was to review pages that were judged in game 2. Hence, we have, in essence, introduced a filtering stage (game 1) before the Book Explorer's Competition (game 2 and 3) in order to reduce the number of books to judge in detail.

The aim of game 1 was to collect book level judgements for the evaluation of the BR task, while page level assessments gathered in games 2 and 3 would be used to evaluate the FBS task.

Table 2. Collected relevance judgements per topic (up to April 15, 2010)

Topic	Judged books (game 1)	Rel. books (game 1)	Judged pages (games 1/2&3)	Rel. pages (games 1/2&3)	Impl. irrel. (pages)
1	61	10	628/0	602/0	1364
2	57	8	55/0	48/0	993
3	106	65	107/235	106/235	1850
5	1763	14	17/26	16/26	25074
6	90	9	192/0	20/0	4104
7	171	58	26/0(26)	25/0(25)	1608
8	471	155	12/0	1/0	9979
9	121	29	23/0(23)	23/0(23)	581
10	172	25	88/0	39/0	4526
11	1104	95	46/0	0/0	18860
13	9	7	0/0	0/0	19
14	195	18	3/0	1/0	3822
15	31	22	0/0	0/0	4
16	79	33	78/0	66/0	1200
Total	4,430	548	1,275/310	947/309	73,984

4.5   Collected Relevance Assessments

We run the 'Read and Play' games for three weeks (ending on March 15, 2010), with weekly prizes of $50 worth of Amazon gift card vouchers, shared between the top three scorers, proportionate to their scores. Additional judgments were collected up to the period of April 15, 2010, with no prizes. Table 2 provides a summary of all the collected relevance assessments. The last column shows the implicit page level judgements, i.e., for pages in the assessment pool that are inside books that were judged irrelevant.

In total, we collected 4,668 book level relevance judgements from 9 assessors in game 1. Assessors were allowed to judge books for any topic, thus some books were judged by multiple assessors. The total number of unique topic-book pair judgements is 4,430.

In game 1, assessors could choose from 4 possible labels: "relevant", "top 10 relevant", "irrelevant" and "unsure". The latter label could be used either to delay a decision on a given book, or when it was not possible to assess the relevance of a book due to language or technical reasons (e.g., the book was unreadable or could not be displayed). Books ranked in the top 10 most relevant books for a topic were labeled with "top 10 relevant". This was, however, seldom assigned, only in 34 cases across 10 topics.

Page level judgements could be contributed in all three games. However, in game 1, pages could only be judged with respect to the whole topic, while in games 2 and 3, pages were judged with respect to the individual topic aspects. The latter is required for the evaluation of the FBS task. For topics with a single aspect, i.e., 7, 9, 12, and 13, page level judgements could be collected in any of the games.

Table 3. Results for the Book Retrieval Task

Run id	MAP	MRR	P10	bpref	Rel.Ret.
p14_BR_BOOKS2009_FUS_TA*	0.1536	0.6217	0.3429	0.3211	200
p14_BR_BOOKS2009_FUS_TITLE*	0.1902	0.7907	0.4214	0.4007	310
p14_BR_BOOKS2009_OK_INDEX_TA*	0.0178	0.1903	0.0857	0.1737	127
p14_BR_BOOKS2009_OK_TOC_TA*	0.0185	0.1529	0.0714	0.1994	153
p14_BR_BOOKS2009_T2_INDEX_TA*	0.0448	0.2862	0.1286	0.1422	80
p14_BR_BOOKS2009_T2_TOC_TA*	0.0279	0.3803	0.0929	0.1164	75
p14_BR_BOOKS2009_OK_TOPIC_TA	0.0550	0.2647	0.1286	0.0749	41
p14_BR_BOOKS2009_T2FB_TOPIC_TA	0.2309	0.6385	0.4143	0.4490	329
p14_BR_BOOKS2009_T2FB_TOPIC_TITLE	0.2643	0.7830	0.4714	0.5014	345
p6_BR_inex09.book.fb.10.50	**0.3471**	**0.8507**	0.4857	**0.5921**	419
p6_BR_inex09.book	0.3432	0.8120	**0.5286**	0.5842	416
p7_BR_to_b_submit*	0.0915	0.4180	0.2000	0.2375	184
p7_BR_to_g_submit	0.1691	0.5450	0.3357	0.3753	325
p7_BR_tw_b3_submit*	0.0639	0.3984	0.1857	0.2015	164
p7_BR_tw_g3_submit	0.1609	0.5394	0.3214	0.3597	319
p7_BR_tw_b5_submit*	0.0646	0.4292	0.2000	0.1866	139
p7_BR_tw_g5_submit	0.1745	0.6794	0.3357	0.3766	326
p7_BR_wo_b3_submit*	0.0069	0.0333	0.0286	0.0422	70
p7_BR_wo_g3_submit	0.0272	0.2102	0.0786	0.1163	140
p7_BR_wo_b5_submit*	0.0108	0.0686	0.0500	0.0798	48
p7_BR_wo_g5_submit	0.0680	0.4779	0.1214	0.2067	173

From the table, it is clear that game 1 proved much more popular than games 2 and 3. There are two principle reasons for this. On the one hand, games 2 and 3 can only start once books filtered through to them from game 1. On the other hand, in game 1, it is enough to find a single relevant page in a book to mark it relevant, while in games 2 and 3, judges need to read and judge a lot more of a book's content.

Out of the 4,430 books 230 was judged by 2 assessors and 4 by 3 judges. Judges only disagreed on 23 out of the 230 double-judged books, and 2 of the 4 triple-judged books.

Due to the very few judgements available for topic aspects, we will only report results for the BR task in the next section.

4.6   Evaluation Measures and Results

For the evaluation of the BR task, we converted the book level assessments into binary judgements. Judgements labeled "relevant" or "top 10 relevant" were mapped to 1, and judgements labeled "irrelevant" or "unsure" were mapped to 0. If multiple assessors judged a book for a topic, a majority vote was used to determine whether a book is relevant or not. Ties were treated as relevant.

Table 3 shows the results for the BR task. Based on participants' descriptions of their retrieval methods, we marked runs that were book-specific in some way, e.g., used back-of-book index, with an * in the table. From these results, it appears that book-specific information is not yet incorporated into the retrieval

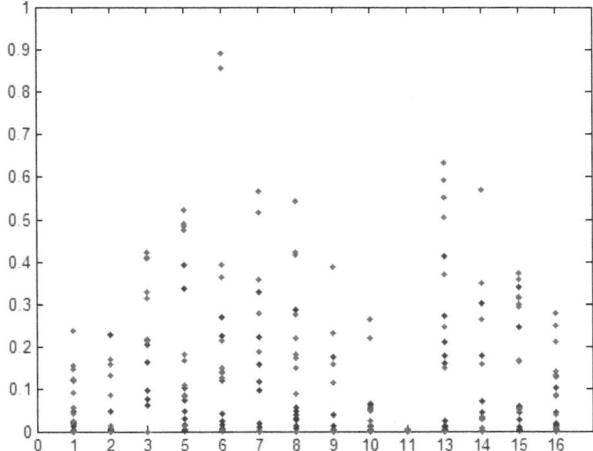

Fig. 4. Distribution of MAP scores across the 14 assessed topics in the BR task. Book-specific approaches are shown as blue dots, while generic IR approaches are shown as red dots.

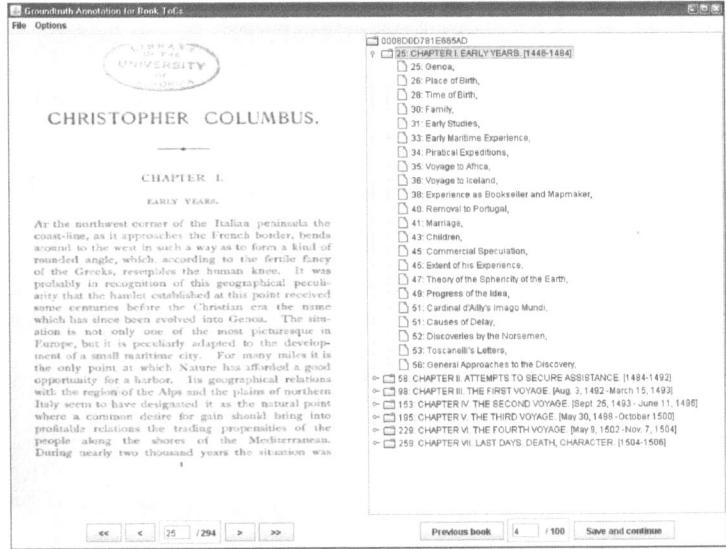

Fig. 5. A screenshot of the ground-truth annotation tool. In the application window, the right-hand side displays the baseline ToC with clickable (and editable) links. The left-hand side shows the current page and allows to navigate through the book. The JPEG image of each visited page is downloaded from the INEX server at www.booksearch.org.uk and is locally cached to limit bandwidth usage.

approaches successfully, but it seems to hurt retrieval effectiveness in the current state of the art. Looking at the per topic results for MAP, see Figure 4, we found that only topic 2 had a book-specific approach as its best performance. For P10, book-specific retrieval strategies obtained best performance for topic 2, and tied with generic retrieval methods on topics 1, 5, 13, and 15. The MRR measure ties the two approaches on all but three topics: ge method is best on topics 1, and 11, and book-specific is best on topic 2. Bpref shows that generic IR methods are superior for all topics. For possible explanations into why book-specific methods do not improve on the traditional IR approaches, please refer to the respective papers, published by the participants of the book track, in the proceedings.

5   The Structure Extraction (SE) Task

The goal of the SE task was to test and compare automatic techniques for extracting structure information from digitized books and building a hyperlinked table of contents (ToC). The task was motivated by the limitations of current digitization and OCR technologies that produce the full text of digitized books with only minimal structure markup: pages and paragraphs are usually identified, but more sophisticated structures, such as chapters, sections, etc., are typically not recognised.

The first round of the structure extraction task, in 2008, ran as a pilot test and permitted to set up appropriate evaluation infrastructure, including guidelines, tools to generate ground-truth data, evaluation measures, and a first test set of 100 books. The second round was run both at INEX 2009 and at the International Conference on Document Analysis and Recognition (ICDAR) 2009 [2]. This round built on the established infrastructure with an extended test set of 1,000 digitized books.

Participants of the task were provided a sample collection of 1,000 digitized books of different genre and styles in DjVu XML format. Unlike the BookML format of the main corpus, the DjVu files only contain markup for the basic structural units (e.g., page, paragraph, line, and word); no structure labels and markers are available. In addition to the DjVu XML files, participants were distributed the PDF of books.

Participants could submit up to 10 runs, each containing the generated table of contents for the 1,000 books in the test set.

A total of 8 runs were submitted by 4 groups (1 run by Microsoft Development Center Serbia (MDCS), 3 runs by Xerox Research Centre Europe (XRCE), 1 run by Noopsis Inc., and 3 runs by the University of Caen).

5.1   Evaluation Measures and Results

For the evaluation of the SE task, the ToCs generated by participants were compared to a manually built ground-truth. This year, the annotation of a minimum number of books was required to gain access to the combined ground-truth set.

To make the creation of the ground-truth set for 1,000 digitized books feasible, we 1) developed a dedicated annotation tool, 2) made use of a baseline annotation as starting point and employed human annotators to make corrections to this, and 3) shared the workload across participants.

The annotation tool was specifically designed for this purpose and developed at the University of Caen, see Figure 5. The tool takes as input a generated ToC and allows annotators to manually correct any mistakes.

Performance was evaluated using recall/precision like measures at different structural levels (i.e., different depths in the ToC). Precision was defined as the ratio of the total number of correctly recognized ToC entries and the total number of ToC entries; and recall as the ratio of the total number of correctly recognized ToC entries and the total number of ToC entries in the ground-truth. The F-measure was then calculated as the harmonic of mean of precision and recall. The ground-truth and the evaluation tool can be downloaded from `http://users.info.unicaen.fr/~doucet/StructureExtraction2009/`.

Table 4. Evaluation results for the SE task (complete ToC entries)

ParticipantID+RunID	Participant	Precision	Recall	F-measure
MDCS	MDCS	41.33%	42.83%	41.51%
XRCE-run1	XRCE	29.41%	27.55%	27.72%
XRCE-run2	XRCE	30.28%	28.36%	28.47%
XRCE-run3	XRCE	28.80%	27.31%	27.33%
Noopsis	Noopsis	9.81%	7.81%	8.32%
GREYC-run1	University of Caen	0.40%	0.05%	0.08%
GREYC-run2	University of Caen	0.40%	0.05%	0.08%
GREYC-run3	University of Caen	0.47%	0.05%	0.08%

The evaluation results are given in Table 4. The best performance ($F = 41.51\%$) was obtained by the MDCS group, who extracted ToCs by first recognizing the page(s) of a book that contained the printed ToC [5]. Noopsis Inc. used a similar approach, although did not perform as well. The XRCE group and the University of Caen relied on title detection within the body of a book.

6 The Active Reading Task (ART)

The main aim of ART is to explore how hardware or software tools for reading eBooks can provide support to users engaged with a variety of reading related activities, such as fact finding, memory tasks, or learning. The goal of the investigation is to derive user requirements and consequently design recommendations for more usable tools to support active reading practices for eBooks. The task is motivated by the lack of common practices when it comes to conducting usability studies of e-reader tools. Current user studies focus on specific content and user groups and follow a variety of different procedures that make comparison, reflection, and better understanding of related problems difficult. ART is hoped to turn into an ideal arena for researchers involved in such efforts with the crucial opportunity to access a large selection of titles, representing different genres, as well as benefiting from established methodology and guidelines for organising effective evaluation experiments.

ART is based on the evaluation experience of EBONI [6], and adopts its evaluation framework with the aim to guide participants in organising and running user studies whose results could then be compared.

The task is to run one or more user studies in order to test the usability of established products (e.g., Amazon's Kindle, iRex's Ilaid Reader and Sony's Readers models 550 and 700) or novel e-readers by following the provided EBONI-based procedure and focusing on INEX content. Participants may then gather and analyse results according to the EBONI approach and submit these for overall comparison and evaluation. The evaluation is task-oriented in nature. Participants are able to tailor their own evaluation experiments, inside the EBONI framework, according to resources available to them. In order to gather user feedback, participants can choose from a variety of methods, from low-effort online questionnaires to more time consuming one to one interviews, and think aloud sessions.

6.1 Task Setup

Participation requires access to one or more software/hardware e-readers (already on the market or in prototype version) that can be fed with a subset of the INEX book corpus (maximum 100 books), selected based on participants' needs and objectives. Participants are asked to involve a minimum sample of 15/20 users to complete 3-5 growing complexity tasks and fill in a customised version of the EBONI subjective questionnaire, allowing to gather meaningful and comparable evidence. Additional user tasks and different methods for gathering feedback (e.g., video capture) may be added optionally. A crib sheet is provided to participants as a tool to define the user tasks to evaluate, providing a narrative describing the scenario(s) of use for the books in context, including factors affecting user performance, e.g., motivation, type of content, styles of reading, accessibility, location and personal preferences.

Our aim is to run a comparable but individualized set of studies, all contributing to elicit user and usability issues related to eBooks and e-reading.

The task has so far only attracted 2 groups, none of whom submitted any results at the time of writing.

7 Conclusions and Plans

The Book Track this year has attracted considerable interest, cow from previous years. Active participation, however, remained a challenge for most of the participants. A reason may be the high initial setup costs (e.g., building infrastructure). Most tasks also require considerable planning and preparations, e.g., for setting up a user study. At the same time, the Structure Extraction task run at ICDAR 2009 (International Conference on Document Analysis and Recognition) has been met with great interest and created a specialist community.

The search tasks, although explored real-world scenarios, were only tackled by a small set of groups. Since the evaluation of the BR and FBS tasks requires a great deal of effort, e.g., developing the assessment system and then collecting relevance judgements, we will be re-thinking the setup of these tasks for INEX 2010. For example, we plan to concentrate on more focused (narrow) topics for which only few pages in the corpus may be relevant. In addition, to improve the quality of the topics, we will look for ways to automate this process, hence also removing the burden from the participants.

To provide real value in improving the test corpus, we plan to run the SE task with the goal to use its results to convert the current corpus to an XML format that contains rich structural and semantic markup, which can then be used in subsequent INEX competitions.

Following the success of running the SE task in parallel at two forums, we will look for possible collaborators, both within and outside of INEX, to run ART next year.

Our plans for the longer term future are to work out ways in which the initial participation costs can be reduced, allowing more of the 'passive' participants to take an active role.

Acknowledgements

The Book Track is supported by the Document Layout Team of Microsoft Development Center Serbia, who developed the BookML format and a tool to convert books from the original OCR DjVu files to BookML.

References

1. Coyle, K.: Mass digitization of books. Journal of Academic Librarianship 32(6), 641–645 (2006)
2. Doucet, A., Kazai, G., Dresevic, B., Uzelac, A., Radakovic, B., Todic, N.: ICDAR 2009 Book Structure Extraction Competition. In: Proceedings of the Tenth International Conference on Document Analysis and Recognition (ICDAR 2009), Barcelona, Spain, July 2009, pp. 1408–1412 (2009)
3. Kantor, P., Kazai, G., Milic-Frayling, N., Wilkinson, R. (eds.): BooksOnline '08: Proceeding of the 2008 ACM workshop on Research advances in large digital book repositories. ACM, New York (2008)
4. Kazai, G., Milic-Frayling, N., Costello, J.: Towards methods for the collective gathering and quality control of relevance assessments. In: SIGIR '09: Proceedings of the 32nd Annual International ACM SIGIR Conference on Research and Development in Information Retrieval. ACM Press, New York (2009)
5. Uzelac, A., Dresevic, B., Radakovic, B., Todic, N.: Book layout analysis: TOC structure extraction engine. In: Geva, S., Kamps, J., Trotman, A. (eds.) INEX 2008. LNCS, vol. 5631, pp. 164–171. Springer, Heidelberg (2009)
6. Wilson, R., Landoni, M., Gibb, F.: The web experiments in electronic textbook design. Journal of Documentation 59(4), 454–477 (2003)

XRCE Participation to the 2009 Book Structure Task

Hervé Déjean and Jean-Luc Meunier

Xerox Research Centre Europe
6 Chemin de Maupertuis, F-38240 Meylan
Firstname.lastname@xrce.xerox.com

Abstract. We present here the XRCE participation to the Structure Extraction task of the INEX Book track 2009. After briefly explaining the four methods used for detecting the book structure in the book body, we explain how we composed them to address the book structure task. We then discuss the Inex evaluation method and propose another measure together with the corresponding software. We then report on each individual method. Finally, we report on our evaluation of the results of all participants.

1 Introduction

We present in this paper our participation to the Structure Extraction task of the INEX Book 2009. Our objective was to experiment with the use of multiple unsupervised methods to realize the task. This article will therefore briefly describe each of them before focusing on the evaluation of our results as well as those of the other participants. We use here the metric we proposed in 2008, whose software implementation is now available at: http://users.info.unicaen.fr/~doucet/StructureExtraction2009/AlternativeResults.html.

Most of the document conversion methods we discuss here are available interactively at http://rossinante.xrce.xerox.com:8090 with inex/colorqube as login/password, and programmatically at https://esp.xrce.xerox.com.

This work is supported by the Large Scale Integrating Project SHAMAN, co-funded under the EU 7th Framework Programme (http://shaman-ip.eu/shaman/).

2 Pre-processing

The first step simply consists in reformatting the XML INEX format into our internal format, mostly renaming elements and adding some attributes (such as unique IDs). This was performed using XSLT technology.

A second step consists in detecting page headers and footers that often introduce noise for our table of contents detector (see [1]).

A third step consists in recognizing the page numbering of the document (see [3]), in order to associate each physical page with zero or one logical page number, the latter being a piece of text. This is again an unsupervised method.

S. Geva, J. Kamps, and A. Trotman (Eds.): INEX 2009, LNCS 6203, pp. 160–169, 2010.
© Springer-Verlag Berlin Heidelberg 2010

3 The Four Methods

The first method aims at parsing the ToC page so as to segment it into entries, each entry being formed by a label and a reference to a page number. Our second method is dedicated at parsing an index page. The third method is our classical Toc detection method (see [2]). The fourth method uses some characteristic page layout in the document body.

None of these methods aims at determining the entry level, so it is arbitrarily set to 1.

3.1 Parsing the ToC Pages

Parsing the ToC pages involves first finding them. For this purpose, we tried first with a simple heuristic that consists in looking for the keyword 'contents' in the few first lines of each page, under the hypothesis of the presence of one ToC, possibly split over consecutive pages at the beginning or end of the book. We look for a series of pages containing this word and tolerate a certain number of misses.

Somehow to our own surprise, this method led to a F1, in term of ToC page retrieval task, in the range 90-95% over the 2008 INEX dataset. However, retrieving the page of the ToC of a document is not enough.

We then need to parse the contiguous pages deemed to be the ToC. The segmentation into paragraphs is unfortunately not adequate, since a valid paragraph may both correspond to multiple entries, as shown below, or to part of one entry. In the latter case, one could argue that the paragraph segmentation is wrong but it is so.

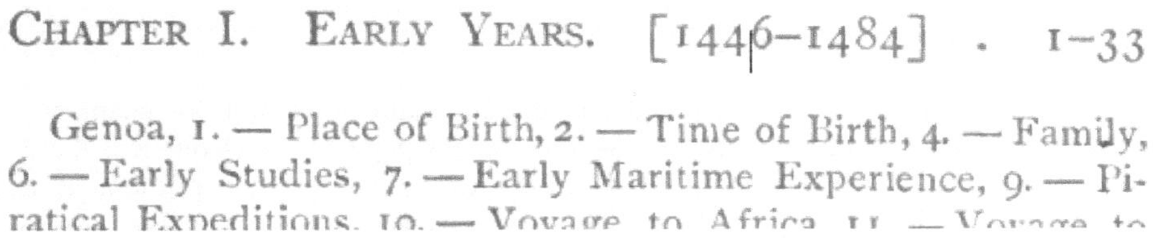

CHAPTER I. EARLY YEARS. [1446–1484] . 1–33

Genoa, 1. — Place of Birth, 2. — Time of Birth, 4. — Family, 6. — Early Studies, 7. — Early Maritime Experience, 9. — Pi- ratical Expeditions. 10. — Voyage to Africa 11 — Voyage to

Fig. 1. Excerpt from the 2008 book #3 (0008D0D781E665AD)

We decided to use the reference to page numbers as ToC entry segmenter, as already mentioned in the literature [4, 5, 6]. Once the ToC is segmented, the page that is referred to is considered as the target of the entry and the entry text becomes the title in the result. Here we take benefit of the recognition of the page numbering, performed in the pre-processing step, and which associates each physical page with zero or one logical page number.

3.2 Parsing the Index Pages

In a similar way, it is possible to look for some index pages, using the 'index' keyword, and to segment it based on the appearance of a page number at the end of the lines. The text in-between two page numbers is deemed to be the title of some part of the document, starting at the indicated page number. This step is performed due to the presence of some books without table of contents, but with an index, the latter being used to build the ground truth.

3.3 "Classical" Detection of the ToCs

The method is detailed in [1], [2] and in this section we will only sketch its outline. The design of this method has been guided by the interest in developing a generic method that uses very intrinsic and general properties of the object known as a table of contents. In view of the large variation in shape and content a ToC may display, we believe that a descriptive approach would be limited to a series of specific collections. Therefore, we instead chose a functional approach that relies on the functional properties that a ToC intrinsically respects. These properties are:

1. Contiguity: a ToC consists of a series of contiguous references to some other parts of the document itself;
2. Textual similarity: the reference itself and the part referred to share some level of textual similarity;
3. Ordering: the references and the referred parts appear in the same order in the document;
4. Optional elements: a ToC entry may include (a few) elements whose role is not to refer to any other part of the document, e.g. decorative text;
5. No self-reference: all references refer outside the contiguous list of references forming the ToC.

Our hypothesis is that those five properties are sufficient for the entire characterization of a ToC, independently of the document class and language.

3.4 Trailing Page Whitespace

Here we are exploiting a widespread convention: main book parts or chapters are often preceded by a page break, i.e. an incomplete page. In consequence, there is some extra blank space between the end of the previous and the start of the next part. So detecting such blank space, also called trailing page whitespace, leads to discover some of the parts of the book. In addition, the title must also be extracted and we currently rely on a heuristic to locate it in the few first lines of the page.

We conjecture that this method has the potential for finding the top level structure of many books while inner structure may remain hidden to it.

4   The Three Runs

We combined in three different ad-hoc ways the methods described earlier and submitted three results.

4.1   First Run

To start with, we applied the keyword heuristic, explained at the beginning of the section 3.1, to all books. Those with a 'contents' keyword were then processed as said in 3.1. Those without 'contents' but with 'index' were processed as said in 3.2. In both case, if the method failed to find resp. a ToC or an Index, the classical method (3.3) was then applied.

The rationale for this process was the assumption that the presence of a 'contents' keyword indicated a ToC. Of course the found ToC might be not parsable based on the page number, for instance because many secondary entries have no associated page number.

768 books were deemed to have a ToC indicated by a keyword 'contents', among which 722 could be segmented based on the occurrence of page numbers. This left 46 documents for the 'classical' method.

21 books had the 'index' keyword but not the 'contents', among which 16 were processed by the 3.2 method. This left 5 more for the 'classical' method, which eventually processed 51 books plus the 211 books with neither 'contents' nor 'index' in them (262 books).

4.2 Second Run

Our second run was an extension of the first one: if the 'classical' method fails to detect a ToC, then we apply the 'trailing space' method (3.4) to structure the book.

We found that among the 262 books processed by the 'classical' method, 50 of them got no result at all, and went then trough the 'trailing space' method.

4.3 Third Run

Our third run was a variation of the previous ones. If the 'contents' keyword is found, we put the method 3.1 and 3.3 in competition by running each of them separately and taking the best output. The best one is supposed to be the one with the longest textual contents, assuming it "explains" better the ToC pages.

For the other books, we applied the 'index' method if the appropriate keyword was found, and in last chance we ran the 'trailing space' method.

5 Evaluation of XRCE results

5.1 Inex Metric

According to the Inex metric, the 3 runs show little difference, with a "complete entries" F1 measure at about 28%.

In more detail, let us look at the first run. The title precision is at 48.23% and the link precision is at 45.22%. This decrease is indicative of the proportion of entries with a valid title that also have a valid link. It means that 94% of the entries with a valid title, have a valid link as well, although the INEX links metric indicate a precision of 45.22%. In our view, this again points at a room for improvement of the INEX links measure. It is desirable in our view to separate the title quality measure from the link quality measure.

5.2 Proposed Link Metric

As initiated in our 2008 paper, we advocate for a complementary metric that would qualify the quality of the links in first intention, rather than conditionally to the title validity.

Such a metric is now available as Python software at:
http://sourceforge.net/projects/inexse/

It computes for each book:

- the precision, recall and F1 measure of the links, by considering a submission as a list of page breaks, ordered by the page number of the breaks. So it counts valid, wrong and missed links. The software actually can display this low-level information per book in a submission.
- for each valid link, it computes the similarity of the proposed title with the ground truth one, using the INEX weighted Levenshtein distance:

$$simil(s1, s2) = 1 - \frac{weightedLevenshtein(s1, s2)}{\max(weight(s1), weight(s2))}$$

  Averaging over all entries with a valid link gives a title accuracy measure for each book.
- In the same manner, a level accuracy per book is computed by considering all entries with a valid link.
- A INEX-like link measure, by matching the title of valid links against the ground truth. So it is similar to the INEX link apart that the link validity is checked before the title validity. If both are valid, the entry is valid. Our experiments validate the closeness of the two measures. A slight difference may occur if several links point to the same page. Since we do not look for the optimal alignment of the corresponding title, the INEX-like measure can be lower.

Each of the measures is averaged on the entire book set to produce the final report.

According to this metric, our 3 runs are also similar to each other, the third one being the better. The table 1 summarizes all these values.

Table 1. XRCE results according to the Inex links measure and to the proposed measure

%	Inex - Links			Proposed link measure				
	Prec.	Rec.	F1	Prec.	Rec.	F1	Title acc.	Level acc.
xrx1	45.22	41.40	42.13	67.1	63.0	62.0	74.6	68.9
xrx2	46.39	42.47	43.15	69.2	64.8	63.8	74.4	69.1
xrx3	45.16	41.70	42.28	69.7	65.7	64.6	74.4	68.8

We observe that the proposed link measure indicates significantly better results, in the range 60-70%, which matches better our feeling when looking at the results.

Those valid links have a textual similarity around 75% with the ground truth. This is why a part of them is invalided by the official INEX links measure. Again, we question this behavior; indeed, should the title "I Introduction" be invalidated because the ground truth specifies "Chapter I Introduction", for instance?

Regarding the levels, the table 2 gives the proportion of each level in the ground truth.

Table 2. Proportion of each level in the ground truth

Level	1	2	3	4	5	6	7
Proportion	39%	41%	16%	4%	1%	0.1%	0.02%

Our methods set by construction all levels to 1, since we did not try to determine the level. We get 7166 valid levels out of 17309 valid links, which means that 41% of the valid links belong to level 1. So our combined method does pretty equally well on all levels. On the other hand, the level accuracy per book is higher on smaller books, this is why the micro-accuracy of 41% is much lower than the macro-accuracy of 68.9% reported in the table 1 above (a book with few entries has the same importance than a larger book with many entries in the macro-accuracy measure).

5.3 Evaluation of the Four XRCE Methods

We will use the proposed measure in the rest of this document.

Using the ground truth, we then evaluated separately all the four methods. We discarded the books 0F7A3F8A17DCDA16 (#40) and E0A9061DCCBA395A (#517) because of massive errors in their respective ground truth. We also decided to only use the ground truth data built collectively in 2009, with the hope to benefit from a higher quality ground truth, thanks to the use of the nice annotation tool provided to us.

This led to a ground truth of 427 books (only 98 books from 2008 were in the 2009 ground truth).

Table 3. The four XRCE methods measured individually

%	Inex–LIKE - Links			(Proposed) link measure				
	Prec.	Rec.	F1	Prec.	Rec.	F1	Title acc.	Level acc.
ToC page parsing	37.2	35.2	35.5	54.8	53.2	52.1	76.3	72.3
Index page parsing	0.0	0.0	0.0	0.5	0.3	0.2	28.5	53.7
Classical ToC	24.6	19.5	20.8	60.3	49.2	50.8	70.4	71.2
Trailing whitespace	22.9	24.4	21.4	56.1	62.6	52.4	59.1	78.8

The first observation is that the use of the index pages, if any, largely fails. Of course, on some documents, like the 491st of the ground truth, the book structuration would benefit from a merge of the ToC and of the index (p15-18).

The second observation is a threshold effect on the title accuracy. It seems that an accuracy below ~75% importantly penalizes the Inex-like result, at almost equal F1 (proposed) measure. Or said differently, given a certain F1 measure, a title accuracy below 75% seems to penalize heavily the INEX F1 measure.

Interestingly, the combination of the 4 methods for the runs leads to a 7 points gain on the F1, reaching 59.3 on these 427 books.

There is room for improvement here since by taking for each book, the max of the F1 of the 4 methods, the averaged F1 reaches 75% with 49% of the books over 0.9 in F1, and 63% over 0.8. On the other hand, 16% of the document would remain below 0.5 in F1. This is the optimal reachable F1 one can obtain by choosing at best which method to trust. Such choice is however difficult to make automatically. Another approach could consist in mixing the results of the four methods for each book, rather than choosing one of them, in some way to be researched.

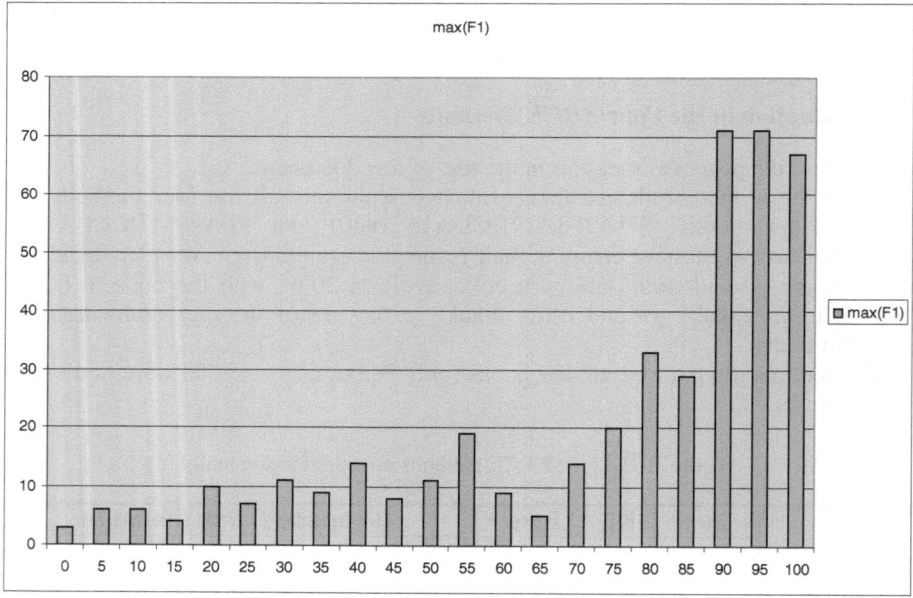

Fig. 2. The histogram of the distribution of the F1 when manually choosing optimally which method to apply to each document. This is theoretical data. *(The vertical scale indicates a number of documents.)*

6   Evaluation of All Participant Results

We again use the proposed new measure to evaluate the results of all the participants, in term of links, titles and levels and show the result in table 4.

This evaluation is performed using the 2009 official ground truth, which include 527 books. For the GREYC result, we observed an unfortunate shift of +1 on all page numbers so we corrected their result before evaluating it under the name GREYC-1.

In term of F1 measure of the links, we observe two groups of result, one in the 65% range the other in the 40% range. The title accuracy ranges from 87% to 42%, with again an important impact on the initial INEX links measure.

As for the level accuracy, MDCS and GREYC do slightly better than the two other participants who simply set all level to 1, gaining about 7% in accuracy, i.e. the proportion of valid level among the valid links. We tend to think that it is indicative of a very difficult task.

Table 4. The results of all participants measured using the proposed metric. Each reported value is an average computed over all books of the ground truth.

Average %	Inex–LIKE - Links			(Proposed) link measure				
	Prec.	Rec.	F1	Prec.	Rec.	F1	Title acc.	Level acc.
MDCS	52.4	54.6	52.8	65.9	70.3	**66.4**	**86.7**	**75.2**
XRX3	44.1	40.8	41.4	69.7	65.7	**64.6**	**74.4**	**68.8**
XRX2	45.3	41.6	42.3	69.2	64.8	63.8	74.4	69.1
XRX1	44.2	40.6	41.3	67.1	63.0	62.0	74.6	68.9
NOOPSIS	15.1	12.0	12.9	46.4	38.0	**39.9**	**71.9**	**68.5**
GREYC-1	10.2	7.2	7.7	59.7	34.2	**38.0**	**42.1**	**73.2**
GREYC	0.0	0.0	0.0	6.7	0.7	1.2	13.9	31.4

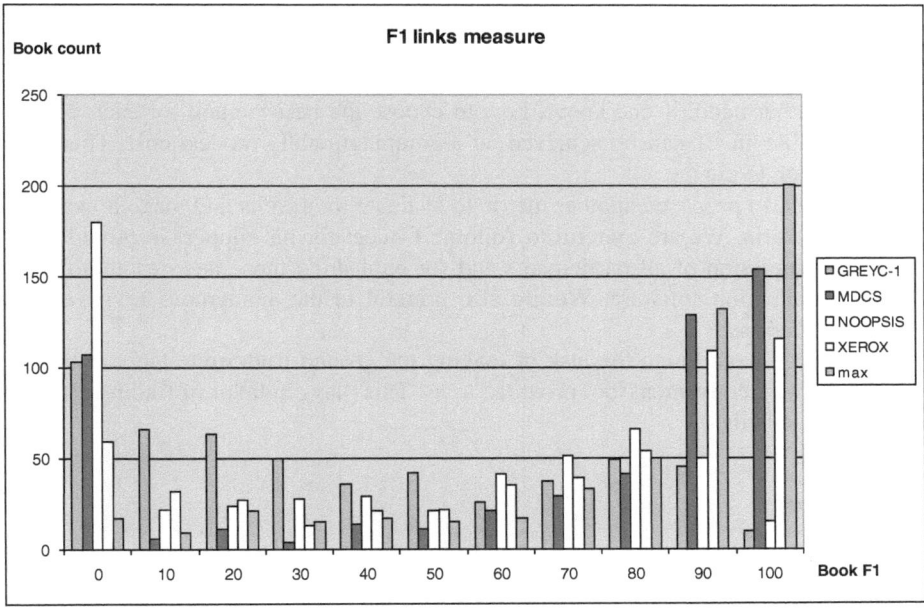

Fig. 3. The histogram of the distribution of the F1 that is computed for each book, for each participant. The orange histogram shows a similar but fictitious distribution obtained by choosing the best F1 among the 4 participants for each book.

We now show in figure 3 the histogram of the link F1 per participant as well as the same histogram computed with the maximum F1 reached by one participant for each book. Such result is achievable is one knows how to choose the best result among the 4 participants for each book.

Choosing optimally the best result per document would lead to a 79% average F1 over the ground truth. Looking at each book, 63% of them would reach over 0.9 in F1, and 73% over 0.8. On the other hand, 20% of the books would remain below 0.5 in F1.

Unfortunately, this measure does not distinguish a non-response from a wrong response, since both get a 0% in precision and recall. Indeed, non-responses are tolerable in many applications while wrong responses are problematic. So we computed the number of non-responses per participant:

GREYC=80   MDCS=101   NOOPSIS=171   XRCE=45

It turns out that the combination "by maximization" leaves only 14 books without answers. It would be interesting to look in more details at those books.

7   Conclusion

We have experimented with combining multiple methods to perform the task, except the level determination for which we have no method in place. Our method for combining was ad-hoc and it provided a limited gain, ~7%, over each individual method. Among those methods, we have found that the method for parsing the index pages was disappointing while the method looking at trailing page whitespace is promising and was under-exploited.

On the other hand, if one knows how to choose the best method for each book, a gain over 20% in F1 can be achieved, at a computationally modest cost. This is an attractive path to pursue.

We have also proposed another metric to evaluate the results and made it available in software form. We are grateful to Antoine Doucet for his support in providing us with the submission of all participants and for publishing those new results together with the evaluation software. We are also grateful to the anonymous reviewers for their contributions.

Finally, we have found the task of making the ground truth quite labor intensive, despite the very convenient tool provided to us. This plays in favor of finding efficient automatic methods.

References

1. Déjean, H., Meunier, J.-L.: Structuring Documents according to their Table of Contents. In: Proceedings of the 2005 ACM symposium on Document engineering, pp. 2–9. ACM Press, New York (2005)
2. Déjean, H., Meunier, J.-L.: On Tables of Contents and how to recognize them. International Journal of Document Analysis and Recognition, IJDAR (2008)

3. Déjean, H., Meunier, J.L.: Versatile page numbering analysis. In: Document Recognition and Retrieval XV, part of the IS&T/SPIE International Symposium on Electronic Imaging, DRR 2008, San Jose, California, USA, January 26-31 (2008)
4. Belaïd, A.: Recognition of table of contents for electronic library consulting. International Journal of Document Analysis and Recognition, IJDAR (2001)
5. Lin, X.: Detection and analysis of table of contents based on content association. International Journal of Document Analysis and Recognition, IJDAR (2005)
6. Gao, L., Tang, Z., Lin, X., Tao, X., Chu, Y.: Analysis of Book Documents' Table of Content Based on Clustering. In: International Conference of Document Analysis and Recognition, ICDAR (2009)

The Book Structure Extraction Competition with the Resurgence Software at Caen University

Emmanuel Giguet and Nadine Lucas

GREYC Cnrs, Caen Basse Normandie University
BP 5186 F-14032 CAEN Cedex France
name.surname@info.unicaen.fr

Abstract. The GREYC Island team participated in the Structure Extraction Competition part of the INEX Book track for the first time, with the Resurgence software. We used a minimal strategy primarily based on top-down document representation. The main idea is to use a model describing relationships for elements in the document structure. Chapters are represented and implemented by frontiers between chapters. Page is also used. The periphery center relationship is calculated on the entire document and reflected on each page. The strong points of the approach are that it deals with the entire document; it handles books without ToCs, and titles that are not represented in the ToC (e. g. preface); it is not dependent on lexicon, hence tolerant to OCR errors and language independent; it is simple and fast.

1 Introduction

The GREYC Island team participated for the first time in the Book Structure Extraction Competition held at ICDAR in 2009 and part of the INEX evaluations [1]. The Resurgence software was modified to this end. This software was designed to handle various document formats, in order to process academic articles (mainly in pdf format) and news articles (mainly in HTML format) in various text parsing tasks [2]. We decided to join the INEX competition because the team was also interested in handling voluminous documents, such as textbooks.

The experiment was conducted over 1 month. It was run from pdf documents to ensure the control of the entire process. The document content is extracted using the pdf2xml software [3]. The evaluation rules were not thoroughly studied, for we simply wished to check if we were able to handle large corpora of voluminous documents.

The huge memory needed to handle books was indeed a serious obstacle, as compared with the ease in handling academic articles. The OCR texts were also difficult to cope with. Therefore, Resurgence was modified in order to handle the corpus. We could not propagate our principles on all the levels of the book hierarchy at a time. We consequently focused on chapter detection.

In the following, we explain the main difficulties, our strategy and the results on the INEX book corpus. We provide corrected results after a few modifications were

S. Geva, J. Kamps, and A. Trotman (Eds.): INEX 2009, LNCS 6203, pp. 170–178, 2010.

made. In the last section, we discuss the advantages of our method and make proposals for future competitions.

2 Our Book Structure Extraction Method

2.1 Challenges

In the first stage of the experiment, the huge memory needed to handle books was found to be indeed a serious hindrance: pdf2xml required up to 8 Gb of memory and Resurgence required up to 2 Gb to parse the content of large books (> 150 Mb). This was due to the fact that the whole content of the book was stored in memory. The underlying algorithms did not actually require the availability of the whole content at a time. They were so designed since they were meant to process short documents.

Therefore, Resurgence was modified in order to load the necessary pages only. The objective was to allow processing on usual laptop computers.

The fact that the corpus was OCR documents also challenged our previous program that detected the structure of electronic academic articles. A new branch in Resurgence had to be written in order to be tolerant to OCR documents.

We had no time to propagate our document parsing principles on all the levels of the book hierarchy at a time. We consequently focused on chapter detection.

2.2 Strategy

Very few principles were tested in this experiment. The strategy in Resurgence is based on document positional representation, and does not rely on the table of contents (ToC). This means that the whole document is considered first. Then document constituents are considered top-down (by successive subdivision), with focus on the middle part (main body). The document is thus the unit that can be broken down ultimately to pages. The main idea is to use a model describing relationships for elements in the document structure. The model is a periphery-center dichotomy. The periphery center relationship is calculated on the entire document and reflected on each page. The algorithm aims at retrieving the book main content bounded by annex material like preface and postface with different layout. It ultimately retrieves the page body in a page, surrounded by margins [2].

However, for this first experiment, we focused on chapter title detection so that the program detects only one level, i. e. chapter titles.

Chapter title detection throughout the document was conducted using a sliding window. It is used to detect chapter transitions. The window captures a four-page context with a look-ahead of one page and look-behind of two pages. The underlying idea is that the chapter begins after a blank, or at least is found in a relatively empty zone at the top of page. The half page fill rate is the simple cue used to decide on chapter transition. The beginning of a chapter is detected by one

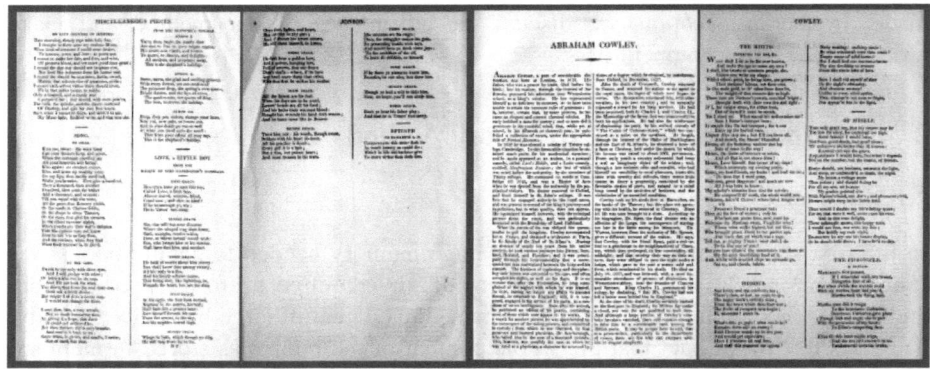

Fig. 1. View of the four-page sliding window to detect chapter beginning. Pattern 1 matches. Excerpt from 2009 book id = 00AF1EE1CC79B277.

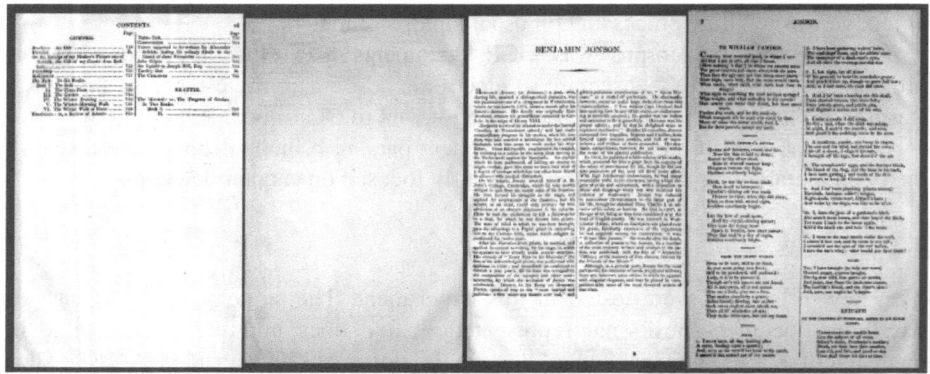

Fig. 2. View of the four-page sliding window to detect chapter announced by a blank page. Pattern 2 matches. Excerpt from 2009 book id= 00AF1EE1CC79B277.

of the two patterns below, where i is the page where a chapter starts. Figure 1 and 2 illustrate the two patterns.

Pattern 1:   - top and bottom of page i-2 equally filled
               - bottom of page i-1 less filled than top of page i-1
               - top of page i less filled than bottom of page i
               - top and bottom of page i+1 equally filled

Pattern 2:   - any content for page i-2
               - empty page i-1
               - top of page i less filled than bottom of page i
               - top and bottom of page i+1 equally filled

Chapter title extraction is made from the first third of the beginning page. The model assumes that the title begins at the top of the page. The end of the title was not carefully looked for. The title is grossly delineated by a constraint rule allowing a number of lines containing at most 40 words.

2.3 Experiment

The program detected only chapter titles. No effort was exerted to find the sub-titles. The three runs were not very different since runs 2 and 3 amount to post-processing of the ToC generated by run 1.

Run 1 was based on minimal rules as stated above.

Run 2 was the same + removing white spaces at the beginning and end of the title (trim)

Run 3 was the same + trim + pruning lower-case lines following a would-be title in higher-case.

2.4 Commented Results

The entire corpus was handled. The results were equally very bad for the three runs. This was due to a page numbering bug where p = p-1. The intriguing value above zero 0,08% came from rare cases where the page contained two chapters (two poems).

Table 1. Book Structure Extraction official evaluation[1]

RunID Participant	F-measure (complete entries)
MDCS Microsoft Development Center Serbia	41,51%
XRCE-run2 Xerox Research Centre Europe	28,47%
XRCE-run1 Xerox Research Centre Europe	27,72%
XRCE-run3 Xerox Research Centre Europe	27,33%
Noopsis Noopsis inc.	8,32%
GREYC-run1 GREYC - University of Caen, France	0,08%
GREYC-run2 GREYC - University of Caen, France	0,08%
GREYC-run3 GREYC - University of Caen, France	0,08%

Table 2. Detailed results for GREYC

	Precision	Recall	F-Measure
Titles	19,83%	13,60%	13,63%
Levels	16,48%	12,08%	11,85%
Links	1,04%	0,14%	0,23%
Complete entries	0,40%	0,05%	0,08%
Entries disregarding depth	1,04%	0,14%	0,23%

[1] http://users.info.unicaen.fr/~doucet/StructureExtraction2009/

The results were recomputed with correction on the unfortunate page number shift in the INEX grid (Table 3).

The alternative evaluation grid suggested by [4, 5], was applied. In table 4, for the GREYC result, the corrected run p= p-1 is computed under the name "GREYC-1"[2].

Table 3. GREYC results with page numbering correction

	precision	recall	F-measure
run-1	10, 41	7,41	7,66
run-2	10,56	7,61	7,85
run-3	11,22	7,61	8,02

Table 4. Alternative evaluation

	XRCE Link-based Measure				
	Links			Accuracy (for valid links)	
	Precision	Recall	F1	*Title*	*Level*
MDCS	65.9	70.3	66.4	86.7	75.2
XRCE-run3	69.7	65.7	64.6	74.4	68.8
XRCE-run2	69.2	64.8	63.8	74.4	69.1
XRCE-run1	67.1	63.0	62.0	74.6	68.9
Noopsis	46.4	38.0	39.9	71.9	68.5
GREYC-1	59.7	34.2	38.0	42.1	73.2
GREYC	6.7	0.7	1.2	13.9	31.4

The results still suffered from insufficient provision made for the evaluation rules. Notably, the title hierarchy is not represented, which impairs recall. Titles were grossly segmented on the right side, which impairs precision. Title accuracy is also very low for the same reason.

However, level accuracy balances the bad results reflected in the F1 measure. The idea behind level accuracy is that good results at a given level are more satisfying than errors scattered everywhere. The accuracy for chapter level, which was the only level we tempted, was 73,2%, second high. It means that few chapter beginnings were missed by Resurgence. Errors reflect both non responses and wrong responses. Our system returned 80 non responses for chapters, out of 527 in the sample, and very few wrong responses. Chapter titles starting on the second half of the page have been missed, as well as some chapters where the title was not very clearly contrasted against the background.

2.5 Corrections after Official Competition

A simple corrective strategy was applied in order to better compare methods. First the bug on page number was corrected.

---

[2] http://users.info.unicaen.fr/~doucet/StructureExtraction2009/AlternativeResults.html

A new feature boosted precision. In a supplementary run (run 4) both page number shift and chapter title detection were amended. The title right end is detected, by calculating the line height disruption: a contrast between the would-be title line height and the rest of the page line height.

These corrections result in a better precision as shown in Table 5 (line GREYC-2) with the XRCE link-based measure. The recall rate is not improved because the subtitles are still not looked for.

Table 5. Corrected run (GREYC-2) with better title extraction compared with previous results

	XRCE Link-based Measure				
	Links			Accuracy (for valid links)	
	Precision	Recall	F1	*Title*	*Level*
GREYC-2	64.3	37.1	41.1	45.1	73.1
GREYC-1	59.7	34.2	38.0	42.1	73.2
GREYC	6.7	0.7	1.2	13.9	31.4

Table 6 reorders the final results of the Resurgence program (GREYC-2) against other participants known performance. The measure is the XRCE Link-based measure.

Table 6. Best alternative evaluation

	XRCE Link-based Measure				
	Links			Accuracy (for valid links)	
	Precision	Recall	F1	*Title*	*Level*
MDCS	65.9	70.3	66.4	86.7	75.2
XRCE-run3	69.7	65.7	64.6	74.4	68.8
GREYC-2	64.3	37.1	41.1	45.1	73.1
Noopsis	46.4	38.0	39.9	71.9	68.5

3 Discussion

The experiment was preliminary. We were pleased to be able to handle the entire corpus and go through evaluation, since only four competitors out of eleven finished [1]. The results were very bad even if small corrections significantly improved them. In addition to the unfortunate page shift, the low recall is due to the fact that the hierarchy of titles was not addressed as mentioned earlier. This will be addressed in the future.

3.1 Reflections on the Experiment

On the scientific side, the importance of quick and light means to handle the corpus was salient. The pdf2xml program for instance returned too much information for our needs and was expensive to run. We wish to contribute to the improvement of the software on these aspects.

Although the results are bad, they showed some strong points of the Resurgence program, based on relative position and differential principles. We intend to further explore this way. The advantages are the following:

- The program deals with the entire document, not only the table of contents;
- It handles books without ToCs, and titles that are not represented in the ToC (e. g. preface);
- It is dependent on typographical position, which is very stable in the corpus;
- It is not dependent on lexicon, hence tolerant to OCR errors and language independent.
- Last, it is simple and fast.

Some examples below underline our point. They illustrate problems that are met in classical "literal" approaches but are avoided by the "positional" solution.

Example 1. Varying forms for the string "chapter" due to OCR errors handled by Resurgence

CHAPTEK
CHAPTEE
CH^PTEE
CHAP TEE
CHA 1 TKR
C II APT Kit
(MI A I TKIl
C II A P T E II
C H A P TEH
C H A P T E R
C II A P T E U
Oil A PTKR

Since no expectations bear on language-dependent forms, chapter titles can be extracted from any language. A reader can detect *a posteriori* that this is being written in French (first series) or in English (second series).

- TABLE DES MATIÈRES
- DEUXIEME PARTIE
- CHAPITRE
- TROISIÈME PARTIE
- QUATRIÈME PARTIE
- PREFACE

- CHAPTER
- TABLE OF CONTENTS
- INTRODUCTION
- APPENDIX

Since no list of expected and memorized forms is used, but position instead, fairly common strings are extracted, such as CHAPTER or SECTION, but also uncommon ones, such as PSALM or SONNET. When chapters have no numbering and no prefix such as *chapter*, they are found as well, for instance a plain title "Christmas Day".

Resurgence did not rely on numbering of chapters: this is an important source of OCR errors, like in the following series. Hence they were retrieved as they were by our robust extractor.

- II
- HI
- IV
- V
- VI

- SECTION VI
- SECTION VII
- SECTION YTIT
- SECTION IX

- SKOTIOX XMI
- SECTION XV
- SECTION XVI

- THE FIRST SERMON
- THE SECOND SERMON
- THE THIRD SERMON
- THE FOURTH SERMON

- CHAPTEE TWELFTH
- CHAPTER THIRTEENTH
- CHAPTER FOURTEENTH

The approach we used was minimal, but reflects an original breakthrough to improve robustness without sacrificing quality.

3.2 Proposals

Concerning evaluation rules, generally speaking, it is unclear whether the ground truth depends on the book or on the ToC. If the ToC is the reference, it is an error to extract prefaces, for instance. The participants using the whole text as main reference would be penalized if they extract the whole hierarchy of titles as it appears in the book, when the ToC represents only higher levels, as is often the case.

Concerning details, it should be clear whether or when the prefix indicating the book hierarchy level (*Chapter, Section*, and so on) and the numbering should be part of the extracted result. As it was mentioned earlier and as it can be seen in Figure 1, the chapter title is not necessarily preceded by such mentions, but in other cases there is no specific chapter title and only a number. The ground truth is not clear either on the extracted title case: sometimes the case differs in the ToC and in the actual title in the book.

It would be very useful to provide results by title depth (level) as suggested by [4], because it seems that providing complete results for one or more level(s) would be more satisfying than missing some items at all levels. It is important to get coherent and comparable text spans for many tasks, such as indexing, helping navigation or text mining.

The reason why the beginning and end of the titles are overrepresented in the evaluation scores is not clear and a more straightforward edit distance for extracted titles should be provided.

There is also a bias introduced by a semi-automatically constructed ground truth. Manual annotation is still to be conducted to improve the ground truth quality, but it is time-consuming. We had technical difficulties to meet that requirement in summer 2009. It might be a better idea to open annotation to a larger audience and for a longer period of time.

It might be a good idea to give the bounding box containing the title as a reference for the ground truth. This solution would solve conflicts between manual annotation and automatic annotation, leaving man or machine to read and interpret the content of the bounding box. It would also alleviate conflicts between ToC-based or text-based approaches.

The corpus provided for the INEX Book track is very valuable, it is the only available corpus offering full books. Although it comprises old printed books only, it is interesting for it provides various examples of layout.

References

1. Doucet, A., Kazai, G.: ICDAR 2009 Book Structure Extraction Competition. In: 10th International Conference on Document Analysis and Recognition ICDAR 2009, Barcelona, Spain, pp. 1408–1412. IEEE, Los Alamitos (2009)
2. Giguet, E., Lucas, N., Chircu, C.: Le projet Resurgence: Recouvrement de la structure logique des documents électroniques. In: JEP-TALN-RECITAL'08 Avignon (2008)
3. Déjean, H.: pdf2xml open source software (2010),
   http://sourceforge.net/projects/pdf2xml/ (last visited March 2010)
4. Déjean, H., Meunier, J.-L.: XRCE Participation to the Book Structure Task. In: Geva, S., Kamps, J., Trotman, A. (eds.) INEX 2008. LNCS, vol. 5631, pp. 124–131. Springer, Heidelberg (2009) doi: 10.1007/978-3-642-03761-0
5. Déjean, H., Meunier, J.-L.: XRCE Participation to the Book Structure Task. In: Kamps, J. (ed.) INEX 2009. LNCS, vol. 6203, pp. 160–169. Springer, Heidelberg (2010)

Ranking and Fusion Approaches for XML Book Retrieval

Ray R. Larson

School of Information
University of California, Berkeley
Berkeley, California, USA, 94720-4600
ray@ischool.berkeley.edu

Abstract. For INEX 2009 UC Berkeley focused on the Book track and all of our submissions were for that track. We tried a variety of different approaches for our Book Track runs, the TREC2 logistic regression probabilistic model used in previous INEX Book Track submissions as well as various fusion approaches including use of the Okapi BM-25 algorithm. This paper focusses on the approaches used in the various runs submitted, and shows the preliminary results from the book search evaluation.

1 Introduction

In this paper we will first discuss the algorithms and fusion operators used in our official INEX 2009 Book Track runs. Then we will look at how these algorithms and operators were used in combination with indexes for various parts of the book contents in our submissions for this track, and finally we will discuss possible directions for future research.

2 The Retrieval Algorithms and Fusion Operators

This section largely duplicates earlier INEX papers in describing the probabilistic retrieval algorithms used for both the Adhoc and Book tracks. Although these are the same algorithms that we have used in previous years for INEX and in other evaluations (such as CLEF), including a blind relevance feedback method used in combination with the TREC2 algorithm, we are repeating the formal description here instead of refering to those earlier papers alone. In addition we will again discuss the methods used to combine the results of searches of different XML components in the collections. The algorithms and combination methods are implemented as part of the Cheshire II XML/SGML search engine [9,8,7] which also supports a number of other algorithms for distributed search and operators for merging result lists from ranked or Boolean sub-queries.

2.1 TREC2 Logistic Regression Algorithm

Once again the principle algorithm used for our INEX runs is based on the *Logistic Regression* (LR) algorithm originally developed at Berkeley by Cooper, et al. [5].

S. Geva, J. Kamps, and A. Trotman (Eds.): INEX 2009, LNCS 6203, pp. 179–189, 2010.
© Springer-Verlag Berlin Heidelberg 2010

The version that we used was the Cheshire II implementation of the "TREC2" [4,3] that provided good thorough retrieval performance in the INEX 2005 evaluation [9]. As originally formulated, the LR model of probabilistic IR attempts to estimate the probability of relevance for each document based on a set of statistics about a document collection and a set of queries in combination with a set of weighting coefficients for those statistics. The statistics to be used and the values of the coefficients are obtained from regression analysis of a sample of a collection (or similar test collection) for some set of queries where relevance and non-relevance has been determined. More formally, given a particular query and a particular document in a collection $P(R \mid Q, D)$ is calculated and the documents or components are presented to the user ranked in order of decreasing values of that probability. To avoid invalid probability values, the usual calculation of $P(R \mid Q, D)$ uses the "log odds" of relevance given a set of S statistics derived from the query and database, such that:

$$\log O(R|C,Q) = log \frac{p(R|C,Q)}{1 - p(R|C,Q)} = log \frac{p(R|C,Q)}{p(\overline{R}|C,Q)}$$

$$= c_0 + c_1 * \frac{1}{\sqrt{|Q_c|} + 1} \sum_{i=1}^{|Q_c|} \frac{qtf_i}{ql + 35}$$

$$+ c_2 * \frac{1}{\sqrt{|Q_c|} + 1} \sum_{i=1}^{|Q_c|} \log \frac{tf_i}{cl + 80}$$

$$- c_3 * \frac{1}{\sqrt{|Q_c|} + 1} \sum_{i=1}^{|Q_c|} \log \frac{ctf_i}{N_t}$$

$$+ c_4 * |Q_c|$$

where C denotes a document component and Q a query, R is a relevance variable, and

$p(R|C,Q)$ is the probability that document component C is relevant to query Q,

$p(\overline{R}|C,Q)$ the probability that document component C is not relevant to query Q, (which is $1.0 - p(R|C,Q)$)

$|Q_c|$ is the number of matching terms between a document component and a query,

qtf_i is the within-query frequency of the ith matching term,

tf_i is the within-document frequency of the ith matching term,

ctf_i is the occurrence frequency in a collection of the ith matching term,

ql is query length (i.e., number of terms in a query like $|Q|$ for non-feedback situations),

cl is component length (i.e., number of terms in a component), and

N_t is collection length (i.e., number of terms in a test collection).

c_k are the k coefficients obtained though the regression analysis.

Assuming that stopwords are removed during index creation, then ql, cl, and N_t are the query length, document length, and collection length, respectively. If the query terms are re-weighted (in feedback, for example), then qtf_i is no longer the original term frequency, but the new weight, and ql is the sum of the new weight values for the query terms. Note that, unlike the document and collection lengths, query length is the relative frequency without first taking the log over the matching terms.

The coefficients were determined by fitting the logistic regression model specified in $\log O(R|C, Q)$ to TREC training data using a statistical software package. The coefficients, c_k, used for our official runs are the same as those described by Chen[1]. These were: $c_0 = -3.51$, $c_1 = 37.4$, $c_2 = 0.330$, $c_3 = 0.1937$ and $c_4 = 0.0929$. Further details on the TREC2 version of the Logistic Regression algorithm may be found in Cooper et al. [4].

2.2   Blind Relevance Feedback

It is well known that blind (also called pseudo) relevance feedback can substantially improve retrieval effectiveness in tasks such as TREC and CLEF. (See for example the papers of the groups who participated in the Ad Hoc tasks in TREC-7 (Voorhees and Harman 1998)[14] and TREC-8 (Voorhees and Harman 1999)[15].)

Blind relevance feedback is typically performed in two stages. First, an initial search using the original queries is performed, after which a number of terms are selected from the top-ranked documents (which are presumed to be relevant). The selected terms are weighted and then merged with the initial query to formulate a new query. Finally the reweighted and expanded query is run against the same collection to produce a final ranked list of documents. It was a simple extension to adapt these document-level algorithms to document components for INEX.

The TREC2 algorithm has been been combined with a blind feedback method developed by Aitao Chen for cross-language retrieval in CLEF. Chen[2] presents a technique for incorporating blind relevance feedback into the logistic regression-based document ranking framework. Several factors are important in using blind relevance feedback. These are: determining the number of top ranked documents that will be presumed relevant and from which new terms will be extracted, how to rank the selected terms and determining the number of terms that should be selected, how to assign weights to the selected terms. Many techniques have been used for deciding the number of terms to be selected, the number of top-ranked documents from which to extract terms, and ranking the terms. Harman [6] provides a survey of relevance feedback techniques that have been used.

Obviously there are important choices to be made regarding the number of top-ranked documents to consider, and the number of terms to extract from those documents. For this year, having no truly comparable prior data to guide us, we chose to use the top 10 terms from 10 top-ranked documents. The terms were chosen by extracting the document vectors for each of the 10 and computing the Robertson and Sparck Jones term relevance weight for each document. This

Table 1. Contingency table for term relevance weighting

	Relevant	Not Relevant	
In doc	R_t	$N_t - R_t$	N_t
Not in doc	$R - R_t$	$N - N_t - R + R_t$	$N - N_t$
	R	$N - R$	N

weight is based on a contingency table where the counts of 4 different conditions for combinations of (assumed) relevance and whether or not the term is in a document. Table 1 shows this contingency table.

The relevance weight is calculated using the assumption that the first 10 documents are relevant and all others are not. For each term in these documents the following weight is calculated:

$$w_t = log \frac{\frac{R_t}{R - R_t}}{\frac{N_t - R_t}{N - N_t - R + R_t}} \tag{1}$$

The 10 terms (including those that appeared in the original query) with the highest w_t are selected and added to the original query terms. For the terms not in the original query, the new "term frequency" (qtf_i in main LR equation above) is set to 0.5. Terms that were in the original query, but are not in the top 10 terms are left with their original qtf_i. For terms in the top 10 and in the original query the new qtf_i is set to 1.5 times the original qtf_i for the query. The new query is then processed using the same TREC2 LR algorithm as shown above and the ranked results returned as the response for that topic.

2.3 Okapi BM-25 Algorithm

The version of the Okapi BM-25 algorithm used in these experiments is based on the description of the algorithm in Robertson [12], and in TREC notebook proceedings [13]. As with the LR algorithm, we have adapted the Okapi BM-25 algorithm to deal with document components :

$$\sum_{j=1}^{|Q_c|} w^{(1)} \frac{(k_1 + 1)tf_j}{K + tf_j} \frac{(k_3 + 1)qtf_j}{k_3 + qtf_j} \tag{2}$$

Where (in addition to the variables already defined):

K is $k_1((1 - b) + b \cdot dl/avcl)$
k_1, b and k_3 are parameters (1.5, 0.45 and 500, respectively, were used),
$avcl$ is the average component length measured in bytes
$w^{(1)}$ is the Robertson-Sparck Jones weight:

$$w^{(1)} = log \frac{(\frac{r + 0.5}{R - r + 0.5})}{(\frac{n_{t_j} - r + 0.5}{N - n_{t_j} - R - r + 0.5})}$$

r is the number of relevant components of a given type that contain a given term,

R is the total number of relevant components of a given type for the query.

Our current implementation uses only the *a priori* version (i.e., without relevance information) of the Robertson-Sparck Jones weights, and therefore the $w^{(1)}$ value is effectively just an IDF weighting. The results of searches using our implementation of Okapi BM-25 and the LR algorithm seemed sufficiently different to offer the kind of conditions where data fusion has been shown to be be most effective [10], and our overlap analysis of results for each algorithm (described in the evaluation and discussion section) has confirmed this difference and the fit to the conditions for effective fusion of results.

2.4   Result Combination Operators

As we have also reported previously, the Cheshire II system used in this evaluation provides a number of operators to combine the intermediate results of a search from different components or indexes. With these operators we have available an entire spectrum of combination methods ranging from strict Boolean operations to fuzzy Boolean and normalized score combinations for probabilistic and Boolean results. These operators are the means available for performing fusion operations between the results for different retrieval algorithms and the search results from different components of a document. We use these operators to implement the data fusion approaches used in the submissions. Data fusion is the approach of combining results from two or more different searches to provide results that are better (it is hoped) than any of the individual searches alone.

For the several of the submitted book search runs we used a merge/reweighting operator based on the "Pivot" method described by Mass and Mandelbrod[11] to combine the results for each type of document component considered. In our case the new probability of relevance for a component is a weighted combination of the initial estimate probability of relevance for the component and the probability of relevance for the entire article for the same query terms. Formally this is:

$$P(R \mid Q, C_{new}) = (X * P(R \mid Q, C_{comp})) + ((1 - X) * P(R \mid Q, C_{art})) \quad (3)$$

Where X is a pivot value between 0 and 1, and $P(R \mid Q, C_{new})$, $P(R \mid Q, C_{comp})$ and $P(R \mid Q, C_{art})$ are the new weight, the original component weight, and article weight for a given query. Although we found that a pivot value of 0.54 was most effective for INEX04 data and measures, we adopted the "neutral" pivot value of 0.4 for all of our 2009 adhoc runs, given the uncertainties of how this approach would fare with the new database.

3   Database and Indexing Issues

For the Book Track data we attempted to use multiple elements or components that were identified in the Books markup as Tables of Contents and Indexes as well as the full text of the book, since the goal of the main Books Adhoc task was to retrieval entire books and not elements, the entire book was retrieved regardless of the matching elements. Obviously this is not an optimal matching strategy, and will inevitably retrieve many false matches, but for the 2009 submissions we were trying to set up a "bag of words" baseline. Other elements were available and used in some of the data fusion submissions. We created the same indexes for the Books and for their associated MARC data that we created last year (the MARC fields are shown in Table 5, for the books themselves we used a single index of the entire document content. We did not use the Entry Vocabulary Indexes used in previous years Book track runs, since their performance was, in general, less effective than using the full contents.

Table 2. Book-Level Indexes for the INEX Book Track 2009

Name	Description	Contents	Vector?
topic	Full content	//document	Yes
toc	Tables of Contents	//section@label="SEC_TOC"	No
index	Back of Book Indexes	//section@label="SEC_INDEX"	No

Table 2 lists the Book-level (/article) indexes created for the INEX Books database and the document elements from which the contents of those indexes were extracted.

Table 3. Components for INEX Book Track 2009

Name	Description	Contents
COMPONENT_PAGE	Pages	//page
COMPONENT_SECTION	Sections	//section

Cheshire system permits parts of the document subtree to be treated as separate documents with their own separate indexes. Tables 3 & 4 describe the XML components created for the INEX Book track and the component-level indexes that were created for them.

Table 3 shows the components and the paths used to define them. The first, refered to as COMPONENT_PAGE, is a component that consists of each identified page of the book, while COMPONENT_SECTION identifies each section of the books, permitting each individual section or page of a book to be retrieved separately. Because most of the areas defined in the markup as "section"s are actually paragraphs, we treat these as if they were paragraphs for the most part.

Table 4 describes the XML component indexes created for the components described in Table 3. These indexes make the individual sections (such as COMPONENT_SECTION) of the INEX documents retrievable by their titles, or by any terms occurring in the section. These are also proximity indexes, so phrase searching is supported within the indexes. Individual paragraphs (COMPONENT_PARAS) are searchable by any of the terms in the paragraph, also

Table 4. Component Indexes for INEX Book Track 2009

Component or Index Name	Description	Contents	Vector?
COMPONENT_SECTION			
para_words	Section Words	* (all)	Yes
COMPONENT_PAGES			
page_words	Page Words	* (all)	Yes

Table 5. MARC Indexes for INEX Book Track 2009

Name	Description	Contents	Vector?
names	All Personal and Corporate names	//FLD[1670]00, //FLD[1678]10, //FLD[1670]11	No
pauthor	Personal Author Names	//FLD[170]00	No
title	Book Titles	//FLD130, //FLD245, //FLD240, //FLD730, //FLD740, //FLD440, //FLD490, //FLD830	No
subject	All Subject Headings	//FLD6..	No
topic	Topical Elements	//FLD6.., //FLD245, //FLD240, //FLD4.., //FLD8.., //FLD130, //FLD730, //FLD740, //FLD500, //FLD501, //FLD502 //FLD505, //FLD520, //FLD590	Yes
lcclass	Library of Congress Classification	//FLD050, //FLD950	No
doctype	Material Type Code	//USMARC@MATERIAL	No
localnum	ID Number	//FLD001	No
ISBN	ISBN	//FLD020	No
publisher	Publisher	//FLD260/b	No
place	Place of Publication	//FLD260/a	No
date	Date of Publication	//FLD008	No
lang	Language of Publication	//FLD008	No

with proximity searching. Individual figures (COMPONENT_FIG) are indexed by their captions.

The indexes used in the MARC data are shown in Table 5. Note that the tags represented in the "Contents" column of the table are from Cheshire's MARC to XML conversion, and are represented as regular expressions (i.e., square brackets indicate a choice of a single character). We did not use the MARC data in our 2009 submitted runs.

3.1   Indexing the Books XML Database

Because the structure of the Books database was derived from the OCR of the original paper books, it is primarily focused on the page organization and layout and not on the more common structuring elements such as "chapters" or "sections". Because this emphasis on page layout goes all the way down to the individual word and its position on the page, there is a very large amount of markup for page with content. For original version of the Books database, there are actually NO text nodes in the entire XML tree, the words actually present on a page are represented as attributes of an empty word tag in the XML. The entire document in XML form is typically multiple megabytes in size. A separate version of the Books database was made available that converted these empty tags back into text nodes for each line in the scanned text. This provided a significant reduction in the size of database, and made indexing much simpler. The primary index created for the full books was the "topic" index containing the entire book content.

We also created page-level "documents" as we did in 2008. As noted above the Cheshire system permits parts of the document subtree to be treated as separate documents with their own separate indexes. Thus, paragraph-level components were extracted from the page-sized documents. Because unique object (page) level indentifiers are included in each object, and these identifiers are simple extensions of the document (book) level identifier, we were able to use the page-level identifier to determine where in a given book-level document a particular page or paragraph occurs, and generate an appropriate XPath for it.

Indexes were created to allow searching of full page contents, and component indexes for the full content of each of individual paragraphs on a page. Because of the physical layout based structure used by the Books collection, paragraphs split across pages are marked up (and therefore indexed) as two paragraphs. Indexes were also created to permit searching by object id, allowing search for specific individual pages, or ranges of pages.

The system problems encountered last year have been (temporarily) corrected for this years submissions. Those problems were caused by the numbers of unique terms exceeding the capacity of the integers used to store them in the indexes. For this year, at least, moving to unsigned integers has provided a temporary fix for the problem but we will need to rethink how statistical summary information is handled in the future – perhaps moving to long integers, or even floating point numbers and evaluating the tradeoffs between precision in the statistics and index size (since moving to Longs could double index size).

4   INEX 2009 Book Track Runs

We submitted nine runs for the Book Search task of the Books track,

As Table 6 shows, a number of variations of algorithms and search elements were tried this year. In Table 6 the first column is the run name (all official submissions had names beginning with "BOOKS09" which has been removed from the name), the second column is a short description of the run, since each topic included not only a title, but one or more "aspects" each of which had an aspect title, column four also indicates whether title only or title and aspects combined were used in the submitted queries. The third column shows which

Table 6. Berkeley Submissions for the INEX Book Track 2009

Name	Description	Algorithm	Aspects?
T2FB_TOPIC_TITLE	Uses topic index with title and blind feedback	TREC2 +BF	No
T2FB_TOPIC_TA	Uses topic index with title and aspects with blind feedback	TREC2 +BF	Yes
T2_INDEX_TA	Uses Back of Book index with title and aspects	TREC2	Yes
T2_TOC_TA	Uses Table of Contents index with title and aspects	TREC2	Yes
OK_TOPIC_TA	Uses Topic index with title and aspects	Okapi BM-25	Yes
OK_TOC_TA	Uses Tables of Contents index with title and aspects	Okapi BM-25	Yes
OK_INDEX_TA	Uses Back of Book indexes with title and aspects	Okapi BM-25	Yes
FUS_TA	Fusion of Topic, Table of Contents, and Back of Book Indexes - title and aspects	TREC2 +BF	Yes
FUS_TITLE	Fusion of Topic, Table of Contents, and Back of Book Indexes - title only	TREC +BF	No

algorithms where used for the run, TREC2 is the TREC2 Logistic regression algorithm described above, "BF" means that blind relevance feedback was used in the run, and OKAPI BM-25 means that the OKAPI algorithm described above was used.

4.1 Preliminary Results for the Book Search Task

The Book track organizers have released preliminary performance data for the Book Search runs submitted. They noted, however in the release that it was based on a relatively small set of judgements (around 3000 for 13 out 16 topics, and that the distribution is quite skewed). So the results presented in Table 7 should be considered preliminary only. They suggest that the BPREF figures in Table 7 are probably best, given the limited data. There are, however, some interesting aspects of the results and implications for future approaches to the Book Search task.

Perhaps the most obvious take-away with the results are that the use of blind feedback appears to obtain the best results. Another possible lesson is that the different aspects included in the topics appear to be more a source of noise than a focusing mechanism, since the runs that used title alone from the topics instead of combining title and aspects worked better. It also appears that using Table of Contents and Back-of-Book-Index alone doesn't perform very well. Regardless of the measure selected, the topic index search with blind feedback using the topic title is the best performing of the runs that Berkeley submitted, and was ranked third among the submitted runs by all participants (using the BPREF measure, based only on the preliminary results).

Table 7. Preliminary Results for Berkeley Submissions

Name	MAP	recip. rank	P10	bpref	num. rel.	num. rel. retr.
T2FB_TOPIC_TITLE	0.2643	0.7830	0.4714	0.5014	548	345
T2FB_TOPIC_TA	0.2309	0.6385	0.4143	0.4490	548	329
FUS_TITLE	0.1902	0.7907	0.4214	0.4007	548	310
FUS_TA	0.1536	0.6217	0.3429	0.3211	548	200
OK_TOPIC_TA	0.0550	0.2647	0.1286	0.0749	548	41
T2_INDEX_TA	0.0448	0.2862	0.1286	0.1422	548	80
T2_TOC_TA	0.0279	0.3803	0.0929	0.1164	548	75
OK_TOC_TA	0.0185	0.1529	0.0714	0.1994	548	153
OK_INDEX_TA	0.0178	0.1903	0.0857	0.1737	548	127

5   Conclusions and Future Directions

In the notebook version of this paper we made a few observations about the submitted runs. We suspected that the fusion of full contents with specific table of contents matching and back of the book indexes would probably perform best. This proved not to be the case, and in the results available the fusion approaches lagged behind the topic index (i.e., full-text search of the book contents). This may be due to the table of contents and back of the book indexes missing relevant items due to the lack of such areas in many of the books. We also hoped, with some confirmation from the preliminary results, that the topic index with blind feedback would provide a good baseline that can be enhanced by matches in tables of contents and back of book indexes. But the actual effectiveness of fusion methods is often slightly less than a single effective method alone. The provision of evaluation results showed that this was once again the case with our Book Search task submissions this year. The fusion of the additional indexes apparently leads to more non relevant items than full-text search alone. However it is worth noting that the weights used to combine indexes have not been tuned for this collection (which requires previous results), so there is still the possibility that some of the fusion approaches may work better than observed, if the appropriate weighting factors for combining them can be found.

References

1. Chen, A.: Multilingual information retrieval using english and chinese queries. In: Peters, C., Braschler, M., Gonzalo, J., Kluck, M. (eds.) CLEF 2001. LNCS, vol. 2406, pp. 44–58. Springer, Heidelberg (2002)
2. Chen, A.: Cross-Language Retrieval Experiments at CLEF 2002. LNCS, vol. 2785, pp. 28–48. Springer, Heidelberg (2003)
3. Chen, A., Gey, F.C.: Multilingual information retrieval using machine translation, relevance feedback and decompounding. Information Retrieval 7, 149–182 (2004)
4. Cooper, W.S., Chen, A., Gey, F.C.: Full Text Retrieval based on Probabilistic Equations with Coefficients fitted by Logistic Regression. In: Text REtrieval Conference (TREC-2), pp. 57–66 (1994)

5. Cooper, W.S., Gey, F.C., Dabney, D.P.: Probabilistic retrieval based on staged logistic regression. In: 15th Annual International ACM SIGIR Conference on Research and Development in Information Retrieval, Copenhagen, Denmark, June 21-24, pp. 198–210. ACM, New York (1992)
6. Harman, D.: Relevance feedback and other query modification techniques. In: Frakes, W., Baeza-Yates, R. (eds.) Information Retrieval: Data Structures & Algorithms, pp. 241–263. Prentice Hall, Englewood Cliffs (1992)
7. Larson, R.R.: A logistic regression approach to distributed IR. In: SIGIR 2002: Proceedings of the 25th Annual International ACM SIGIR Conference on Research and Development in Information Retrieval, Tampere, Finland, August 11-15, pp. 399–400. ACM, New York (2002)
8. Larson, R.R.: A fusion approach to XML structured document retrieval. Information Retrieval 8, 601–629 (2005)
9. Larson, R.R.: Probabilistic retrieval, component fusion and blind feedback for XML retrieval. In: Fuhr, N., Lalmas, M., Malik, S., Kazai, G. (eds.) INEX 2005. LNCS, vol. 3977, pp. 225–239. Springer, Heidelberg (2006)
10. Lee, J.H.: Analyses of multiple evidence combination. In: SIGIR '97: Proceedings of the 20th Annual International ACM SIGIR Conference on Research and Development in Information Retrieval, Philadelphia, July 27-31, pp. 267–276. ACM, New York (1997)
11. Mass, Y., Mandelbrod, M.: Component ranking and automatic query refinement for xml retrieval. In: Fuhr, N., Lalmas, M., Malik, S., Szlávik, Z. (eds.) INEX 2004. LNCS, vol. 3493, pp. 73–84. Springer, Heidelberg (2005)
12. Robertson, S.E., Walker, S.: On relevance weights with little relevance information. In: Proceedings of the 20th annual international ACM SIGIR conference on Research and development in information retrieval, pp. 16–24. ACM Press, New York (1997)
13. Robertson, S.E., Walker, S., Hancock-Beauliee, M.M.: TREC-7: ad hoc, filtering, vlc and interactive track. In: Text Retrieval Conference (TREC-7), November 9-1 (Notebook), pp. 152–164 (1998)
14. Voorhees, E., Harman, D. (eds.): The Seventh Text Retrieval Conference (TREC-7). NIST (1998)
15. Voorhees, E., Harman, D. (eds.): The Eighth Text Retrieval Conference (TREC-8). NIST (1999)

OUC's Participation in the 2009 INEX Book Track

Michael Preminger, Ragnar Nordlie, and Nils Pharo

Oslo University College

Abstract. In this article we describe the Oslo University College's participation in the INEX 2009 Book track. This year's tasks have been featuring complex topics, containing aspects. These lend themselves to use in both the book retrieval and the focused retrieval tasks. The OUC has submitted retrieval results for both tasks, focusing on using the Wikipedia texts for query expansion, as well as utilizing chapter division information in (a number of) the books.

1 Introduction

In recent years large organizations like national libraries, as well as multinational organizations like Microsoft and Google have been investing labor, time and money in digitizing books. Beyond the preservation aspects of such digitization endeavors, they call on finding ways to exploit the newly available materials, and an important aspect of exploitation is book and passage retrieval.

The INEX Book Track, which has now been running for three years, is an effort aiming to develop methods for retrieval in digitized books. One important aspect here is to test the limits of traditional methods of retrieval, designed for retrieval within "documents" (such as news-wire), when applied to digitized books. One wishes to compare these methods to book-specific retrieval methods.

One of the aims of the 2009 Book Track experiments[1] was to explore the potential of query expansion using Wikipedia texts to improve retrieval performance. Another aim, which this paper only treats superficially, is to compare book specific retrieval to generic retrieval for both (whole) book retrieval and focused retrieval. In the short time Wikipedia has existed, its use as a source of knowledge and reference has increased tremendously even if its credibility as a trustworthy resource is sometimes put to doubt [2]. This combination of features would make it interesting to use digitized books as a resource with which one can verify or support information found in the Wikipedia.

The most interesting part of this year's topics, which also constitutes the essence of this years task, is, no doubt, the Wikipedia text that is supplied with each aspect. The first thing coming to mind is, of course, using the Wikipedia texts for query expansion, which could intuitively provide a precision enhancing device. The Wikipedia texts are quite long, and the chances of zero hits using the entire text as a query are quite significant (particularly if using logical AND to combine the terms). Query expansion needs thus be approached with caution.

S. Geva, J. Kamps, and A. Trotman (Eds.): INEX 2009, LNCS 6203, pp. 190–199, 2010.

Whereas a query text (even a test query) is said to originally be formulated by the user, a Wikipedia article does not origin with the user, so that there may be elements in the article that the user would not have endorsed, and thus are unintentional. Used uncritically in a query, those parts may reduce experienced retrieval performance.

A measure to counter this effect would be either using only parts of the Wikipedia text that (chances are that) the user would knowingly endorse, or to use the topic title to process the Wikipedia text, creating a version of the latter that is closer to the user's original intention, while still benefitting from the useful expansion potential the text entails.

In investigations involving book retrieval, [3] have experimented with different strategies of retrieval based on query length and document length. Their conclusion has been that basing one's book retrieval on collating results obtained from searching in *book pages* as basic retrieval units (shorter documents), performed better than using *the book as a whole* as a basic retrieval unit. Moreover, manually adding terms to the query improved page level (shorter document) retrieval, but did not seem to improve retrieval of longer documents (whole books).

At the OUC, we wished to pursue this observation, and pose the following questions:

- Can the Wikipedia page partly play the same role which manually added terms would play in a batch retrieval situation (laboratory setting)?
- In case it does, would it also benefit users in real life situations?
- What kind of usage of the Wikipedia text would, on average, provide the better retrieval?

2   A Brief Analysis of a Wikipedia Topic Text

A Wikipedia text may vary in length. Even if we assume the text is very central to – and a good representative of – the user's information need, we hypothesize that using the entire text uncritically as a query text or expansion device, would be hazardous.

Being a collaborative resource, the Wikipedia community has a number of rules that contributors are asked to adhere to "using common sense" [4]. This means that Wikipedia has a realtively free style. An examination of a number of articles treating various subjects indicates that they resemble each other in structure, starting off with a general introduction, followed by a table of contents into the details of the discussed subject. Given that it is the general topic of the article which is of interest to the user (or the writer of the article for that matter), This makes the beginning of an article quite important as a source for formulating the search.

However, a glance at the Wikipedia texts supplied with [5] (both on topic and aspect level) leaves the impression that using even the initial sentences or sections uncritically may result in poor retrieval, or no retrieval at all.

In the beginning of a Wikipedia article, there often occurs a "to be" (is, are, were a.s.o) or a "to have" ("has", "had" a.s.o.) inflection. The occurrence is not necessarily in the first sentence[1], but relatively early in the document. The idea is to use this occurrence as an entry point to the important part of the text[2].

Our hypothesis is that on both sides of such an occurrence, one generally finds words that could be used to compose meaningful queries. There are also grounds to assume that the user, who hunts for book references to this Wikipedia article the way he or she conceives its contents, has read this part of the text and approves of it as representative before proceeding to the rest of the article. Hence this part may arguably be a good representative of the text as seen by the user.

3   Extracting Important Terms from a Wikipedia Article

A way of testing the idea mentioned above is, for each applicable Wikipedia text, to locate the first salient occurrence of a "to be" or "to have" inflection (see also footnotes 1 and 2, and define a window of both preceding and succeeding text (omitting stop-words). The length of the window may vary (See Figure 1). the content of this window is either used as the query or is added to an existing query as an expansion text.

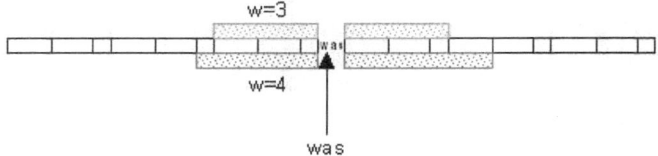

Fig. 1. Using windows of varying widths around an entry word in a Wikipedia text

3.1   Word Occurrences in Windows

Using this year's topics' Wikipedia texts as a sample, we have tried to analyze the occurrences of words in the samples, based on some parameters. The most important parameter was the length of the window. Based on the window length, it was interesting to categorize the distribution of expansion terms into nominals, verbs, adjectives etc.

Experiments performed by [6] indicate that nouns have an advantage when using only certain parts of speech in retrieval, in the sense that using only nouns in queries entails very little loss in precision recall. This would call for experimenting with Wikipedia text extracting only nouns. In this paper we are not pursuing this, merely observing the number of nouns in the queries we are composing. The purpose is to see how the presence of particular types of words in the query contribute to the retrieval quality, and what role Wikipedia can play here.

---

[1] It is sometimes omitted in favor of phrases like "refers to", or another similar combination. This observation may be important to follow up in future research.

[2] On a too early occurrence, the second or third occurrence might be considered as an alternative entry.

4   Book Retrieval

4.1   Generic Retrieval

Books seen as traditional documents are, on average, very long. This means that word may co-occur in line with their co-occurrence in a query, without necessarily being very relevant to the query. For this reason we have experimented with two types of queries. The one that looks at the book as a single large chunk of text, and the other that looks at each page as a chunk of text and combines the performances of the query against all pages.

We choose to regard this as a type of "generic retrieval" for the sake of the current discussion, although it does incorporate page division. In practice, page is not a part of the semantic structure, as page boundaries are mostly a function of physical attributes (font size, page size, illustration size etc.). Moreover, we feel that the former type of retrieval evaluation (evaluating the retrieval of a book as a large chunk of text) might possibly provide results that are unrealistically good, due to an (again, possible) overly high correspondence with the method books are pre-selected by for relevance assessments and assessed (online) thereafter.

4.2   Book-Specific Retrieval

One of the research objectives of this year's book track is to compare the performance of generic retrieval methods with more book-specific retrieval methods.

There are probably a large number of possibilities of utilizing book structure in the collection. We have chosen to identify chapter titles with the help of the TOC entries of the book[3]. In addition to the indication of being TOC sections (or lines or words), the marker elements also have references to the page where the referred chapter begins. The chapter names are available and can be used to boost the chances of certain pages (title pages or inner pages of the chapter) to be retrieved as response to queries that include these words.

We create an index for which we identify words in chapter titles, section titles and the like, so we can enhance their significance at query time, and then try to run the expanded queries against this index as well. In practice we identify words constituting chapter or section titles in the TOC section of the book, find the physical location of their respective chapter and prepend them, specially tagged, to the chapter. This tagging facilitates different weighting of these words related to the rest of the text. Different weighting strategies can then be tested.

For book retrieval it is not important where in the book the title words increase in weight, as they will increase the chances of retrieving this book as a response to a query featuring these words. For focused retrieval we have the limitation that we do not have an explicit chapter partition of the book, only a page partition. One choice of handling this is to identify all pages belonging to a partition, and adding the title words (with enhanced weights) to the text of each page. Within

---

[3] Results may suffer due to the fact that only some 36000 of the 50000 books indeed feature these markup attributes "refid" and "link" of the page and marker element respecively.

the title page (first page of the chapter) the same title words can be given a different relative weight than for the pages inside the chapter.

5   Focused Retrieval

In our experiments, focused retrieval follows along similar lines as book retrieval, just that here the purpose is to retrieve book pages rather than books. We have submitted a number of runs participating in the effort to improve the quality of the test collection. As page-relevance assessments for focused runs were not available by the submission deadline of this paper we choose to defer further analysis regarding this task to a later stage of the research.

6   Runs

We have been running comparable experiments for book and focused retrieval, using the same Indri (http://www.lemurproject.org) index. We were using Indri'support for retrieval by extents. We added chapter titles (enclosed in elements we named *titleinfront* and *titleinside* respectively) in all the pages that constituted a chapter title page or a chapter content page.

In both book retrieval and focused retrieval we have experimented with generic as well as book specific retrieval as described above, using the code pattern as described in table 1.

Table 1. Code pattern of run submissions to the book track. Rows represent the composition of the query. For columns, *book specific* refers to retrieval where chapter title and chapter front page are weighted higher than the rest of the text. No weighting for *generic* retrieval. Grey cells represent the runs analyzed in this paper.

	book retrieval		focused retrieval	
	generic	book specific	generic	book specific
title only	book_to_g	book_to_b	focused_to_g	focused_to_b
title and wiki	book_tw_g#	book_tw_b#	focused_tw_g#	focused_tw_b#
wiki only	book_wo_g#	book_wo_b#	focused_wo_g#	focused_wo_b#

The main partition follows the line of book vs. focused retrieval, so that parallel generic and book specific runs can be compared. The first row features runs with queries involving topic titles only. The second row has runs where the topic is composed of terms from the topic title and the Wikipedia text, and the third row represents queries composed of the Wikipedia texts only. The hash character is a place holder for the size of the window as described in Section 3, where applicable. In this submission we have experimented with window sizes of 3 and 5 for each run type (as represented by a table cell). This gives a total of 20 runs. The choice of 3 and 5 is somewhat arbitrary as hypothetically salient choices. The choice of 10 was later added for the sake of control. More extensive experimentation with different combinations will be beneficial.

7   Results

7.1   Word Distribution in Queries

In table 2 we are listing the proportion of nouns[4] in the queries.

Table 2. Average percentage of nouns in the topic queries

	to_g	tw_g3	tw_g5	tw_g10
nouns	44	99	145	227
total	55	148	213	376
%nouns	80,00	66,89	68,08	60,37

7.2   Retrieval Performance

Below we show retrieval performance results. The results are shown for some of the runs we submitted, and are based on a relatively small number of relevance assessments. Therefore conclusions are drawn with some caution. The results are shown separately for generic retrieval (Figure 2) and for book-specific - structure supported retrieval (Figure 3). Precision at 5 retrieved documents is shown on the plot for each of the runs.

Generic retrieval. In sub-figures 2(a) and 2(b) we are showing precision-recall curves of some of the runs, based on (a) each book as a long chunk of text - and (b) combination of single page performances, respectively. The most important finding is that the increased involvement of Wikipedia terms seems to improve retrieval performance at the low recall region in the runs labeled (a), while it seems to deteriorates the performance of the runs labeled (b). The increase in (a) is not linearly proportional to the number of terms added, and we see that terms added by a window of 5 draw the curve down at the low recall region. This may be due to an interaction between the basic query terms (topic title) and the Wikipedia text. The low number of relevant judgments may also be a factor here. The tendency as a whole, and the difference between the two modes may, again, be explained by the fact that a group of many terms will tend to co-occur in a book as a long chunk of text, but will have a lower probability of co-occurrence in a page. The correlation with the percentage of nouns in the queries (table 2) is not clear, and is difficult to judge on the basis of the current results.

Book-specific (structure-supported) retrieval. In sub-figures 3(a) and 3(b) we are showing precision-recall curves of some of the runs, based on each book as a long chunk of text - and combination of single page performances, respectively. Here the increased involvement of Wikipedia terms does not seem to improve retrieval performance, and the higher the involvement, the worse the performace gets.

---

[4] Based on the definition in [6], "A noun is any person, place or thing", both person name, place name and general nouns are included when nouns are counted.

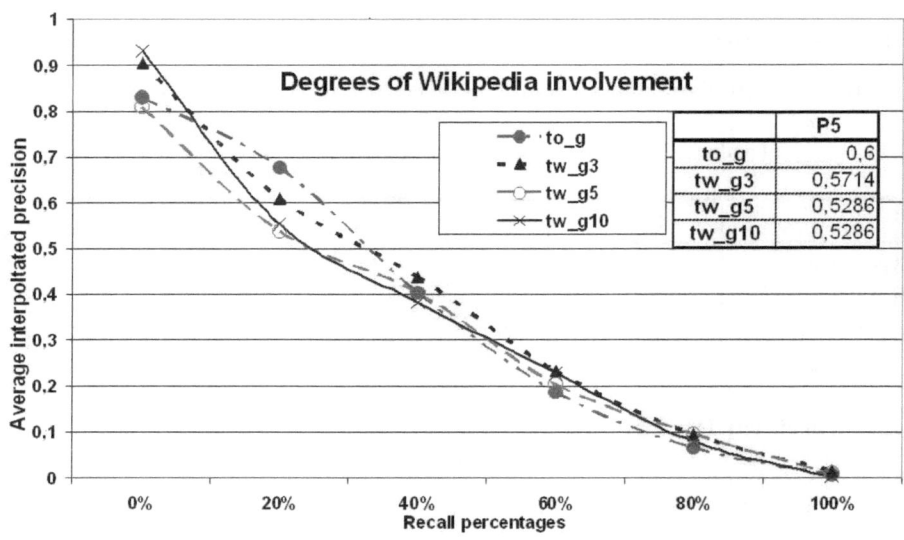

(a) Each book regarded as a large chunk of text

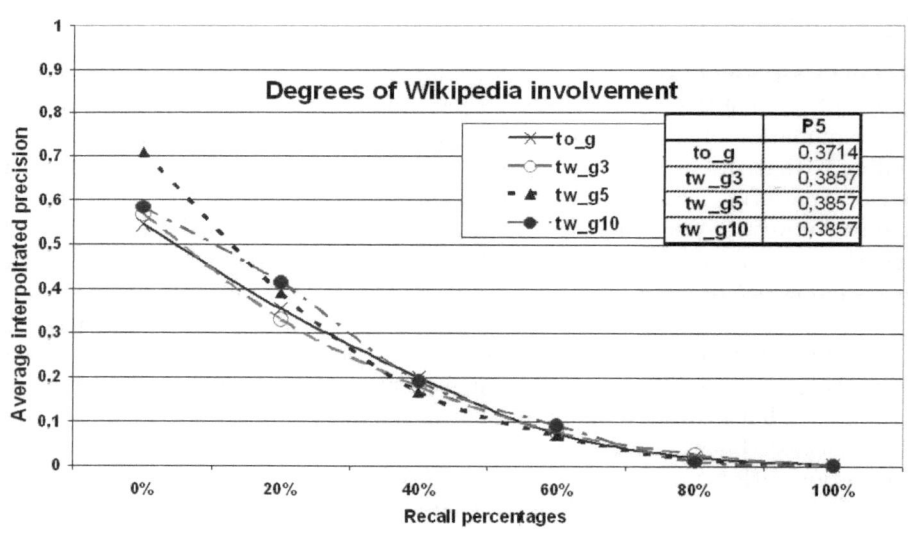

(b) Each book regarded as a combination of pages

Fig. 2. Precision-recall curves of generic book retrieval. Precision at 5 retrieved documents is given for all plotted runs.

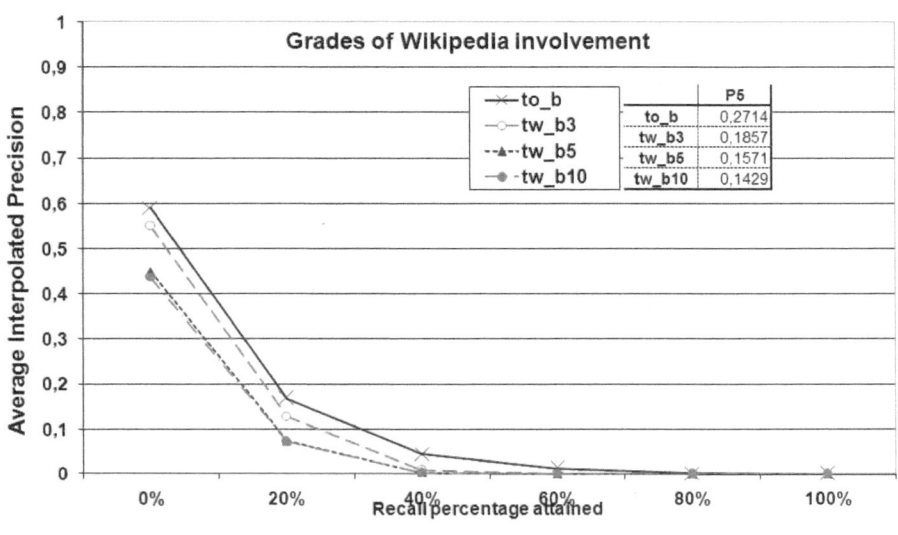

(a) Each book regarded as a large chunk of text

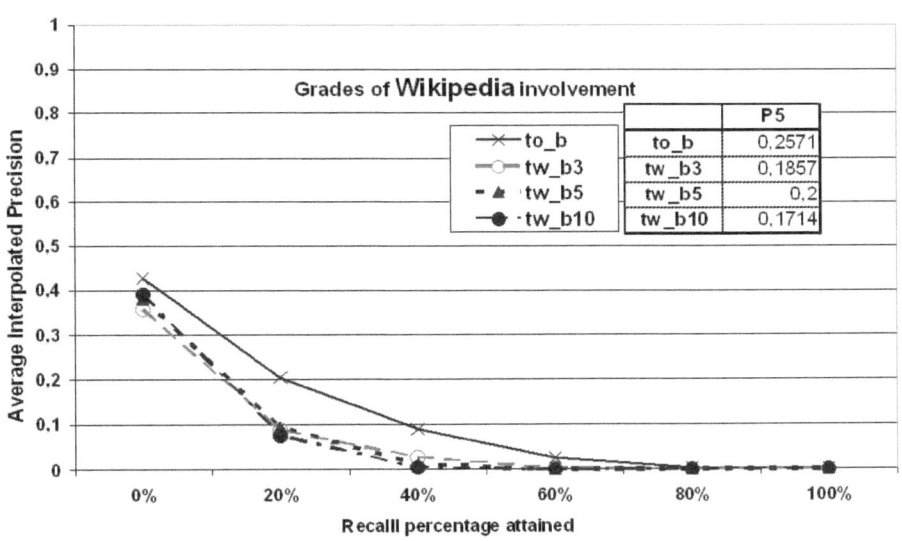

(b) Each book regarded as a combination of pages

Fig. 3. Precision-recall curves of book-specific (structure supported) book retrieval. Precision at 5 retrieved documents is given for all plotted runs.

Also here, looking at a book as a long chunk of text is more beneficial for the results than combining the results from each page. As the reviewer pointed out, both here and in the previous group of results there seems to be a contradiction between our results and those of [3], as the latter observed increased performance when combining (merging) pages. Here it is important to point out that the method of merging pages differs for the two contributions, as we combine all the pages in the books when ranking, while [3] retrieves the 3000 pages with highest score, ranks a book by the proportions of these pages belonging to it. This latter ranking method to a lesser extent takes into account the contribution of non-relevant pages.

With regard to the questions we posed in the introduction, the results seem to cast doubt on whether Wikipedia terms can substitute manually assigned terms as in [3]'s setting, as Wikipedia terms seem to have a different effect in our experiments than manually assigned terms had in the former, in both main retrieval modes. A general observation that we make about adding Wikipedia terms to a query is that the positive effect, if present, seems to be exhausted beyond a relatively short window (in our experiments 5 terms). this indicates that such query expansion should be done with caution.

Even if the potential is there, we are doubtful whether the best results attained in the current setting actually indicates better experienced retrieval on the side of the user. More research and experimentation will be needed to establish that.

8   Conclusion

The results we obtain indicate that Wikipedia article texts do have potential as retrieval aid for digitized books that are topically relevant for the subject of the articles. There is, however, little doubt that more experiments in laboratory conditions along this line of research, as well as conditions resembling real life usage of Wikipedia in combination with digitized books, will be necessary in order to approach combinations and query expansion settings that will be beneficial in real life situations. It is doubtful whether the best results we get in the current experiments also predict the user-experienced retrieval performance.

Trying to find book-specific (structure supported) methods for book retrieval, the page as a unit seems to have some disadvantages. The page is quite a little partition, and, in addition page breaks are often a consequence of physical attributes of the book rather than a conscious or structural choice taken by the author. It would be interesting to repeat our book-specific experiments with larger, structurally more coherent partitions of books.

References

1. Kazai, G., Koolen, M., Landoni, M.: Summary of the book track. In: Geva, S., Kamps, J., Trotman, A. (eds.) INEX 2009. LNCS, vol. 6203, pp. 145–159. Springer, Heidelberg (2010)
2. Luyt, B., Tan, D.: Improving wikipedia's credibility: References and citations in a sample of history articles. Journal of the American Society for Information Science and Technology 43(3) (2010)

3. Wu, M., Scholer, F., Thom, J.A.: The impact of query length and document length on book search effectiveness. In: Geva, S., Kamps, J., Trotman, A. (eds.) INEX 2008. LNCS, vol. 5631, pp. 172–178. Springer, Heidelberg (2009)
4. unknown (2010),
   http://en.wikipedia.org/wiki/Wikipedia:Policies_and_guidelines (retrieved March 5, 2010)
5. Kazai, G., Koolen, M.: Inex book search topics for 2009 (2009)
6. Chowdhury, A., McCabe, M.C.: Improving information retrieval systems using part of speech tagging. Technical report, ISR, Institute for Systems Research (1998)

Overview of the INEX 2009 Efficiency Track

Ralf Schenkel[1,2] and Martin Theobald[2]

[1] Saarland University, Saarbrücken, Germany
[2] Max Planck Institute for Informatics, Saarbrücken, Germany

Abstract. This paper presents an overview of the Efficiency Track that was run
for the second time in 2009. This track is intended to provide a common forum for
the evaluation of both the effectiveness and efficiency of XML ranked retrieval
approaches on *real data* and *real queries*. The Efficiency Track significantly extends the Ad-Hoc Track by systematically investigating different types of queries
and retrieval scenarios, such as classic ad-hoc search, high-dimensional query expansion settings, and queries with a deeply nested structure (with all topics being
available in both the NEXI-style CO and CAS formulations, as well as in their
XPath 2.0 Full-Text counterparts). The track received 68 runs submitted by 4 participating groups using 5 different systems. The best systems achieved interactive
retrieval times for ad-hoc search, with a result quality comparable to the best runs
in the AdHoc track.

1 Introduction

The Efficiency Track was run for the second time in 2009, with its first incarnation at
INEX 2008 [2]. It is intended to provide a common forum for the evaluation of both
the effectiveness and efficiency of XML ranked retrieval approaches on *real data* and
real queries. The Efficiency Track significantly extends the Ad-Hoc Track by systematically investigating different types of queries and retrieval scenarios, such as classic
ad-hoc search, high-dimensional query expansion settings, and queries with a deeply
nested structure (with all topics being available in both the NEXI-style CO and CAS
formulations, as well as in their XPath 2.0 Full-Text counterparts).

2 General Setting

2.1 Test Collection

The Efficiency Track uses the INEX-Wikipedia collection[1] that has been introduced in
2009, an XML version of English Wikipedia articles with semantic annotations. With
almost 2.7 million articles, more than a billion elements and an uncompressed size of
approximately 50 GB, this collection is a lot larger than the old Wikipedia collection
used in previous years (and for last year's Efficiency track). The collection has an ir-
regular structure with many deeply nested paths, which turned out to be challenging for
most systems.

---

[1] Available from
http://www.mpi-inf.mpg.de/departments/d5/software/inex/

S. Geva, J. Kamps, and A. Trotman (Eds.): INEX 2009, LNCS 6203, pp. 200–212, 2010.

2.2   Topic Types

One of the main goals to distinguish the Efficiency Track from traditional Ad-Hoc retrieval is to cover a broader range of query types than the typical NEXI-style CO or CAS queries, which are mostly using either none or only very little structural information and only a few keywords over the target element of the query. Thus, two natural extensions are to extend Ad-Hoc queries with high-dimensional query expansions and/or to increase the amount of structural query conditions without sacrificing the IR aspects in processing these queries (with topic `description` and `narrative` fields providing hints for the human assessors or allowing for more semi-automatic query expansion settings, see Figure 1). The Efficiency Track focuses on the following types of queries (also coined "topics" in good IR tradition), each representing different retrieval challenges:

- **Type (A) Topics:** 115 topics (ids 2009-Eff-001—2009-Eff-115) were taken over from the Ad-hoc Track. These topics represent classic, Ad-Hoc-style, focused passage or element retrieval, with a combination of NEXI CO and CAS queries.
- **Type (B) Topics:** Another 115 topics (ids 2009-Eff-116—2009-Eff-230) were generated by running Rocchio-based blind feedback on the results of the article-only AdHoc reference run. These CO topics are intended to simulate high-dimensional query expansion settings with up to 101 keywords, which cannot be evaluated in a conjunctive manner and are expected to pose a major challenge to any kind of search engine. Relevance assessments for these topics can be taken over from the corresponding adhoc topics; a reference run (using TopX2) with the expanded topics was submitted to the adhoc track to make sure that results with the expanded topics were also present in the result pools.
- **Type (C) Topics:** These topics were planned to represent high-dimensional, structure-oriented retrieval settings over a DB-style set of CAS queries, with deeply nested structure but only a few keyword conditions. The focus of the evaluation should have been on execution times. Unfortunately, we did not get any proposals for type (C) topics by the track participants.

2.3   Topic Format

The Efficiency Track used an extension of the topic format of the Adhoc Track. All adhoc fields were identical to the corresponding adhoc topics for type (A) topics. For the type (B) topics, the expanded keyword queries were put into the `title` field and corresponding NEXI queries of the form `//*[keyword1, keyword2, ...]` were generated for the `castitle` field. All topics contained an additional `xpath_title` field with an XPath FullText expression that should be equivalent to the NEXI expression in the `castitle` field. This Xpath FT expression was automatically created from the NEXI expression by replacing `about()` predicates with corresponding `ftcontains()` expressions and connecting multiple keywords within such an expression by `ftand`. Figure 1 shows the representation of topic 2007-Eff-0578 as an example for a type (A) topic. After releasing the topics it turned out that `ftand` enforces a conjunctive evaluation of the predicate, which was not the original intension and especially for the type (B) topics would lead to many topics with empty result set. The topics were therefore updated and all occurrences of ftand replaced by the less strict ftor.

```
<topic ct_no="242" id="2009-Eff-057" type="A">
  <title>movie Slumdog Millionaire directed by Danny Boyle</title>
  <castitle>
    //movie[about(.,"Slumdog Millionaire")]//director[about(.,"Danny Boyle")]
  </castitle>
  <xpath_title>
    //movie[. ftcontains ("Slumdog Millionaire")]
    //director[. ftcontains ("Danny Boyle")]
  </xpath_title>
  <phrasetitle> "Slumdog Millionaire" "Danny Boyle" </phrasetitle>
  <description>
    Retrieve information about the movie Slumdog Millionaire
    directed by Danny Boyle.
  </description>
  <narrative>
    The relevant texts must contain information on: the movie, the awards
    it has got or about the casts and crew of the movie. The criticisms of
    the movie are relevant as well. The other movies directed by Danny Boyle
    are not relevant here. Information about the making of the movie and
    about the original novel and its author is also relevant. Passages or
    elements about other movies with the name "millionaire" as a part of
    them are irrelevant. Information about the game show "Who wants to be a
    millionaire" or any other related game shows is irrelevant.
  </narrative>
</topic>
```

Fig. 1. Example topic (2009-Eff-057)

2.4 Tasks

Adhoc Task. The Efficiency Track particularly encourages the use of top-k style query engines. In the AdHoc task, participants were asked to create top-15, top-150, and top-1500 results with their systems and to report runtimes, using the different title field, including the NEXI CO, CAS, or XPATH titles, or additional keywords from the narrative or description fields. Following the INEX AdHoc Track, runs could be sumbitted in either Focused (i.e., non-overlapping), Thorough (incl. overlap), or Article retrieval mode:

- **Article:** Here, results are always at the article level, so systems may consider the XML or the plain text version of documents. Results are always free of overlap by definition.
- **Thorough:** The Thorough mode represents the original element-level retrieval mode used in INEX 2003-2005. Here, any element identified as relevant should be returned. Since removing overlap may mean a substantial burden for a system, this setting intentionally allows overlapping results, so query processing times can be clearly distinguished from the time needed to remove overlapping results.
- **Focused:** Focused (i.e., overlap-free) element and/or passage retrieval typically is favorable from a user point-of-view and therefore replaced the Thorough retrieval as primary retrieval mode in the Ad-hoc Track in 2006. Here, the reported runtimes include the time needed to remove overlap, which may give rise to interesting comparisons between systems following both Thorough and Focused retrieval strategies.

Budget-Constrained Task. This novel task in 2009 asked participants to retrieve results within a fixed budget of runtime (one of 10ms, 100ms, 1000ms and 10000ms), simulating interactive retrieval situations. Standard top-k algorithms cannot easily be used with such a constraint or may return arbitrarily bad results. However, we did not get any submissions for this task, probably because it required specific modifications to the systems which was too much effort for the participants.

3 Run Submissions

The submission format for all Efficiency Track submissions was defined by a DTD similar to the DTD used by the INEX AdHoc track up to 2008, but extended by fields to report efficiency statistics. The DTD is depicted in Appendix A; the different fields have the following meanings:

- Each *run* submission *must* contain the following information:
    - `participant-id` - the INEX participant id
    - `run-id` - your run id
    - `task` - either `adhoc` or one of the budget-constrained tasks
    - `type` - either `focused`, `thorough`, or `article`
    - `query` - either automatic or manual mode
    - `sequential` - queries being processed sequentially or in parallel (independent of whether distribution is used)

- Furthermore, each *run* submission *should* contain some basic system and retrieval statistics:
    - `no_cpu` - the number of CPUs (cores) in the system (sum over all nodes for a distributed system)
    - `ram` - the amount of RAM in the system in GB (sum over all nodes for a distributed system)
    - `no_nodes` - the number of nodes in a cluster (only for a distributed system)
    - `hardware_cost` - estimated hardware cost
    - `hardware_year` - date of purchase of the hardware
    - `topk` - top-k run or not (if it is a top-k run, there may be at most k elements per topic returned)
    - `index_size_bytes` - the overall index size in bytes
    - `indexing_time_sec` - the indexing time in seconds to create the indexes used for this run

- Each *run* submission *should* also contain the following brief system descriptions (keywords), if available:
    - `general_description` - a general system and run description
    - `ranking_description` - the ranking strategies used
    - `indexing_description` - the indexing structures used
    - `caching_description` - the caching hierarchies used

– Each *topic* element in a run submission *must* contain the following elements:
  - `topic_id` - the id of the topic
  - `total_time_ms` - the total processing time in milliseconds: this should include the time for parsing and processing the query but does not have to consider the extraction of resulting file names or element paths (needed to create the above format for the run submission)

– Furthermore, each *topic* element of a run submission *should* contain the following elements:
  - `cpu_time_ms` - the CPU time spent on processing the query in milliseconds
  - `io_time_ms` - the total I/O time spent on physical disk accesses in milliseconds
  - `io_bytes` - the number of I/O bytes needed for processing the query. For a distributed system, this should contain the entire amount of bytes spent on network communication.

Particularly interesting for the Efficiency Track submissions is the `runtime` field, of course. This can optionally be split into `cpu_time` and `io_time`, which has been done only by a single participant. We therefore focus on actual wallclock running times as efficiency measure.

4   Metrics

To assess the quality of the retrieved results, the Efficiency Track applied the same metrics and tools used in the Ad-Hoc track. Runs were evaluated with the interpolated precision metric [1]. Like the Adhock Track, we are mainly interested in early precision, so we focus on iP[0.01], but also report mean average iP values (MAiP). All runs were first converted to the submission format of the AdHoc Track and then evaluated with the standard tools from that track.

5   Participants

An overall amount of 68 runs was submitted by 4 participating groups using 5 different systems. Here are short system descriptions submitted by the participants.

Max-Planck-Institut Informatik – TopX 2.0 [Part.ID 10]. The TopX 2.0 system provides a compressed object-oriented storage for text-centric XML data with direct access to customized inverted files and C++-based implementation of a structure-aware top-k algorithm.

Max-Planck-Institut Informatik – Proximity-Enhanced Top-K [Part.ID 10]. Following our work on proximity-based XML retrieval at INEX 2008, we developed a proximity-aware indexing framework at the article level that uses inverted lists for both terms and term pairs and includes an automated pruning step that cuts the size of these lists to a very short prefix with the highest-scoring entries. We evaluate queries through a merge join of the corresponding term and term pair lists, yielding low CPU overhead and fast answer times.

University of Frankfurt – Spirix [Part.ID 16]. Spirix is a Peer-to-Peer (P2P) search engine for Information Retrieval of XML documents. The underlying P2P protocol is based on a Distributed Hash Table (DHT). Due to the distributed architecture of the system, efficiency aspects have to be considered in order to minimize bandwidth consumption and communication overhead. Spirix is a top-k search engine aiming at efficient selection of posting lists and postings by considering structural information, e.g. taking advantage of CAS queries. As collections in P2P systems are usually quite heterogeneous, no underlying schema is assumed but schema-mapping methods are of interest to detect structural similarity.

University of Otago [Part.ID 4]. We submitted whole document runs. We examined two parameters. One was dynamic pruning of tf-ordered postings lists - this effects the time taken to compute the BM25 score for each document. Computed rsvs were held in an array of accumulators, and the second parameter affected the sorting of this array. An efficient select-top-k-from-n algorithm was used, the second parameter was k.

University of Konstanz – BaseX [Part.ID 304]. BaseX is a native XML database and XPath/XQuery processor, including support for the latest XQuery Full Text recommendation. As we put our main focus on efficiency and generic evaluation of all types of XQuery requests and input documents, our scoring model is based on a classical TF/IDF implementation. Additional scoring calculations are performed by XQFT (ftand, ftor, ftnot) and XQuery operators (union, location steps, ...). A higher ranking is given to those text nodes which are closer to the location steps of the input query than others. We decided to stick with conjunctive query evaluation (using 'ftand' instead of 'ftor' in the proposed topic queries), as a change to the disjunctive mode would have led to too many changes, which could not have been reasonably implemented in the remaining time frame. Next, we decided to not extend the proposed queries with stemming, stop words or thesaurus options. As a consequence, many queries might return less results than the TopX reference engine (and sometimes no results at all). To give a realistic picture, we have included both the total time for accessing indexes as well as traversing the inverted specified location paths in our final performance results.

6   Results

Table 1 summarizes important system parameters as they were delivered in the runs' headers, grouping runs with similar id prefixes and settings. The majority of all runs was submitted for the Focused or Article subtask, only four runs were submitted for the Thorough subtask. Detailed effectiveness (iP, MAiP) and efficiency (average of wall-clock runtimes, cpu and io times in milliseconds) results of all submitted runs can be found in Appendix B and C.

As we received only four Thorough runs (all from the same participant), we consider only the effectiveness of these runs. Figure 2 shows the precision-recall chart for these Thorough runs and, as a reference, the plot for the (focused) run Eff-20 which has the highest Thorough MAiP among all Efficiency (and in fact also AdHoc) runs. It is evident from the chart that there is a significant gap between the four submitted runs and the best Focused run.

Table 1. Run parameters as taken from the submission headers

Part.ID	Run prefix	#CPU	RAM	#Nodes	Hardw.Cost	Year
4	Eff	8	8	1	3000 NZD	2008
10	MPI-eff	8	32	1	3000 EUR	2008
10	TopX2-09	4	16	1	5000 EUR	2005
16	Spirix			no information		
304	BaseX	2	32		no information	

Thorough evaluation for type (A) topics

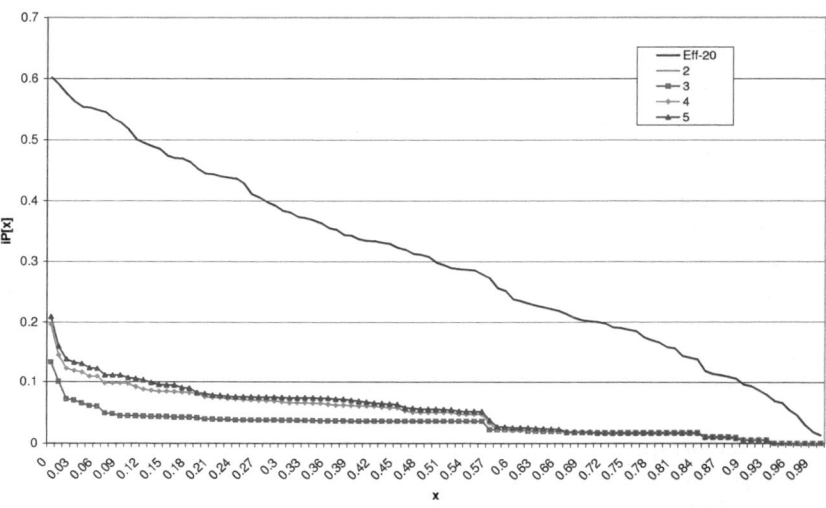

Fig. 2. Precision/recall plots for thorough runs, type (A) topics

The Efficiency Track aims at investigating the trade-off of efficiency and effectiveness, so we depict the results for Focused and Article runs in charts that show, for each run, its efficiency (measured as runtime) and its effectiveness (measured through the official measure iP[0.01] or MAiP). Figures 3 and 4 depict these plots for the type (A) topics, where each run is labeled by the system that generated it. Regarding efficiency, average running times per topic varied from 8.8 ms to 50 seconds. It is important to notice that absolute runtimes across systems are hardly comparable due to differences in the hardware setup and caching. Both the Otago runs and the MPI-prox runs clearly show that the dominant part of retrieval time is spent in IO activity, so improving or reducing IO access could be a promising way to improve efficiency. Most runs (with the exception of a few TopX2 runs) were article-only runs, and like last year these runs generally yielded very good efficiency results. Only TopX2 generated element-level results using the CAS NEXI queries, and it is evident that the additional structure in the queries increases processing time. Spirix used the CAS NEXI queries as well, but returned only article-level results. Overall effectiveness results were generally comparable to the Ad-hoc Track, with the best runs achieving a MAiP value of 0.301 and interpolated (early) precision values of 0.589 at 1% recall (iP[0.01]) and 0.517 at 10% recall (iP[0.10]), respectively.

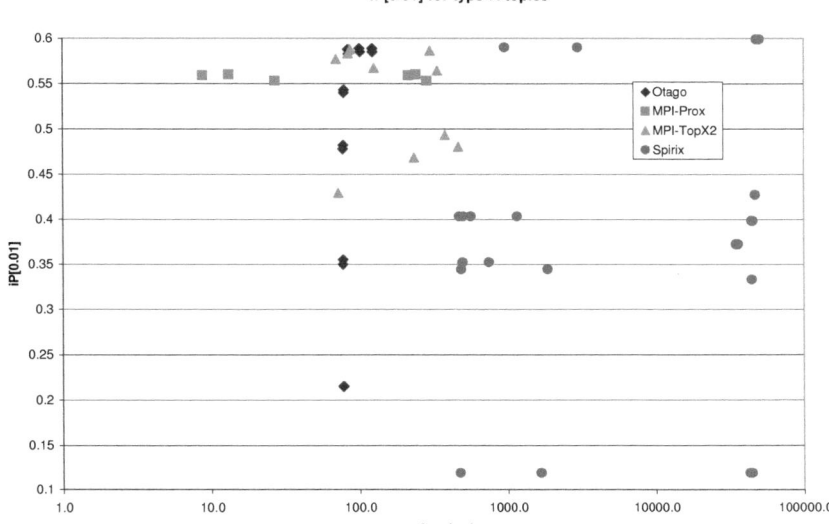

Fig. 3. Runtime vs. early precision plots for Focused and Article runs, type (A) topics

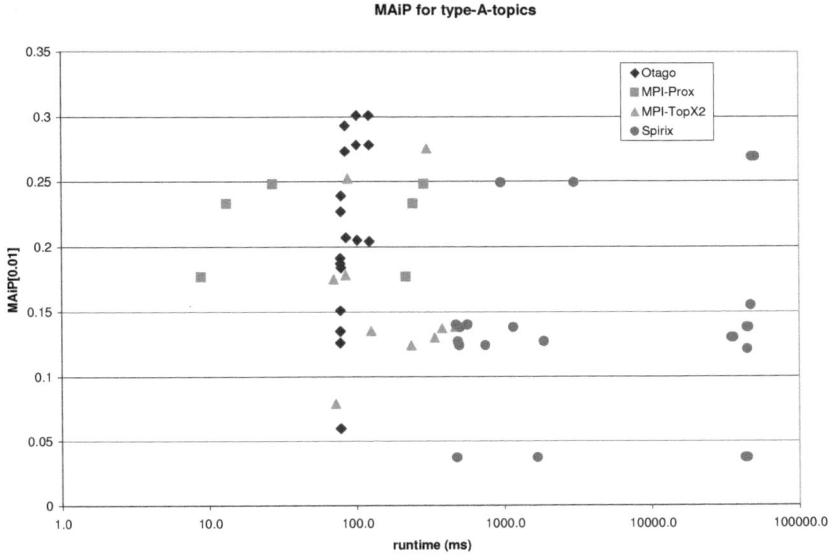

Fig. 4. Runtime vs. MAiP plots for Focused and Article runs, type (A) topics

Figures 5 and 6 show similar charts for the type (B) topics. Here, average running times per topic varied from 367.4 ms to 250 seconds, which is much larger than for the type (A) topics. This is clearly caused by the much higher number of keywords in these

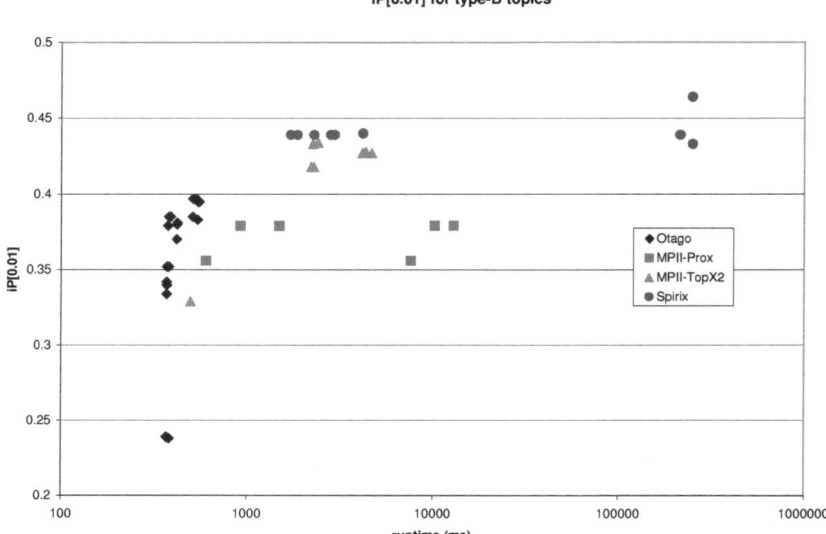

Fig. 5. Runtime vs. early precision plots for Focused and Article runs, type (B) topics

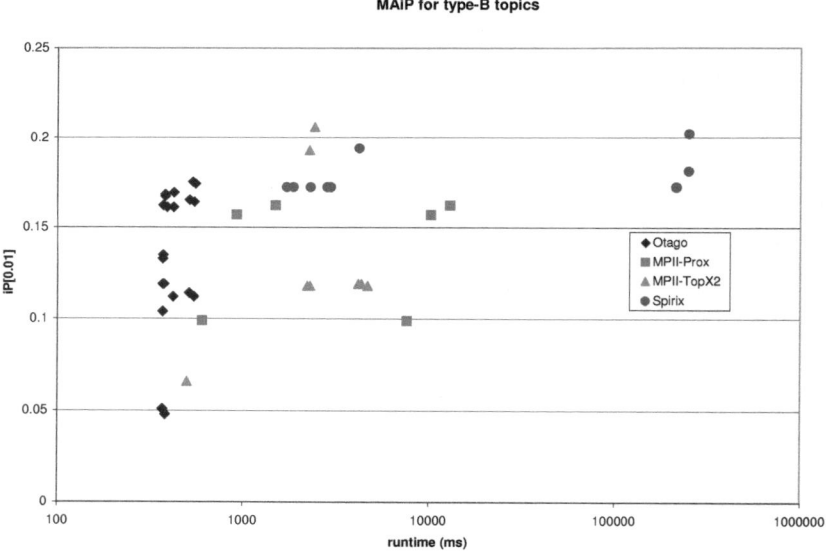

Fig. 6. Runtime vs. MAiP plots for Focused and Article runs, type (B) topics

topics. Result quality on the type (B) topics was generally slightly worse compared to the type (A) topics, which was probably caused by the extreme query expansion used to generate them, leading to typical problems such as topic drift. This effect is not simply caused by missing assessments: Article-level results for type (B) topics from TopX2 were submitted to the AdHoc track (using their original query ids), and iP[0.01] dropped from 0.6090 (rank 5) to 0.4593 (rank 42) with otherwise unchanged settings. A possible explanation for this is that the original queries are already quite verbose, and there is limited information redundancy in the Wikipedia encyclopedia, hence blind feedback may have introduced some overfitting here.

7   Conclusions

This paper gave an overview of the INEX 2009 Efficiency Track that provided a platform for comparing retrieval efficiency of different systems. The Efficiency Track has demonstrated that a number of systems are able to achieve very good result quality within very short processing times that allow for interactive retrieval, at least for the short type (A) topics. The majority of the submitted runs retrieved only article-level results, using the provided content-and-structure queries merely as hints for finding relevant articles. For the first time, runs generated with an XQuery FullText system were submitted.

Given the interactive retrieval times achieved by a number of systems, the efficiency problem for article-level retrieval on the INEX Wikipedia collection seems to be solved. A future Efficiency Track needs to introduce new challenges, which could come from a much larger collection (such as ClueWeb), from more complex queries with more structural conditions, or from a combination of both.

References

1. Kamps, J., Pehcevski, J., Kazai, G., Lalmas, M., Robertson, S.: INEX 2007 Evaluation Measures. In: Fuhr, N., Kamps, J., Lalmas, M., Trotman, A. (eds.) INEX 2007. LNCS, vol. 4862, pp. 24–33. Springer, Heidelberg (2008)
2. Theobald, M., Schenkel, R.: Overview of the inex 2008 efficiency track. In: Geva, S., Kamps, J., Trotman, A. (eds.) INEX 2008. LNCS, vol. 5631, pp. 179–191. Springer, Heidelberg (2009)

A   DTD for Efficiency Track Submissions

```
<!ELEMENT efficiency-submission (topic-fields,
                                general_description,
                                ranking_description,
                                indexing_description,
                                caching_description,
                                topic+)>
<!ATTLIST efficiency-submission
  participant-id CDATA #REQUIRED
  run-id         CDATA #REQUIRED
  task           (adhoc | budget10 | | budget100 | budget1000 | budget10000) #REQUIRED
  type           (focused | thorough | article) #REQUIRED
  query          (automatic | manual) #REQUIRED
  sequential     (yes|no) #REQUIRED
```

```
    no_cpu          CDATA #IMPLIED
    ram             CDATA #IMPLIED
    no_nodes        CDATA #IMPLIED
    hardware_cost   CDATA #IMPLIED
    hardware_year   CDATA #IMPLIED
    topk            (15 | 150 | 1500) #IMPLIED
    index_size_bytes CDATA #IMPLIED
    indexing_time_sec CDATA #IMPLIED
>
<!ELEMENT topic-fields EMPTY>
<!ATTLIST topic-fields
    co_title        (yes|no) #REQUIRED
    cas_title       (yes|no) #REQUIRED
    xpath_title     (yes|no) #REQUIRED
    text_predicates (yes|no) #REQUIRED
    description     (yes|no) #REQUIRED
    narrative       (yes|no) #REQUIRED
>
<!ELEMENT general_description  (#PCDATA)>
<!ELEMENT ranking_description  (#PCDATA)>
<!ELEMENT indexing_description (#PCDATA)>
<!ELEMENT caching_description  (#PCDATA)>
<!ELEMENT topic (result*)>
<!ATTLIST topic
    topic-id        CDATA #REQUIRED
    total_time_ms   CDATA #REQUIRED
    cpu_time_ms     CDATA #IMPLIED
    io_time_ms      CDATA #IMPLIED
    io_bytes        CDATA #IMPLIED
>
<!ELEMENT result (file, path, rank, rsv?)>
<!ELEMENT file   (#PCDATA)>
<!ELEMENT path   (#PCDATA)>
<!ELEMENT rank   (#PCDATA)>
<!ELEMENT rsv    (#PCDATA)>
```

B   Performance Summaries for All Runs for Type (A) Topics

Part.ID	Run ID	T	iP[0.00]	iP[0.01]	iP[0.05]	iP[0.10]	MAiP	avg total [ms]	avg CPU [ms]	avg IO [ms]	k	query
					Focused/Article							
4	Eff-1	A	0.220	0.215	0.177	0.141	0.060	77.8	20.4	57.3	15	CO
4	Eff-2	A	0.350	0.350	0.315	0.285	0.126	77.1	20.4	56.7	15	CO
4	Eff-3	A	0.480	0.478	0.433	0.359	0.151	77.2	20.0	57.2	15	CO
4	Eff-4	A	0.549	0.540	0.500	0.443	0.184	78.1	21.0	57.1	15	CO
4	Eff-5	A	0.597	0.583	0.532	0.468	0.207	84.5	26.9	57.6	15	CO
4	Eff-6	A	0.598	0.585	0.532	0.475	0.205	101.3	43.0	58.2	15	CO
4	Eff-7	A	0.598	0.585	0.532	0.475	0.204	122.2	64.9	57.3	15	CO
4	Eff-8	A	0.220	0.215	0.177	0.142	0.060	77.7	20.4	57.3	150	CO
4	Eff-9	A	0.355	0.355	0.328	0.305	0.135	76.9	19.9	57.0	150	CO
4	Eff-10	A	0.484	0.482	0.438	0.377	0.187	77.4	20.3	57.2	150	CO
4	Eff-11	A	0.552	0.543	0.513	0.474	0.227	78.3	21.3	57.0	150	CO
4	Eff-12	A	0.600	0.588	0.553	0.510	0.273	83.8	26.9	56.9	150	CO
4	Eff-13	A	0.602	0.589	0.553	0.517	0.278	100.0	42.7	57.3	150	CO
4	Eff-14	A	0.602	0.589	0.553	0.517	0.278	122.2	64.9	57.3	150	CO
4	Eff-15	A	0.220	0.215	0.177	0.142	0.060	76.9	20.3	56.6	1500	CO
4	Eff-16	A	0.355	0.355	0.328	0.305	0.135	77.1	20.2	56.9	1500	CO
4	Eff-17	A	0.484	0.482	0.438	0.377	0.191	77.4	20.1	57.3	1500	CO
4	Eff-18	A	0.552	0.543	0.513	0.474	0.239	78.5	20.9	57.6	1500	CO
4	Eff-19	A	0.600	0.588	0.553	0.510	0.293	83.6	26.8	56.9	1500	CO
4	Eff-20	A	0.602	0.589	0.553	0.517	0.301	100.3	42.7	57.6	1500	CO
4	Eff-21	A	0.602	0.589	0.553	0.517	0.301	121.7	64.3	57.4	1500	CO
10	MPI-eff-1500-1810	A	0.566	0.553	0.532	0.464	0.248	27.1			1500	CO
10	MPI-eff-1500-1810-cold	A	0.566	0.553	0.532	0.464	0.248	287.0			1500	CO
10	MPI-eff-150-610	A	0.574	0.560	0.531	0.466	0.233	13.2			150	CO
10	MPI-eff-150-610-cold	A	0.574	0.560	0.531	0.466	0.233	242.5			150	CO
10	MPI-eff-15-210	A	0.575	0.559	0.511	0.400	0.177	8.8			15	CO
10	MPI-eff-15-210-cold	A	0.575	0.559	0.511	0.400	0.177	216.5			15	CO
10	TopX2-09-Ar-Fo-15-Hot	A	0.598	0.583	0.494	0.397	0.178	84.0			15	CO
10	TopX2-09-ArHeu-Fo-1500-Hot	A	0.597	0.586	0.530	0.475	0.275	301.2			1500	CO
10	TopX2-09-ArHeu-Fo-150-Hot	A	0.598	0.588	0.531	0.474	0.252	87.2			150	CO
10	TopX2-09-ArHeu-Fo-15-Hot	A	0.589	0.577	0.482	0.398	0.175	69.8			15	CO
10	TopX2-09-CAS-Fo-15-Cold	F	0.546	0.480	0.423	0.355	0.138	467.7			15	CAS
10	TopX2-09-CAS-Fo-15-Hot	F	0.545	0.493	0.418	0.350	0.137	379.2			15	CAS
10	TopX2-09-CASHeu-Fo-15-Hot	F	0.525	0.468	0.358	0.304	0.124	234.9			15	CAS
10	TopX2-09-CO$-Fo-15-Hot	F	0.645	0.567	0.406	0.285	0.135	125.6			15	CAS
10	TopX2-09-CO-Fo-15-Hot	F	0.641	0.564	0.405	0.291	0.130	338.2			15	CAS
10	TopX2-09-COHeu-Fo-15-Hot	F	0.507	0.429	0.306	0.196	0.079	71.8			15	CAS
16	Spirix09RX01	A	0.621	0.599	0.551	0.499	0.269	50063.5			1500	CAS
16	Spirix09RX02	A	0.435	0.427	0.379	0.319	0.155	46820.5			150	CAS
16	Spirix09RX03	A	0.398	0.372	0.335	0.293	0.130	34663.6			150	CAS
16	Spirix09RX03	A	0.398	0.372	0.335	0.293	0.130	34663.6			150	CAS
16	Spirix09RX04	A	0.402	0.398	0.370	0.295	0.138	44191.5			150	CAS
16	Spirix09RX05	A	0.368	0.333	0.311	0.268	0.121	44352.2			150	CAS
16	Spirix09RX06	A	0.119	0.119	0.119	0.107	0.037	42878.4			150	CAS
16	Spirix09RX07	A	0.604	0.590	0.535	0.497	0.249	959.8			1500	CAS
16	Spirix09RX08	A	0.405	0.403	0.369	0.309	0.140	563.4			150	CAS
16	Spirix09RX09	A	0.354	0.352	0.334	0.283	0.124	496.2			150	CAS
16	Spirix09RX10	A	0.405	0.403	0.386	0.313	0.138	502.6			150	CAS
16	Spirix09RX11	A	0.354	0.344	0.330	0.278	0.127	483.9			150	CAS
16	Spirix09RX12	A	0.119	0.119	0.118	0.099	0.037	474.9			150	CAS
16	Spirix09RX13	A	0.604	0.590	0.535	0.497	0.249	2986.3			1500	CAS
16	Spirix09RX14	A	0.405	0.403	0.369	0.309	0.140	470.5			150	CAS
16	Spirix09RX15	A	0.354	0.352	0.334	0.283	0.124	746.5			150	CAS
16	Spirix09RX16	A	0.405	0.403	0.386	0.313	0.138	1156.6			150	CAS
16	Spirix09RX17	A	0.354	0.344	0.330	0.278	0.127	1863.0			150	CAS
16	Spirix09RX18	A	0.119	0.119	0.118	0.099	0.037	1675.5			150	CAS
16	Spirix09RX19	A	0.621	0.599	0.551	0.499	0.269	47857.1			1500	CAS
16	Spirix09RX20	A	0.435	0.427	0.379	0.319	0.155	46712.3			150	CAS
16	Spirix09RX21	A	0.398	0.372	0.335	0.293	0.130	35746.8			150	CAS
16	Spirix09RX22	A	0.402	0.398	0.370	0.295	0.138	45072.0			150	CAS
16	Spirix09RX23	A	0.368	0.333	0.311	0.268	0.121	44285.8			150	CAS
16	Spirix09RX24	A	0.119	0.119	0.119	0.107	0.037	44256.9			150	CAS
					Thorough							
304	2	T	0.133	0.101	0.061	0.045	0.032	11504.4			1500	XPath
304	3	T	0.133	0.101	0.061	0.045	0.032	2553.3			1500	XPath
304	4	T	0.197	0.144	0.109	0.097	0.049	2510.0			1500	XPath
304	5	T	0.209	0.160	0.123	0.107	0.054	2726.7			1500	XPath

C   Performance Summaries for All Runs for Type (B) Topics

Part.ID	Run ID	T	iP[0.00]	iP[0.01]	iP[0.05]	iP[0.10]	MAiP	avg total [ms]	avg CPU [ms]	avg IO [ms]	k	query
4	Eff-1	A	0.240	0.238	0.167	0.139	0.048	380.2	32.0	348.3	15	CO
4	Eff-2	A	0.336	0.334	0.322	0.285	0.104	367.5	32.1	335.5	15	CO
4	Eff-3	A	0.350	0.342	0.326	0.293	0.119	367.4	33.4	334.0	15	CO
4	Eff-4	A	0.379	0.379	0.339	0.316	0.119	374.0	41.7	332.2	15	CO
4	Eff-5	A	0.377	0.370	0.334	0.311	0.112	418.0	89.7	328.3	15	CO
4	Eff-6	A	0.392	0.385	0.335	0.298	0.114	511.6	184.9	326.7	15	CO
4	Eff-7	A	0.390	0.383	0.334	0.299	0.112	543.0	217.0	326.0	15	CO
4	Eff-8	A	0.240	0.239	0.179	0.149	0.051	367.2	32.1	335.1	150	CO
4	Eff-9	A	0.340	0.340	0.330	0.295	0.133	367.5	32.1	335.4	150	CO
4	Eff-10	A	0.356	0.352	0.336	0.321	0.162	370.1	33.4	336.7	150	CO
4	Eff-11	A	0.385	0.385	0.350	0.340	0.161	387.6	42.0	345.6	150	CO
4	Eff-12	A	0.386	0.380	0.335	0.335	0.161	419.4	90.0	329.4	150	CO
4	Eff-13	A	0.403	0.397	0.357	0.331	0.165	512.5	185.1	327.5	150	CO
4	Eff-14	A	0.401	0.395	0.356	0.330	0.164	543.5	216.6	326.9	150	CO
4	Eff-15	A	0.240	0.239	0.179	0.149	0.051	368.3	31.8	336.5	1500	CO
4	Eff-16	A	0.340	0.340	0.330	0.296	0.135	369.5	32.3	337.1	1500	CO
4	Eff-17	A	0.356	0.352	0.336	0.321	0.167	378.7	33.3	345.5	1500	CO
4	Eff-18	A	0.385	0.385	0.350	0.341	0.168	378.2	41.8	336.4	1500	CO
4	Eff-19	A	0.386	0.381	0.354	0.335	0.169	421.8	90.1	331.7	1500	CO
4	Eff-20	A	0.403	0.397	0.357	0.331	0.175	533.3	184.9	348.4	1500	CO
4	Eff-21	A	0.401	0.395	0.356	0.330	0.174	551.8	217.5	334.3	1500	CO
10	MPI-eff-1500-1810	A	0.391	0.379	0.337	0.316	0.162	1492.3			1500	CO
10	MPI-eff-1500-1810-cold	A	0.391	0.379	0.337	0.316	0.162	12979.9			1500	CO
10	MPI-eff-150-610	A	0.391	0.379	0.338	0.315	0.157	922.7			150	CO
10	MPI-eff-150-610-cold	A	0.391	0.379	0.338	0.315	0.157	10235.3			150	CO
10	MPI-eff-15-210	A	0.374	0.356	0.304	0.272	0.099	604.3			15	CO
10	MPI-eff-15-210-cold	A	0.374	0.356	0.304	0.272	0.099	7630.4			15	CO
10	TopX2-09-Ar-TOP15-Hot	A	0.440	0.427	0.362	0.315	0.119	4163.2			15	CO
10	TopX2-09-ArHeu-TOP1500-Hot	A	0.443	0.434	0.381	0.358	0.206	2412.0			1500	CO
10	TopX2-09-ArHeu-TOP150-Hot	A	0.443	0.433	0.381	0.358	0.193	2260.2			150	CO
10	TopX2-09-ArHeu-TOP15-Hot	F	0.431	0.418	0.344	0.303	0.118	2205.4			15	CO
10	TopX2-09-CAS-TOP15-Cold	F	0.442	0.428	0.363	0.316	0.119	4352.3			15	CAS
10	TopX2-09-CAS-TOP15-Hot	F	0.440	0.427	0.357	0.315	0.118	4685.1			15	CAS
10	TopX2-09-CASHeu-TOP15-Hot	F	0.431	0.418	0.344	0.303	0.118	2293.1			15	CAS
10	TopX2-09-COHeu-TOP15-Hot	F	0.364	0.329	0.221	0.175	0.066	497.8			15	CO
16	Spirix09RX13	A	0.450	0.440	0.412	0.382	0.194	4199.2			1500	CAS
16	Spirix09RX14	A	0.456	0.439	0.412	0.365	0.172	1712.1			150	CAS
16	Spirix09RX15	A	0.456	0.439	0.412	0.365	0.172	1859.6			150	CAS
16	Spirix09RX16	A	0.456	0.439	0.412	0.365	0.172	2301.6			150	CAS
16	Spirix09RX17	A	0.456	0.439	0.412	0.365	0.172	2953.0			150	CAS
16	Spirix09RX18	A	0.456	0.439	0.412	0.365	0.172	2823.9			150	CAS
16	Spirix09RX19	A	0.444	0.433	0.406	0.390	0.202	250640.3			1500	CAS
16	Spirix09RX20	A	0.484	0.464	0.421	0.383	0.181	249851.5			150	CAS
16	Spirix09RX21	A	0.456	0.439	0.412	0.365	0.172	215376.4			150	CAS
16	Spirix09RX22	A	0.456	0.439	0.412	0.365	0.172	214315.3			150	CAS
16	Spirix09RX23	A	0.456	0.439	0.412	0.365	0.172	214068.9			150	CAS
16	Spirix09RX24	A	0.456	0.439	0.412	0.365	0.172	215025.8			150	CAS

Index Tuning for Efficient Proximity-Enhanced Query Processing

Andreas Broschart[1,2] and Ralf Schenkel[1,2]

[1] Max-Planck-Institut für Informatik, Saarbrücken, Germany
[2] Saarland University, Saarbrücken, Germany
{abrosch,schenkel}@mpi-inf.mpg.de

Abstract. Scoring models that make use of proximity information usually improve result quality in text retrieval. Considering that index structures carrying proximity information can grow huge in size if they are not pruned, it is helpful to tune indexes towards space requirements and retrieval quality. This paper elaborates on our approach used for INEX 2009 to tune index structures for different choices of result size k. Our best tuned index structures provide the best CPU times for type A queries among the Efficiency Track participants, still providing at least BM25 retrieval quality. Due to the number of query terms, Type B queries cannot be processed equally performant. To allow for comparison as to retrieval quality with non-pruned index structures, we also depict our results from the Adhoc Track.

1 Introduction

Proximity-enhanced scoring models are known to improve result quality in text retrieval. In the last decade a number of scoring models which integrate content and proximity scores have been proposed, among them the scoring model by Büttcher et al. [2] which we use in a modified version for the Adhoc Track. Schenkel et al. [3] found that index structures for proximity can grow prohibitively large, if they are not pruned. As index pruning is a lossy operation, it risks result quality which translates into lower precision values. For our contribution to the Efficiency Track, therefore, we prune our index to tolerable size levels considering precision at the same time.

2 Scoring Model

The scoring model we use in INEX 2009 corresponds to the one used in INEX 2008 [1], this time retrieving article elements only. The score for a document d is computed by a variant of [2] that uses a linear combination of a standard BM25-based score and a proximity score, which is itself computed by plugging the accumulated proximity scores into a BM25-like scoring function:

$$score_{\text{Büttcher}}(d,q) = score_{\text{BM25}}(d,q) + score_{\text{proximity}}(d,q)$$

where b=0.5 and k_1=1.2, respectively in $score_{\text{BM25}}(d,q)$ and

S. Geva, J. Kamps, and A. Trotman (Eds.): INEX 2009, LNCS 6203, pp. 213–217, 2010.

$$score_{\text{proximity}}(d,q) = \sum_{t \in q} min\{1, idf(t)\} \frac{acc(d,t) \cdot (k_1 + 1)}{acc(d,t) + k_1}$$

where $k_1 = 1.2$.

To compute $acc(d,t)$, we consider every query term occurrence. The accumulation function $acc(d,t)$ is defined as

$$acc(d,t) = \sum_{x \in pos(d,t)} \sum_{t' \in q, t' \neq t} idf(t') \underbrace{\sum_{y \in pos(d,t')} \frac{1}{(pos(d,t') - pos(d,t))^2}}_{acc(d,t,t')}$$

where $x \in pos(d,t)$ means that term t occurs at position x, i.e. as x^{th} term in document d.

3   Adhoc Track Results

For our contribution we removed all tags from the XML documents in the Official INEX 2009 collection and worked on their textual content only. The last two runs have been submitted to INEX 2009, the first is the non-submitted baseline:

- MPII-COArBM': a CO run that considered the stemmed terms in the title of a topic (including the terms in phrases, but not their sequence) except terms in negations and stop words. We restricted the collection to the top-level article elements and computed the 1,500 articles with the highest $score_{\text{BM25}}$ value as described in our last year's INEX contribution [1]. Note that this approach corresponds to standard document-level retrieval. This run is the actual non-submitted baseline to enable a comparison to the submitted runs which all use proximity information.
- MPII-COArBP: a CO run which aims to retrieve the 1,500 articles with the highest $score_{\text{BM25}} + score_{\text{proximity}}$ where $score_{\text{proximity}}$ is calculated based on all possible stemmed term pairs in the title of a topic (including the terms in phrases, but not their sequence) except terms in negations and stop words.
- MPII-COArBPP: a CO run which is similar to MPII-COArBP but calculates the $score_{\text{proximity}}$ part based on a selection of stemmed term pairs. Stemmed term pairs are selected as follows: we consider all stemmed tokens in phrases that occur both in the phrasetitle and in the title and are no stop words. The modified phrases in the phrasetitle are considered one at a time to combine term pairs usable to calculate $score_{\text{proximity}}$. If the phrasetitle is empty we use approach MPII-COArBP.

The results in Table 1 show that computing our proximity score with a subset of term pairs based on information taken from the phrasetitles (MPII-COArBPP) doesn't improve the iP values compared to using all term pairs (MPII-COArBP). As expected MPII-COArBP leads to a slight improvement over MPII-COArBM'.

Table 1. Results for the Adhoc Track: interpolated precision at different recall levels (ranks for iP[0.01] are in parentheses) and mean average interpolated precision

run	iP[0.00]	iP[0.01]	iP[0.05]	iP[0.10]	MAiP
MPII-COArBM'	0.5483	0.5398	0.5112	0.4523	0.2392
MPII-COArBP	0.5603	0.5516(26)	0.5361	0.4692	0.2575
MPII-COArBPP	0.5563	0.5477(28)	0.5283	0.4681	0.2566

4   Efficiency Track Results

This section describes our effort in INEX 2009 to tune our index structures for efficient query processing, taking into account the expected retrieval quality and index size.

As in [3] we use

1. text index lists (TL) which contain for each term t, a list of the form $(d, score_{BM25}(d,t))$ and
2. combined index lists (CL) which contain, for each term pair (t_1, t_2), a list of $(d, acc(d, t_1, t_2), score_{BM25}(d, t_1), score_{BM25}(d, t_2))$ entries.

As full, non-pruned indexes will grow huge, we aim at pruning the index structures after a fixed number of entries per list. The final pruned index is used as input to a merge join which avoids overhead costs of threshold algorithms such as book-keeping of candidates.

To measure retrieval quality one usually compares the retrieval results with a set of relevance assessments. As at the time of tuning we didn't have any relevance assessments, for each number of results k (top-15, top-150, and top-1500), we first built up a groundtruth as a substitute. That groundtruth consists of the first k results obtained through processing the non-pruned TLs and CLs. Note that this corresponds to the k highest scoring results of MPII-COArBP.

In our recent studies on the TREC .GOV2 collection we found that it was reasonable to use an overlap of 75% between the top-k documents obtained by query processing on pruned TLs and CLs and the top-k documents of the groundtruth. This is enough to achieve the retrieval quality of non-pruned TLs (i.e., BM25 retrieval quality) using pruned TLs and CLs. (Note that the overlap is computed by the amount of overlapping documents and is not based on the number of characters returned.)

The optimization process is supported by Hadoop, an Open Source MapReduce framework, which distributes the evaluation and indexing workload across a cluster of 10 server nodes in the same network. The optimization proceeds as follows: For all list lengths L ranging between 10 and 20,000 (step size of 100) we estimate the index size first by hashcode based sampling 1% of all terms and term pairs. We restrict the processing of the query load to those that fit the index size constraint set to 100 GB in our experiments. The shortest list length that fulfills the overlap and the index size constraint is considered the optimal list length L_{opt}. We prefer short list lengths, since we process the pruned lists in a merge join which reads the relevant pruned index structures completely.

Figure 1 depicts the pruning process on text as well as combined index lists and the query processing. Unpruned TLs that are ordered by descending $score_{BM25}$

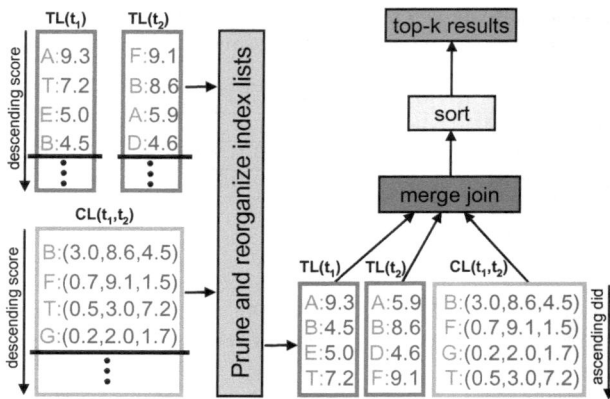

Fig. 1. Pruning TLs and CLs plus query processing

and unpruned CLs that are ordered by descending $acc(d, t_1, t_2)$ score are cut off after L_{opt} entries. They are ordered by ascending did to enable merge join processing of the pruned TLs and CLs. As the results of the merge join are ordered by ascending did, they have to be reordered by aggregated score over all dimensions. Finally, the k results with the highest aggregated score are returned.

Table 2 shows the results of the tuned index structures for type A queries. For performance reasons, tuning was carried out using the type A queries only, type B queries use the same pruned indexes. MPII-eff-k depicts the optimal list lengths for different choices of k, the average cold and warm cache running times and interpolated precision values at different recall levels. While measuring the cold cache running times, we have emptied the cache after each query execution, not just after each batch. To collect the warm cache running times, in a first round we fill the cache by processing the complete query load and measure the running times in the second round. The difference between the cold and warm cache running times can be considered as I/O time.

Table 2. Efficiency Track results, type A queries

run	L_{opt}	$\varnothing t_{warm}$[ms]	$\varnothing t_{cold}$[ms]	iP[0.00]	iP[0.01]	iP[0.05]	iP[0.10]	MAiP
MPII-eff-15	210	8.8	216.5	0.575	0.559	0.511	0.400	0.177
MPII-eff-150	610	13.2	242.5	0.574	0.560	0.531	0.466	0.233
MPII-eff-1500	1810	27.1	287.0	0.566	0.553	0.532	0.464	0.248

Queries are processed using the pruned index structures which have been reordered by docid to enable for merge join query processing. As the pruned index is created by Hadoop and stored in a MapFile accessed by Hadoop in a non-optimized way during query execution, we think that there's still room for performance improvements. It turns out that already very short list prefixes are sufficient to lead to a result quality comparable to MPII-COArBP at early recall levels (until iP[0.01]) and to MPII-COArBM' at later recall levels.

Table 3. Efficiency Track results, type B queries

run	L_{opt}	$\varnothing t_{warm}$[ms]	$\varnothing t_{cold}$[ms]	iP[0.00]	iP[0.01]	iP[0.05]	iP[0.10]	MAiP
MPII-eff-15	210	604.3	7,630.4	0.374	0.356	0.304	0.272	0.099
MPII-eff-150	610	922.7	10,235.3	0.391	0.379	0.338	0.315	0.157
MPII-eff-1500	1810	1,492.3	12,979.9	0.391	0.379	0.337	0.316	0.162

Table 3 shows the results of the tuned index structures for type B queries. It is clear that in our setting type B queries that consist of partly more than 100 keywords cannot be executed as fast as type A queries. Many thousands of possible single pair lists per query have to be fetched from harddisk first before the evaluation can start.

5 Conclusion

This paper has presented an approach to perform index pruning in a retrieval-quality aware manner to realize performance improvements and smaller indexes at the same time. Our best tuned index structures provide the best CPU times for type A queries among all Efficiency Track participants while still providing at least BM25 retrieval quality. Type B queries which consist of partly more than 100 keywords cannot be processed equally performant as too many lists have to be fetched from harddisk before they can be evaluated.

References

1. Broschart, A., Schenkel, R., Theobald, M.: Experiments with proximity-aware scoring for xml retrieval at inex 2008. In: Geva, S., Kamps, J., Trotman, A. (eds.) INEX 2008. LNCS, vol. 5631, pp. 29–32. Springer, Heidelberg (2009)
2. Büttcher, S., Clarke, C.L.A., Lushman, B.: Term proximity scoring for ad-hoc retrieval on very large text collections. In: SIGIR, pp. 621–622 (2006)
3. Schenkel, R., Broschart, A., won Hwang, S., Theobald, M., Weikum, G.: Efficient text proximity search. In: Ziviani, N., Baeza-Yates, R.A. (eds.) SPIRE 2007. LNCS, vol. 4726, pp. 287–299. Springer, Heidelberg (2007)

TopX 2.0 at the INEX 2009 Ad-Hoc and Efficiency Tracks

Distributed Indexing for Top-k-Style Content-And-Structure Retrieval

Martin Theobald[1], Ablimit Aji[2], and Ralf Schenkel[3]

[1] Max Planck Institute for Informatics, Saarbrücken, Germany
[2] Emory University, Atlanta, USA
[3] Saarland University, Saarbrücken, Germany

Abstract. This paper presents the results of our INEX 2009 Ad-hoc and Efficiency track experiments. While our scoring model remained almost unchanged in comparison to previous years, we focused on a complete redesign of our XML indexing component with respect to the increased need for scalability that came with the new 2009 INEX Wikipedia collection, which is about 10 times larger than the previous INEX collection. TopX now supports a CAS-specific *distributed* index structure, with a completely *parallel* execution of all indexing steps, including parsing, sampling of term statistics for our element-specific BM25 ranking model, as well as sorting and compressing the index lists into our final inverted block-index structure. Overall, TopX ranked among the top 3 systems in both the Ad-hoc and Efficiency tracks, with a maximum value of 0.61 for iP[0.01] and 0.29 for MAiP in focused retrieval mode at the Ad-hoc track. Our fastest runs achieved an average runtime of 72 ms per CO query, and 235 ms per CAS query at the Efficiency track, respectively.

1 Introduction

Indexing large XML collections for Content-And-Structure (CAS) retrieval consumes a significant amount of time. In particular inverting (i.e., sorting) index lists produced by our XML parser constitutes a major bottleneck in managing very large XML collections such as the 2009 INEX Wikipedia collection, with 55 GB of XML sources and more than 1 billion XML elements. Thus, for our 2009 INEX participation, we focused on a complete redesign of our XML indexing component with respect to the increased need for scalability that came with the new collection. Through distributing and further splitting the index files into multiple smaller files for sorting, we managed to break our overall indexing time down to less than 20 hours on a single-node system and less than 4 hours on a cluster with 16 nodes, for a complete CAS index with individual tag-term pairs as keys to inverted lists.

With TopX being a native XML engine, our basic index units are inverted lists over combined tag-term pairs, where the occurrence of each term in an XML element is propagated "upwards" the XML tree structure, and the term is bound to the tag name of each element that contains it (see [8]). Content-Only (CO) queries are treated as CAS

S. Geva, J. Kamps, and A. Trotman (Eds.): INEX 2009, LNCS 6203, pp. 218–228, 2010.
© Springer-Verlag Berlin Heidelberg 2010

queries with a virtual "$*$" tag. Term frequencies (TF) and element frequencies (EF) are computed for each tag-term pair in the collection individually. In summary, we used the same XML-specific extension to BM25 (generally known as EBM25) as in last years also for the 2009 Ad-hoc track and Efficiency tracks. For the EF component, we precompute an individual element frequency for each distinct tag-term pair, capturing the amount of tags under which the term appears in the entire collection. Because of the large size of the new Wikipedia collection, we approximate these collection-wide statistics by sampling over only a subset of the collection before computing the actual scores. New for 2009 was also the introduction of a static decay factor for the TF component to make the scoring function favor smaller elements rather than entire articles (i.e., the root of the documents), in order to obtain more diverse results in focused element retrieval mode (used in our two best Ad-hoc runs `MPII-COFoBM` and `MPII-COBIBM`).

2 Scoring Model

Our XML-specific extension to the popular Okapi BM25 [5] scoring model, as we first introduced it for XML ranking in 2005 [9], remained largely unchanged also in our 2009 setup. It is very similar to later Okapi extensions in [4,6]. Notice that regular text retrieval with entire documents as retrieval units is just a special case of the below ranking function, which in principle computes a separate Okapi model for each element type individually.

Thus, for content scores, we make use of collection-wide element statistics that consider the *full-content* of each XML element (i.e., the recursive concatenation of all its descendants' text nodes in its XML subtree) as a bag of words:

1) the *full-content term frequency*, $ftf(t, n)$, of term t in an element node n, which is the number of occurrences of t in the full-content of n;
2) the *tag frequency*, N_A, of tag A, which is the number of element nodes with tag A in the entire corpus;
3) the *element frequency*, $ef_A(t)$, of term t with regard to tag A, which is the number of element nodes with tag A that contain t in their full-contents in the entire corpus.

The score of a tag-term pair of an element node n with tag name A with respect to a content condition of the form `//A[about(., t)]` (in NEXI [10] syntax), where A either matches the tag name A or is the tag wildcard $*$, is then computed by the following BM25-based formula:

$$score(n, //\texttt{A[about(., t)]}) = \frac{(k_1 + 1)\, ftf(t, n)}{K + ftf(t, n)} \cdot \log\left(\frac{N_A - ef_A(t) + 0.5}{ef_A(t) + 0.5}\right)$$

$$\text{with} \quad K = k_1\left((1 - b) + b\frac{\sum_{t'} ftf(t', n)}{avg\{\sum_{t'} ftf(t', n') \mid n' \text{ with tag } A\}}\right)$$

For 2009, we used values of $k_1 = 2.0$ and $b = 0.75$ as Okapi-specific tuning parameters, thus changing k_1 from the default value of 1.25 (as often used in text retrieval) to 2.0 in comparison to 2008 (see also [1] for tuning BM25 on INEX data). Consequently, for an `about` operator with multiple terms, the score of an element satisfying this tag constraint is computed as the sum over all the element's content scores, i.e.:

$$score(n, //\texttt{A[about(.,} \ t_1 \ldots t_m \texttt{)])}) = \sum_{i=1}^{m} score(n, //\texttt{A[about(.,} \ t_i \texttt{)])})$$

Moreover, for queries with multiple support elements (like for example in the query `//A//B[about(., t)]`), we assign a small and constant score mass c for each supporting tag condition that is matched by the XML element (like for the tag `A` in this example). This structural score mass is then aggregated with the content scores, again using summation. In our INEX 2009 setup (just like in 2008), we have set $c = 0.01$. Note that our notion of tag-term pairs enforces a strict matching of a query's target element, while content conditions and support elements can be relaxed (i.e., be skipped on-the-fly) by the non-conjunctive query processor [8]. Also, content scores are normalized to $[0, 1]$.

2.1 2009 Extensions

Decay Factor for Term Frequencies. New for 2009 was the introduction of a static decay factor for the ftf component to make the scoring function favor smaller elements rather than entire articles (i.e., the root of the documents), in order to obtain more diverse results in focused element retrieval mode. With respect to the fairly deep structure of the new 2009 collection, we chose a relatively high decay factor of 0.925. That is, in addition to summing up the ftf values of each tag-term pair among the children of an XML element (recursively upwards to the root), we also multiply the ftf value from each of the child nodes by 0.925 before propagating these values upwards.

Sampling for Element Frequencies. With very large, diverse XML collections and very many distinct (but mostly infrequent) tag-term pairs, exact element frequencies as needed for the ef component cannot easily be kept in memory anymore. An additional difficulty in a distributed indexing setting is that these statistics need to be shared among peers. Therefore, we introduced a *sampling phase* for these combined tag-term frequencies, which however generates approximate statistics only. During the sampling phase, we uniform-randomly scan (and keep) only a fixed amount of tag-term statistics in memory from the XML parser output, for each of the distributed nodes in the network individually. Tag-term pairs for which no statistics are kept in memory after this sampling phase are smoothed by an ef value of 1 when materializing the above scoring function.

3 Distributed Indexing

Indexing an XML collection with TopX consists of a 3-pass process: 1) parsing the XML documents (using a standard SAX parser) and hashing individual tag-term pairs and navigational tags into a distributed file storage; 2) sampling these files for the BM25-specific ef statistics for all tag-term pairs and materializing the BM25 model; and 3) sorting these BM25-scored files to obtain their final inverted block structure and compressing the blocks into a more compact binary format. While we are keeping all intermediate index files of steps 1 and 2 in a simple (GZip'ed) ASCII format, our final block-index structure created in step 3 is stored in a customized (compressed) binary format as described in [7], which can be decompressed much faster than a GZip format.

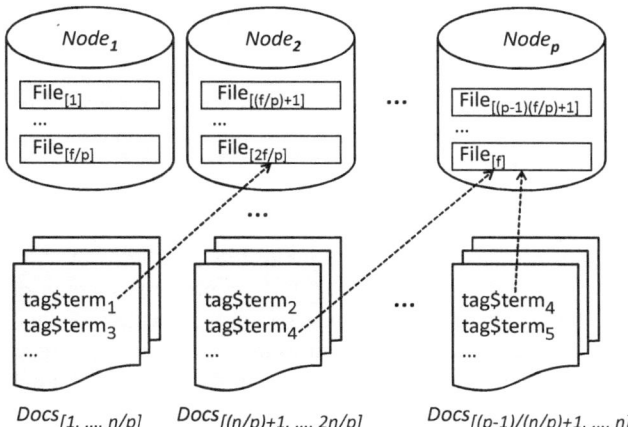

Fig. 1. Two-level hashing of tag-term pairs onto network nodes and index files

The basic hashing phase is illustrated in Figure 1. We are given a collection of n XML documents yielding m distinct tag-term pairs, f files to hold our inverted index, and p distributed nodes (e.g., a compute cluster or peers in a network). Before we start the actual indexing phase, the document collection is partitioned into n/p equally sized chunks, and the chunks are distributed over the p nodes. During indexing, every compute node is used for parsing and storing the index files at the same time, i.e., every node has write access to every other node in the network. Let $hash(t_i)$ be the hash code of tag-term pair t_i for all $i = 1, \ldots m$, then $(hash(t_i) \mod f)$ denotes the *file identifier* where the inverted list for t_i is stored. Moreover, $(hash(t_i) \mod f \mod p)$ then denotes the *node identifier* at which the file containing t_i is located. This ensures that all tag-term pairs from the entire collection that share the same hash code are stored on the same node and in the same index file. The reason to further partition the index into $f \geq p$ files is that, on each compute node, we can sort multiple such files concurrently in a multi-threaded fashion. Multiple smaller files can be sorted in parallel more efficiently than a single large file. Thus, every index file contains at least one but possibly more inverted lists. For our INEX 2009 indexing experiments, we used $f = 256$ index files which were distributed over $p = 16$ nodes.

This simple two-level hashing allows for a MapReduce-like [2] but more specialized form of distributed indexing. After the initial parsing and hashing phase (corresponding to the Map phase in MapReduce), all files needed for materializing the above scoring function are readily available per node, and thus the sampling and scoring can be kept perfectly parallel (corresponding to the Reduce phase in MapReduce). Since all nodes can operate independently in the second phase, this approach allows for a substantially more lightweight Reduce phase than in a classical MapReduce setting. The only data structure that is finally shared across all nodes is a dictionary (see [7]) that maps a tag-term key from a query back to a file (including the byte-offset within that file) as entry point to the respective inverted list in the distributed storage. Our dictionary maps a 64-bit hash key computed from the combined tag-term string of t_i onto a 64-bit value whose upper 8 bits encode the node id, whose middle 12 bits encode the file id that contains the corresponding inverted list, and whose lower 44 bits encode the byte offset within that file in order to mark the beginning of the inverted list. In this 64-bit setting, we can address up to $2^8 = 256$ nodes with up to $2^{12} = 4,096$ files, each file with a maximum size of $2^{44} = 16$ Terabytes of compressed index data. Of course, the same hash function needs to be used for both indexing and query processing.

This mapping step works particularly well for a CAS-based index, as it is using tag-term pairs as keys to access the inverted lists. In particular, the resulting index files are more uniform in length as compared to a plain (i.e., CO-style) text index because of the much larger amount of distinct tag-term keys in the CAS case. As such, this method scales very well to many distributed index files and nodes in a network. Analogously, the same hashing step is performed for inverted tag lists, which results in a distinct set of inverted files for navigational tag conditions as needed by TopX (see [7]) to resolve structural constraints. These files are less uniform in size (due to the much lower amount of distinct tag keys in the collection), but they are anyway much more compact and can benefit from a higher cache locality.

4   Query Processing

TopX is a top-k engine for XML with non-conjunctive XPath evaluations. It supports dynamic top-k-style index pruning for both CO and CAS queries. In dynamic pruning, the traversal of index list scans at query processing time can be pruned early, i.e., when no more candidate elements from the intermediate candidate queue can make it into the top-k list anymore. Also, since TopX was designed as a native XML engine based on element retrieval, relevant passages are identified based on the XML elements that embrace them.

In its current version, query processing in TopX is *multi-threaded* but *not distributed*. After a query is decomposed into its basic tag-term conditions, an index scan is issued for each of the basic tag-term pairs in a separate thread. Each of these scan threads block-wisely accesses an inverted list for such a tag-term condition, merges the list's index objects into the processors' internal candidate queue (a main-memory data structure which is shared among threads), and iteratively checks for the top-k stopping condition (see [8,7] for details). Although processing an individual query is not distributed, the index files that the processor accesses may well stay distributed across an entire cluster of storage nodes. Also, multiple query processors may be run in parallel within the cluster, each processing a different query in this case. In the case of a distributed storage, the multi-threading within each processor may still yield quite substantial runtime benefits, since parallel index scans to different disks are supported much better over such a network infrastructure (e.g., over a shared file system such as NFS) than over a single (local) disk, and repeated network latencies may be largely avoided. Further multi-threading improvements are achieved by *prefetching* also the next block of an index list from disk (asynchronously) already while processing the current block in memory.

4.1 Retrieval Modes

Article Mode. In Article mode, all CO queries (including `//*` queries) are rewritten into `/article` CAS conditions. Thus we only return entire `article` elements as target element of the query. This mode conforms to a regular document retrieval mode with our BM25 model collapsing into a document-based scoring model (however including the per element-level decay factor of the TF component). Article mode can thus be seen as a simplest-possible form of focused element retrieval, as entire articles are always guaranteed to be overlap-free.

CO Mode. In CO mode, any target element of the collection is allowed as result. CO queries are processed by TopX equivalently to queries with `//*` as structural condition and exactly one `about` operator with no further structural constraints as filter predicate. The `*` is treated like a virtual tag name and is fully materialized into a distinct set of corresponding `*`-term pairs directly at indexing time. That is, a `*`-term pair is always generated in addition to any other tag-term pair, with individual TF and EF statistics for these `*` tags, which roughly doubles the index size. As a new extension in our 2009 experiments, we cut off very small elements of less than 24 terms (as a form of static index pruning) from these CO-related index lists. At query processing time, a CO query can be processed by directly accessing these precomputed (inverted) CO lists by attaching a virtual * tag to each query term.

CAS Mode. In CAS mode, TopX allows for the formulation of arbitrary path queries in the NEXI [10] or XPath 2.0 Full-Text [10] syntax. As an optimization, leading `*` tags are rewritten as `article` tags, for CAS queries that were otherwise led by a `//*` step.

4.2 Optimizations

Focused, Best-In-Context, and Thorough Modes. We performed experiments with Focused, Best-In-Context and Thorough runs. Element overlap in the *Focused* and *Best-In-Context* retrieval modes is eliminated on-the-fly during query processing by comparing the pre- and post-order labels [9] of elements already contained in the top-k list with the pre- and post-order label of each element that is about to be inserted into the top-k list. During query processing, this top-k list is a constantly updated queue. Thus, if an element is a descendant of another parent element that already is in the top-k queue, and the descendant has a higher score than its parent, then the parent is removed from the queue and the descendant is inserted; if otherwise an element is a parent of one or more descendants in the top-k queue, and the parent has a higher score than all the descendants, then all the descendants are removed from the top-k queue and the parent is inserted. In *Thorough* retrieval mode, this overlap check is simply omitted.

Caching. Entire index lists (or their prefixes in case of dynamic top-k pruning) can be cached by the TopX engine, and the decoded and decompressed data structures for each index list can directly be reused by the engine for subsequent queries. Thus, when running over a *hot cache*, only joins and XPath evaluations are carried out over these cached index lists at query processing time, while physical disk accesses can be almost completely avoided for query conditions that were already processed in a previous query.

5   Results

We next present the results for our Ad-hoc and Efficiency track experiments. While indexing was performed on a distributed cluster of 16 nodes, the final runs were performed on a single 3Ghz AMD Opteron Quad-Core machine with 16 GB RAM and a RAID5 storage, with all index files being copied to the local storage. All queries were run sequentially.

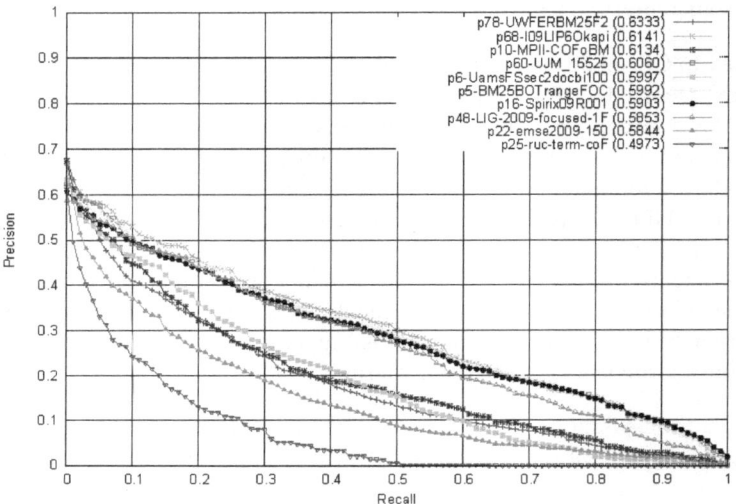

Fig. 2. iP[x] for top runs in Focused mode, Ad-hoc track

5.1   Ad-Hoc Track

Figures 2, 3 and 4 depict the iP[x] and gP[x] plots of participants' runs which were submitted to the Ad-Hoc track, as functions of the recall x for Focused, Best-In-Context and Thorough modes, respectively (see also [3] for an overview of current INEX metrics). We see that at iP[0.01] (Focused and Thorough) and gP[0.01] (Best-In-Context), the best TopX runs rank at positions 3, 3 and 7, respectively, when grouping the runs by participant id (runs denoted as p10-MPII-COFoBM and p10-MPII-COBIBM). For all retrieval modes, we generally observe a high early (interpolated or generalized) precision rate (i.e., for iP[0.01] and gP[0.01], each at a recall of 1%), which is an excellent behavior for a top-k engine. Both our top runs used a true focused element retrieval mode, with CO queries being rewritten into //* CAS queries, i.e., any result element type was allowed as result. Both our 2009 CO runs achieved higher iP[0.01] (and MAiP) values than our Article runs, as opposed to 2008.

5.2   Efficiency Track

Figure 5 depicts the iP[x] plots for all Focused type A runs submitted to the Efficiency track. The best TopX run (TopX2-09-ArHeu-Fo-150-Hot) is highlighted.

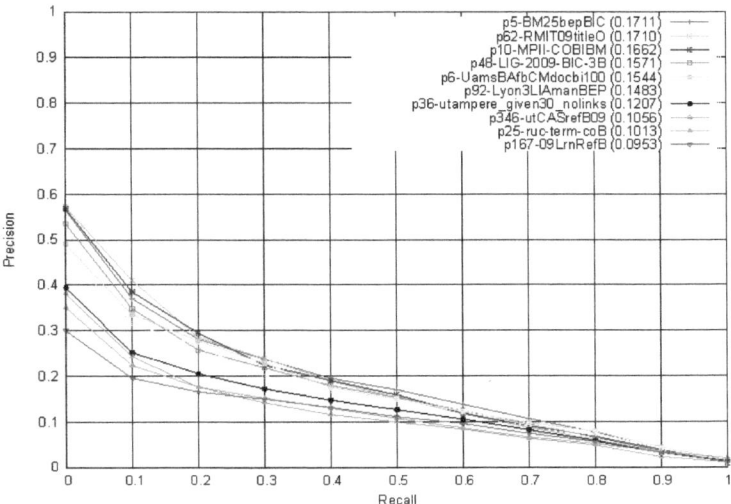

Fig. 3. gP[x] for top runs in Best-In-Context mode, Ad-hoc track

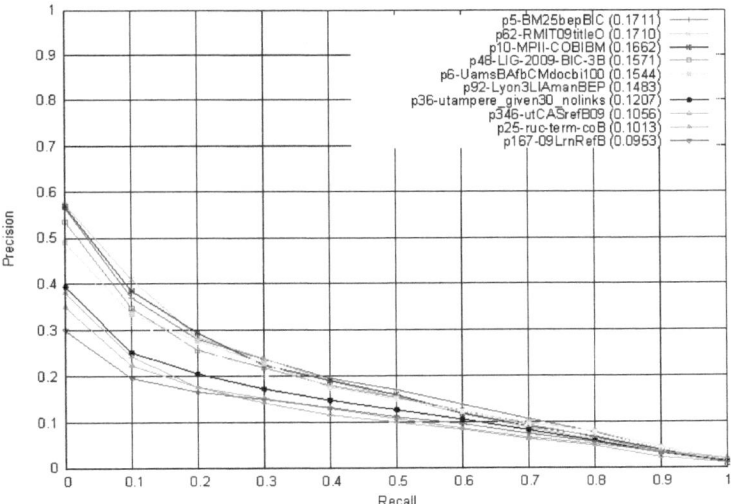

Fig. 4. iP[x] for top runs in Thorough mode, Ad-hoc track

Figures 6 and 7 depict the iP[0.01] plots of all participants' Focused runs submitted to the Efficiency track, in comparison to their runtime. For type A topics, our fastest runs achieved an average of 72 ms per CO query, and 235 ms per CAS query at the Efficiency track, respectively (denoted as `TopX2-09-ArHeu-Fo-15-Hot` and `TopX2-09-CASHeu-Fo-15-Hot`). TopX ranks among the top engines, with a very good retrieval quality vs. runtime trade-off (compare also Tables 1 and 2). Our fastest

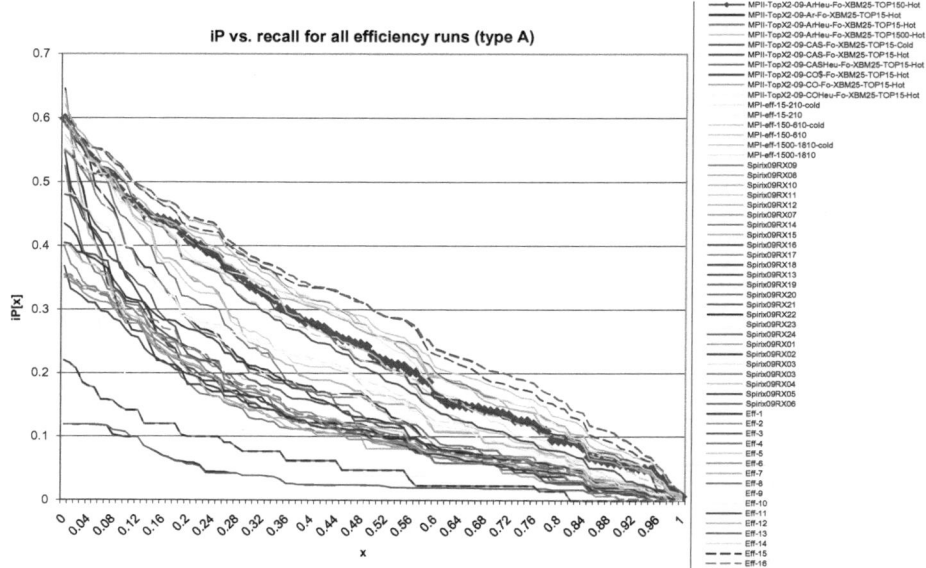

Fig. 5. iP[x] for all Efficiency runs, type A queries

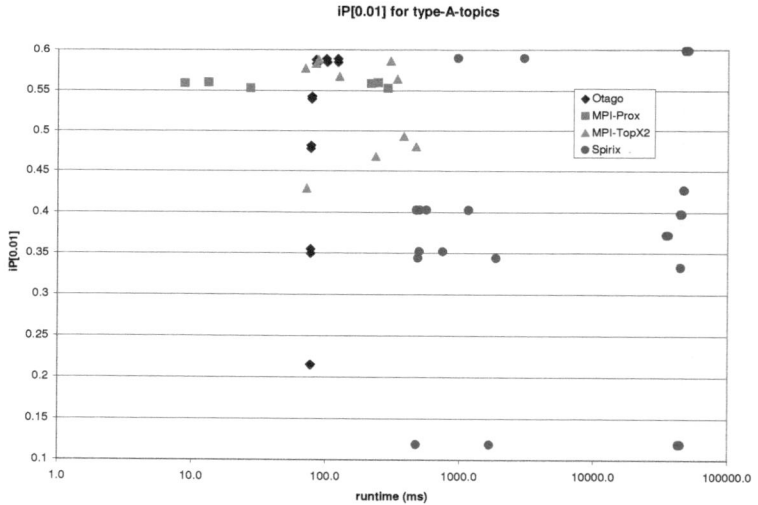

Fig. 6. Runtime vs. iP[0.01] for all Efficiency runs, type A queries

runs operated over a hot cache and employed a heuristic top-k stopping condition (denoted as *Heu*, thus terminating query evaluations after the first block of elements was read and merged from each of the inverted lists that are related to the query (see [7]). Unsurprisingly, our best Article runs on type A topics (70 ms) were slightly faster than

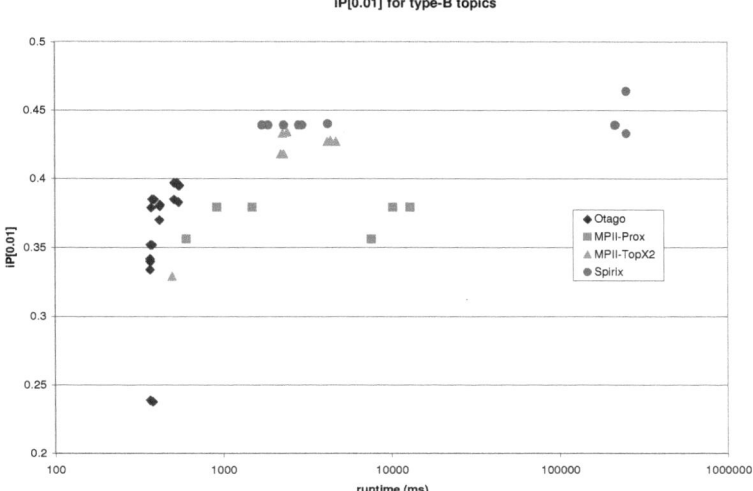

Fig. 7. Runtime vs. iP[0.01] for all Efficiency runs, type B queries

Table 1. Summary of all TopX Efficiency runs, type A queries

Part.ID	Run ID	Type	iP[0.00]	iP[0.01]	iP[0.05]	iP [0.10] [ms]	MAiP [ms]	avg total [ms]	k	Mode
10	TopX2-09-Ar-Fo-15-Hot	Article	0.598	0.583	0.494	0.397	0.178	84.0	15	CO
10	TopX2-09-ArHeu-Fo-1500-Hot	Article	0.597	0.586	0.530	0.475	0.275	301.2	1500	CO
10	TopX2-09-ArHeu-Fo-150-Hot	Article	0.598	0.588	0.531	0.474	0.252	87.2	150	CO
10	TopX2-09-ArHeu-Fo-15-Hot	Article	0.589	0.577	0.482	0.398	0.175	69.8	15	CO
10	TopX2-09-CAS-Fo-15-Cold	Focused	0.546	0.480	0.423	0.355	0.138	467.7	15	CAS
10	TopX2-09-CAS-Fo-15-Hot	Focused	0.545	0.493	0.418	0.350	0.137	379.2	15	CAS
10	TopX2-09-CASHeu-Fo-15-Hot	Focused	0.525	0.468	0.358	0.304	0.124	234.9	15	CAS
10	TopX2-09-CO\$-Fo-15-Hot	Focused	0.645	0.567	0.406	0.285	0.135	125.6	15	CO
10	TopX2-09-CO-Fo-15-Hot	Focused	0.641	0.564	0.405	0.291	0.130	338.2	15	CO
10	TopX2-09-COHeu-Fo-15-Hot	Focused	0.507	0.429	0.306	0.196	0.079	71.8	15	CO

Table 2. Summary of all TopX Efficiency runs, type B queries

Part.ID	Run ID	Type	iP[0.00]	iP[0.01]	iP[0.05]	iP [0.10] [ms]	MAiP [ms]	avg total [ms]	k	Mode
10	TopX2-09-Ar-TOP15-Hot	Article	0.440	0.427	0.362	0.315	0.119	4163.2	15	CO
10	TopX2-09-ArHeu-TOP1500-Hot	Article	0.443	0.434	0.381	0.358	0.206	2412.0	1500	CO
10	TopX2-09-ArHeu-TOP150-Hot	Article	0.443	0.433	0.381	0.358	0.193	2260.2	150	CO
10	TopX2-09-ArHeu-TOP15-Hot	Focused	0.431	0.418	0.344	0.303	0.118	2205.4	15	CO
10	TopX2-09-CAS-TOP15-Cold	Focused	0.442	0.428	0.363	0.316	0.119	4352.3	15	CAS
10	TopX2-09-CAS-TOP15-Hot	Focused	0.440	0.427	0.357	0.315	0.118	4685.1	15	CAS
10	TopX2-09-CASHeu-TOP15-Hot	Focused	0.431	0.418	0.344	0.303	0.118	2293.1	15	CAS
10	TopX2-09-COHeu-TOP15-Hot	Focused	0.364	0.329	0.221	0.175	0.066	497.8	15	CO

our best CO runs (72 ms), and about a factor of 3 faster than our CAS runs (235 ms), at a comparable result quality as in the Ad-hoc track. For type B topics, containing CO topics with partly more than 90 keywords, average runtimes were more than an order of magnitude worse than for type A topics. Altogether, TopX was the only engine to submit other than Article runs to the Efficiency track.

6   Conclusions

TopX was one out of only a few engines to consider CAS queries in the Ad-hoc and Efficiency tracks over the large 2009 Wikipedia collection, and even the only engine in the Efficiency track that processed CAS queries in an actual element-retrieval mode. In our ongoing work, we already started looking into further XPath Full-Text operations, including phrase matching and proximity-based ranking for both CO and CAS queries, as well as top-k support for more complex XQuery constructs. Our long-term goal is to make TopX a comprehensive open-source indexing and search platform for the W3C XPath 2.0 and XQuery 1.0 Full-Text standards. While we believe that our distributed indexing strategy already scales to future XML indexing needs (with many Terabytes potential index size), making also the search process truly distributed remains a challenging topic for future work.

References

1. Clarke, C.L.A.: Controlling overlap in content-oriented XML retrieval. In: Baeza-Yates, R.A., Ziviani, N., Marchionini, G., Moffat, A., Tait, J. (eds.) SIGIR, pp. 314–321. ACM, New York (2005)
2. Dean, J., Ghemawat, S.: MapReduce: Simplified data processing on large clusters. In: OSDI 2004, pp. 137–150 (2004)
3. Kamps, J., Pehcevski, J., Kazai, G., Lalmas, M., Robertson, S.: INEX 2007 evaluation measures. In: Fuhr, N., Kamps, J., Lalmas, M., Trotman, A. (eds.) INEX 2007. LNCS, vol. 4862, pp. 24–33. Springer, Heidelberg (2008)
4. Lu, W., Robertson, S.E., MacFarlane, A.: Field-weighted XML retrieval based on BM25. In: Fuhr, N., Lalmas, M., Malik, S., Kazai, G. (eds.) INEX 2005. LNCS, vol. 3977, pp. 161–171. Springer, Heidelberg (2006)
5. Robertson, S.E., Walker, S., Hancock-Beaulieu, M., Gatford, M., Payne, A.: Okapi at TREC-4. In: TREC (1995)
6. Robertson, S.E., Zaragoza, H., Taylor, M.: Simple BM25 extension to multiple weighted fields. In: Grossman, D., Gravano, L., Zhai, C., Herzog, O., Evans, D.A. (eds.) CIKM, pp. 42–49. ACM, New York (2004)
7. Theobald, M., AbuJarour, M., Schenkel, R.: TopX 2.0 at the INEX 2008 Efficiency Track. In: Geva, S., Kamps, J., Trotman, A. (eds.) INEX 2008. LNCS, vol. 5631, pp. 224–236. Springer, Heidelberg (2009)
8. Theobald, M., Bast, H., Majumdar, D., Schenkel, R., Weikum, G.: TopX: efficient and versatile top-k query processing for semistructured data. VLDB J. 17(1), 81–115 (2008)
9. Theobald, M., Schenkel, R., Weikum, G.: An efficient and versatile query engine for TopX search. In: Böhm, K., Jensen, C.S., Haas, L.M., Kersten, M.L., Larson, P.-Å., Ooi, B.C. (eds.) VLDB, pp. 625–636. ACM, New York (2005)
10. Trotman, A., Sigurbjörnsson, B.: Narrowed Extended XPath I (NEXI). In: Fuhr, N., Lalmas, M., Malik, S., Szlávik, Z. (eds.) INEX 2004. LNCS, vol. 3493, pp. 16–40. Springer, Heidelberg (2005)

Fast and Effective Focused Retrieval

Andrew Trotman[1], Xiang-Fei Jia[1], and Shlomo Geva[2]

[1] Computer Science, University of Otago, Dunedin, New Zealand
[2] Queensland University of Technology, Brisbane, Australia

Abstract. Building an efficient and an effective search engine is a very challenging task. In this paper, we present the efficiency and effectiveness of our search engine at the INEX 2009 Efficiency and Ad Hoc Tracks. We have developed a simple and effective pruning method for fast query evaluation, and used a two-step process for Ad Hoc retrieval. The overall results from both tracks show that our search engine performs very competitively in terms of both efficiency and effectiveness.

1 Introduction

There are two main performance issues in Information Retrieval (IR); effectiveness and efficiency. In the past, the research was mainly focused on effectiveness. Only until recent years, efficiency is getting more research focus under the trend of larger document collection sizes. In this paper, we present our approaches towards efficient and effective IR and show our submitted results at the INEX 2009 Efficiency and Ad Hoc Tracks. We have developed a simple and effective pruning method for fast query evaluation, and used a two-step process for Ad Hoc retrieval. The overall results from both tracks show that our search engine performs very competitively in terms of both efficiency and effectiveness.

In Section 2, IR efficiency issues are discussed. Section 3 explains how we achieve fast indexing and searching for large document collections. Experiments and results are shown in Section 4 and 5. Section 6 discusses our runs in the Ad Hoc Track. The last section provides the conclusion and future work.

2 Background

Inverted files [1,2] are the most widely used index structures in IR. The index has two parts: a dictionary of unique terms extracted from a document collection and a list of postings (a pair of <document number, term frequency>) for each of the dictionary terms.

When considering efficiency issues, IR search engines are very interesting because search engines are neither purely I/O-intensive nor solely CPU-intensive. To serve a query, I/O is needed in order to read dictionary terms as well as postings lists from disk. Then postings lists are processed using a ranking function and intermediate results are stored in accumulators. At the end, the accumulators are sorted and the top results are returned. There are two obvious questions; (1) How do we reduce the I/O required for reading dictionary terms and posting lists, and (2) how do we minimise the processing and sorting.

S. Geva, J. Kamps, and A. Trotman (Eds.): INEX 2009, LNCS 6203, pp. 229–241, 2010.

When considering effectiveness of Focused Retrieval, it is necessary to consider whether to index documents, elements or passages. This leads to the question of how effectiveness is affected by these index types — we have experimented using document index and post processing to focus.

2.1   Disk I/O

The dictionary has a small size and can be loaded into memory at start-up. Due to their large size, postings must be compressed and stored on disk. Various compression algorithms have been developed, including Variable Byte, Elias gamma, Elias delta, Golomb and Binary Interpolative. Trotman [3] concludes that Variable Byte coding provides the best balance between the compression ratio and the CPU cost for decompression. Anh & Moffat [4,5] construct word-aligned binary codes, which are effective at compression and fast at decompression. We are experimenting with these compression algorithms.

Caching can also be used to reduce disk I/O. There are two levels of caching; system-level and application-level. At the system-level, operating systems provide general purpose I/O caching algorithms. For example, the Linux kernel provides several I/O caching algorithms [6]. At the application-level, caching is more effective since the application can deploy specialised caching algorithms [7]. We are experimenting with caching approaches.

For IR search engines, there are two ways of caching at the application-level. The first solution is to cache query results, which not only reduces disk I/O but also avoids re-evaluation of queries. However, queries tend to have low frequency of repetition [8]. The second is to cache raw postings lists. The challenge is to implement a efficient replacement algorithm in order to keep the postings in memory. We are also experimenting with caching algorithms.

Since the advent of 64-bit machine with vast amount of memory, is has become feasible to load both the dictionary and the compressed postings of a whole-document inverted file into main memory, thus eliminating all disk I/O. For Focused Retrieval a post process of the documents can be a second step. If the documents also fit into memory, then no I/O is needed for Focused Retrieval. This is the approach we are taking, however our experiments in this paper were performed without caching.

2.2   Query Pruning

The processing of postings and subsequent sorting of the accumulators can be computationally expensive, especially when queries contain frequent terms. Frequent terms appear in many documents in a collection and have low similarity scores due to having a low Inverse Document Frequency (IDF). Processing the postings for these terms not only takes time, but also has little impact on the final ranking results.

The purpose of query pruning is to eliminating any unnecessary evaluation while still maintaining good precision. In order to best approximate original results, query pruning requires that (1) every term is assigned a weight [9,10],

(2) query terms are sorted in decreasing order of their weights (such as IDF), (3) the postings are sorted in decreasing order of their weights (such as TF). Partial similarity scores are obtained when some stop condition is met. Either partial or the whole postings list of a query term might be pruned.

Harman & Candeka [11] experimented with a static pruning algorithm in which complete similarity scores are calculated by processing all query terms and postings of the terms. But only a limited number of accumulators, those above a given threshold, are sorted and returned. A dynamic pruning algorithm developed by Buckley and Lewit [12] keeps track of the top $k+1$ partial similarity scores in the set of accumulators, and stops the query evaluation when it is impossible to alter the top-k documents. The algorithm tries to approximate the upper-bound of the top k candidates.

Moffat & Zobel [13] developed two pruning algorithms; the *quit* algorithm is similar to the top-k algorithm and stops processing query terms when a none-zero number of accumulators exceeds a constant value. While the *continue* algorithm continues to process query terms when the stopping condition is met, but only updates documents already in the set of accumulators.

Persin et al. [14,15] argue that a single stopping condition is not efficient enough to maintain fair partial similarity scores. They introduced both a global and a local threshold. The global threshold determines if a new document should be inserted into the set of accumulators, while the local threshold checks if existing accumulators should be updated. The global threshold is similar to the *quit* algorithm, while the combination of the global and local thresholds is like the *continue* algorithm. However, there are two differences; (1) the quit algorithm keeps adding new documents into the set of accumulators until reaching a stopping condition, while the global threshold algorithm adds a new document into the set of accumulators only if the partial similarity score of the document is above the predefined global threshold. (2) The local threshold algorithm only updates the set of accumulators when a partial similarity score is above the local threshold, while the continue algorithm has no condition to update the accumulators.

Anh et al. [16] introduced impact ordering, in which the postings for a term are ordered according to their overall contribution to the similarity scores. They state that Persin et al. [14,15] defined term-weighting as a form of TF-IDF (the global threshold is the IDF and the local threshold is the TF), while Anh et al. used normalised TF-IDF. The term impact is defined as $w_{d,t}/W_d$ where $w_{d,t}$ is the document term weight and W_d is the length of the document vector.

In this paper, we present a simple but effective static pruning method, which is similar to the *continue* algorithm.

3   Efficiency

3.1   Indexer

Memory management is a challenge for fast indexing. Efficient management of memory can substantially reduce indexing time. Our search engine has a memory management layer above the operating system. The layer pre-allocates large chunks of memory. When the search engine requires memory, the requests are served from the pre-allocated pool, instead of calling system memory allocation functions. The sacrifice is that some portion of pre-allocated memory might be wasted. The memory layer is used both in indexing and in query evaluation. As we show in our results, only a very small portion of memory is actually wasted.

The indexer uses hashing with a collision binary tree for maintaining terms. We tried several hashing functions including Hsieh's super fast hashing function. By default, the indexer uses a very simple hashing function, which only hashes the first four characters of a term and its length by referencing a pre-defined look-up table. A simple hashing function has less computational cost, but causes more collisions. Collisions are handled by a simple unbalanced binary tree. We will examine the advantages of various hashing and chaining algorithms in future work.

Postings lists can vary substantially in length. The indexer uses various sizes of memory blocks chained together. The initial block size is 8 bytes and the resize factor is 1.5 for the subsequent blocks.

In order to reduce the size of the inverted file, we always use 1 byte to store term frequencies. This limits term frequencies to a maximum value of 255. Truncating term frequencies could have an impact on long documents. But we assume long documents are rare in a collection and terms with high frequencies in a document are more likely to be common words.

As shown in Figure 1, the index file has four levels of structure. Instead of using the pair of <document number, term frequency> for postings, we group documents with the same term frequency together and store the term frequency at the beginning of each group. By grouping and impacting order documents according to term frequency, during query evaluation we can easily process documents with potential high impacts first and prune the less important documents at the end of the postings list. The difference of document ids in each group are then stored in increasing order and each group ends with a zero. Postings are compressed with Variable Byte coding.

The dictionary of terms is split into two parts. The first level stores the first four bytes of a term string, the length of the term string and the position to locate the second level structure. Terms with the same prefix (the first four bytes) are stored in a term block in the second level. The term block stores the statistics for the terms, including collection frequency, document frequency, offset to locate the postings list, the length of the postings list stored on disk, the uncompressed length of the postings list, and the position to locate the term suffix which is stored at the end of the term block.

At the very end of the index file, the small footer stores the location of the first level dictionary and other values for the management of the index.

3.2   Query Evaluation

At start-up, only the the first-level dictionary is loaded into memory. To process a query term, two disk reads have to be issued; The first reads the second-level

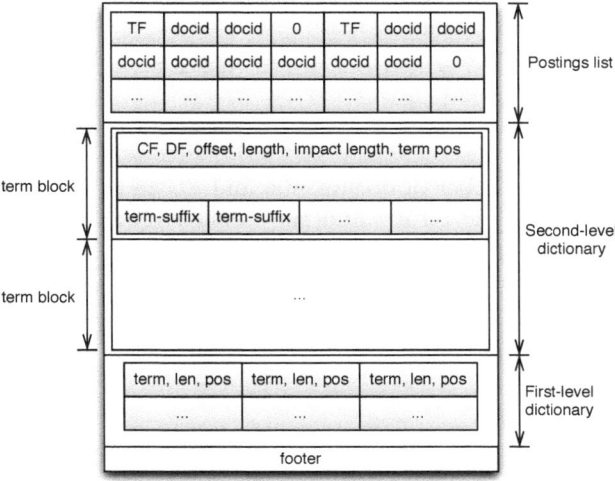

Fig. 1. The index structures

dictionary. Then the offset in that structure is used to locate postings, since we do not use caching in these experiments. The current implementation has no disk I/O caching. We simply deploy the general purpose caching provided by the underlying operating system.

An array is used to store the accumulators. We used fixed point arithmetic on the accumulators because it is faster than the floating point.

We have implemented a special version of quick sort algorithm [17] for fast sorting of the accumulators. One of the features of the algorithm is partial sorting; It will return the top-k documents by partitioning and then only sorting the top partition. Pruning accumulators using partial sorting is similar to that of Harman & Candeka [11]. A command line option (lower-k) to our search engine is used to specify how many top documents to return.

We have also developed a method for static pruning of postings. A command line option (upper-K) is used to specify a value, which is the number of document ids (in the postings list of a term) to be processed. The upper-K value is only a hint. The search engine will always finish processing all postings with the same TF at the Kth postings. The combined use of both the lower-k and upper-K methods is similar to the *continue* algorithm.

When upper-K is specified at the command line, the whole postings list of a term is decompressed, even though only partial postings will be processed (this is left for future work). Moffat and Anh [13,18] have developed methods for partial decompression for Variable Byte compressed lists. However, these methods do not come without a cost; Extra housekeeping data must be inserted into the postings lists, thus increasing the size of the index. Further more, there is also a computational cost in keeping track of the housekeeping data.

A modified BM25 is used for ranking. This variant does not result in negative IDF values and is defined thus:

$$RSV_d = \sum_{t \in q} log\left(\frac{N}{df_t}\right) \cdot \frac{(k_1 + 1)\, tf_{td}}{k_1\left((1 - b) + b \times \left(\frac{L_d}{L_{avg}}\right)\right) + tf_{td}}$$

Here, N is the total number of documents, and df_t and tf_{td} are the number of documents containing the term t and the frequency of the term in document d, and L_d and L_{avg} are the length of document d and the average length of all documents. The empirical parameters k_1 and b have been set to 0.9 and 0.4 respectively by training on the previous INEX Wikipedia collection.

4   Experiments

We conducted our experiments on a system with dual quad-core Intel Xeon E5410 2.3 GHz, DDR2 PC5300 8 GB main memory, Seagate 7200 RPM 500 GB hard drive, and running Linux with kernel version 2.6.30.

The collection used in the INEX 2009 Efficiency Track is the INEX 2009 Wikipedia collection [19]. The collection was indexed using the default parameters as discussed in Section 3. No words were stopped and stemming was not used. The indexing took about 1 hour and 16 minutes. The memory layer allocated a total memory of 5.3 GB with a utilisation of 97%. Only 160 MB of memory was allocated but never used. Table 1 shows a summary of the document collection.

The INEX 2009 Efficiency Track used two types of topics, with both types having 115 queries. Type A Topics are short queries and the same as the INEX 2009 Ad Hoc topics. Type B Topics are expansions of topics in Type A and intended as long queries. Both topics allow *focused*, *thorough* and *article* query evaluations. Our search engine does not natively support focused retrieval yet, but we instead use a post-process. We only evaluated the topics for *article* Content-Only. We used the BM25 ranking model as discussed in previous section. The k_1 and b values were 0.9 and 0.4 respectively.

We experimented only sorting the top-k documents using the lower-k parameter with k = 15, 150 and 1500 as required by the Efficiency Track. Query terms do not have to be sorted in descending order of term frequency since our pruning method does not prune query terms. We also experimented pruning of postings using the

Table 1. Summary of INEX 2009 Wikipedia Collection

Collection Size	50.7 GB
Documents	2666190
Average Document Length	881 words
Unique Words	11393924
Total Worlds	2348343176
Postings Size	1.2 GB
Dictionary Size	369 MB

upper-K parameter. For each iteration of the lower-k, we specified the upper-K of 1, 15, 150, 1500, 15000, 150000, 1500000. In total we submitted 21 runs.

The disk cache was flushed before each run. No caching mechanism was deployed except that provided by the Linux operating system.

5   Results

This section talks about the evaluation and performance of our 21 submitted runs, obtained from the official Efficiency Track.

Table 2 shows a summary of the runs evaluated on the Type A topics. The first column shows the run-id. The interpolated Precision (iP) reflects the evaluations of top-k documents at points of 0%, 1%, 5% and 10%. The overall performance is shown as Mean Average interpolated Precision (MAiP). The average run time, consisting of the CPU and I/O, is the total time taken for the runs. The last two columns show the lower-k and upper-K parameters. In terms of MAiP, the best runs are Eff-21, Eff-20 and Eff-19 with a value of 0.3, 0.3 and 0.29 respectively.

Figure 2(a) shows the Precision-Recall graph of our 21 runs for Type A topics. Except the Eff-1, Eff-8 and Eff-15 runs, all other runs achieved a very good early precision. Bad performance of the three runs was caused by pruning too many postings (a too small value for upper-K) regardless the number of top-k documents retrieved.

Table 2. A summary of the runs for Type A topics

run-id	iP[0.00]	iP[0.01]	iP[0.05]	iP[0.01]	MAiP	Total time	CPU	I/O	Lower-k	Upper-K
Eff-01	0.22	0.22	0.18	0.14	0.06	77.8	20.4	57.3	15	1
Eff-02	0.35	0.35	0.32	0.29	0.13	77.1	20.4	56.7	15	15
Eff-03	0.48	0.48	0.43	0.36	0.15	77.2	20.0	57.2	15	150
Eff-04	0.55	0.54	0.5	0.44	0.18	78.1	21.0	57.1	15	1500
Eff-05	0.6	0.58	0.53	0.47	0.21	84.5	26.9	57.6	15	15000
Eff-06	0.6	0.59	0.53	0.48	0.21	101.3	43.0	58.2	15	150000
Eff-07	0.6	0.59	0.53	0.48	0.2	122.2	64.9	57.3	15	1500000
Eff-08	0.22	0.22	0.18	0.14	0.06	77.7	20.4	57.3	150	1
Eff-09	0.36	0.36	0.33	0.31	0.14	76.9	19.9	57.0	150	15
Eff-10	0.48	0.48	0.44	0.38	0.19	77.4	20.3	57.2	150	150
Eff-11	0.55	0.54	0.51	0.47	0.23	78.3	21.3	57.0	150	1500
Eff-12	0.6	0.59	0.55	0.51	0.27	83.8	26.9	56.9	150	15000
Eff-13	0.6	0.59	0.55	0.52	0.28	100.0	42.7	57.3	150	150000
Eff-14	0.6	0.59	0.55	0.52	0.28	122.2	64.9	57.3	150	1500000
Eff-15	0.22	0.22	0.18	0.14	0.06	76.9	20.3	56.6	1500	1
Eff-16	0.36	0.36	0.33	0.31	0.14	77.1	20.2	56.9	1500	15
Eff-17	0.48	0.48	0.44	0.38	0.19	77.4	20.1	57.3	1500	150
Eff-18	0.55	0.54	0.51	0.47	0.24	78.5	20.9	57.6	1500	1500
Eff-19	0.6	0.59	0.55	0.51	0.29	83.6	26.8	56.9	1500	15000
Eff-20	0.6	0.59	0.55	0.52	0.3	100.3	42.7	57.6	1500	150000
Eff-21	0.6	0.59	0.55	0.52	0.3	121.7	64.3	57.4	1500	1500000

The relationship between the MAiP measures and the lower-k and upper-K parameters is plotted in Figure 3(a) using data from Table 2. When upper-K has values of 150 and 1500, MAiP measures are much better than the upper-K 15. In terms of lower-k, MAiP measures approach constant at a value of 15000.

To have a better picture of the total time cost, we plotted the time costs of all runs in Figure 4(a) using data from Table 2. Regardless of the values used for lower-k and upper-K, the same number of postings were retrieved from disk, thus causing all runs to have the same amount of disk I/O. The figure also shows that the CPU usage is high when upper-K has a value greater than 1500.

We used the same measures for Type B topics. Table 3 shows the summary of averaged measures for Type B topics. The best runs are Eff-20, Eff-21, Eff-13 with an MAiP measure of 0.18, 0.17 and 0.17 respectively. An interesting observation is that the best run (Eff-20) does not has the highest upper-k value. Processing fewer postings not only saves time, but also improves precision. The best MAiP in Type A is 0.3 while only 0.18 in Type B. We are investigating why.

Figure 2(b) shows the Precision-Recall graph for Type B topics. The Eff-1, Eff-8 and Eff-15 runs also achieved low precision at the early stage. All other runs received good early precision. Figure 3(b) shows the MAiP measures using various lower-k and upper-K values. It shows a similar pattern to that of Figure 3(a). However, good performance is seen when upper-K has a value of 150, rather than the 15000 for Type A topics.

Table 3. A summary of the runs for Type B topics

run-id	iP[0.00]	iP[0.01]	iP[0.05]	iP[0.10]	MAiP	Total time	CPU	I/O	Lower-k	Upper-K
Eff-01	0.24	0.24	0.17	0.14	0.05	380.22	31.97	348.25	15	1
Eff-02	0.34	0.33	0.32	0.29	0.1	367.53	32.07	335.46	15	15
Eff-03	0.35	0.34	0.33	0.29	0.12	367.44	33.41	334.03	15	150
Eff-04	0.38	0.38	0.34	0.32	0.12	373.95	41.72	332.23	15	1500
Eff-05	0.38	0.37	0.33	0.31	0.11	418.02	89.73	328.29	15	15000
Eff-06	0.39	0.39	0.34	0.3	0.11	511.56	184.9	326.66	15	150000
Eff-07	0.39	0.38	0.33	0.3	0.11	542.98	216.97	326.02	15	1500000
Eff-08	0.24	0.24	0.18	0.15	0.05	367.21	32.08	335.13	150	1
Eff-09	0.34	0.34	0.33	0.3	0.13	367.51	32.14	335.37	150	15
Eff-10	0.36	0.35	0.34	0.32	0.16	370.12	33.43	336.69	150	150
Eff-11	0.39	0.39	0.35	0.34	0.16	387.61	41.98	345.63	150	1500
Eff-12	0.39	0.38	0.35	0.34	0.16	419.43	90.03	329.39	150	15000
Eff-13	0.4	0.4	0.36	0.33	0.17	512.54	185.07	327.47	150	150000
Eff-14	0.4	0.4	0.36	0.33	0.16	543.53	216.59	326.94	150	1500000
Eff-15	0.24	0.24	0.18	0.15	0.05	368.33	31.84	336.49	1500	1
Eff-16	0.34	0.34	0.33	0.3	0.14	369.46	32.33	337.13	1500	15
Eff-17	0.36	0.35	0.34	0.32	0.17	378.73	33.23	345.5	1500	150
Eff-18	0.39	0.39	0.35	0.34	0.17	378.19	41.77	336.42	1500	1500
Eff-19	0.39	0.38	0.35	0.34	0.17	421.83	90.11	331.72	1500	15000
Eff-20	0.4	0.4	0.36	0.33	0.18	533.32	184.88	348.44	1500	150000
Eff-21	0.4	0.4	0.36	0.33	0.17	551.8	217.52	334.28	1500	1500000

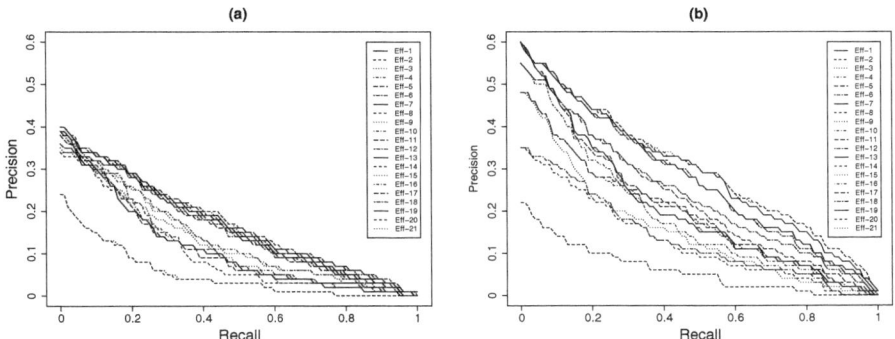

Fig. 2. Precision-Recall plot for (a) Type A and (b) Type B topics

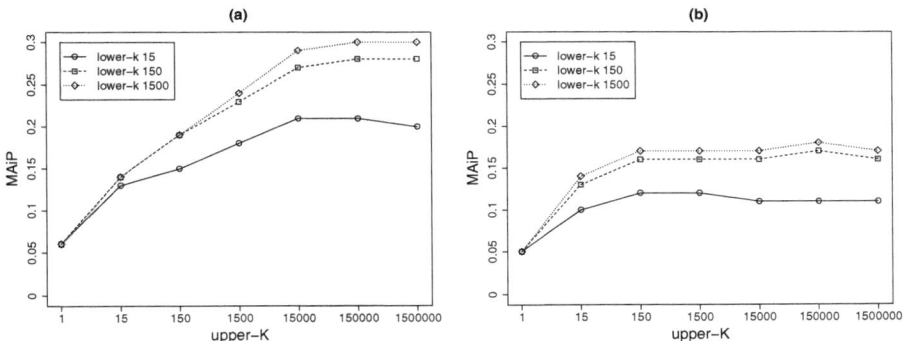

Fig. 3. MAiP measures for (a) Type A and (b) Type B topics

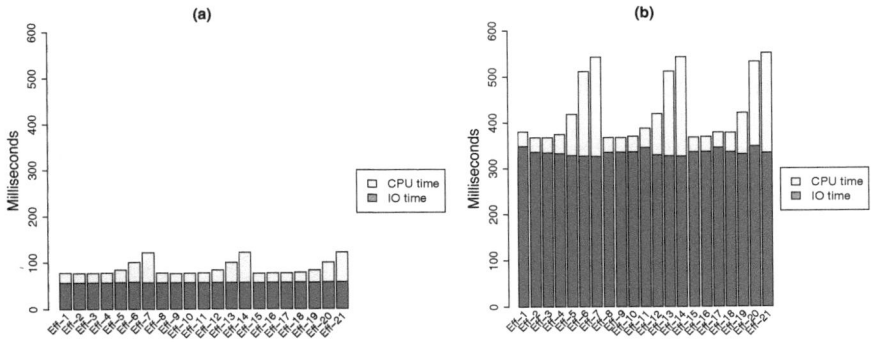

Fig. 4. Total runtime for (a) Type A and (b) Type B topics

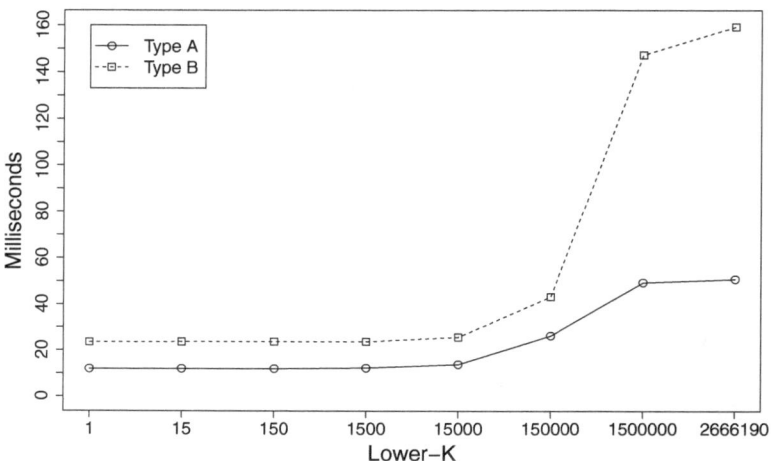

Fig. 5. Times taken for sorting accumulators

The time cost for Type B queries is plotted in Figure 4(b). All runs used the same amount of time for I/O, and have different CPU cost due to various values used for the lower-k and upper-K parameters. The lower-k again has no effect on the CPU cost, and values of 1500 or above for upper-K causes more CPU usage. It took a much longer time for I/O, due to more terms, when compared with I/O cost in Type A.

As shown in both Figure 4(a) and 4(b), the runtime is dominated by the I/O. This leads use to consider that storing the whole index in memory is important.

The submitted runs used small values for the lower-k parameter. In order to see the impact of lower-k, we evaluated both topic sets using only lower-k with a value of 1, 15, 150, 1500, 15000, 150000, 1500000 and 2666190. The number 2666190 is the total number of documents in the collection. As shown in Figure 5, the time taken for sorting the accumulators increases when lower-k has a value above 15000. The sorting times increase from 13.72 ms and 25.57 for Type A and B topics (when lower-k is 15000) to 50.81 ms and 159.39 ms (when lower-k is 2666190) respectively.

6 Ad Hoc

We also used our search engine in the ad hoc track. The whole-document results were extracted and submitted as the REFERENCE run. We then took the reference run and ran a post-process to focus the top 50 results. Our rationale is that document retrieval should rank document from mostly about a topic is mostly not about a topic. If this is the case then focusing should be a fast and relatively simple post process.

6.1 Query Evaluation

Three sets of runs were submitted: Those starting BM25 were based on the reference run and generated from the topic title field (CO) using the BM25 search engine. Those starting ANTbigram were an experiment into phrase searching (and are not discussed here).

For structure searching (CO+S/CAS) we indexed all those tags in a document as special terms. If the path /A/B/C were present in the document then we indexed the document as containing tags A, B, and C. Searching for these tags did not take into consideration the path, only the presence of the tag; that is, /C/B/A would match /A/B/C. Ranking of tags was done with BM25 because the special terms were treated as ordinary terms during ranking. We call this technique Bag-Of-Tags.

Runs containing BOT in their name were generated from the CAS title using the Bag-Of-Tags approach. All search terms and tag names from the paths were included and the queries were ranked using BM25.

The post processing step was not done by the search engine — we leave that for future work. Several different techniques were used:

1. No Focusing (ARTICLE)
2. Deepest enclosing ancestor of all search terms (ANCESTOR)
3. Enclosing element range between first and last occurrence of a search term (RANGE/BEP)
4. All non-overlapping elements containing a search term
5. All overlapping elements containing a search term (THOROUGH)

Runs were submitted to the BIC task (1 & 2), RIC (1-4), Focused (1-4) and thorough (1-5) tasks.

6.2 Results

In the BIC task our run BM25bepBIC placed first. It used BM25 to rank documents and then placed the Best Entry Point at the start of the first element that contained the first occurrence of any search term. Our second best run placed third (RMIT placed second) and it used the ancestor approach.

In the RIC task our runs placed first through to ninth. Our best run was BM25RangeRIC which simply trimmed all those elements from the start and end of the document that did not contain any occurrences of the search terms. The next most effective run was BM25AncestorRIC which chose the lowest common ancestor of the range (and consequently more non-relevant material). Of note, the REFERENCE run placed third – that is, whole document retrieval was very effective.

In the Focused task all our Bag-Of-Tags (CAS) runs placed better than our CO runs. Our best run placed ninth and used ranges, the ancestor run placed tenth and the article run placed eleventh.

In the thorough task our best run, BM25thorough, placed sixth with the Bag-of-Tags placing seventh. We have not concentrated on the focused track.

7   Conclusion and Future Work

In this paper, we introduced our search engine, discussed our design and implementation. We also demonstrated the initial evaluation on the INEX 2009 Efficiency and Ad Hoc Tracks. Our best runs for Type A topics have an MAiP measure of 0.3 and runtime of 100 milliseconds, an MAiP measure of 0.18 and runtime of 533 milliseconds for Type B topics. Compared with the overall results from the Efficiency Track, we believe that our results are very competitive.

Our ad hoc experiments have shown that the approach of finding relevant documents and then post-processing is an effective way of building a Focused Retrieval search engine for the in-Context tasks (where we placed first). They also show that ignoring the structural hints present in a query is reasonable.

Our Focused and Thorough results were not as good as our in-Context runs (we placed respectively fifth and third institutionally). Our experiments here suggest that the Bag-of-Tags approach is effective with our BOT runs performing better than ignoring structural hints in the Focused task and comparably to ignoring the hints in the Focused task. In the Focused task we found that ranges are more effective than common ancestor and (because they are better excluders of non-relevant material). In future work we will be concentrating on increasing our performance in these two tasks.

Of particular interest to us, our runs did not perform best when measured as whole-document retrieval. Our focusing were, however, effective. LIG, RMIT University, and University of Amsterdam bettered our REFERENCE run and we are particularly interested in their approaches and how they might be applied to whole document retrieval (so that we may better our own runs).

In the future, we will continue to work on pruning for more efficient query evaluation. We are also interested in other techniques for improving efficiency without loss of effectiveness, including compression, caching and multi-threading on multi-core architectures.

References

1. Zobel, J., Moffat, A., Ramamohanarao, K.: Inverted files versus signature files for text indexing. ACM Trans. Database Syst. 23(4), 453–490 (1998)
2. Zobel, J., Moffat, A.: Inverted files for text search engines. ACM Comput. Surv. 38(2), 6 (2006)
3. Trotman, A.: Compressing inverted files. Inf. Retr. 6(1), 5–19 (2003)
4. Anh, V.N., Moffat, A.: Inverted index compression using word-aligned binary codes. Inf. Retr. 8(1), 151–166 (2005)
5. Anh, V.N., Moffat, A.: Improved word-aligned binary compression for text indexing. IEEE Transactions on Knowledge and Data Engineering 18(6), 857–861 (2006)
6. Bovet, D.P., Cesati, M.: Understanding the linux kernel, 3rd edn. (November 2005)
7. Jia, X., Trotman, A., O'Keefe, R., Huang, Z.: Application-specific disk I/O optimisation for a search engine. In: PDCAT '08: Proceedings of the 2008 Ninth International Conference on Parallel and Distributed Computing, Applications and Technologies, Washington, DC, USA, pp. 399–404. IEEE Computer Society, Los Alamitos (2008)

8. Baeza-Yates, R., Gionis, A., Junqueira, F., Murdock, V., Plachouras, V., Silvestri, F.: The impact of caching on search engines. In: SIGIR '07: Proceedings of the 30th annual international ACM SIGIR conference on Research and development in information retrieval, pp. 183–190. ACM, New York (2007)
9. Salton, G., Buckley, C.: Term-weighting approaches in automatic text retrieval, pp. 513–523 (1988)
10. Lee, D.L., Chuang, H., Seamons, K.: Document ranking and the vector-space model. IEEE Softw. 14(2), 67–75 (1997)
11. Harman, D., Candela, G.: Retrieving records from a gigabyte of text on a minicomputer using statistical ranking. Journal of the American Society for Information Science 41, 581–589 (1990)
12. Buckley, C., Lewit, A.F.: Optimization of inverted vector searches, pp. 97–110 (1985)
13. Moffat, A., Zobel, J.: Self-indexing inverted files for fast text retrieval. ACM Trans. Inf. Syst. 14(4), 349–379 (1996)
14. Persin, M.: Document filtering for fast ranking, pp. 339–348 (1994)
15. Persin, M., Zobel, J., Sacks-Davis, R.: Filtered document retrieval with frequency-sorted indexes. J. Am. Soc. Inf. Sci. 47(10), 749–764 (1996)
16. Anh, V.N., de Kretser, O., Moffat, A.: Vector-space ranking with effective early termination, pp. 35–42 (2001)
17. Bentley, J.L., Mcilroy, M.D.: Engineering a sort function (1993)
18. Anh, V.N., Moffat, A.: Compressed inverted files with reduced decoding overheads, pp. 290–297 (1998)
19. Schenkel, R., Suchanek, F., Kasneci, G.: YAWN: A semantically annotated wikipedia xml corpus (March 2007)

Achieving High Precisions with Peer-to-Peer Is Possible!

Judith Winter and Gerold Kühne

University of Applied Science, Frankfurt, Germany
winter@fb3.fh-frankfurt.de, gkuehne@cs.uni-frankfurt.de

Abstract. Until previously, centralized stand-alone solutions had no problem coping with the load of storing, indexing and searching the small test collections used for evaluating search results at INEX. However, searching the new large-scale Wikipedia collection of 2009 requires much more resources such as processing power, RAM, and index space. It is hence more important than ever to regard efficiency issues when performing XML-Retrieval tasks on such a big collection. On the other hand, the rich markup of the new collection is an opportunity to exploit the given structure and obtain a more efficient search. This paper describes our experiments using distributed search techniques based on XML-Retrieval. Our aim is to improve both effectiveness and efficiency; we have thus submitted search results to both the Efficiency Track and the Ad Hoc Track. In our experiments, the collection, index, and search load are split over a peer-to-peer (P2P) network to gain more efficiency in terms of load balancing when searching large-scale collections. Since the bandwidth consumption between searching peers has to be limited in order to achieve a scalable, efficient system, we exploit XML-structure to reduce the number of messages sent between peers. In spite of mainly aiming at efficiency, our search engine SPIRIX resulted in quite high precisions and made it into the top-10 systems (focused task). It ranked 7 at the Ad Hoc Track (59%) and came first in terms of precision at the Efficiency Track (both categories of topics). For the first time at INEX, a P2P system achieved an official search quality comparable with the top-10 centralized solutions!

Keywords: XML-Retrieval, Large-Scale, Distributed Search, INEX, Efficiency.

1 Introduction

1.1 Motivation

For years, we have discussed at INEX the need for a test collection providing tags with more semantic meaning than offered by the Wikipedia collection used at INEX from year 2006 to 2008 [11]. Now, we can finally make use of rich markup that includes a wide variety of tags with real semantic meaning, other than the pure structural tags such as paragraph and section that were used before. However, the new collection has also grown in size, from 4.6 GB to more than 50 GB (not including the images contained in articles). Building an index on such a large-scale collection and

S. Geva, J. Kamps, and A. Trotman (Eds.): INEX 2009, LNCS 6203, pp. 242–253, 2010.
© Springer-Verlag Berlin Heidelberg 2010

searching it, now requires much more resources such as computing power, memory and disk space for indexing and retrieval. While with previous collections, centralized stand-alone solutions could cope well with the index and search load, it is now more important than ever to develop efficient systems able to manage large-scale search. Splitting the resource consumption over a set of computers can distribute the search load and thus gain more efficiency of the search system [5].

We have developed a distributed search engine for XML-Retrieval that is based on a peer-to-peer system. We use a set of autonomous and equal computers (= peers) that are pooled together to share resources in a self-organized manner without using a central control [8]. Due to this self-organization of the system, it provides the potential to realize fault-tolerance and robustness. It may scale to theoretically unlimited numbers of participating nodes, and can bring together otherwise unused resources [2].

Owners of objects such as documents often use P2P networks to share their collections with others without giving up full control over their documents. Many users do not want to store their collections on central servers where the documents might be subject to censorship or control by an authority. They prefer to store them locally on their own machines such that it is the owner's sole decision, who can access which documents when. Not only the ownership argument but also privacy and security issues are reasons for people not wanting to store their information in central systems. P2P networks are emerging infrastructures for distributed computing, real-time communication, ad-hoc collaboration, and information sharing in large-scale distributed systems [8].

Applications include E-Business, E-Commerce, E-Science, and Digital Libraries, that are used by large, diverse, and dynamic sets of users. There are a number of P2P solutions that use information retrieval techniques to improve effectiveness and efficiency while searching large-scale collections. The general idea of P2P search and criteria for the classification of such approaches are presented in [6]. An overview of how to look up data in structured P2P systems can be found in [2]. A summary of approaches for IR in unstructured P2P networks is given in [12].

However, none of the existing P2P solutions has ever applied or evaluated the use of XML-Retrieval methods such as developed in the INEX community. Our system is still the only P2P solution for XML-Retrieval.

In this paper, we concentrate on the technical advantages of P2P systems: their potential to distribute the search load over the participating peers. Consequently, both the Wikipedia collection and the corresponding index are distributed over a P2P network. On this basis, we have performed experiments that exploit the new rich structure of the collection in order to achieve more effectiveness while ranking (Ad Hoc Track) and to achieve more efficiency while routing CAS-queries in the P2P network (Efficiency Track). Our top results achieved high precisions in both tracks. In the Ad Hoc Track (focused task), we gained rank 7. In the Efficiency Track, we ranked highest in both tasks (type A and type B topics).

1.2 Comparing Structural Similarity in Ranking and Routing

We applied different structural similarity functions for comparing the structural similarity of the hints given by users in CAS-queries with the structure extracted of the indexed articles. In general, there are four groups of functions that can be used and that differ in the thoroughness they achieve in analyzing the similarity: *perfect match, partial match, fuzzy match,* and *baseline (flat)* [3].

We developed the following formula as a representative of the *partial match* type of strategies [3], i.e. where one of the compared structures has to be a sub-sequence of the other and the overlapping ratio is measured.

$$Sim_1(s_q, s_{ru}) = \begin{cases} \left(\dfrac{1+|s_q|}{1+|s_{ru}|}\right)^\alpha, & \text{if } s_q \text{ is sub-sequence of } s_{ru} \\ \beta \cdot Sim_1(s_{ru}, s_q), & \text{if } s_{ru} \text{ is sub-sequence of } s_q \\ 0, & \text{else} \end{cases} \qquad \textbf{(ArchSim)}$$

s_q represents the structural condition in the query and s_{ru} stands for the structure of the search term found in the collection. $|s_x|$ is the number of words in s_x. Both parameters α and β allow for finer tuning of the calculated similarity value. We also implemented a small tag dictionary that contains values for the similarity between known tags. For example, *<author>* and *<writer>* are rather similar and a precise similarity value can be assigned to these tags.

The *fuzzy match* type of functions takes into account gaps or wrong sequences in the different tags. As a representative of the class of functions based on cost calculation for transforming the query structure into the target one, we used a method based on the *Levenstein Edit Distance* to compute the similarity between two strings by counting the operations *delete, insert* and *replace* needed to transform one string into another [10]. This method allows similar measuring of the difference between XML structures by considering every tag as a single character. We implemented and evaluated this method with a suitable normalization and, as above, an enhancement with a tag dictionary was also applied (**Sim₂=PathSim**).

Another approach for the similarity analysis of XML structures within the scope of *fuzzy* type strategies is the definition of a number of factors describing specific properties of the compared structures and combining them in a single function. Five such factors are proposed in [4]: semantic completeness (SmCm), *semantic correctness (SmCr),* structural completeness (StCm), *structural correctness (StCr),* and *structural cohesion (StCh).* They can be used for a thorough analysis and representation of the different similarity aspects between two XML structures. We used the arithmetic mean to compute the *SmCr* and measured the similarities between tags. We used these factors to construct combined similarity functions. In order to compare these functions, we built a small heterogeneous document collection with search terms occurring in many different XML contexts, resulting in a number of structures to be compared and ranked. Several functions were tested with a number of parameters. We achieved the best ranking results with the following formula:

$$Sim_3 = \alpha \cdot \left(\frac{\omega_1 \, SmCm + \omega_2 \, SmCr}{2}\right) + \beta \cdot \left(\frac{\omega_3 \, StCm + \omega_4 \, StCr + \omega_5 \, StCh}{3}\right) \qquad \textbf{(FineSim)}$$

All similarity factors were normalized and parameter boundaries were set such that the resulting single similarity value remains within the interval [0,1]. Parameters α and β provide an opportunity to shift the weight of the similarity between the two classes of factors – *semantic* and *structural*. The five parameters ω_i can be used for further fine-tuning.

The three developed functions were compared with the *perfect match strategy* where exact match of structures is measured as well as the *baseline (flat)*, where no structure was taken into account (CO-run).

3 System Used for Experiments

3.1 SPIRIX, a Distributed Solution

For our experiments, we used *SPIRIX*, a P2P search engine for XML-Retrieval that is based on a structured P2P network (DHT). Figure 1 describes the architecture of each SPIRIX peer.

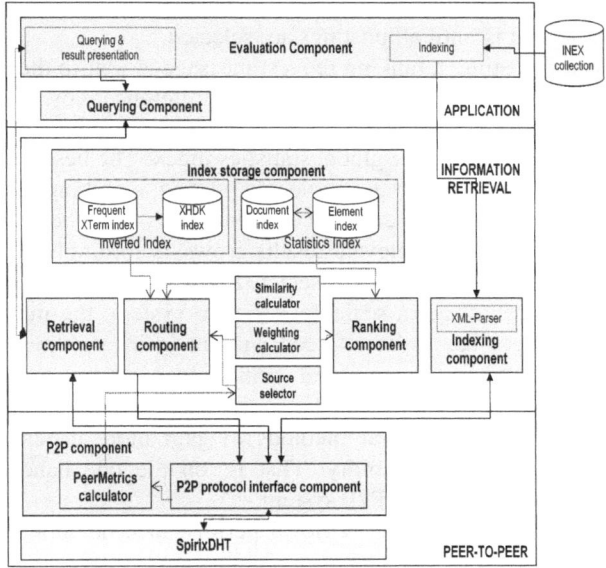

Fig. 1. SPIRIX and its components

When used for INEX experiments, the whole collection is indexed by one single peer. However, the information extracted by the *indexing component* is distributed over all participating peers such that each peer is responsible for storing, maintaining and –at querying time- providing parts of the global index. The indexed information consists of XTerms, which are tuples of XML structure and terms. For each XTerm or combination of XTerms (so called XHDKs), the according posting list are distributed into the global index, using the P2P protocol *SpirixDHT* which is based on Chord [9]. Term statistics about the indexed documents and their elements are also extracted and distributed in form of vectors, where each component of a vector represents the weight of an XTerm in a potentially relevant retrieval unit. All extracted information can be identified by a unique key and will be stored on the peer assigned to the hash value of this key. Posting lists are stored in the distributed *inverted index*, term statistics are stored in the *statistics index* (split into statistics for documents and statistics for elements).

At querying time, topics are prepared by the local *evaluation component* of the evaluating/querying peers. For this, each topic is converted into an XTerm-format and split into keys (XTerms or XTerm-combinations). For each key, a message is sent (using *SpirixDHT*) to the peer assigned to this key, i.e. holding the key's posting list. From this posting list, the best postings are selected, using a combination of ranking algorithms (*weighting calculator*), structural similarity functions (*similarity calculator*), and peer metrics (*source selector* and *peerMetrics calculator*). If the topic consists of several keys, the according posting lists have to be merged with each other by processing them in a pipelined manner: the shortest posting list is selected, and then sent to the peer assigned to the key with the next shortest posting list etc. until all keys are processed, that is until all postings are selected.

For each selected posting, a ranking request message is sent to the peer assigned to the term statistics of the document or element referenced by the posting. This document or element will be ranked by the local *ranking component*, using statistics information of the local part of the global statistics index. The best results will be sent back to the evaluating peer. As for each selected posting, a ranking request message is sent, the number of ranked documents and elements grow with the number of selected postings. The chance of finding relevant results grows, accordingly. To achieve a high precision, enough postings have to be selected. However, each message produces network traffic. In order to get a scalable, efficient system, the number of selected postings / sent messages has to be reduced without losing too much effectiveness. XML-Retrieval algorithms and structural similarity functions to judge similarity between CAS-queries and posting list keys are used to select adequate postings, taking advantage of XML-Retrieval methods to gain more efficiency in the P2P system and to guarantee its scalability. That is, on the one hand XML-Retrieval techniques are used to improve the P2P system.

How can, on the other hand, the retrieval performance be supported by the P2P design, especially when searching a large-size collection?

First of all, the global index is split over all participating peers such that each peer has to take care of only a fraction of the total index. This not only reduces the disk space consumed but also allows for a bigger portion of the local index to be hold in memory of each peer. At querying time, the processing of the posting lists is performed in parallel on all the peers that hold adequate posting lists. Those peers share the load of selecting the best postings in respect to computing power necessary for the selection algorithms, the memory used for this task, and the disk I/O for reading the posting lists. Also, the ranking process is performed in parallel: its load is split over all the peers that hold term statistics and thus get a ranking request message with the request to perform a local ranking. Hence, both the process of selecting postings and executing the relevance computations are performed in parallel. This is a big advantage when dealing with large-scale systems such as the new Wikipedia collection of INEX 2009.

3.2 Technical Details of SPIRIX

SPIRIX features element retrieval, can answer CO and CAS topics, indexes structure, does not support phrases, performs stemming and stopword removal, and punishes negative terms.

For participation at the Ad Hoc track, we aimed at high precision. Many parameters that influence the balance between effectiveness and efficiency were thus set to values that aim at effectiveness. For instance, this includes global statistics. For the selection of postings, we decided for a compromise of 1500 postings respectively 150 postings.

Structure size in index: For each term, a separate posting list for each of its structures is stored. This allows efficient selecting of postings according to structural similarity between hints in CAS topics and the stored structure of an XTerm. To reduce the variety of structures, several methods were tried such as:

♦ Stemming of tags (Snowball stemmer),
♦ Deleting stopword tags (e.g. conversionwarnings),
♦ Deleting of tags at specific positions to limit the structure size to 20 bytes (tags are deleted until a path size of 20 bytes is achieved).

In comparison to a full structure index, we saved 40% of the index size (without compression) and reduced the amount of tags per structure to 3.3 tags on average. The total amount of different structures was reduced to 1539. However, evaluation showed that we can achieve a significant improvement on early precision measures when using structural similarity for ranking and routing – but only with the full structure index. With the reduced index, there is almost no improvement (less than 1%). We assume that the reduction to 20 bytes retarded our similarity functions from detecting the correct similarity between structures. Further experiments have to be conducted to analyze, which of the used techniques can reduce the structure variety without losing precision. For the experiments submitted to INEX this year, we used two different indexes: one without structure and one with full structure.

Early termination (Selection of 1500 respectively 150 postings): Due to our system architecture, using all postings from a posting list is not efficient as this leads to many ranking request messages. However, precision increases with the amount of selected postings (until a collection specific point). Thus, the best 1500 respectively 150 documents were selected from each query term's posting list. Note, that this is done on separate peers and thus without merging – we lose precision for multi term queries when good documents are on positions > 1500 respectively 150. Our techniques to support multi term queries (XHDK-algorithms) have not been used for the described INEX experiments in order to focus on routing and ranking algorithms.

Global statistics: SPIRIX is based on a DHT which enables collecting and storing of global statistics. Usually, we estimate these statistics from the locally stored part of the distributed document index that contains randomly hashed samples from the collection. This estimation technique saves the messages necessary for distributing and accessing global statistics at the cost of losing precision depending on the estimations. Thus, for the INEX runs the exact global statistics were used.

Indexing of structure: XML-structure is considered both in the routing and in the ranking. It is extracted while parsing the XML-document at indexing and stored together with the extracted terms.

Element retrieval: Retrieval results can be either whole documents (Wikipedia articles) or XML elements out of these documents. In the indexing, elements are treated as documents, that is they are indexed independently by extracting their statistics and storing these in the retrieval unit index. However, we believe in a strong correlation between the relevance of an XML element and its parent document, as shown in related work. Therefore, the parent document influences the retrieval of elements: when ranking, the score of an element is computed based on the stored statistics but smoothed by the score of its parent document. In the routing process, only evidence of the parent document and its best retrieval unit are used for the decision, which documents plus their containing elements are to be ranked.

Other technical details: Phrases are not supported. Terms that appear together in a phrase are regarded independently. Stemming and removal of stopwords are used. All elements with less than 15 words are also removed. Results that contain negative terms (specified by the user by "not" respective "-") are still ranked. However, their relevance is reduced by the weight of all negative terms multiplied with a negativity factor.

4 Submitted Runs

We submitted two runs for the Ad Hoc Track (Focused Task) to compare two different ranking functions without the use of structural hints, e.g. we used CO-queries. We furthermore submitted several runs for the Efficiency Track where CO- and CAS-queries were applied for ranking or routing. Here, we evaluated four different structural similarity functions and compared the results with a run, where no structural similarity was taken into account (CO-run). Table 1 and 2 give an overview of the submitted runs.

5 Evaluation of INEX Experiments

5.1 Ad Hoc Track Results

Our runs achieved 59.03% and 58.91% precision at recall level 1% (iP[0.01]) of the Ad Hoc Track (focused). In the official list of the top-10 systems, we thus are on rank seven, with the best performing system being the solution developed by the University of Waterloo (63.33%).

Table 1. CO-Runs submitted by University of Applied Science, Frankfurt

Name (run#)	Task	Ranking function	Struct. Sim. function: Ranking	Routing function	#Postings per peer	iP[0.01]
Spirix09R001 (#872)	Ad Hoc: Focused	BM25-adaption	Flat	-	1500	0.5903
Spirix09R002 (#873)	Ad Hoc: Focused	tf*idf-adaption	Flat	-	1500	0.5891

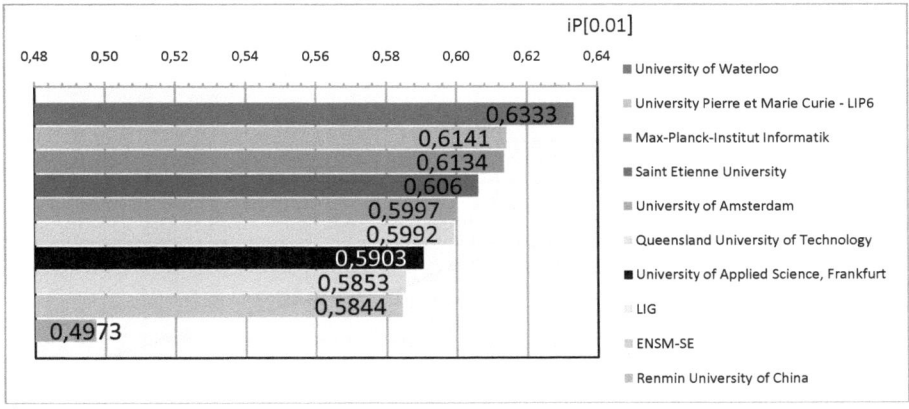

Fig. 2. Rank 7 in the top-10 systems of Ad Hoc Track (Focused)

Our best performing run was Spirix09R001, where an adaption of BM25 was applied based on the use of our XTerms. The Recall-Precision graph in figure 3 shows a curve descending slower than that of most other participants. From what we can see at the official INEX webpage results, this is the case for only the best 5 systems. As only the iP[0.01] values have been released, we cannot make an exact statement on this but we expect that the MAiP of SPIRIX might be significantly higher than that of other search engines. SPIRIX seems to be very good in identifying 1% of the relevant results but is able to find good results at higher recall levels, too.

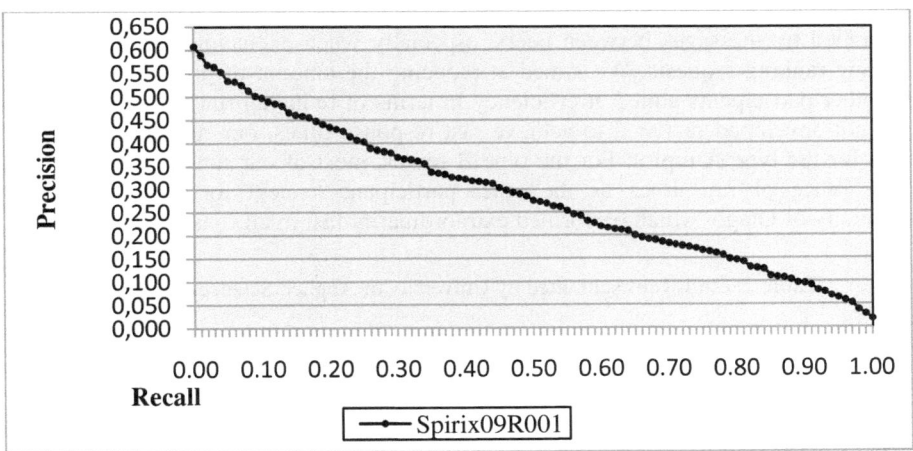

Fig. 3. Recall-Precision graph of the best performing run Spirix09R001: relatively flat curve

Figure 4 shows a comparison of both SPIRIX runs for early recall levels. Run Spirix09R001, where a BM25-adaption was applied, performs only slightly better than run Spirix09R002, where a Robertson TF*IDF-adaption was applied. The differences are not significant and can hardly be seen. The same applies for higher recall levels and for MAiP.

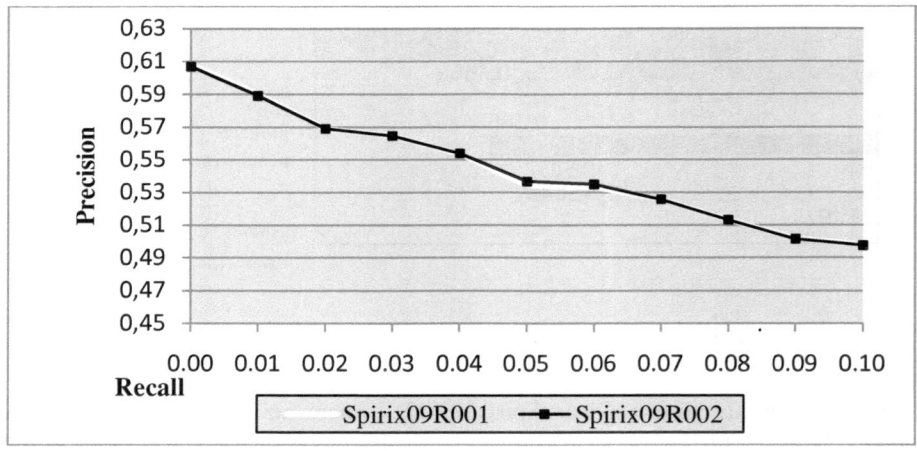

Fig. 4. SPIRIX runs of the Ad Hoc track: different ranking functions

5.2 Efficiency Track Results

At the Efficiency track, we achieved the highest precision in both tasks, which is 59.9% for the type A topics and 46.4% for the type B topics. Our notion of efficiency is to get the P2P system to scale. This is achieved by reducing the bandwidth consumed by messages between peers, especially when exchanging posting lists or sending ranking requests. We aimed at reducing the amount of postings to be used. All other participants aimed at efficiency in terms of reducing run times. As SPIRIX was not developed or tuned to achieve fast response times, our run times are rather slow for the type A topics. For the type B topics, most of our runs have been in the same range of run times as the other participants except for the system from university of Otago, which performed extraordinarily fast in all runs and tasks.

Table 2. CAS-Runs submitted by University of Applied Science, Frankfurt

Name	Task	Ranking function	Struct. sim. function: Ranking	Routing function	Struct. similar. function: Routing	# po- stings per peer	iP [0.01]	Avg- time
Spirix09 RX01	Eff.: typeA	BM25	Flat	BM25E	Flat	1500	0.599	50063
...RX02	Eff.: typeA	BM25E	Flat	BM25E	Flat	150	0.427	46820
...RX03	Eff.: typeA	BM25E	PathSim	BM25E	PathSim	150	0.372	34663
...RX04	Eff.: typeA	BM25E	FineSim	BM25E	FineSim	150	0.372	34663
...RX05	Eff.: typeA	BM25E	ArchSim	BM25E	ArchSim	150	0.398	44191
...RX06	Eff.: typeA	BM25E	Strict	BM25E	Strict	150	0.333	44352
...RX07	Eff.: typeA	BM25	Flat	Baseline	Flat	1500	0.119	42878

Table 2. (*continued*)

Name	Task	Ranking function	Struct. sim. function: Ranking	Routing function	Struct. similar. function: Routing	# po-stings per peer	iP [0.01]	Avg-time
...RX08	Eff.: typeA	BM25E	Flat	Baseline	Flat	150	0.59	959
...RX09	Eff.: typeA	BM25E	PathSim	Baseline	Flat	150	0.352	496
...RX10	Eff.: typeA	BM25E	FineSim	Baseline	Flat	150	0.403	502
...RX11	Eff.: typeA	BM25E	ArchSim	Baseline	Flat	150	0.344	483
...RX12	Eff.: typeA	BM25E	Strict	Baseline	Flat	150	0.119	474
...RX13	Eff.:typeA+B	BM25	Flat	Baseline	Flat	1500	0.59	2986
...RX14	Eff.:typeA+B	BM25E	Flat	BM25E	Flat	150	0.403	470
...RX15	Eff.:typeA+B	BM25E	PathSim	BM25E	Flat	150	0.352	746
...RX16	Eff.:typeA+B	BM25E	FineSim	BM25E	Flat	150	0.403	1156
...RX17	Eff.:typeA+B	BM25E	ArchSim	BM25E	Flat	150	0.344	1863
...RX18	Eff.:typeA+B	BM25E	Strict	BM25E	Flat	150	0.119	1675
...RX19	Eff.:typeA+B	BM25	Flat	BM25E	Flat	1500	0.599	47857
...RX20	Eff.:typeA+B	BM25E	Flat	BM25E	Flat	150	0.427	46712
...RX21	Eff.:typeA+B	BM25E	PathSim	BM25E	PathSim	150	0.372	35746
...RX22	Eff.:typeA+B	BM25E	FineSim	BM25E	FineSim	150	0.398	45072
...RX23	Eff.:typeA+B	BM25E	ArchSim	BM25E	ArchSim	150	0.333	44285
...RX24	Eff.:typeA+B	BM25E	Strict	BM25E	Strict	150	0.119	44256

For some of our experiments, the use of structure in the routing process showed a clear increase of precision in comparison to our runs without exploiting XML-structure. For example, this could be noticed when analyzing the use of our adaption of BM25E [7] in the routing process. Figure 5 compares run SpirixRX01 with SpirixRX07. Both runs use the same techniques for ranking but apply different routing algorithms. The RX01 run takes advantage of the indexed structure (stored in the posting list) by applying a BM25E-adaption while RX07 ignores structure and thus achieves a precision of only 59% instead of 59.9%.

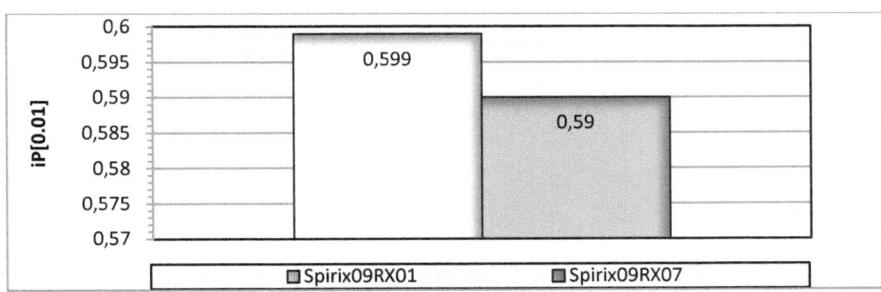

Fig. 5. Run RX01 used BM25E as routing function and thus outperformed Run RX07

As shown in figure 6 and 7, our runs ranked best in terms of precision. Figure 6 shows the precision iP[0.01] of the top runs in regard of the task where type A topics where queried. Our best runs, e.g. RX01, achieved 59.9% precision at this task.

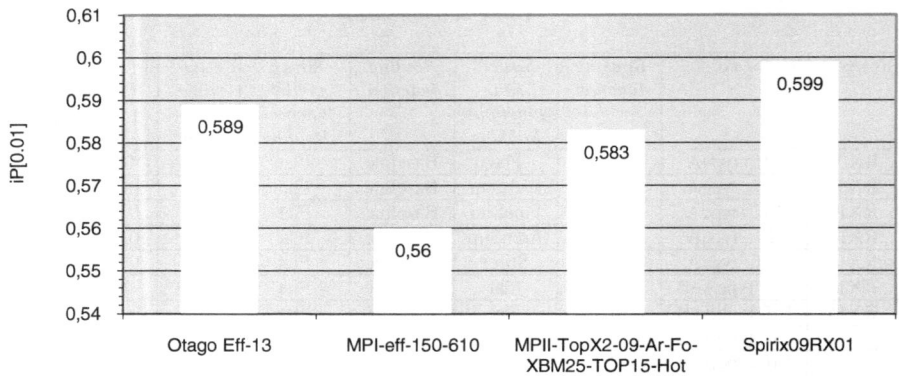

Fig. 6. Top runs for type A topics

Figure 7 displays iP[0.01] for the top runs, this time for the type B topics. Again, SPIRIX performed best, e.g. with run RX20 and iP[0.01] of 46.4%.

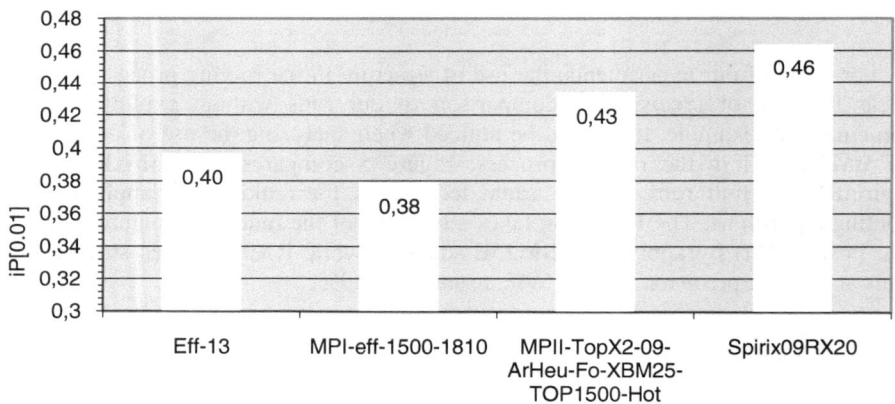

Fig. 7. Top runs for type B topics

6 Discussion

Our system SPIRIX gained high precisions at both the Ad Hoc Track and the Efficiency Track. A precision of 59.03% (iP[0.01]) was achieved at the Ad Hoc track (focused task), where the best performing system achieved 63.33%. SPIRIX thus resulted in rank 7 (considering the best run for each group). At the Efficiency track, we achieved the highest precision in both tasks, which is 59.9% for the type A topics and 46.4% for the type B topics.

These results were achieved with a P2P system, where communication overhead has to be taken care of and thus not all postings of the inverted index can be considered. Several XML-Retrieval techniques are used to select adequate postings in the routing process. For instance, our selection algorithms are based on BM25- and BM25E-adaptions as well as on several structural similarity functions. The parallelisation of both the posting selection and the ranking process is a big advantage when dealing with large-scale collections like the new Wikipedia collection, e.g. by being able to process more postings in memory at the same time.

The INEX results show that using XML-Retrieval methods in order to gain a scalable, efficient P2P system does not necessarily reduce the search quality but that distributed systems such as SPIRIX can achieve a search quality comparable with that of the top-10 centralized stand-alone systems. Achieving high precisions in P2P settings is possible!

References

[1] Baeza-Yates, R., Castillo, C., Junqueira, F., Plachouras, V., Silvestri, F.: Challenges on Distributed Web Retrieval. In: IEEE Int. Conf. on Data Engineering (ICDE'07), Turkey (2007)
[2] Balakrishnan, H., Kaashoek, F., Karger, D., Morris, R., Stoica, I.: Looking Up Data in P2P Systems. Communications of the ACM 46(2) (2003)
[3] Carmel, D., Maarek, Y., Mandelbrod, M., Mass, Y., Soffer, A.: Searching XML Documents via XML Fragments. In: Proc. of the 26th Int. ACM SIGIR, Toronto, Canada (2003)
[4] Ciaccia, P., Penzo, W.: Adding Flexibility to Strucuture Similarity Queries on XML Data. In: Andreasen, T., et al. (eds.) FQAS 2002. LNCS (LNAI), vol. 2522. Springer, Heidelberg (2002)
[5] Moffat, A., Webber, W., Zobel, J., Baeza-Yates, R.: A pipelined architecture for distributed text query evaluation. In: Springer Science + Business Media, LLC 2007 (2007)
[6] Risson, J., Moors, T.: Survey of research towards robust peer-to-peer networks – search methods. In: Technical Report UNSW-EE-P2P-1-1, Uni. of NSW, Australia (2004)
[7] Robertson, S., Zaragoza, H., Taylor, M.: Simple BM25 extension to multiple weighted fields. In: Proc. of CIKM'04. ACM Press, New York (2004)
[8] Steinmetz, R., Wehrle, K. (eds.): Peer-to-Peer Systems and Applications. LNCS, vol. 3485. Springer, Heidelberg (2005)
[9] Stoica, I., Morris, R., Liben-Nowell, D., Karger, D., Kaashoek, F., Dabek, F., Balakrishnan, H.: Chord - A Scalable Peer-to-peer Lookup Protocol for Internet Applications. IEEE/ACM Transactions on Networking 11(1) (2003)
[10] Vinson, A., Heuser, C., Da Silva, A., De Moura, E.: An Approach to XML Path Matching. In: WIDM'07, Lisboa, Portugal, November 9 (2007)
[11] Winter, J., Jeliazkov, N., Kühne, G.: Aiming For More Efficiency By Detecting Structural Similarity. In: Geva, S., Kamps, J., Trotman, A. (eds.) INEX 2008. LNCS, vol. 5631, pp. 237–242. Springer, Heidelberg (2009)
[12] Zeinalipour-Yazti, D., Kalogeraki, V., Gunopulos, D.: Information Retrieval in Peer-to-Peer Networks. IEEE CiSE Magazine, Special Issue on Web Engineering (2004)

Overview of the INEX 2009 Entity Ranking Track

Gianluca Demartini[1], Tereza Iofciu[1], and Arjen P. de Vries[2]

[1] L3S Research Center
Leibniz Universität Hannover
Appelstrasse 9a D-30167 Hannover, Germany
{demartini,iofciu}@L3S.de
[2] CWI & Delft University of Technology
The Netherlands
arjen@acm.org

Abstract. In some situations search engine users would prefer to retrieve entities instead of just documents. Example queries include "Ital-
ian Nobel prize winners", "Formula 1 drivers that won the Monaco Grand
Prix", or "German spoken Swiss cantons". The XML Entity Ranking
(XER) track at INEX creates a discussion forum aimed at standardiz-
ing evaluation procedures for entity retrieval. This paper describes the
XER tasks and the evaluation procedure used at the XER track in 2009,
where a new version of Wikipedia was used as underlying collection; and
summarizes the approaches adopted by the participants.

1 Introduction

Many search tasks would benefit from the ability of performing typed search, and
retrieving entities instead of 'just' web pages. Since 2007, INEX has organized
a yearly XML Entity Ranking track (INEX-XER) to provide a forum where
researchers may compare and evaluate techniques for engines that return lists
of entities. In entity ranking (ER) and entity list completion (LC), the goal is
to evaluate how well systems can rank entities in response to a query; the set
of entities to be ranked is assumed to be loosely defined by a generic category,
implied in the query itself (for ER), or by some example entities (for LC). This
year we adopted the new Wikipedia document collection containing annotations
with the general goal of understanding how such semantic annotations can be
exploited for improving Entity Ranking.

Entity ranking concerns triples of type <query, category, entity>. The category (i.e., the entity type), specifies the type of 'objects' to be retrieved. The
query is a free text description that attempts to capture the information need.
The Entity field specifies example instances of the entity type. The usual information retrieval tasks of document and element retrieval can be viewed as special
instances of this more general retrieval problem, where the category membership
relates to a syntactic (layout) notion of 'text document', or 'XML element'. Expert finding uses the semantic notion of 'people' as its category, where the query

S. Geva, J. Kamps, and A. Trotman (Eds.): INEX 2009, LNCS 6203, pp. 254–264, 2010.

would specify 'expertise on T' for finding experts on topic T. While document retrieval and expert finding represent common information needs, and therefore would require specific technologies to be developed, the XER track challenges participants to develop generic ranking methods that apply to entities irrespective of their type: e.g., actors, restaurants, museums, countries, etc.

In this paper we describe the INEX-XER 2009 track running both the ER and the LC tasks, using selected topics from the previous years over the new INEX Wikipedia collection. For evaluation purpose we adopted a stratified sampling strategy for creating the assessment pools, using xinfAP as the official evaluation metric [6].

The remainder of the paper is organized as follows. In Section 2 we present details about the collection used in the track and the two different search tasks. Next, in Section 3 we briefly summarize the approaches designed by the participants. In Section 4 we summarize the evaluation results computed on the final set of topics for both the ER and LC tasks. As this year we used a selection of topics from the past INEX-XER campaigns, in Section 5 we provide an initial comparison of the new test collection with the previous ones. Finally, in Section 6, we conclude the paper.

2   INEX-XER Setup

2.1   Data

The INEX-XER 2009 track uses the new Wikipedia 2009 XML data based on a dump of the Wikipedia taken on 8 October 2008 and annotated with the 2008-w40-2 version of YAGO [5], as described in [4]. The Wikipedia pages and links are annotated with concepts from the WordNet thesaurus. Available annotations could be exploited to find relevant entities. Category information about the pages loosely defines the entity sets. The entities in such a set are assumed to loosely correspond to those Wikipedia pages that are labeled with this category (or perhaps a sub-category of the given category). Obviously, this is not perfect as many Wikipedia articles are assigned to categories in an inconsistent fashion. There are no strict restrictions in Wikipedia for category assignment, thus categories can represent the type of an entity as well as a label for it. Retrieval methods should handle the situation that the category assignments to Wikipedia pages are not always consistent, and also far from complete. The intended challenge for participants is therefore to exploit the rich information from text, structure, links and annotations to perform the entity retrieval tasks.

2.2   Tasks

This year's INEX-XER track consists of two tasks, i.e., entity ranking (with categories), and entity list completion (with examples). Entity list completion is a special case of entity ranking where a few examples of relevant entities are provided instead of the category information as relevance feedback information.

Entity Ranking. The motivation for the entity ranking (ER) task is to return entities that satisfy a topic described in natural language text. Given preferred categories, relevant entities are assumed to loosely correspond to those Wikipedia pages that are labeled with these preferred categories (or perhaps sub-categories of these preferred categories). Retrieval methods need to handle the situation where the category assignments to Wikipedia pages are not always consistent or complete. For example, given a preferred category 'art museums and galleries', an article about a particular museum such as the 'Van Gogh Museum' may not be labeled by 'art museums and galleries' at all, or, be labeled by a sub-category like 'art museums and galleries in the Netherlands'. Therefore, when searching for "art museums in Amsterdam", correct answers may belong to other categories close to this category in the Wikipedia category graph, or may not have been categorized at all by the Wikipedia contributors. The category 'art museums and galleries' is only an indication of what is expected, not a strict constraint.

List Completion. List completion (LC) is a sub-task of entity ranking which considers relevance feedback information. Instead of knowing the desired category (entity type), the topic specifies a number of correct entities (instances) together with the free-text context description. Results consist again of a list of entities (Wikipedia pages). If we provide the system with the topic text and a number of entity examples, the task of list completion refers to the problem of completing the partial list of answers. As an example, when ranking 'Countries' with topic text 'European countries where I can pay with Euros', and entity examples such as 'France', 'Germany', 'Spain', then 'Netherlands' would be a correct completion, but 'United Kingdom' would not.

2.3 Topics

Based on the topics from the previous two INEX-XER years, we have set up a collection of 60 XER topics for both the ER and LC tasks, with 25 from 2007 and 35 topics form 2008. The <categories> part is provided exclusively for the Entity Ranking Task. The <entities> part is given to be used exclusively for the List Completion Task.

2.4 The INEX-XER 2009 Test Collection

The judging pools for the topics have been based on all submitted runs, using a stratified sampling strategy [6]. As we aimed at performing relevance judgments on 60 topics (as compared to 49 in 2008), we adopted a less aggressive sampling strategy that would make the judging effort per topic lower. We used the following strata and sampling rates for the pool construction of INEX-XER 2009:

- $[1, 8]$ 100%
- $[9, 31]$ 70%
- $[32, 50]$ 30%
- $[51, 100]$ 10%

where $[i, j]$ indicates the interval of retrieved results considered (i.e., from rank i to rank j) followed by the sampling rate for the interval. The resulting pools contained on average 312 entities per topic (as compared to 400 in 2008 and 490 in 2007).

All 60 topics have been re-assessed by INEX-XER 2009 participants on the new collection. As last year, from the originally proposed ones, topics with less than 7 relevant entities (that is, 104, and 90) and topics with more than 74 relevant entities (that is, 78, 112, and 85) have been excluded [3]. The final set consists of 55 genuine XER topics with assessments.

Out of the 55 XER topics, 3 topics have been excluded for the LC task (i.e., 143, 126, and 132). The reason is that example entities for these topics were not relevant as the underlying Wikipedia collection has changed. After this selection, 52 List Completion topics are part of the final set and are considered in the evaluation.

2.5   Not-An-Entity Annotations

An additional difference from the relevance judgments performed during last INEX-XER is the possibility for the assessor to mark a retrieved result as not being an entity. This choice is intended for those Wikipedia pages that do not represent an entity and, thus, would be irrelevant to any XER query. Examples include "list-of" or "disambiguation" pages.

Differentiating between a non-relevant and not-an-entity result does not influence the evaluation of INEX-XER systems as both judgments are considered a wrong result for XER tasks.

3   Participants

At INEX-XER 2009 five groups submitted runs for both the ER and LC tasks. We received a total of 16 ER runs and 16 LC runs. In the following we report a short description of the approaches used, as reported by the participants.

Waterloo. Our two runs for each task is based on Clarke et al.'s question answering technique that uses redundancy [1]. Specifically, we obtained top scoring passages from each article in the corpus using topic titles (for ER task) and topic titles+examples (for LC task). For LC task, we estimated the categories of entities to return as the union of categories in the examples. Within each top scoring passage, we located candidate terms that have a Wikipedia page that fall under the desired categories. We ranked the candidate terms by the number of distinct passages that contain the term.

AU-CEG (Anna University,Chennai). In our approach, we have extracted the Entity Determining Terms (EDTs), Qualifiers and prominent n-grams from the query. As a second step, we strategically exploit the relation between the extracted terms and the structure and connectedness of the corpus to retrieve links which are highly probable of being entities and then use a recursive mechanism for retrieving relevant documents through the Lucene Search. Our ranking mechanism combines various approaches that make use of category information, links, titles and WordNet information, initial description and the text of the document.

PITT team (School of Information Sciences, University of Pittsburgh). As recent studies indicate that named entities exist in queries and can be useful for retrieval, we also notice the ubiquitous existence of entities in entity ranking queries. Thus, we try to consider entity ranking as the task of finding entities related to existing entities in a query. We implement two generative models, i.e., MODEL1EDR and MODEL1EDS, both of which try to capture entity relations. These two models are compared with two baseline generative models: MODEL1D, which estimates models for each entity using Wikipedia entity documents; MODEL1E, which interpolates entity models in MODEL1D with entity category models.

UAms (Turfdraagsterpad). We rank entities by combining a document score, based on a language model of the document contents, with a category score, based on the distance of the document categories to the target categories. We extend our approach from last year by using Wordnet categories and by refining the categories we use as target categories.

UAms (ISLA). We propose a novel probabilistic framework for entity retrieval that explicitly models category information in a theoretically transparent manner. Queries and entities are both represented as a tuple: a term-based plus a category-based model, both characterized by probability distributions. Ranking of entities is then based on similarity to the query, measured in terms of similarities between probability distributions.

Discussion. It is possible to notice that a common behavior of participants this year was to identify entity mentions in the text of Wikipedia articles, passages, or queries. They then applied different techniques (e.g., detect entity relations, exploit category information) to produce a ranked list of Wikipedia articles that represents the retrieved entities. The best performing approach exploited a probabilistic framework ranking entities using similarity between probability distributions.

4   Results

The five groups submitted 32 runs to the track. The evaluation results for the ER task are presented in Table 1, those for the LC task in Table 2, both reporting xinfAP [6]. We can see that best effectiveness values are higher than 0.5 xinfAP which shows improvement over past years. It must be noticed that high effectiveness values may be due to the adoption of topics already used in past campaigns. The best performing group based their runs on a probabilistic framework for ranking entities using similarity measures between probability distributions.

As we considered all the runs during the pooling phase and as some groups submitted more runs than others, we performed an analysis of possible bias in

Table 1. Evaluation results for ER runs at INEX XER 2009

Run	xinfAP
2_UAmsISLA_ER_TC_ERreltop:	0.517
4_UAmsISLA_ER_TC_ERfeedbackSP:	0.505
1_AU_ER_TC_mandatoryRun.txt:	0.270
3_UAmsISLA_ER_TC_ERfeedbackS:	0.209
2_UAmsISLA_ER_TC_ERfeedback:	0.209
1_TurfdraagsterpadUvA_ER_TC_base+asscats:	0.201
3_TurfdraagsterpadUvA_ER_TC_base+asscats+prfcats:	0.199
2_TurfdraagsterpadUvA_ER_TC_base+prfcats:	0.190
1_UAmsISLA_ER_TC_ERbaseline:	0.189
4_TurfdraagsterpadUvA_ER_TC_base:	0.171
1_PITT_ER_T_MODEL1EDS:	0.153
1_PITT_ER_T_MODEL1EDR:	0.146
1_PITT_ER_T_MODEL1ED:	0.130
1_PITT_ER_T_MODEL1D:	0.129
1_Waterloo_ER_TC_qap:	0.095
5_TurfdraagsterpadUvA_ER_TC_asscats:	0.082

Table 2. Evaluation results for LC runs at INEX XER 2009

Run	xinfAP
5_UAmsISLA_LC_TE_LCexpTCP:	0.520
3_UAmsISLA_LC_TE_LCreltop:	0.504
6_UAmsISLA_LC_TE_LCexpTCSP:	0.503
1_UAmsISLA_LC_TE_LCexpTC:	0.402
1_UAmsISLA_LC_TE_LCtermexp:	0.358
2_UAmsISLA_LC_TEC_LCexpTCS:	0.351
3_UAmsISLA_LC_TE_LCexpT:	0.320
1_AU_LC_TE_mandatoryRun.txt:	0.308
2_UAmsISLA_LC_TE_LCbaseline:	0.254
4_UAmsISLA_LC_TE_LCexpC:	0.205
4_TurfdraagsterpadUvA_LC_TE_base+wn20cats:	0.173
3_TurfdraagsterpadUvA_LC_TE_base+wiki20cats+wn20cats:	0.165
2_TurfdraagsterpadUvA_LC_TE_base+wiki20cats+prfcats:	0.160
5_TurfdraagsterpadUvA_LC_TE_base+wiki20cats:	0.157
1_TurfdraagsterpadUvA_LC_TE_base+wiki20cats:	0.156
1_Waterloo_LC_TE:	0.100

the pool. Figure 1 shows the *pool coverage* (i.e., the number of entities retrieved by the run which are present in the pool and, therefore, have been judged) as compared with the total number of retrieved entities. Figure 2 shows the *pool unique contribution* (the number of entities in the pool which were sampled only in this run) for each run submitted to INEX-XER 2009. The runs having worse coverage from the pool are also those that contribute most unique entities. This means that such runs are "different" from others in the sense that they retrieve different entities. Interestingly, runs from the AU-CEG group contributed a big

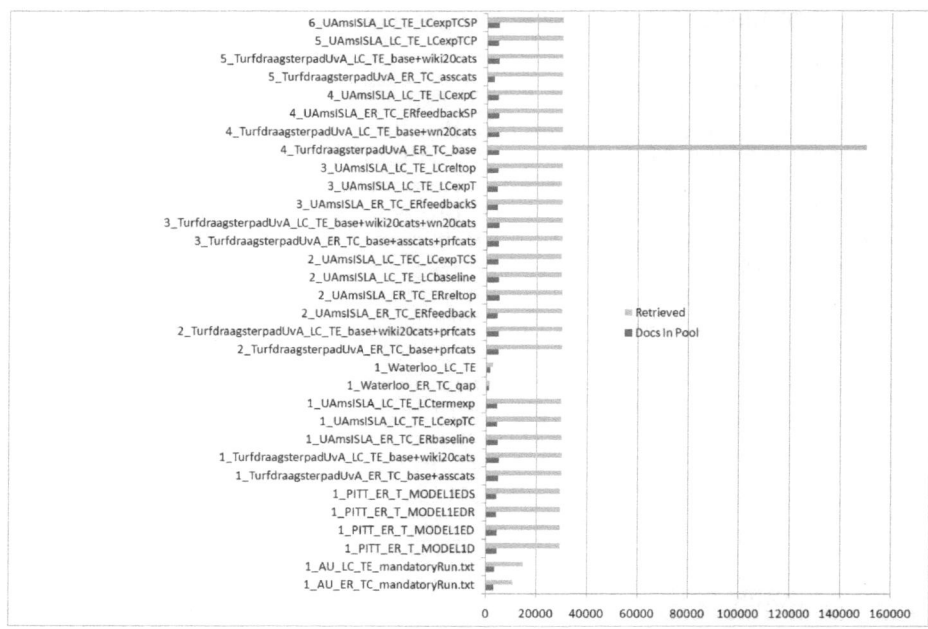

Fig. 1. Pool coverage: number of entities retrieved by the runs and present in the pool

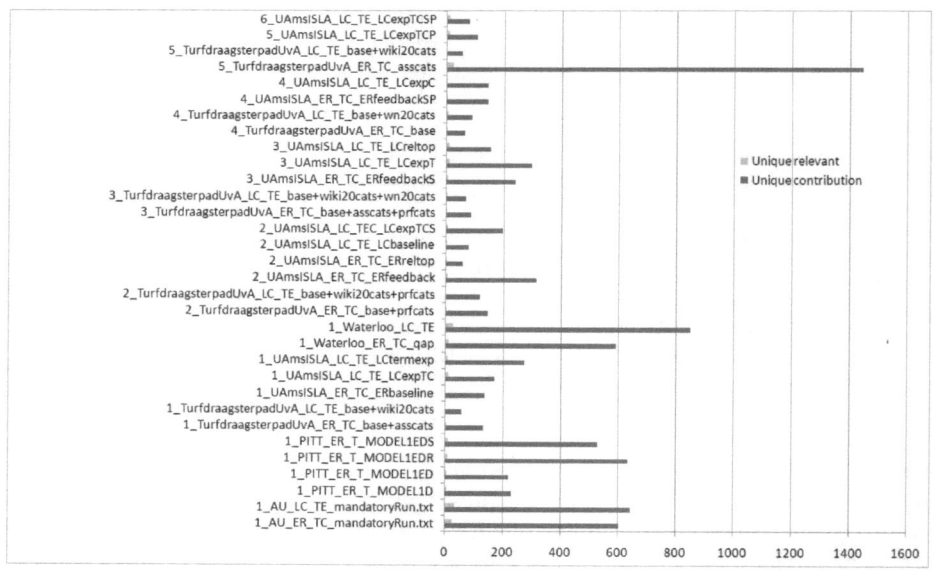

Fig. 2. Pool Unique Contribution: number of (relevant) entities sampled only in this run

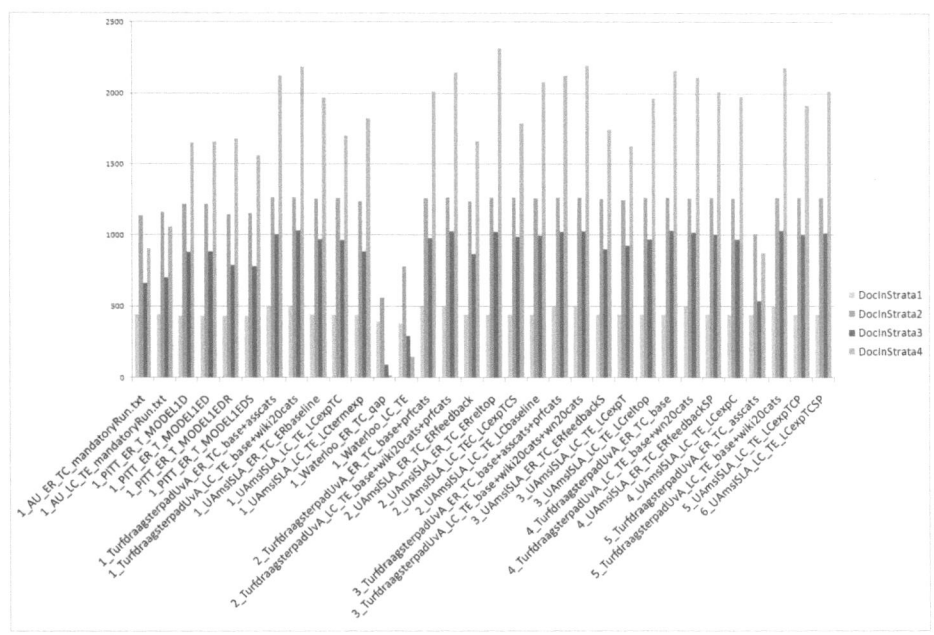

Fig. 3. Per-stratum pool coverage: number of entities retrieved by runs in different strata and present in the pool

number of unique relevant entities. We can see in Figure 3 that a relatively high proportion of retrieved entities belong to strata 1 and 2, which guarantees a fair evaluation. However, as some runs did not retrieve up to 8 results for a topic and as some systems did not run all the topics, not all runs have an equal number of entities covered in stratum 1 (which considers a complete sampling). For the Waterloo runs for example, only few entities have been sampled due to the low number of entities retrieved per topic (see also Figure 1).

5   Comparison with Previous INEX-XER Collections

At INEX-XER 2009 we used a selected set of topics from previous years while using the newer and annotated Wikipedia collection. This allows us to perform some comparisons with previous collections.

Comparison on the number of relevant entities. Figure 4 shows the number of entities judged relevant for each topic at INEX-XER 2009 as well as in the previous years. While we would expect to find the same number of relevant entities while re-judging the same topic, we must take into account that the new Wikipedia is bigger and contains more up-to-date information. Thus, we expect the number of relevant entities to be greater or equal to that in the past year. This is not the case for 12 topics out of 60. The highest difference can be seen for

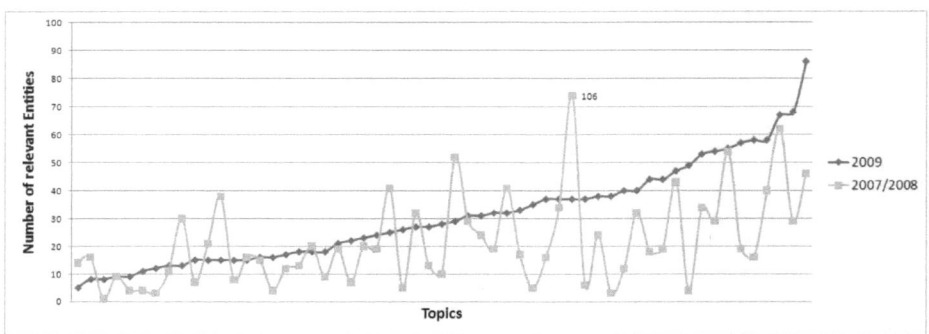

Fig. 4. Number of relevant entities per topic compared to previous years

Table 3. Comparison of samples and judgments between INEX-XER 2009 and previous years

	common					agreement		disagreement			
Year	S-co	S-past	S-2009	R-past	R-2009	R-co	I-co	Rpast-I09	R09-Ipast	UniRelPast	UniRel09
2007	57.86	490	295.27	16.36	26.18	10	41.86	3.86	2.55	2.5	13.64
2008	79.24	400.03	314.55	26.09	31.64	16.91	54	4.94	3.24	4.24	11.48

topic 106 "Noble English person from the Hundred Years' War". By a manual inspection of the judgments on this topic, we observed a high disagreement between assessors in the two different campaigns.

Preliminary Comparison on Samples and Judgments. Based on the titles of the sampled pages we compared the pool and assessments against the previous years. As the 2007 and 2008 topics have been assessed on the old smaller corpus, we made a comparison between the two datasets based on the entity title performing a simple textual comparison. Thus minor changes in the title of an entity in the two collections would lead to the entity not being identified as the same in the two collections. Table 3 shows the comparison results for the final set of 55 topics assessed in 2007/2008 and 2009. We show the following indicators in the table (average values per topic):

- S-co: the number of entities that have been sampled in both campaigns, based on entity title comparison;
- S-past: the total number of entities that have been sampled in the past campaign;
- S-2009: the total number of entities that have been sampled at INEX-XER 2009;
- R-past: the total number of relevant entities in the past campaign;
- R-2009: the total number of relevant entities at INEX-XER 2009;
- R-co: the number of entities assessed as relevant in both campaigns;
- I-co: the number of entities assessed as not-relevant in both campaigns;

- Ryear1-Iyear2: the number of entities assessed as relevant in year1 and as not-relevant in year2;
- UniRelYear: the number of entities that were both sampled and assessed as relevant only in the respective year.

For the set of 22 topics assessed both in 2007 and 2009, from the entities sampled in both years (S-co), 17% were relevant in both years, and 72% were not relevant (the agreement between the assessments being of 89%). On the other hand, 6.7% entities were relevant in 2007 and assessed as not relevant in 2009, and 4.4% the other way around, thus amounting to a disagreement of 11%. Additionally, on average 2.5 entities relevant in 2007 have not been sampled in 2009 (UniRelPast), and 13.64 entities not sampled in 2007 have been sampled and are relevant in 2009 (UniRel09).

For the set of 33 topics assessed both in 2008 and 2009, from the entities sampled in both years, 21% were relevant in both years, and 68% were not relevant (the agreement between the assessments being of 89%). On the other hand, 6.2% entities were relevant in 2008 and assessed as not relevant in 2009, and 4.1% the other way around, thus amounting to a disagreement of 10%. Additionally, on average 4.2 entities relevant in 2008 have not been sampled in 2009 (UniRelPast), and 11 entities not sampled in 2008 have been sampled and are relevant in 2009 (UniRel09).

In conclusion, we observe that for both sets of topics, the agreement between assessments (R-co + I-co) is much larger than the disagreement (Rpast-I09 + R09-Ipast).

We have not explicitly forbidden the use of previous years assessments for training and tuning 2009 systems, and indeed, a group (i.e., ISLA) had used them. For this reason we analyzed the effect of these runs on the pool. We removed from the pool the results uniquely contributed by these runs and analyzed how this affects the number of common relevant results between two corresponding years. As it can be seen in Table 4 the main effect of the system using previous years assessments for training was in the sampling of less non-relevant entities. The sampling of relevant entities (R-co) was affected only a little.

Table 4. Analysis of usage of previous assessments for training and tuning in INEX-XER 2009

2007-2009			
	S-co	R-co	I-Co
all runs	57.86	10	41.86
runs - ISLA	53.77	9.73	36.18
2008-2009			
	S-co	R-co	I-Co
all runs	79.24	16.91	54
runs - ISLA	73.79	16.64	44.45

6   Conclusions and Further Work

After the first two years of the XER Track at INEX 2007 and INEX 2008 [2,3], INEX-XER 2009 created additional evaluation material for IR systems that retrieve entities instead of documents. The new aspect in INEX XER 2009 is the use of the new annotated Wikipedia collection, re-using topics developed for the two previous years of this track. The track created a set of 55 XER topics with relevance assessments for the ER task and 52 for the LC task.

This has been the last year for INEX-XER which has built, over three years, a set of 55 ER topics with relevance assessments for two different document collections. Moreover, we have observed, over three campaigns, an improvement in term of effectiveness and more advanced techniques being used by Entity Retrieval systems participating to the track.

Acknowledgements. This work is partially supported by the EU Large-scale Integrating Projects OKKAM[1] - Enabling a Web of Entities (contract no. ICT-215032), LivingKnowledge[2] - Facts, Opinions and Bias in Time (contract no. 231126), VITALAS (contract no. 045389), and the Dutch National project MultimediaN.

References

1. Charles, L.A., Clarke, G.V., Lynam, T.R.: Exploiting redundancy in question answering. In: SIGIR '01: Proceedings of the 24th annual international ACM SIGIR conference on Research and development in information retrieval, pp. 358–365. ACM, New York (2001)
2. de Vries, A.P., Vercoustre, A.-M., Thom, J.A., Craswell, N., Lalmas, M.: Overview of the INEX 2007 Entity Ranking Track. In: Fuhr, N., Kamps, J., Lalmas, M., Trotman, A. (eds.) INEX 2007. LNCS, vol. 4862, pp. 245–251. Springer, Heidelberg (2008)
3. Demartini, G., de Vries, A.P., Iofciu, T., Zhu, J.: Overview of the INEX 2008 Entity Ranking Track. In: Geva, S., Kamps, J., Trotman, A. (eds.) INEX 2008. LNCS, vol. 5631, pp. 243–252. Springer, Heidelberg (2009)
4. Schenkel, R., Suchanek, F.M., Kasneci., G.: YAWN: A semantically annotated Wikipedia XML corpus. In: Symposium on Database Systems for Business, Technology and the Web of the German Society for Computer science, BTW 2007 (2007)
5. Suchanek, F.M., Kasneci, G., Weikum, G.: Yago: A Core of Semantic Knowledge. In: 16th international World Wide Web conference (WWW 2007). ACM Press, New York (2007)
6. Yilmaz, E., Kanoulas, E., Aslam, J.A.: A simple and efficient sampling method for estimating AP and NDCG. In: Myaeng, S.-H., Oard, D.W., Sebastiani, F., Chua, T.-S., Leong, M.-K. (eds.) SIGIR, pp. 603–610. ACM, New York (2008)

---

[1] http://fp7.okkam.org/
[2] http://livingknowledge-project.eu/

Combining Term-Based and Category-Based Representations for Entity Search

Krisztian Balog, Marc Bron, Maarten de Rijke, and Wouter Weerkamp

ISLA, University of Amsterdam, Science Park 107, 1098 XG Amsterdam, The Netherlands
{k.balog,m.bron,derijke,w.weerkamp}@uva.nl

Abstract. We describe our participation in the INEX 2009 Entity Ranking track. We employ a probabilistic retrieval model for entity search in which term-based and category-based representations of queries and entities are effectively integrated. We demonstrate that our approach achieves state-of-the-art performance on both the entity ranking and list completion tasks.

1 Introduction

Our aim for the INEX 2009 Entity Ranking track was to evaluate a recently proposed probabilistic framework for entity retrieval that explicitly models category information in a theoretically transparent manner [2]. Information needs and entities are represented as tuples: each consists of a term-based model plus a category-based model, both characterized by probability distribution over words. Ranking of entities is then based on similarity to the query, measured in terms of similarity between probability distributions. In our participation, our focus is on two core steps: (1) entity modeling and (2) query modeling and expansion. Moreover, we seek to answer how well parameter settings trained on the 2007 and 2008 editions of the Entity Ranking track perform on the 2009 setup. We find that our probabilistic approach is indeed very effective on the 2009 edition of the tasks, and delivers top performance on both tasks.

The remainder of this paper is organized as follows. We introduce our retrieval model in Section 2. Next, we discuss the submitted runs (Section 3), followed by their results and a discussion of these results (Section 4). We conclude in Section 5.

2 Modeling Entity Ranking

In this section we present a probabilistic retrieval framework for the two tasks that have been formulated within the Entity Ranking track. In the *entity ranking* task we are given a query (Q) and a set of target categories (C) and have to return entities. For *list completion* we need to return entities given a query (Q), a set of similar entities (E), and (optionally also) a set of target categories (C).[1]

[2] recently proposed a probabilistic retrieval model for entity search, in which term-based and category-based representations of queries and entities are effectively

---

[1] We use q to denote the information need provided by the user (i.e., all components), and Q for the textual part of the query.

S. Geva, J. Kamps, and A. Trotman (Eds.): INEX 2009, LNCS 6203, pp. 265–272, 2010.
© Springer-Verlag Berlin Heidelberg 2010

integrated. With the exception of the formula used for weighting terms for query expansion, we present the original approach unchanged.

The remainder of this section is organized as follows. In §2.1 we introduce our general scheme for ranking entities. This is followed by a discussion of the two main components of our framework: entity and query models in §2.2 and §2.3, respectively.

2.1 Modeling Entity Ranking

We rank entities e according to their probability of being relevant given the query q: $P(e|q)$. Instead of estimating this probability directly, we apply Bayes' rule and rewrite it to:

$$P(e|q) \propto P(q|e) \cdot P(e), \tag{1}$$

where $P(q|e)$ expresses the probability that query q is generated by entity e, and $P(e)$ is the *a priori* probability of e being relevant, i.e., the entity prior.

Each entity is represented as a pair: $\theta_e = (\theta_e^T, \theta_e^C)$, where θ_e^T is a distribution over terms and θ_e^C is a distribution over categories. Similarly, the query is also represented as a pair: $\theta_q = (\theta_q^T, \theta_q^C)$, which is then (optionally) refined further, resulting in an expanded query model that is used for ranking entities.

The probability of an entity generating the query is estimated using a mixture model:

$$P(q|e) = \lambda \cdot P(\theta_q^T|\theta_e^T) + (1 - \lambda) \cdot P(\theta_q^C|\theta_e^C), \tag{2}$$

where λ controls the interpolation between the term-based and category-representations. The estimation of $P(\theta_q^T|\theta_e^T)$ and $P(\theta_q^C|\theta_e^C)$ requires a measure of the difference between two probability distributions. We opt for the Kullback-Leibler divergence—also known as the relative entropy. The term-based similarity is estimated as follows:

$$P(\theta_q^T|\theta_e^T) \propto -KL(\theta_q^T||\theta_e^T) = -\sum_t P(t|\theta_q^T) \cdot \frac{P(t|\theta_q^T)}{P(t|\theta_e^T)}, \tag{3}$$

where the probability of a term given an entity model ($P(t|\theta_e^T)$) and the probability of a term given the query model ($P(t|\theta_q^T)$) remain to be defined. Similarly, the category-based component of the mixture in Eq. 2 is calculated as:

$$P(\theta_q^C|\theta_e^C) \propto -KL(\theta_q^C||\theta_e^C) = -\sum_c P(c|\theta_q^C) \cdot \frac{P(c|\theta_q^C)}{P(c|\theta_e^C)}, \tag{4}$$

where the probability of a category according to an entity's model ($P(c|\theta_e^C)$) and the probability of a category according to the query model ($P(c|\theta_q^C)$) remain to be defined.

2.2 Modeling Entities

Term-based representation To estimate $P(t|\theta_e^T)$ we smooth the empirical entity model with the background collection to prevent zero probabilities. We employ Bayesian smoothing using Dirichlet priors which has been shown to achieve superior performance on a variety of tasks and collections [7, 5] and set:

$$P(t|\theta_e^T) = \frac{n(t,e) + \mu^T \cdot P(t)}{\sum_t n(t,e) + \mu^T}, \tag{5}$$

where $n(t,e)$ denotes the number of times t occurs in the document, $\sum_t n(t,e)$ is the total number of term occurrences, i.e., the document length, and $P(t)$ is the background model (the relative frequency of t in the collection). Since entities correspond to Wikipedia articles, this representation of an entity is identical to constructing a smoothed document model for each Wikipedia page, in a standard language modeling approach [6, 4]. Alternatively, the entity model can be expanded with terms from related entities, i.e., entities sharing the categories or entities linking to or from the Wikipedia page [3]. To remain focused, we do not explore this direction here.

Category-based representation Analogously to the term-based representation detailed above, we smooth the maximum-likelihood estimate with a background model. We employ Dirichlet smoothing again and use the parameter μ^C to avoid confusion with μ^T:

$$P(c|\theta_e^C) = \frac{n(c,e) + \mu^C \cdot P(c)}{\sum_c n(c,e) + \mu^C}. \tag{6}$$

In Eq. 6, $n(c,e)$ is 1 if entity e is assigned to category c, and 0 otherwise; $\sum_c n(c,e)$ is the total number of categories to which e is assigned; $P(c)$ is the background category model and is set using a maximum-likelihood estimate:

$$P(c) = \frac{\sum_e n(c,e)}{\sum_c \sum_e n(c,e)}, \tag{7}$$

where $\sum_c \sum_e n(c,e)$ is the number of category-entity assignments in the collection.

Entity priors. We use uniform entity priors, i.e., all pages in the collection are equally likely to be returned.

2.3   Modeling Queries

In this subsection we introduce methods for estimating and expanding query models. This boils down to estimating the probabilities $P(t|\theta_q^T)$ and $P(c|\theta_q^C)$ as discussed in §2.1.

Term-based representation. The term-based component of the baseline query model is defined as follows:

$$P(t|\theta_q^T) = P_{bl}(t|\theta_q^T) = \frac{n(t,Q)}{\sum_t n(t,Q)}, \tag{8}$$

where $n(t,Q)$ stands for the number of times term t occurs in query Q.

The general form we use for expansion is a mixture of the baseline (subscripted with *bl*) defined in Eq. 8 and an expansion (subscripted with *ex*):

$$P(t|\theta_q^T) = (1 - \lambda^T) \cdot P_{bl}(t|\theta_q^T) + \lambda^T \cdot P_{ex}(t|\theta_q^T). \tag{9}$$

Given a set of feedback entities FB, the expanded query model is constructed as follows:

$$P_{ex}(t|\theta_q^T) = \frac{P_{K_T}(t|FB)}{\sum_{t'} P_{K_T}(t'|FB)}, \tag{10}$$

where $P_{K_T}(t|FB)$ is estimated as follows. First, $P(t|FB)$ is computed according to Eq. 11. Then, the top K_T terms with the highest $P(t|FB)$ value are taken to form $P_{K_T}(t|FB)$, by redistributing the probability mass in proportion to their corresponding $P(t|FB)$ values:

$$P(t|FB) = \frac{1}{|FB|} \sum_{e \in FB} \frac{s(t,e)}{\sum_t s(t,e)} \tag{11}$$

and

$$s(t,e) = \log\left(\frac{n(t,e)}{P(t) \cdot \sum_t n(t,e)}\right), \tag{12}$$

where $\sum_t n(t,e)$ is the total number of terms, i.e., the length of the document corresponding to entity e. (This is the same as the *EXP* query model generation method using example documents from [1], with the simplification that all feedback documents are assumed to be equally important.)

The set of feedback entities, FB, is defined in two ways: for the entity ranking task, it is the top N relevant entities according to a ranking obtained using the initial (baseline) query. For the list completion task, the set of example entities provided with the query are used as the feedback set ($FB = E$).

Category-based representation. Our baseline model uses the keyword query (Q) to infer the category-component of the query model (θ_q^C), by considering the top N_c most relevant categories given the query; relevance of a category is estimated based on matching between the name of the category and the query, i.e., a standard language modeling approach on top of an index of category names:

$$P(c|\theta_q^C) = P_q(c|\theta_q^C) = \begin{cases} P(Q|\theta_c)/\sum_{c \in N_c} P(Q|\theta_c), & \text{if } c \in \text{top } N_c \\ 0, & \text{otherwise.} \end{cases} \tag{13}$$

Note that this method does not use the category information provided with the query. To use target category information, we set $n(c,q)$ to 1 if c is a target category, and $\sum_c n(c,q)$ to the total number of target categories provided with the topic statement. Then, we put

$$P_c(c|\theta_q^C) = \frac{n(c,q)}{\sum_c n(c,q)}. \tag{14}$$

To combine the two methods (categories relevant to the query and categories provided as input), we put:

$$P(c|\theta_q^C) = \frac{1}{2} P_q(c|\theta_q^C) + \frac{1}{2} P_c(c|\theta_q^C). \tag{15}$$

(For the sake of simplicity, each model contributes half of the probability mass.)

Expansion of the category-based component is performed similarly to the term-based case; we use a linear combination of the baseline (either Eq. 13 or Eq. 15) and expanded components:

$$P(c|\theta_q^C) = (1 - \lambda^C) \cdot P_{bl}(c|\theta_q^C) + \lambda^C \cdot P_{ex}(c|\theta_q^C). \qquad (16)$$

Given a set of feedback entities FB, the expanded query model is constructed as follows:

$$P_{ex}(c|\theta_q^C) = \frac{P_{K_C}(c|FB)}{\sum_{c'} P_{K_C}(c'|FB)}, \qquad (17)$$

where $P_{K_C}(c|FB)$ is calculated similarly to the term-based case: first, $P(c|FB)$ is calculated according to Eq. 18 (where, as before, $n(c, e)$ is 1 if e belongs to c). Then, the top K_C categories with the highest $P(c|FB)$ value are selected, and their corresponding probabilities are renormalized, resulting in $P_{K_C}(c|FB)$.

$$P(c|FB) = \frac{1}{|FB|} \sum_{e \in FB} \frac{n(c, e)}{\sum_t n(c, e)}. \qquad (18)$$

The set of feedback entities is defined as before (the top N entities obtained using blind relevance feedback for entity ranking, and the example entities E for list completion).

2.4 Heuristic Promotion

This year's topics are a selection of topics from prior editions of the track (from 2007 and 2008). We experiment with making use of the relevance assessments available from prior years; we view this information as click-through data. For so-called repeat queries such data can easily be collected in an operational system, thus assuming its availability is not unreasonable.

We incorporate this information into our ranking by mixing the original query likelihood probability ($P(q|e)$) with query likelihood based on click-through data ($P_c(q|e)$):

$$P'(q|e) = (1 - \alpha) \cdot P(q|e) + \alpha \cdot P_c(q|e). \qquad (19)$$

For the sake of simplicity we set α in Eq. 19 to 0.5.

Let $c(q, e)$ denote the number of clicks entity e has received for query q; we set $c(q, e)$ to 1 if the entity was judged relevant for the query on the 2007 or 2008 topic sets, otherwise we set it to 0. We use a maximum likelihood estimate to obtain $P_c(q|e)$:

$$P_c(q|e) = \frac{c(q, e)}{\sum_{e'} c(q, e')}, \qquad (20)$$

where the term $\sum_{e'} c(q, e')$ denotes the total number of clicks that query q has received (in our setting: the total number of entities judged as relevant for q).

Note that the underlying Wikipedia crawl has changed (from $\sim$659K to $\sim$2,666K documents), therefore this method is expected to increase early precision, but does not affect recall.

3 Runs

Parameter settings. Using the 2007 and 2008 editions of the Entity Ranking track as training material, we set the parameters of our models as follows.

- Importance of the term-based vs. the category-based component (Eq. 2): $\lambda = 0.7$
- Number of categories obtained given the query (Eq. 13): $N_c = 15$
- Number of feedback entities: $N = 3$
- Number of feedback terms (Eq. 17): $K_T = 35$
- Weight of feedback terms (Eq. 9): $\lambda^T = 0.7$
- Number of feedback categories (Eq. 17): $K_C = \infty$ (not limited)
- Weight of feedback categories (Eq. 16): $\lambda^C = 0.3$

Entity ranking. Table 1 summarizes the 4 runs we submitted for the entity ranking task. The baseline query models are estimated in the same manner for all runs: using Eq. 8 for the term-based component ($P(t|\theta_q^T)$) and Eq. 15 for the category-based component ($P(c|\theta_q^C)$). Runs that employ feedback estimate the term-based and category-based components of expanded query models using Eq. 10 and Eq. 17, respectively. The last two runs apply blind feedback only on a selection of topics; on those that were helped by this technique in prior editions of the task. Note that expansion always takes place in both the term-based and category-based components.

Table 1. Entity ranking runs. Feedback values are: N=no feedback, B=blind feedback, TB=topic-dependent blind feedback (only for topics that were helped by blind feedback on the 2007/2008 topic set.)

RunID (UAmsISLA_ER_...)	Feedback	Promotion	xinfAP
TC_ERbaseline	N	N	0.1893
TC_ERfeedback	B	N	0.2093
TC_ERfeedbackS	TB	N	0.2094
TC_ERfeedbackSP	TB	Y	0.5046

Table 2. List completion runs. (*Topics that were helped by using example entities on the 2007/2008 topic set do not use input category information (i.e., use Eq. 13 for constructing $P_{bl}(c|\theta_q^C)$); the remainder of the topics use the input category information (i.e., $P_{bl}(c|\theta_q^C)$ is estimated using Eq. 15).)

RunID (UAmsISLA_LC_...)	Expansion Term	Expansion Cat.	Promotion	xinfAP
TE_LCexpT	Y	N	N	0.3198
TE_LCexpC	N	Y	N	0.2051
TE_LCexpTC	Y	Y	N	0.4021
TE_LCexpTCP	Y	Y	Y	0.5204
TEC_LCexpTCS*	Y	Y	N	0.3509
TEC_LCexpTCSP*	Y	Y	Y	0.5034

List completion. Table 2 summarizes the 6 runs we submitted for the list completion task. The baseline query models are estimated as follows: using Eq. 8 for the term-based component and Eq. 13 for the category-based component. The first four runs use only example entities (E), while the last two runs, that employ selective blind feedback (`TEC_LCexpTCS` and `TEC_LCexpTCSP`), also use input category information (C) for constructing the category-based component of the baseline query model (Eq. 15). We make use of example entities (E) in the feedback phase, by expanding the term-based and/or category-based component of query models with information extracted from examples (using Eq. 10 and Eq. 17 for term-based and category-based expansion, respectively).

4 Results and Discussion

Tables 1 and 2 present the results for the entity ranking (ER) and list completion (LC) tasks. Our main findings are as follows. First, feedback improves performance on both tasks; not surprisingly, explicit feedback (using example entities) is more beneficial than blind feedback. As to term-based vs. category-based feedback, the former is more effective on the LC task (`TE_LCexpT` vs. `TE_LCexpC`). This finding is especially interesting, as in previous work category-based feedback was found to be more advantageous [2]. This is probably due to the different term importance weighting scheme that we employed in the current work, which seems to be capable of sampling more meaningful terms from example entities. Combining term-based and category-based feedback improves over each method individually (`TE_LCexpTC` vs. `TE_LCexpC` and `TE_LCexpTC` vs. `TE_LCexpT`); this is in line with findings of [2], although the degree of improvement is more substantial here.

We made use of information from prior editions of the track in various ways. Applying blind feedback only to a subset of topics, that are likely to be helped, does not lead to meaningful score differences (`TC_ERfeedback` vs. `TC_ERfeedbackS`). Using explicit category information for a selected subset of topics hurt performance on the LC task (`TE_LCexpTC` vs. `TEC_LCexpTCS`); this is rather unexpected and suggests that the category assignments underlying Wikipedia has changed in the new dump. Finally, we promoted entities which were previously known to be relevant given the query. This heuristic proved to be very effective for both tasks (see `TC_ERfeedbackS` vs. `TC_ERfeedbackSP`, `TE_LCexpTC` vs. `TE_LCexpTCP`, and `TEC_LCexpTCS` vs. `TEC_LCexpTCSP`).

5 Conclusions

We have described our participation in the INEX 2009 Entity Ranking track. Building on earlier work [2], we employed a probabilistic modeling framework for entity search, in which term-based and category-based representations of queries and entities are effectively integrated.

We submitted 4 runs for the entity ranking and 6 runs for the list completion tasks. Our main focus was on evaluating the effectiveness of our recently proposed entity retrieval framework on the new Wikipedia collection. We demonstrated that our approach is robust, and that it delivers very competitive performance on this year's platform too. We experimented with various ways of making use of information from prior editions of the task, and found that using known relevant pages as click-through data has a very positive effect on retrieval performance.

One change we implemented to the original approach concerned the way in which the importance of feedback terms, sampled from example entities, is estimated. Using more discriminative terms proved advantageous, which suggests that there is more to be gained by developing alternative methods for estimating the importance of terms and categories to be sampled from example entities.

Acknowledgements. This research was supported by the European Union's ICT Policy Support Programme as part of the Competitiveness and Innovation Framework Programme, CIP ICT-PSP under grant agreement nr 250430, by the DuOMAn project carried out within the STEVIN programme which is funded by the Dutch and Flemish Governments under project nr STE-09-12, and by the Netherlands Organisation for Scientific Research (NWO) under project nrs 612.066.512, 612.061.814, 612.061.815, 640.004.802.

References

[1] Balog, K., Weerkamp, W., de Rijke, M.: A few examples go a long way: constructing query models from elaborate query formulations. In: SIGIR '08: Proceedings of the 31st annual international ACM SIGIR conference on Research and development in information retrieval, pp. 371–378. ACM Press, New York (2008)

[2] Balog, K., Bron, M., de Rijke, M.: Category-based query modeling for entity search. In: 32nd European Conference on Information Retrieval (ECIR 2010), March 2010, pp. 319–331. Springer, Heidelberg (2010)

[3] Fissaha Adafre, S., de Rijke, M., Tjong Kim Sang, E.: Entity retrieval. In: Recent Advances in Natural Language Processing (RANLP 2007), September 2007, Borovets, Bulgaria (2007)

[4] Lafferty, J., Zhai, C.: Document language models, query models, and risk minimization for information retrieval. In: SIGIR '01: Proceedings of the 24th annual international ACM SIGIR conference on Research and development in information retrieval, pp. 111–119. ACM, New York (2001)

[5] Losada, D., Azzopardi, L.: An analysis on document length retrieval trends in language modeling smoothing. Information Retrieval 11(2), 109–138 (2008)

[6] Song, F., Croft, W.B.: A general language model for information retrieval. In: CIKM '99: Proceedings of the eighth international conference on Information and knowledge management, pp. 316–321. ACM, New York (1999)

[7] Zhai, C., Lafferty, J.: A study of smoothing methods for language models applied to information retrieval. ACM Transactions on Information Systems 22(2), 179–214 (2004)

Focused Search in Books and Wikipedia: Categories, Links and Relevance Feedback

Marijn Koolen[1], Rianne Kaptein[1], and Jaap Kamps[1,2]

[1] Archives and Information Studies, Faculty of Humanities, University of Amsterdam
[2] ISLA, Faculty of Science, University of Amsterdam

Abstract. In this paper we describe our participation in INEX 2009 in the Ad Hoc Track, the Book Track, and the Entity Ranking Track. In the Ad Hoc track we investigate focused link evidence, using only links from retrieved sections. The new collection is not only annotated with Wikipedia categories, but also with YAGO/WordNet categories. We explore how we can use both types of category information, in the Ad Hoc Track as well as in the Entity Ranking Track. Results in the Ad Hoc Track show Wikipedia categories are more effective than WordNet categories, and Wikipedia categories in combination with relevance feedback lead to the best results. Preliminary results of the Book Track show full-text retrieval is effective for high early precision. Relevance feedback further increases early precision. Our findings for the Entity Ranking Track are in direct opposition of our Ad Hoc findings, namely, that the WordNet categories are more effective than the Wikipedia categories. This marks an interesting difference between ad hoc search and entity ranking.

1   Introduction

In this paper, we describe our participation in the INEX 2009 Ad Hoc, Book, and Entity Ranking Tracks. Our aims for this year were to familiarise ourselves with the new Wikipedia collection, to continue the work from previous years, and to explore the opportunities of using category information, which can be in the form of Wikipedia's categories, or the enriched YAGO/WordNet categories.

The rest of the paper is organised as follows. First, Section 2 describes the collection and the indexes we use. Then, in Section 3, we report our runs and results for the Ad Hoc Track. Section 4 briefly discusses our Book Track experiments. In Section 5, we present our approach to the Entity Ranking Track. Finally, in Section 6, we discuss our findings and draw preliminary conclusions.

2   Indexing the Wikipedia Collection

In this section we describe the index that is used for our runs in the ad hoc and the entity ranking track, as well as the category structure of the collection. The collection is based, again, on the Wikipedia but substantially larger and with

S. Geva, J. Kamps, and A. Trotman (Eds.): INEX 2009, LNCS 6203, pp. 273–291, 2010.

longer articles. The original Wiki-syntax is transformed into XML, and each article is annotated using "semantic" categories based on YAGO/Wikipedia. We used Indri [15] for indexing and retrieval.

2.1 Indexing

Our indexing approach is based on earlier work [1, 4, 6, 12–14].

- *Section index*: We used the `<section>` element to cut up each article in sections and indexed each section as a retrievable unit. Some articles have a leading paragraph not contained in any `<section>` element. These leading paragraphs, contained in `<p>` elements are also indexed as retrievable units. The resulting index contains no overlapping elements.
- *Article index*: We also build an index containing all full-text articles (i.e., all wikipages) as is standard in IR.

For all indexes, stop-words were removed, and terms were stemmed using the Krovetz stemmer. Queries are processed similar to the documents. In the ad hoc track we use either the CO query or the CAS query, and remove query operators (if present) from the CO query and the about-functions in the CAS query.

2.2 Category Structure

A new feature in the new Wikipedia collection is the assignment of WordNet labels to documents [11]. The WordNet categories are derived from Wikipedia categories, but are designed to be conceptual. Categories for administrative purposes, such as 'Article with unsourced statements', categories yielding non-conceptual information, such as '1979 births' and categories that indicate merely thematic vicinity, such as 'Physics', are not used for the generation of WordNet labels, but are excluded by hand and some shallow linguistic parsing of the category names. WordNet concepts are matched with category names and the category is linked to the most common concept among the WordNet concepts. It is claimed this simple heuristic yields the correct link in the overwhelming majority of cases.

A second method which is used to generate WordNet labels, is based on information in lists. For example, If all links but one in a list point to pages belonging to a certain category, this category is also assigned to the page that was not labelled with this category. This is likely to improve the consistency of annotation, since annotation in Wikipedia is largely a manual effort.

We show the most frequent category labels of the two category structures in Table 1. Many of the largest categories in Wikipedia are administrative categories. The category *Living people* is the only non-administrative label in this list. The largest WordNet categories are more semantic, that is, they describe what an article is about. The list also shows that many Wikipedia articles are about entities such as persons and locations.

2.3 Comparing the Category Structures

We first analyse the difference between the two category structures by comparing the number of categories assigned to each article in Table 2. In total, over 2.5

Table 1. The most frequent categories of the Wikipedia and WordNet structure

Wikipedia	Wordnet		
Living people	307,317	person	438,003
All disambiguation pages	143,463	physical entity	375,216
Disambiguation pages	103,954	causal agent	373,697
Articles with invalid date parameter in template	77,659	entity	245,049
All orphaned articles	34,612	location	155,304
All articles to be expanded	33,810	region	146,439
Year of birth missing (living people)	32,503	artifact	131,248
All articles lacking sources	21,084	player	109,427

Table 2. The distribution of Wikipedia and WordNet categories over articles

cats/article	N	Min	Max	Med.	Mean	St.dev
Wikipedia	2,547,560	1	72	3	3.50	2.82
WordNet	2,033,848	1	41	3	3.98	3.18

Table 3. The distribution of articles over Wikipedia and WordNet categories

articles/cat	N	Min	Max	Med.	Mean	St.dev
Wikipedia	346,396	1	307,317	5	26	643
WordNet	5,241	1	438,003	57	1,546	12,087

million articles have at least one Wikipedia category and just over 2 million articles have at least one WordNet category. We see that most articles have up to 3 or 4 Wikipedia or WordNet categories. The highest number of categories assigned is somewhat higher for Wikipedia (72) than for WordNet (41). There seem to be no big differences between the distributions of the two category structures.

In Table 3 we show statistics of the number of articles assigned to each category. The most salient difference is the total number of categories. There are 346,396 Wikipedia categories and only 5,241 WordNet categories. As a direct result of this and the statistics of Table 2, most of the WordNet categories are much bigger than the Wikipedia categories. On average, a Wikipedia category has 26 articles, while a WordNet category has 1,546 articles. The median size of both structures is much smaller, indicating a skewed distribution, but we observe the same pattern. 50% of the WordNet categories have at least 57 articles, while 50% of the Wikipedia categories has at most 5 articles. The Wikipedia category structure is thus more fine-grained than the WordNet structure.

3 Ad Hoc Track

For the INEX 2009 Ad Hoc Track we had two main aims. Investigating the value of element level link evidence, and the relative effectiveness of the Wikipedia and WordNet category structures available in the new INEX 2009 Wikipedia collection.

In previous years [2], we have used local link degrees as evidence of topical relevance. We took the top 100 retrieved articles, and computed the link degrees using all the links between those retrieved articles. This year, instead of looking at all local links between the top 100 retrieved articles, we consider only the links occurring in the retrieved elements. A link from article A to article B occurring in a section of article A that is not retrieved is ignored. This link evidence is more focused on the search topic and possibly leads to less infiltration. Infiltration occurs when important pages with many incoming links are retrieved in the top 100 results. Because of their high global in-degree, they have a high probability of having links in the local set. The resulting local link degree is a consequence of their query-independent importance and pushes these documents up the ranking regardless of their topical relevance. If we use only the relevant text in document to derive link evidence, we reduce the chance of picking up topically unrelated link evidence.

The new INEX Wikipedia collection has markup in the form of YAGO elements including WordNet categories. Most Wikipedia articles are manually categorised by the Wikipedia contributors. The category structure can be used to generate category models to promote articles that belong to categories that best match the query. We aim to directly compare the effectiveness of category models based on the Wikipedia and WordNet categorisations for improving retrieval effectiveness.

We will first describe our approach and the official runs, and finally per task, we present and discuss our results.

3.1   Approach

We have four baseline runs based on the indexes described in the previous section:

Article: Run on the article index with linear length prior and linear smoothing $\lambda = 0.15$.

Section: Run on the section index with linear length prior and linear smoothing $\lambda = 0.15$.

Article RF: Run on the article index with blind relevance feedback, using 50 terms from the top 10 results.

Section RF: Run on the section index with blind relevance feedback, using 50 terms from the top 10 results.

These runs have up to 1,500 results per topic. All our official runs for all four tasks are based on these runs. To improve these baselines, we explore the following options.

Category distance: We determine two target categories for a query based on the top 20 results. We select the two most frequent categories to which the top 20 results are assigned and compute a category distance score using parsimonious language models of each category. This technique was successfully employed on the INEX 2007 Ad hoc topics by Kaptein et al. [8]. In the new collection, there are two sets of category labels. One based on the *Wikipedia* category structure and one based on the *WordNet* category labels.

CAS filter: For the CAS queries we extracted from the CAS title all semantic target elements, identified all returned results that contain a target element in the xpath and ranked them before all other results by adding a constant c to the score per matching target element. Other than that, we keep the ranking in tact. A result that matches two target elements gets $2c$ added to its score, while a result matching one target element gets $1c$ added to its score. In this way, results matching n target elements are ranked above results matching $n - 1$ target elements. This is somewhat similar to co-ordination level ranking of content-only queries, where documents matching n query terms are ranked above documents matching $n - 1$ query terms. Syntactic target elements like `<article>`, `<sec>`, `<p>` and `<category>` are ignored.

Link degrees: Both incoming and outgoing link degrees are useful evidence in identifying topical relevance [5, 10]. We use the combined $indegree(d) + outdegree(d)$ as a document "prior" probability $P_{link}(d)$. Local link evidence is not query-independent, so $P_{link}(d)$ is not an actual *prior* probability. We note that for runs where we combine the article or section text score with a category distance score, we get a different score distribution. With these runs we use the link evidence more carefully by taking the log of the link degree as $P_{link}(d)$. In a standard language model, the document prior is incorporated as $P(d|q) = P_{link}(d) \cdot P_{content}(q|d)$, where $P_{content}(q|d)$ is the standard language model score.

Focused Link degrees: We also constructed a focused local link graph based on the retrieved elements of the top 100 articles. Instead of using all links between the top 100 articles, we only use the outgoing links from sections that are retrieved for a given topic. The main idea behind this is that link anchors appearing closer to the query terms are more closely related to the search topic. Thus, if for an article a_i in the top 100 articles only section s_j is retrieved, we use only the links appearing in section s_j that point to other articles in the top 100. This local link graph is more focused on the search topic, and potentially suffers less from infiltration of important but off-topic articles. Once the focused local link graph is constructed, we count the number of incoming + outgoing links as the focused link prior $P_{foc_link}(d)$.

Article ranking: Based on [4], we use the article ranking of an article index run and group the elements returned by a section index run as focused results.

Cut-off(n): When we group returned elements per article for the Relevant in Context task, we can choose to group all returned elements of an article, or only the top ranked elements. Of course, further down the results list we find less relevant elements, so grouping them with higher ranked elements from the same article might actually hurt precision. We set a cut-off at rank n to group only the top returned elements by article.

3.2   Runs

Combining the methods described in the previous section with our baseline runs leads to the following official runs.

For the Thorough Task, we submitted two runs:

UamsTAdbi100: an article index run with relevance feedback. The top 100 results are re-ranked using the link degree prior $P_{link}(d)$.

UamsTSdbi100: a section index run with relevance feedback. We cut off the results list at rank 1500 and re-rank the focused results of the top 100 articles using the link prior $P_{link}(d)$. **However, this run is invalid, since it contains overlap due to an error in the xpaths.**

For the Focused Task, we submitted two runs:

UamsFSdbi100CAS: a section index run combined with the Wikipedia category distance scores. The results of the top 100 articles are re-ranked using the link degree prior. Finally, the CAS filter is applied to boost results with target elements in the xpath.

UamsFSs2dbi100CAS: a section index run combined with the Wikipedia category distance scores. The results of the top 100 articles are re-ranked using the focused link degree prior $P_{foc_link}(d)$.

For the Relevant in Context Task, we submitted two runs:

UamsRSCMACMdbi100: For the article ranking we used the article text score combined with the manual category distance score as a baseline and re-ranked the top 100 articles with the log of the local link prior $P_{link}(d)$. The returned elements are the top results of a combination of the section text score and the manual category distance score, grouped per article.

UamsRSCWACWdbi100: For the article ranking we used the article text score combined with the WordNet category distance score as a baseline and re-ranked the top 100 with the log of the local link prior $P_{link}(d)$. The returned elements are the top results of a combination of the section text score and the WordNet category distance score, grouped per article.

For the Best in Context Task, we submitted two runs:

UamsBAfbCMdbi100: an article index run with relevance feedback combined with the Wikipedia category distance scores, using the local link prior $P_{link}(d)$ to re-rank the top 100 articles. The Best-Entry-Point is the start of the article.

UamsBAfbCMdbi100: a section index run with relevance feedback combined with the Wikipedia category distance scores, using the focused local link prior $P_{foc_link}(d)$ to re-rank the top 100 articles. Finally, the CAS filter is applied to boost results with target elements in the xpath. The Best-Entry-Point is the start of the article.

3.3 Thorough Task

Results of the Thorough Task can be found in Table 4. The official measure is MAiP. For the Thorough Task, the article runs are vastly superior to the

Table 4. Results for the Ad Hoc Track Thorough and Focused Tasks (runs labeled "UAms" are official submissions)

Run id	MAiP	iP[0.00]	iP[0.01]	iP[0.05]	iP[0.10]
UamsTAdbi100	0.2676	0.5350	0.5239	0.4968	0.4712
UamsFSdocbi100CAS	0.1726	0.5567	0.5296	0.4703	0.4235
UamsFSs2dbi100CAS	0.1928	**0.6328**	0.5997	0.5140	0.4647
UamsRSCMACMdbi100	0.2096	0.6284	**0.6250**	0.5363	0.4733
UamsRSCWACWdbi100	0.2132	0.6122	0.5980	0.5317	0.4782
Article	0.2814	0.5938	0.5880	0.5385	0.4981
Article + Cat(Wiki)	0.2991	0.6156	0.6150	**0.5804**	**0.5218**
Article + Cat(WordNet)	0.2841	0.5600	0.5499	0.5203	0.4950
Article RF	0.2967	0.6082	0.5948	0.5552	0.5033
Article RF + Cat(Wiki)	**0.3011**	0.6006	0.5932	0.5607	0.5177
Article RF + Cat(WordNet)	0.2777	0.5490	0.5421	0.5167	0.4908
$(Article + CAT(Wiki)) \cdot P_{link}(d)$	0.2637	0.5568	0.5563	0.4934	0.4662
$(Article + CAT(WordNet)) \cdot P_{link}(d)$	0.2573	0.5345	0.5302	0.4924	0.4567
Section	0.1403	0.5525	0.4948	0.4155	0.3594
Section $\cdot P_{link}(d)$	0.1727	0.6115	0.5445	0.4824	0.4155
Section $\cdot P_{foc\_link}(d)$	0.1738	0.5920	0.5379	0.4881	0.4175
Section + Cat(Wiki)	0.1760	0.6147	0.5667	0.5012	0.4334
Section + Cat(WordNet)	0.1533	0.5474	0.4982	0.4506	0.3831
Section + Cat(Wiki) $\cdot P_{art\_link}(d)$	0.1912	0.6216	0.5808	0.5220	0.4615
Section + Cat(Wiki) $\cdot P_{foc\_link}(d)$	0.1928	**0.6328**	0.5997	0.5140	0.4647
Section RF	0.1493	0.5761	0.5092	0.4296	0.3623
Section RF + Cat(Wiki)	0.1813	0.5819	0.5415	0.4752	0.4186
Section RF + Cat(WordNet)	0.1533	0.5356	0.4794	0.4201	0.3737
Section RF $\cdot P_{art\_link}(d)$	0.1711	0.5678	0.5327	0.4774	0.4174

section level runs. The MAiP score for the baseline Article run is more than twice as high as for the Section run. Although the Section run can be more easily improved by category and link information, even the best Section run comes nowhere near the Article baseline. The official article run UamsTAdbi100 is not as good as the baseline. This seems a score combination problem. Even with log degrees as priors, the link priors have a too large impact on the overall score. The underlying run is already a combination of the expanded query and the category scores. Link evidence might correlate with either of the two or both and lead to over use of the same information. Standard relevance feedback improves upon the baseline. The Wikipedia category distances are even more effective. The WordNet category distances are somewhat less effective, but still lead to improvement for MAiP. Combining relevance feedback with the WordNet categories hurts performance, whereas combining feedback with the Wikipedia categories improves MAiP. The link prior has a negative impact on performance of article level runs. The official run *UamsTAdbi100* is based on the *Article RF* run, but with the top 100 articles re-ranked using the local link prior. With the link evidence added, MAiP goes down considerably.

On the section runs we see again that relevance feedback and link and category information can improve performance. The Wikipedia categories are more effective than the WordNet categories and than the link degrees. The link priors also lead to improvement. On both the *Section* and *Section + Cat(Wiki)* runs, the focused link degrees are slightly more effective than the article level link degrees. For the section results, link and category evidence are complementary to each other.

For the Thorough Task, there seems to be no need to use focused retrieval techniques. Article retrieval is more effective than focused retrieval. Inter-document structures such as link and category structures are more effective.

3.4   Focused Task

We have no overlapping elements in our indexes, so no overlap filtering is done. Because the Thorough and Focused Tasks use the same measure, the Focused results are also shown in Table 4. However, for the Focused Task, the official measure is iP[0.01]. Even for the Focused Task, the article runs are very competitive, with the *Article + Cat(Wiki)* run outperforming all section runs. Part of the explanation is that the first 1 percent of relevant text is often found in the first relevant article. In other words, the iP[0.01 score of the article runs is based on the first relevant article in the ranking, while for the section runs, multiple relevant sections are sometimes needed to cover the first percent of relevant text. As the article run has a very good document ranking, it also has a very good precision at 1 percent recall.

The Wikipedia categories are very effective in improving performance of both the article and section index runs. They are more effective when used without relevance feedback. The link priors have a negative impact on the *Article + Cat(Wiki)* run. Again, this might be explained by the fact that the article run already has a very good document ranking and the category and link information are possibly correlated leading to a decrease in performance if we use both. However, on the *Section + Cat(Wiki)* run the link priors have a very positive effect. For comparison, we also show the official Relevant in Context run *UamsRSC-MACMdbi100*, which uses the same result elements as the *Section + Cat(Wiki)* run, but groups them per article and uses the $(Article + Cat(Wiki)) \cdot P_{link}(d)$ run for the article ranking. This improves the precision at iP[0.01]. The combination of the section run and the article run gives the best performance. This is in line with the findings in [4]. The article level index is better for ranking the first relevant document highly, while the section level index is better for locating the relevant text with the first relevant article.

In sum, for the Focused Task, our focused retrieval approach fails to improve upon standard article retrieval. Only in combination with a document ranking based on the article index does focused retrieval lead to improved performance. The whole article seems be the right level of granularity for focused retrieval with this set of Ad Hoc topics. Again, inter-document structure is more effective than the internal document structure.

Table 5. Results for the Ad Hoc Track Relevant in Context Task (runs labeled "UAms" are official submissions)

Run id	MAgP	gP[5]	gP[10]	gP[25]	gP[50]
UamsRSCMACMdbi100	0.1771	0.3192	0.2794	0.2073	0.1658
UamsRSCWACWdbi100	0.1678	0.3010	0.2537	0.2009	0.1591
Article	0.1775	0.3150	0.2773	0.2109	0.1621
Article RF	0.1880	0.3498	0.2956	0.2230	0.1666
Article + Cat(Wiki)	0.1888	0.3393	0.2869	**0.2271**	0.1724
Article + Cat(WordNet)	0.1799	0.2984	0.2702	0.2199	0.1680
Article RF + Cat(Wiki)	**0.1950**	**0.3528**	**0.2979**	0.2257	**0.1730**
Article RF + Cat(WordNet)	0.1792	0.3200	0.2702	0.2180	0.1638
Section	0.1288	0.2650	0.2344	0.1770	0.1413
Section $\cdot P_{art\_link}(d)$	0.1386	0.2834	0.2504	0.1844	0.1435
Section $\cdot P_{foc\_link}(d)$	0.1408	0.2970	0.2494	0.1823	0.1434
Section + Cat(Wiki)	0.1454	0.2717	0.2497	0.1849	0.1407
Section + Cat(Wiki) $\cdot P_{art\_link}(d)$	0.1443	0.2973	0.2293	0.1668	0.1392
Section + Cat(Wiki) $\cdot P_{foc\_link}(d)$	0.1451	0.2941	0.2305	0.1680	0.1409

3.5 Relevant in Context Task

For the Relevant in Context Task, we group result per article. Table 5 shows the results for the Relevant in Context Task. A simple article level run is just as effective for the Relevant in Context task as the much more complex official runs *UamsRSCMACMdbi100* and *UamsRSCWACWdbi100*, which use the *Article + Cat(Wiki)·log($P_{link}(d)$)* run for the article ranking, and the *Section + Cat(Wiki)* and *Section + Cat(WordNet)* respectively run for the top 1500 sections.

Both relevance feedback and category distance improve upon the baseline article run. The high precision of the *Article RF* run shows that expanding the query with good terms from the top documents can help reducing the amount of non-relevant text in the top ranks and works thus as a precision device. Combining relevance feedback with the Wikipedia category distance gives the best results. The WordNet categories again hurt performance of the relevance feedback run.

For the *Section* run, the focused link degrees are more effective than the article level link degrees. The Wikipedia categories are slightly more effective than the link priors for MAgP, while the link priors lead to a higher early precision. The combination of link and category evidence is less effective than either individually.

Again, the whole article is a good level of granularity for this task and the 2009 topics. Category information is very useful to locate articles focused on the search topic.

3.6 Best in Context Task

The aim of the Best in Context task is to return a single result per article, which gives best access to the relevant elements. Table 6 shows the results for the Best in Context Task. We We see the same patterns as for the previous Tasks.

Table 6. Results for the Ad Hoc Track Best in Context Task (runs labeled "UAms" are official submissions)

Run id	MAgP	gP[5]	gP[10]	gP[25]	gP[50]
UamsBAfbCMdbi100	0.1543	0.2604	0.2298	0.1676	0.1478
UamsBSfbCMs2dbi100CASart1	0.1175	0.2193	0.1838	0.1492	0.1278
UamsTAdbi100	0.1601	0.2946	0.2374	0.1817	0.1444
Article	0.1620	0.2853	0.2550	0.1913	0.1515
Article RF	0.1685	**0.3203**	**0.2645**	0.2004	0.1506
Article + Cat(Wiki)	0.1740	0.2994	0.2537	**0.2069**	**0.1601**
Article + Cat(WordNet)	0.1670	0.2713	0.2438	0.2020	0.1592
Article RF + Cat(Wiki)	**0.1753**	0.3091	0.2625	0.2001	0.1564
Article RF + Cat(WordNet)	0.1646	0.2857	0.2506	0.1995	0.1542

Relevance feedback helps, so do Wikipedia and WordNet categories. Wikipedia categories are more effective than relevance feedback, WordNet categories are less effective. Wikipedia categories combined with relevance feedback gives further improvements, WordNet combined with feedback gives worse performance than feedback alone. Links hurt performance. Finally, the section index is much less effective than the article index.

The official runs fail to improve upon a simple article run. In the case of *UamsBAfbCMdbi100*, the combination of category and link information hurts the *Article RF* baseline, and in the case of *UamsBSfbCMs2dbi100CASart1*, the underlying relevance ranking of the *Section RF + Cat(Wiki)* run is simply much worse than that the *Article* run.

In summary, we have seen that relevance feedback and the Wikipedia category information can both be used effectively to improve focused retrieval. The WordNet categories can lead to improvements in some cases, but are less effective than Wikipedia categories. This is probably caused by the fact that the WordNet categories are much larger and thus have less discriminative power.

Although the difference is small, focused link evidence based on element level link degrees is slightly more effective than article level degrees. Link information is very effective for improving the section index results, but hurts the article level results when used in combination with category evidence. This might be a problem of combining the score incorrectly and requires further analysis. We leave this for future work.

With this year's new Wikipedia collection, we see again that document retrieval is a competitive alternative to element retrieval techniques for focused retrieval performance. The combination of article retrieval and element retrieval can only marginally improve performance upon article retrieval in isolation. This suggests that, for the Ad Hoc topics created at INEX, the whole article is a good level of granularity and that there is little need for sub-document retrieval techniques. Structural information such as link and category evidence also remain effective in the new collection.

4   Book Track

In the INEX 2009 Book Track we participated in the Book Retrieval and Focused Book Search tasks. Continuing our efforts of last year, we aim to find the appropriate level of granularity for Focused Book Search. During last year's assessment phase, we noticed that it is often hard to assess the relevance of an individual page without looking at the surrounding pages. If humans find it hard to assess individual pages, than it is probably hard for IR systems as well. In the assessments of last year, it turned out that relevant passages often cover multiple pages [9]. With larger relevant passages, query terms might be spread over multiple pages, making it hard for a page level retrieval model to assess the relevance of individual pages.

Therefore, we wanted to know if we can better locate relevant passages by considering larger book parts as retrievable units. Using larger portions of text might lead to better estimates of their relevance. However, the BookML markup only has XML elements on the page level. One simple option is to divide the whole book in sequences of n pages. Another approach would be to use the logical structure of a book to determine the retrievable units. The INEX Book corpus has no explicit XML elements for the various logical units of the books, so as a first approach we divide each book in sequences of pages. We created indexes using 3 three levels of granularity:

Book index : each whole book is indexed as a retrievable unit.
Page index : each individual page is indexed as a retrievable unit.
5-Page index : each sequence of 5 pages is indexed as a retrievable unit. That
is, pages 1–5, 6–10, etc., are treated as individual text units.

We submitted six runs in total: two for the Book Retrieval (BR) task and four for the Focused Book Search (FBS) task. The 2009 topics consist of an overall topic statement and one or multiple sub-topics. In total, there are 16 topics and 37 sub-topics. The BR runs are based on the 16 overall topics. The FBS runs are based on the 37 sub-topics.

Book : a standard Book index run. Up to 1000 results are returned per topic.
Book RF : a Book index run with Relevance Feedback (RF). The initial queries
are expanded with 50 terms from the top 10 results.
Page : a standard Page index run.
Page RF : a Page index run with Relevance Feedback (RF). The initial queries
are expanded with 50 terms from the top 10 results.
5-page : a standard 5-Page index run.
5-Page RF : a 5-Page index run with Relevance Feedback (RF). The initial
queries are expanded with 50 terms from the top 10 results.

The impact of feedback. In Table 7 we see the impact of relevance feedback on the number of retrieved pages per topic and per book. Because we set a limit of 5,000 on the number of returned results, the total number of retrieved pages does not change, but the number of books from which pages are returned goes

Table 7. The impact of feedback on the number of results per topic

Run	pages	books	pages/book
Page	5000	2029	2.46
Page RF	5000	1602	3.12
5Page	24929	2158	11.55
5Page RF	24961	1630	15.31
Book	–	1000	–
Book RF	–	1000	–

Table 8. Results of the INEX 2009 Book Retrieval Task

Run id	MAP	MRR	P10	Bpref	Rel.	Rel. Ret.
Book	0.3640	0.8120	0.5071	0.6039	494	377
Book RF	0.3731	0.8507	0.4643	0.6123	494	384

down. Relevance feedback using the top 10 pages (or top 10 5-page blocks) leads to more results from a single book. This is unsurprising. With expansion terms drawn from the vocabulary of a few books, we find pages with similar terminology mostly in the same books. On the book level, this impact is different. Because we already retrieve whole books, feedback can only changes the set of book returned. The impact on the page level also indicates that feedback does what it is supposed to do, namely, find more results similar to the top ranked results.

At the time of writing, there are only relevance assessments at the book level, and only for the whole topics. The assessment phase is still underway, so we show results based on the relevance judgements as off 15 March 2010 in Table 8. The *Book* run has an MRR of 0.8120, which means that for most of the topics, the first ranked result is relevant. This suggests that using full text retrieval on long documents like books is an effective method for locating relevance. The impact of relevance feedback is small but positive for MRR and MAP, but negative for P@10. It also helps finding a few more relevant books.

We will evaluate the page level runs once page-level and aspect-level judgements are available.

5 Entity Ranking

In this section, we describe our approach to the Entity Ranking Track. Our goals for participation in the entity ranking track are to refine last year's entity ranking method, which proved to be quite effective, and to explore the opportunities of the new Wikipedia collection. The most effective part of our entity ranking approach last year was combining the documents score with a category score, where the category score represents the distance between the document categories and the target categories. We do not use any link information, since last year this only lead to minor improvements [7].

5.1   Category Information

For each target category we estimate the distances to the categories assigned to the answer entity, similar to what is done in Vercoustre et al. [16]. The distance between two categories is estimated according to the category titles. Last year we also experimented with a binary distance, and a distance between category contents, but we found the distance estimated using category titles the most efficient and at the same time effective method.

To estimate title distance, we need to calculate the probability of a term occurring in a category title. To avoid a division by zero, we smooth the probabilities of a term occurring in a category title with the background collection:

$$P(t_1, ..., t_n | C) = \sum_{i=1}^{n} \lambda P(t_i | C) + (1 - \lambda) P(t_i | D)$$

where C is the category title and D is the entire wikipedia document collection, which is used to estimate background probabilities. We estimate $P(t|C)$ with a parsimonious model [3] that uses an iterative EM algorithm as follows:

E-step:
$$e_t = t f_{t,C} \cdot \frac{\alpha P(t|C)}{\alpha P(t|C) + (1 - \alpha) P(t|D)}$$

M-step:
$$P(t|C) = \frac{e_t}{\sum_t e_t}, \text{ i.e. normalize the model}$$

The initial probability $P(t|C)$ is estimated using maximum likelihood estimation. We use KL-divergence to calculate distances, and calculate a category score that is high when the distance is small as follows:

$$S_{cat}(C_d | C_t) = -D_{KL}(C_d | C_t) = -\sum_{t \in D} \left(P(t|C_t) * \log \left(\frac{P(t|C_t)}{P(t|C_d)} \right) \right)$$

where d is a document, i.e. an answer entity, C_t is a target category and C_d a category assigned to a document. The score for an answer entity in relation to a target category $S(d|C_t)$ is the highest score, or shortest distance from any of the document categories to the target category.

For each target category we take only the shortest distance from any answer entity category to a target category. So if one of the categories of the document is exactly the target category, the distance and also the category score for that target category is 0, no matter what other categories are assigned to the document. Finally, the score for an answer entity in relation to a query topic $S(d|QT)$ is the sum of the scores of all target categories:

$$S_{cat}(d|QT) = \sum_{C_t \in QT} \underset{C_d \in d}{\operatorname{argmax}} S(C_d | C_t)$$

Besides the category score, we also need a query score for each document. This score is calculated using a language model with Jelinek-Mercer smoothing without length prior:

$$P(q_1, ..., q_n | d) = \sum_{i=1}^{n} \lambda P(q_i | d) + (1 - \lambda) P(q_i | D)$$

Finally, we combine our query score and the category score through a linear combination. For our official runs both scores are calculated in the log space, and then a weighted addition is made.

$$S(d|QT) = \mu P(q|d) + (1 - \mu) S_{cat}(d|QT)$$

We made some additional runs using a combination of normalised scores. In this case, scores are normalised using a min-max normalisation:

$$S_{norm} = \frac{S - Min(S_n)}{Max(S_n) - Min(S_n)}$$

A new feature in the new Wikipedia collection is the assignment of YAGO/ WordNet categories to documents as described in Section 2.2. These WordNet categories have some interesting properties for entity ranking. The WordNet categories are designed to be conceptual, and by exploiting list information, pages should be more consistently annotated. In our official runs we have made several combinations of Wikipedia and WordNet categories.

5.2 Pseudo-relevant Target Categories

Last year we found a discrepancy between the target categories assigned manually to the topics, and the categories assigned to the answer entities. The target categories are often more general, and can be found higher in the Wikipedia category hierarchy. For example, topic 102 with title 'Existential films and novels' has as target categories 'films' and 'novels,' but none of the example entities belong directly to one of these categories. Instead, they belong to lower level categories such as '1938 novels,' 'Philosophical novels,' 'Novels by Jean-Paul Sartre' and 'Existentialist works' for the example entity 'Nausea (Book).' The term 'novels' does not always occur in the relevant document category titles, so for those categories the category distance will be overestimated. In addition to the manually assigned target categories, we have therefore created a set of pseudo-relevant target categories. From our baseline run we take the top n results, and assign k pseudo-relevant target categories if they occur at least 2 times as a document category in the top n results. Since we had no training data available we did a manual inspection of the results to determine the parameter settings, which are $n = 20$ and $k = 2$ in our official runs. For the entity ranking task we submitted different combinations of the baseline document score, the category score based on the assigned target categories, and the category score based on the pseudo-relevant target categories. For the list completion task, we follow a similar procedure to assign target categories, but instead of using pseudo-relevant results, we use the categories of the example entities. All categories that occur at least twice in the example entities are assigned as target categories.

5.3 Results

Before we look at at the results, we take a look at the categories assigned by the different methods. In Table 9 we show a few example topics together with the

Table 9. Target Categories

Topic	olympic classes dinghie sailing	Neil Gaiman novels	chess world champions
Assigned	dinghies	novels	chess grandmasters world chess champions
PR	dinghies sailing	comics by Neil Gaiman fantasy novels	chess grandmasters world chess champions
Wikipedia	dinghies sailing at the olympics boat types	fantasy novels novels by Neil Gaiman	chess grandmasters chess writers living people world chess champion russian writers russian chess players russian chess writers 1975 births soviet chess players people from Saint Petersburg
Wordnet	specification types	writing literary composition novel written communication fiction	entity player champion grandmaster writer chess player person soviet writers

categories as assigned ("Assigned") by each method. As expected the pseudo-relevant target categories ("PR") are more specific than the manually assigned target categories. The number of common Wikipedia categories in the example entities ("Wikipedia") can in fact be quite large. More categories is in itself not a problem, but also non relevant categories such as '1975 births' and 'russian writers' and very general categories such as 'living people' are added as target categories. Finally, the WordNet categories ("WordNet") contain less detail than the Wikipedia categories. Some general concepts such as 'entity' are included. With these kind of categories, a higher recall but smaller precision is expected.

The official results of the entity ranking runs can be found in Table 10. The run that uses the official categories assigned during topic creation performs best, and significantly better than the baseline when we consider Average Precision (xinfAP). The pseudo-relevant categories perform a bit worse, but still significantly better than the baseline. Combining the officially assigned categories and the pseudo-relevant categories does not lead to any additional improvements. Looking at the NDCG measure the results are unpredictable, and do not correlate well to the AP measure. In addition to the official runs, we created some additional runs using min-max normalisation before combining scores. For each combinations, only the best run is given here with the corresponding λ.

In our official list completion runs we forgot to remove the example entities from our result list. The results reported in Table 11 are therefore slightly better than the official results. For all runs we use $\lambda = 0.9$. We see that the run based on the WordNet categories outperforms the runs using the Wikipedia categories, although the differences are small. Again the AP results, do not correspond well to the NDCG measure.

Table 10. Results Entity Ranking

Run	AP	NDCG
Base	0.171	0.441
Off. cats ($\lambda = 0.9$)	0.201•	0.456°
Off. cats norm. ($\lambda = 0.8$)	**0.234•**	**0.501•**
Prf cats ($\lambda = 0.9$)	0.190°	0.421°
Off. cats ($\lambda = 0.45$) + Prf cats ($\lambda = 0.45$)	0.199•	0.447⁻

Table 11. Results List Completion

Run	AP	NDCG
Base	0.152	0.409
Wiki ex. cats	0.163•	0.402⁻
Wiki ex. + prf cats	0.168°	0.397°
WordNet ex. cats	**0.181°**	**0.418⁻**
Wiki + Wordnet ex. cats	0.173•	0.411⁻

Compared to previous years the improvements from using category information are much smaller. In order to gain some information on category distributions within the retrieval results, we analyse the relevance assessment sets of the current and previous years. We show some statistics in Table 12.

Table 12. Relevance assessment sets statistics

Year	07	08	09	09
Cats	Wiki	Wiki	Wiki	WordNet
Avg. # of pages	301	394	314	314
Avg. % relevant pages	0.21	0.07	0.20	0.20
Pages with majority category of all pages:				
all pages	0.232	0.252	0.254	0.442
relevant pages	0.364	0.363	0.344	0.515
non-relevant pages	0.160	0.241	0.225	0.421
Pages with majority category of relevant pages:				
all pages	0.174	0.189	0.191	0.376
relevant pages	0.608	0.668	0.489	0.624
non-relevant pages	0.068	0.155	0.122	0.317
Pages with target category:				
all pages	0.138	0.208	0.077	
relevant pages	0.327	0.484	0.139	
non-relevant pages	0.082	0.187	0.064	

When we look at the Wikipedia categories, the most striking difference with the previous years is the percentage of pages belonging to the target category. In the new assessments less pages belong to the target category. This might be caused by the extension of the category structure. In the new collection there are more categories, and the categories assigned to the pages are more refined than before. Also less pages belong to the majority category of the relevant pages, another sign that the categories assigned to pages have become more diverse. When we compare the WordNet to the Wikipedia categories, we notice that the WordNet categories are more focused, i.e. more pages belong to the same categories. This is in concordance with the previously calculated numbers of the distribution of articles over Wikipedia and WordNet categories, and vice versa in Section 2.2.

We are still investigating if there are other reasons that explain why the performance does not compare well to the performance in previous years. Also we expect some additional improvements from optimising the normalisation and combination of scores.

6   Conclusion

In this paper we discussed our participation in the INEX 2009 Ad Hoc, Book, and the Entity Ranking Tracks.

For the Ad Hoc Track we conclude focused link evidence outperforms local link evidence on the article level for the Focused Task. Focused link evidence leads to high early precision. Using category information in the form of Wikipedia categories turns out to be very effective, and more valuable than WordNet category information. These inter-document structures are more effective than document internal structure. Our focused retrieval approach can only marginally improve an article retrieval baseline and only when we keep the document ranking of the article run. For the INEX 2009 Ad Hoc topics, the whole article level seems a good level of granularity.

For the Book Track, using the full text of books gives high early precision and even good overall precision, although the small number of judgements might lead to an over-estimated average precision. Relevance feedback seems to be very effective for further improving early precision, although it can also help finding more relevant books. The Focused Book Search Task still awaits evaluation because there are no page-level relevance judgements yet.

Considering the entity ranking task we can conclude that in the new collection using category information still leads to significant improvements, but that the improvements are smaller because the category structure is larger and categories assigned to pages are more diverse. WordNet categories seem to be a good alternative to the Wikipedia categories. The WordNet categories are more general and consistent categories.

This brings up an interesting difference between ad hoc retrieval and entity ranking. We use the same category distance scoring function for both tasks, but for the former, the highly specific and noisy Wikipedia categories are more effective, while for the latter the more general and consistent WordNet categories are more effective. Why does ad hoc search benefit more from the more specific Wikipedia categories? And why does entity ranking benefit more from the more general WordNet categories? Does the category distance in the larger Wikipedia category structure hold more focus on the topic and less on the entity type? And vice versa, are the more general categories of the WordNet category structure better for finding similar entities but worse for keeping focus on the topical aspect of the search query? These questions open up an interesting avenue for future research.

Acknowledgments. Jaap Kamps was supported by the Netherlands Organization for Scientific Research (NWO, grants # 612.066.513, 639.072.601, and 640.001.-501). Rianne Kaptein was supported by NWO under grant # 612.066.513 Marijn Koolen was supported by NWO under grant # 640.001.501.

References

[1] Fachry, K.N., Kamps, J., Koolen, M., Zhang, J.: Using and detecting links in Wikipedia. In: Fuhr, N., Kamps, J., Lalmas, M., Trotman, A. (eds.) INEX 2007. LNCS, vol. 4862, pp. 388–403. Springer, Heidelberg (2008)

[2] Fachry, K.N., Kamps, J., Koolen, M., Zhang, J.: Using and detecting links in Wikipedia. In: Fuhr, N., Kamps, J., Lalmas, M., Trotman, A. (eds.) INEX 2007. LNCS, vol. 4862, pp. 388–403. Springer, Heidelberg (2008)

[3] Hiemstra, D., Robertson, S., Zaragoza, H.: Parsimonious language models for information retrieval. In: Proceedings of the 27th Annual International ACM SIGIR Conference on Research and Development in Information Retrieval, pp. 178–185. ACM Press, New York (2004)

[4] Kamps, J., Koolen, M.: The impact of document level ranking on focused retrieval. In: Geva, S., Kamps, J., Trotman, A. (eds.) INEX 2008. LNCS, vol. 5631, pp. 140–151. Springer, Heidelberg (2009)

[5] Kamps, J., Koolen, M.: Is wikipedia link structure different? In: Proceedings of the Second ACM International Conference on Web Search and Data Mining (WSDM 2009). ACM Press, New York (2009b)

[6] Kamps, J., Koolen, M., Sigurbjörnsson, B.: Filtering and clustering XML retrieval results. In: Fuhr, N., Lalmas, M., Trotman, A. (eds.) INEX 2006. LNCS, vol. 4518, pp. 121–136. Springer, Heidelberg (2007)

[7] Kaptein, R., Kamps, J.: Finding entities in Wikipedia using links and categories. In: Geva, S., Kamps, J., Trotman, A. (eds.) INEX 2008. LNCS, vol. 5631, pp. 273–279. Springer, Heidelberg (2009)

[8] Kaptein, R., Koolen, M., Kamps, J.: Using Wikipedia categories for ad hoc search. In: Proceedings of the 32nd Annual International ACM SIGIR Conference on Research and Development in Information Retrieval. ACM Press, New York (2009)

[9] Kazai, G., Milic-Frayling, N., Costello, J.: Towards methods for the collective gathering and quality control of relevance assessments. In: SIGIR '09: Proceedings of the 32nd international ACM SIGIR conference on Research and development in information retrieval, pp. 452–459. ACM, New York (2009), http://doi.acm.org/10.1145/1571941.1572019

[10] Koolen, M., Kamps, J.: What's in a link? from document importance to topical relevance. In: Azzopardi, L., Kazai, G., Robertson, S., Rüger, S., Shokouhi, M., Song, D., Yilmaz, E. (eds.) ICTIR 2009. LNCS, vol. 5766, pp. 313–321. Springer, Heidelberg (2009)

[11] Schenkel, R., Suchanek, F., Kasneci, G.: YAWN: A semantically annotated wikipedia xml corpus. In: 12th GI Conference on Databases in Business, Technology and Web (BTW 2007) (March 2007)

[12] Sigurbjörnsson, B., Kamps, J.: The effect of structured queries and selective indexing on XML retrieval. In: Fuhr, N., Lalmas, M., Malik, S., Kazai, G. (eds.) INEX 2005. LNCS, vol. 3977, pp. 104–118. Springer, Heidelberg (2006)

[13] Sigurbjörnsson, B., Kamps, J., de Rijke, M.: An Element-Based Approach to XML Retrieval. In: INEX 2003 Workshop Proceedings, pp. 19–26 (2004)

[14] Sigurbjörnsson, B., Kamps, J., de Rijke, M.: Mixture models, overlap, and structural hints in XML element retreival. In: Fuhr, N., Lalmas, M., Malik, S., Szlávik, Z. (eds.) INEX 2004. LNCS, vol. 3493, pp. 196–210. Springer, Heidelberg (2005)

[15] Strohman, T., Metzler, D., Turtle, H., Croft, W.B.: Indri: a language-model based search engine for complex queries. In: Proceedings of the International Conference on Intelligent Analysis (2005)

[16] Vercoustre, A.-M., Pehcevski, J., Thom, J.A.: Using Wikipedia categories and links in entity ranking. In: Fuhr, N., Kamps, J., Lalmas, M., Trotman, A. (eds.) INEX 2007. LNCS, vol. 4862, pp. 321–335. Springer, Heidelberg (2008)

A Recursive Approach to Entity Ranking and List Completion Using Entity Determining Terms, Qualifiers and Prominent n-Grams

Madhu Ramanathan, Srikant Rajagopal, Venkatesh Karthik,
Meenakshi Sundaram Murugeshan, and Saswati Mukherjee

Department of Computer Science and Engineering,
College of Engineering, Guindy,
Anna University, Chennai-60025, India
{strmmadhu,r.srikant.k,coolvenks.kar}@gmail.com,
{msundar_26@,msaswati}@yahoo.com

Abstract. This paper presents our approach for INEX 2009 Entity Ranking track which consists of two subtasks *viz.* Entity Ranking and List Completion. Retrieving the correct entities according to the user query is a three-step process *viz.* extracting the required information from the query and the provided categories, extracting the relevant documents which may be either prospective entities or intermediate pointers to prospective entities by making use of the structure available in the Wikipedia Corpus and finally ranking the resultant set of documents. We have extracted the Entity Determining Terms (EDTs), Qualifiers and prominent n-grams from the query, strategically exploited the relation between the extracted terms and the structure and connectedness of the corpus to retrieve links which are highly probable of being entities and then used a recursive mechanism for retrieving relevant documents through the Lucene Search. Our ranking mechanism combines various approaches that make use of category information, links, titles and WordNet information, initial description and the text of the document.

Keywords: Entity Ranking, List Completion, Entity Determining Terms (EDTs), Qualifiers, Prominent n-grams, Named Qualifiers, Wikipedia tags.

1 Introduction

Search Engines are widely used to retrieve relevant information from the World Wide Web. However, the task of identifying the necessary information from the relevant documents is left to the user. Research on Question Answering (QA) and Ad Hoc retrieval addresses this problem by analyzing the documents and locating the information according to the user's need. The Text Retrieval Conference (TREC) pioneered this research through the Question Answering track [1] and promotes three types of Questions *viz.* factoid, list and complex questions. TREC's Enterprise track [2] has a task similar to the Entity Ranking called the Expert Search where a set of expert names (people) are to be identified for a given topic.

S. Geva, J. Kamps, and A. Trotman (Eds.): INEX 2009, LNCS 6203, pp. 292–302, 2010.

The Entity Ranking track and the List completion task [3] started in 2007 as part of the INEX are aimed at exploring methodologies for retrieving relevant list of documents corresponding to the entities (answers) using the Wikipedia XML corpus. In the Entity Ranking task, the query and the category are provided and we are required to retrieve all the entities that match them. In the List Completion task, the query and a few sample entities are given and the remaining entities should be extracted and ranked. The challenge lies in handling different XML tags to filter the exact entities from the set of all related documents and to rank them. The XML corpus consists of 2,666,190 articles collected from Wikipedia. The two subtasks are Entity Ranking (ER) and List Completion (LC). Figure 1 shows an example INEX Entity Ranking topic from INEX 2009.

```
<inex_topic topic_id="139">
<title> Films directed by Akira Kurosawa </title>
<description> find the list of movies directed by Akira Kurosawa
</description>
<narrative>The expected answers are movies directed by the Japanese
 director Akira Kurosawa
</narrative>
<categories>
<category>japanese films </category>
<entities>
<entity id="477031">Sanshiro Sugata </entity>
<entity id="187603">Rashomon (film) </entity>
<entity id="75984">Ran (film) </entity></entities>
</inex_topic>
```

Fig. 1. A sample topic from INEX 2009's ER track

Here, the category "Japanese films" and the title "Films directed by Akira Kurosawa" can be used to identify the relevant entities in the Entity Ranking task, whereas, the title and the example entities "Sanshiro Sugata", "Rashomon (film)" and "Ran (film)" can be used for the List Completion task.

The articles in the Wikipedia corpus are well-organized in such a way that each article starts with an overview which we call Initial Descriptions (IDES) and contains several paragraphs of relevant information labeled with subtitles along with many links to other Wikipedia articles and concludes with references. The documents containing related information are grouped into categories and each document may fall under one or more such categories. In addition, the Wikipedia 2009 corpus contains WordNet tags for titles and links indicating the genre under which the article falls.

Our approach stresses on the importance of extracting the Entity Determining Terms (EDTs), Qualifiers and Prominent n-grams in arriving at a clear distinction between entities and non-entities. We use Lucene to find the initial set of relevant documents and then use a recursive expansion mechanism so as to arrive at a set that encompasses nearly all the relevant documents. For ranking the retrieved documents, we have combined various approaches that rely on the category matching, title keyword matching, WordNet confidence factor and synonyms, expansion of links present at relevant positions and paragraph expansion.

In Section 2 Related Work is explained. Section 3 describes our approach and Section 4 shows the results that we have submitted and Section 5 concludes the paper.

2   Related Work

Using the Part-of-Speech (POS) information for finding the focus of the question given by TREC is explored in [4]. Here the authors have used the predefined patterns to find the question types. Using Named Entity Recognition (NER) for query processing in List Question Answering is shown in [5]. The authors have classified the web pages into four categories *Viz.* collection page, topic page, relevant page and irrelevant page.

The information available from Wikipedia can be used for building a Named-Entity recognition system, which is shown in [6]. In this paper, authors have shown how category labels can be extracted from Wikipedia. Ranking sentences by giving importance to the proper names is discussed in [7]. Assigning different weights for WordNet synonyms and stemmed terms to improve the ranking is also explored. Wikipedia article names can be used to form effective queries and the effectiveness of the Initial Descriptions (IDES) is also explored [8]. The role of Wikipedia categories in ranking the entities is shown in [9].

3   Our Approach

Discovering entities by simply expanding all the links in the top 'n' documents retrieved by Lucene Search would be a rather rudimentary way to discover entities. The structure of the given document needs to be thoroughly understood to discover heuristics as to what to search in which part of the document. For example, in a Wikipedia article on "Akira Kurosawa" the links present in a paragraph named as "Movies" are more probable of being entities than the links in a paragraph named "Early Life" for the query given in Figure 1.

Similarly, if an XML page which has the same title as that given in the query is found, almost all links in the page have good candidature of being entities. Close examination of the structure of the pages may reveal that many such nuances could be found. Further, once all possible candidates for entities have been listed, further heuristics like category information and the WordNet tag information could be used to filter the entities and rank them.

However, in order to mine for the clues scattered in the corpus and to frame heuristics, the right terms need to be searched for at the right locations. This in turn, implies that the required terms need to be correctly segregated into the relevant term groups. This is exactly the approach that this paper proposes. In order to achieve a good precision, we make use of several ranking heuristics. Further, to increase our recall, the entire procedure is made recursive to extract more entities from the available set of prospective entities.

In our proposed approach we try to accomplish both the tasks of Entity Ranking and List Completion by doing the following steps:

1) Query Processing.
2) Extracting relevant documents.
3) Filtering actual entities.
4) Entity Ranking

3.1 Query Processing

The main difference between the classical document retrieval task and the Entity Ranking task is that the former should return all the documents containing relevant information whereas the latter should return only documents that are entities, eliminating all other documents. Hence, in order to distinguish between documents that are entities from those that are not, we extract the "Entity Determining terms (EDTs)". The EDTs are defined as the fundamental terms used to describe the entities which form hypernyms for the query. For the example given in Figure 1, the EDT would be "films". The actual entities for the query would form a subset under the set of documents that are represented by the EDT. For the example in Figure 1, entities for "Films directed by Akira Kurosawa" form a subset under the set of entities denoted by the EDT "films".

For the ER track, the query categories and the words in the query title are used to determine EDTs. The EDTs that are extracted from the query categories and the query title are called Category EDTs and Query EDTs respectively. The category information given along with the query often refers to some category which is higher in the category hierarchy as in "novels" for "Neil Gaiman novels" and "writers" for "Nordic authors who are known for children's literature". Hence, the common nouns present in the query category could be considered as EDTs as they are more generic terms depicting the given query. We have used the Stanford parser for extracting noun phrases and Stanford POS tagger for extracting common nouns. Once the category EDTs are taken from the query category, query EDTs need to be extracted. The common noun in the first noun phrase of the query gives the query EDT. An observation of this year's and previous year's topics shows us that in most cases only the first noun phrase contains the EDT and the remaining noun phrases merely add additional details to the entity type mentioned in the first noun phrase. For example, in "Nordic authors who are known for children's literature", only the first noun phrase "Nordic authors" contains the EDT "authors", while the other noun phrases just provide an additional description for the "authors" entity type. The query EDT may or may not match with the category EDTs. If a query EDT matches any category EDT, it is ignored; else, it is included in the set of EDTs.

Although this method works fine for the given queries, it tends to fail for a different syntactic construction of the query. A more generic mechanism of query EDT extraction (which has not been implemented as part of our system) would be to extract the various noun phrases in the query, get the common nouns from each noun phrase and measure the semantic relatedness between each of those common nouns and the category EDTs. Here we use the Normalized Google Distance (NGD) [10] to calculate semantic relatedness between two terms or nouns. NGD gives the distance between two terms. The smaller this value more related the terms are. NGD is given by:

$$NGD = \frac{Max - log_{10}\ C1C2}{log_{10}\ M - Min} \tag{1}$$

$$Max = Maximum(\log_{10} C1, \log_{10} C2) \tag{2}$$

$$Min = Minimum(\log_{10} C1, \log_{10} C2) \tag{3}$$

where, M is the total number of pages indexed by Google ($\sim 10^{17}$), C1 is number of hits returned when "term1" is searched and C2 is the number of hits returned when "term2" is searched. C1C2 is the number of hits returned when "term1 term2" is searched. Note that term1 and term2 may be unigrams or n-grams. This method would be better than checking if the common noun is a synonym of the category EDT, from any standard lexicon, since, this method can handle n-grams and also spans a wider coverage of terms compared to a lexicon.

For the LC task, where explicit query categories are not given, the category EDTs are extracted as follows. The categories of the sample entities are put in a set C. The common nouns from each of those categories are put in a set N. If a noun occurs more than once in N, then it is added to EDT. Also, if a noun N_i is semantically related to any one noun in the existing EDT list, N_i is added to EDT. Here again, Semantic Relatedness can be calculated by NGD or any similar measure. Once the category EDTs are extracted by the above mentioned algorithm, the query EDTs can be extracted by one of the two procedures described earlier.

Though the EDTs help us to determine whether a document is a possible entity or not, more information is required to determine whether the entity document meets the description specified in the query. In this example, the EDTs specify that the entity should be a Japanese film or a movie in general, but do not indicate anywhere that it should be directed by Akiro Kurosawa. We call such terms that describe the entities as "Qualifiers". From the query, we remove the stop words and EDTs and take all the remaining terms as Qualifiers and we stem the required terms. If a subset of those qualifiers occurs as a compound noun (sequence of nouns) in the query, instead of treating them as separate qualifiers, they are together considered as a single compound qualifier called a "Named Qualifier". The qualifiers and named qualifiers are used for filtering and ranking the entities. For the example given in Fig. 1 the qualifiers would be "direct" and "Akira Kurasawa" and the named qualifiers would be the subset "Akiro Kurasawa".

In natural language, a group of terms together may give a separate meaning when compared with the meaning of each of the terms individually. Such combinations of terms are called n-grams, where 'n' represents the number of terms in the meaningful unit. We call these as Meaningful n-grams. Each of the articles in the Wikipedia corpus discusses about a particular topic and has a name associated with it. The list that we have used in our approach is the expanded list comprising of Wikipedia article names and WordNet synonyms which we call "Wiki Names List". Each of the named qualifiers is checked for its presence in the "Wiki Names List". If a match is found, such a named qualifier is called a "Prominent n-gram". The articles on these prominent n-grams have the highest probability of having the links to the possible entities and can be used in entity extraction and ranking. In our example, the prominent n-gram would be "Akiro Kurosawa". The following table illustrates the extraction of EDTs, Qualifiers, Named Qualifiers and Prominent n-grams from a few sample queries:

Table 1. EDTs, Qualifiers, named Qualifiers and Prominent n-grams for sample queries

Query	Categories	EDTs	Simple Qualifiers	Named Qualifiers	Prominent N Grams
Magazines about indie-music	Music Magazines	Magazines		Indie-music	Indie-music
Nordic authors known for children's literature	Writers	Writers, authors	Nordic	children's literature	
Tom Hanks movies where he plays a leading role	Movies, films	Films, movies	Plays, leading, role	Tom Hanks	Tom Hanks
Computer systems that have a recursive acronym for their name.	Computer Systems, acronyms	Systems, acronyms	Name, computer	Recursive acronym	Recursive acronym
Hanseatic league in Germany in Netherland circle	Cities, cities in Germany	Cities		Hanseatic league, Germany, Netherland circle	Hanseatic league, Germany

It is to be noted that in the last query, there are no query EDTs. All EDTs are category EDTs.

3.2   Extracting Relevant Documents

We have used Lucene to index the Wikipedia corpus. Once the query processing is done, we retrieve the top 'n' documents using the given query as such, to extract the initial set of relevant documents. Though this initial set may contain a few entities, it may not encompass the set of all entities. In order to overcome this problem, we make use of a combination of heuristics viz. Initial Category Expansion, Prominent n-gram expansion, Title query match, Partial title query match, Document category expansion and Paragraph expansion. All these techniques rely on the document titles, category information, proximity information and the prominent n-grams.

In the *Initial Category Expansion* technique, we make use of the categories given for the ER task and the intersection of categories of the sample entities for the LC task and perform a category search that extracts all the documents that fall under these categories. These documents are added to the already existing set of relevant documents. The main reason for performing the Initial Category Expansion is to increase recall. Expanding the query category ensures that the entire pool of prospective entities is identified and put in the set of relevant documents. The pool may later be filtered by other approaches to increase precision. Also, this may lead to intermediate pages that may result in entities after recursion. However, not all entities come under these categories and hence we go in for the other techniques also.

As explained earlier, the articles with the prominent n-grams as their title have a high probability of containing links to the actual entities. Hence, all the links in such documents are added to the list of relevant documents. However, we need to verify if the documents thus obtained are actual entities or not, which is taken care of in step 3.3. This procedure is called *Prominent N-gram Expansion*.

In certain cases, the query itself occurs as the title of a document. For example: Q63 Hugo awarded best novels. In this case there is a XML document with the title of "Hugo Award for best novels". Hence, in *Title query match*, the entries in the relevant documents list is checked for such exact title query matches. If such a match exists, then the links in that document are added to the list of possible entities. Instead of checking for the match of the entire title, if the title of the document contains at least one EDT and one or more qualifiers, then the links of such a document are also included. This procedure is called *Partial Title Query Match*. It is a bit less stringent than Title Query Match.

The initial category given in the query may not encompass all the entities and we need some means of getting all the entities through related categories. For this we perform *Document Category Expansion* where we look for the presence of the EDTs and Qualifiers in the category information provided in the list of documents already extracted through the above methods. Those document categories that have one or more EDTs and qualifiers are also expanded as in Initial category expansion and the extracted documents are added to the relevant document list. Whereas, the Initial Category Expansion is done only initially, the Document Category Expansion is applied recursively, along with the other expansion techniques, on the already retrieved set of relevant documents. This ensures that category information is used to search for entities, not only in the beginning with the initial set of relevant documents returned by Lucene, but also at every level of recursion with the new sets of relevant documents added during the previous level of recursion.

In the List Completion task, apart from the above methods, we make use of the *Paragraph Expansion technique* which is similar to a proximity search where the paragraph size determines the window size. Each of the retrieved documents is checked for paragraphs that contain all the sample entities along with the qualifiers. This implies that the paragraph has a high probability of containing the other entities too and so all the links in that paragraph are added to the relevant documents list.

These techniques when applied recursively to the set of relevant documents helps in deeper exploration for prospective entities. In other words, the recursion is primarily used to boost the recall. The precision does not get affected by increasing the level of recursion because the next step employs techniques using the EDTs and WordNet information to filter the actual entities alone. Increasing the level of recursion beyond a certain threshold may result in stagnation. Recursion till stagnation would ensure complete prospective entity exploration.

3.3 Filtering Actual Entities

Though the previous step populates the list with all the possible entities one has to verify if each of them is an entity or just an article related to entities, because the prominent n-gram expansion and the title query match expansion techniques may yield a few links that do not point to real entities. The document's categories and its WordNet tags are the two main features that enable the distinction of an entity from a non-entity. We follow two mechanisms to separate out the actual entities, namely Category verification and WordNet verification. In category verification, each of the documents obtained as a result of the previous step is taken and the document's categories are explored to check for the presence of one or more EDTs. Similarly, in WordNet verification, each of the documents is checked for the presence of atleast one EDT in the WordNet tags. If any one of the above tests is true then the document is added to the Final Entity List, otherwise it is considered as a non-entity and discarded. This is where the true power of the EDTs can be realized. They are critical in determining entities and distinguishing them from non-entities.

3.4 Entity Ranking

The previous step yields the set of actual entities. However, they have to be ranked according to the actual degree of relevance to the query. This ranking of the retrieved documents is done using the WordNet tags, category terms and the locality of query terms in the paragraphs.

The INEX 2009 Wikipedia corpus contains the WordNet tags added to the titles and links in the document. The WordNet tag contains the WordNet id, WordNet term and the precision percentage which indicates how related the WordNet term is to the word that is tagged. We consider the WordNet terms of the title of the document alone for ranking purposes and count for the number of EDTs that are present in the WordNet tags. The *WordNet Fitness* is assigned a value proportional to this count after scaling.

To enhance the ranking further, we have used the category information available for the articles in Wikipedia. Each Wikipedia article falls under a few categories. Apart from checking for the presence of EDTs in the categories we look for the number of qualifier terms present in the categories. The ratio of the maximum number of qualifiers present in a document category to the total number of qualifiers is taken as the *Category Fitness*.

In addition, for the List Completion task, we used the categories of the given example entities as reference set (E_e). This set is compared against the set of categories the retrieved document belongs to (R_e). The ratio of the match is used to find the similarity between the retrieved entity and the example entities. This is added to the category fitness score.

$$Category\ Fitness = \frac{Max\ no.\ of\ qualifiers\ in\ the\ document's\ categories}{total\ number\ of\ qualifiers} \quad (4)$$

$$Category\ Match\ Score = \frac{Cat(E_e) \cap Cat(R_e)}{Cat(R_e)} \quad (5)$$

Apart from these, we used a third technique called *Paragraph Fitness* similar to a proximity search where each paragraph in a document is checked for the presence of EDTs and qualifiers. Depending upon the number of qualifiers present the paragraph fitness is increased. If the paragraph being considered is the initial description paragraph then a greater weight is assigned since the relevancy of information in the IDES is greater when compared to the other paragraphs. This is same as the proximity search except that the window size is not fixed but depends upon the size of the paragraph.

The final rank is then taken as the sum of the WordNet fitness, category fitness and the paragraph fitness and the entities are ranked based on this score.

4 Evaluation

The INEX 2009 topics are selected from the previous editions INEX ER track. A sample Entity Ranking and List Completion results retrieved by our approach for the topic in Figure 1 are shown below.

Entity Ranking

    139 0 WP1624660 Dersu Uzala (1975 film)
    139 0 WP1942885 Madadayo
    139 0 WP235331 Red Beard
    139 0 WP2553318 Scandal (1950 film)
    139 0 WP477031 Sanshiro Sugata
    139 0 WP6716962 After the Rain (film)
    139 0 WP180241 Sanjuro
    139 0 WP187603 Rashomon (film)

List Completion

    139 0 WP1624660 Dersu Uzala (1975 film)
    139 0 WP235331 Red Beard
    139 0 WP2553318 Scandal (1950 film)
    139 0 WP31371 Seven Samurai
    139 0 WP477031 Sanshiro Sugata
    139 0 WP180241 Sanjuro
    139 0 WP187603 Rashomon (film)
    139 0 WP75984 Ran (film)

Table 2 shows the average precision value given by xinfAP and the average recall values obtained when each of the expansion or matching technique was used separately on the Lucene search results for the Entity Ranking Task. Similar results were obtained for List Completion Task also.

Table 2. Average Precision (xinfAP) and average Recall of each expansion or matching technique for the Entity Ranking Task

Technique	xinfAP	Average Recall
No Expansion (Lucene results alone)	0.1702	0.2420
Initial Category Expansion (ICE)	0.1885	0.3279
Title Query Match (TQM)	0.1990	0.3087
Document Category Expansion (DCE)	0.2098	0.3549
Partial Title Query Match (PTQM)	0.2287	0.3976
Prominent N-Gram Expansion (PNE)	0.2036	0.3081
All expansion techniques combined	0.2696	0.5807

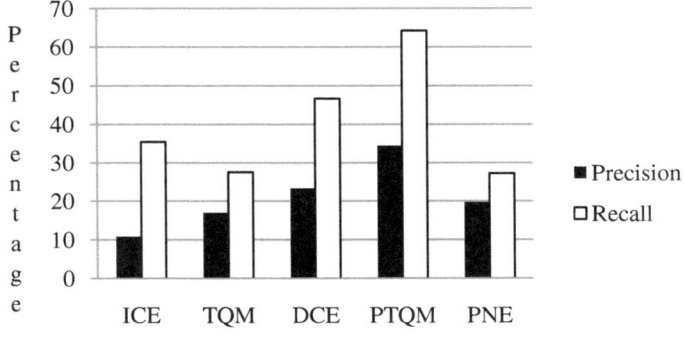

Expansion/Match technique

Fig. 1. The percentage increase in xinfAP and average recall brought by the Expansion/ Matching technique over the Lucene results for the Entity Ranking Task

The information in Table 2 is represented in Figure 1. The percentage increase is shown in the figure instead of the actual average precision/recall values. It can be seen that Partial Title Query matching was the most successful technique to boost both the precision (33% boost) as well as the recall (65% boost). For our mandatory runs, we had used all the expansion and matching techniques over the Lucene results. We had not used recursion in our mandatory run and the average accuracy was reported as 0.2696 for entity ranking and 0.3081 for List completion. We fine-tuned our system by changing the weights for various fitness measures for filtering the actual entities (described in Section 3.4) from which it was observed that maximum precision is obtained when Category fitness, WordNet fitness and Paragraph fitness are assigned decreasing weights in that order. With the tuned parameters, we performed runs by changing the level of recursion. Figure 2 shows the results. As explained earlier, after the second level of recursion the precision and recall begin to stagnate and even drop slightly. Recursion has brought in a 5% increase in precision and 8% increase in recall.

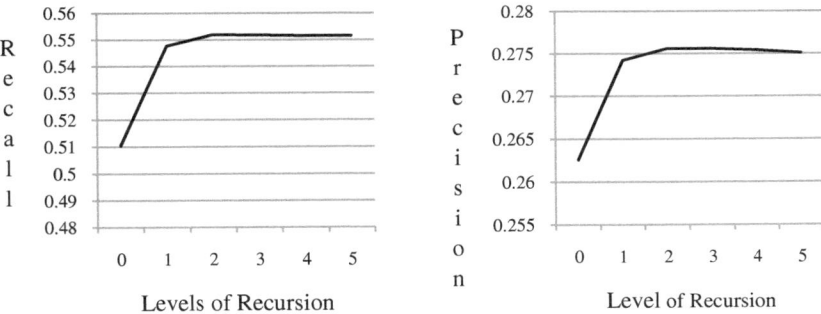

Fig. 2. The recall and Precision Vs. the level of recursion (Entity Ranking)

5 Conclusion

The method we have deployed relies heavily on the Entity Determining terms, qualifiers and the prominent n-grams. We described how differentiating the EDTs and qualifiers in the query and using them for the appropriate expansion and matching techniques would make the task of identifying entities and intermediate pages, much easier. In other words, looking for the right terms at the right places has proved to be a simple, yet highly effective mechanism of mining for entities. We showed a method for EDT and qualifier extraction and also suggested a generic method that would work on all types of queries, regardless of their syntactic composition. Since the category hierarchy has not been specified, the scope of expanding the EDTs is reduced. However, the WordNet tagging introduced in the 2009 Wikipedia corpus provides a better distinction between entities and non-entities. Though our method provides better accuracy and recall in majority of the queries, for those that have some form of reference to time periods as in "movies shot after 1999", it fails. Better results for can be obtained by introducing NLP techniques to semantically process such queries.

References

1. Voorhees, E.M.: Overview of the TREC 2001 Question Answering Track. In: Proceedings of the 10th Text Retrieval Conference (2001)
2. Craswell, N., de Vries, A.P., Soboroff, I.: Overview of the TREC 2005 Enterprise Track. In: Proceedings of the 14th Text Retrieval Conference (2005)
3. de Vries, A.P.: Overview of the INEX 2007 Entity Ranking Track. In: Fuhr, N., Kamps, J., Lalmas, M., Trotman, A. (eds.) INEX 2007. LNCS, vol. 4862, pp. 245–251. Springer, Heidelberg (2008)
4. Chen, J., Diekema, A., Taffett, M.D., McCracken, N., Ozgencil, N.E., Yilmazel, O., Liddy, E.D.: Question Answering: CNLP at the TREC 10 Question Answering Track. In: Proceedings of the 10th Text Retrieval Conference (2001)
5. Yang, H., Chua, T.-S.: Web-based list question answering. In: Proceedings of the 20th International Conference on Computational Linguistics (2004)
6. Kazama, J., Torisawa, K.: Exploiting Wikipedia as External Knowledge for Named Entity Recognition. In: Proceedings of the 2007 Joint Conference on Empirical Methods in Natural Language Processing and Computational Natural Language Learning, EMNLP-CoNLL (2007)
7. Hermjakob, U., Hovy, E.H., Lin, C.-Y.: Knowledge-Based Question Answering. In: Proceedings of the 6th World Multiconference on Systems, Cybernatics and Informatics (SCI 2002), Orlando, FL, U.S.A., July 14-18 (2002)
8. Murugeshan, M.S., Mukherjee, S.: An n-Gram and Initial Description Based Approach for Entity Ranking Track. In: Fuhr, N., Kamps, J., Lalmas, M., Trotman, A. (eds.) INEX 2007. LNCS, vol. 4862, pp. 293–305. Springer, Heidelberg (2008)
9. Thom, J.A., Pehcevski, J., Vercoustre, A.-M.: Use of Wikipedia Categories in Entity Ranking. In: Proceedings of the 12th Australasian Document Computing Symposium (ADCS 2007), Melbourne, Australia, December 10 (2007)
10. Cilibrasi, R., Vitanyi, P.: The Google Similarity Distance. IEEE Trans. Knowledge and Data Engineering 19(3), 370–383 (2007)

Overview of the INEX 2009 Interactive Track

Nils Pharo[1], Ragnar Nordlie[1], Norbert Fuhr[2],
Thomas Beckers[2], and Khairun Nisa Fachry[3]

[1] Faculty of Journalism, Library and Information Science, Oslo University College, Norway
{nils.pharo,ragnar.nordlie}@jbi.hio.no
[2] Department of Computer Science and Applied Cognitive Science,
University of Duisburg-Essen, Germany
norbert.fuhr@uni-due.de, tbeckers@is.inf.uni-due.de
[3] Archives and Information Studies, University of Amsterdam, The Netherlands
k.n.fachry@uva.nl

Abstract. In the paper we present the organization of the INEX 2009 interactive track. For the 2009 experiments the iTrack has gathered data on user search behavior in a collection consisting of book metadata taken from the online bookstore Amazon and the social cataloguing application LibraryThing. Thus the data are more structured than in previous years' experiments, consisting of traditional bibliographic metadata, user-generated tags and reviews and promotional texts and reviews from publishers and professional reviewers. Through monitoring searches based on three different task types the experiment aims at studying how users interact with highly structured data. We describe the methods used for data collection and the tasks performed by the participants. Some preliminary results of the interaction analysis are reported.

1 Introduction

The INEX interactive track (iTrack) is a cooperative research effort run as part of the INEX Initiative for the Evaluation of XML retrieval [1]. The overall goal of INEX is to experiment with the potential of using XML to retrieve relevant parts of documents. In recent years, this has been done through the provision of a test collection of XML-marked Wikipedia articles. The main body of work within the INEX community has been the development and testing of retrieval algorithms. Interactive information retrieval (IIR) [2] aims at investigating the relationship between end users of information retrieval systems and the systems they use. This aim is approached partly through the development and testing of interactive features in the IR systems and partly through research on user behavior in IR systems. In the INEX iTrack the focus over the years has been on how end users react to and exploit the potential of IR systems that facilitate the access to *parts* of documents in addition to the full documents.

The INEX interactive track (iTrack) was run for the first time in 2004 [3], repeated in 2005 [4], in 2006/2007 [5] (due to technical problems the tasks scheduled for 2006 were actually run in early 2007), and in 2008 [14]. Although there has been variations in task content and focus, some fundamental premises has been in force throughout:

S. Geva, J. Kamps, and A. Trotman (Eds.): INEX 2009, LNCS 6203, pp. 303–311, 2010.
© Springer-Verlag Berlin Heidelberg 2010

- a common subject recruiting procedure
- a common set of user tasks and data collection instruments such as interview guides and questionnaires
- a common logging procedure for user/system interaction
- an understanding that collected data should be made available to all participants for analysis

This has ensured that through a manageable effort, participant institutions have had access to a rich and comparable set of data on user background and user behavior, of sufficient size and level of detail to allow both qualitative and quantitative analysis. This has already been the source of a number of papers and conference presentations ([6], [7], [8], [9], [10], [11], [12], [15]).

In 2009, it was felt that although the "common effort" quality of the previous years was valuable and still held potential as an efficient way of collecting user behavior data, the Wikipedia collection had exhausted its potential as a source for studies of user interaction with XML-coded documents. We decided to base the experiments on a new data collection with richer structure and more semantic markup than has previously been available, and have created a collection based on a crawl of 2.7 million records from the book database of the online bookseller Amazon.com, consolidated with corresponding bibliographic records from the cooperative book cataloguing tool LibraryThing (a more specific description of the database is given below). The records present book descriptions on a number of levels: formalized author, title and publisher data; subject descriptions and user tags; book cover images; full text reviews and content descriptions. The database intended to enable investigation of research questions concerning, for instance

- What is the basis for judgments on relevance in a richly structured and diverse material? What fields / how much descriptive text do users make use of / chose to see to be able to judge relevance?
- How do users understand and make use of structure (e.g. representing different levels of description, from highly formalized bibliographic data to free text with varying degrees of authority) in their search development?
- How do users construct and change their queries during search (sources of terms, use and understanding of tags, query development strategies ..)?

2 Tasks

For the 2009 iTrack the experiment was designed with two categories of tasks constructed by the track organizers, from each of which the searchers were instructed to select one of three alternative search topics. In addition the searchers were invited to perform one semi-self-generated task. The two categories of tasks were intended to reflect the most common purposes a searcher would have for visiting a database of primarily bibliographic data, a broad, explorative task and a narrower, more specific, purpose-driven task. The self-selected task was intended to force the searcher to perform a more quality-driven search than the two others.

2.1 The Broad Tasks

These task were designed to investigate thematic exploration, aiming to provide data on query development, metadata type preference and navigation patterns. The tasks were as follows:

1. You are considering to start studying sociology. In order to prepare for the course you would like to get acquainted with some good and recent introductory texts within the field as well as some of its classics.
2. You are interested in taking a course on environmental friendly energy. In order to prepare for the course you would like to get acquainted with some good introductory texts on the field.
3. You are considering to start studying existentialism. In order to prepare for the course you would like to get acquainted with some good introductory texts within the field as well as some of its classics.

2.2 The Narrow Tasks

These tasks represent relatively narrow topical queries where the purpose was to allow us to study the basis for relevance decisions and compare the searchers' preference of different document representations. The following tasks were provided:

1. Find trustworthy books discussing the conspiracy theories which developed after the 9/11 terrorist attacks in New York.
2. Find books which present documentation of the specific health and/or beauty effects of consuming olive oil.
3. The Kabbalah is an esoteric religious tradition which has inspired works of fiction. Find novels where the plot is inspired by the Kabbalah, and a factual treatment of the origins and development of this tradition.

2.3 The Semi Self-selected Task

For one of the courses you are currently attending, you need an additional textbook. You have only money for one book (assuming they all have about the same price). You are free to select the course topic yourself.

3 Participating Groups

Due to unfortunate delays in the preparation of the experimental system, the experiments were launched late in the INEX 2009 research cycle, and only 3 research groups were able to submit experiment data by the deadline for this report: Oslo University College, University of Glasgow, and University of Duisburg-Essen. Data from a total of 123 searches performed by 41 test subjects were collected, in addition to 36 searches by 12 subjects using Duisburg's alternative system (see below).

4 Research Design

4.1 Search System

The experiments were conducted on a java-based retrieval system built within the Daffodil framework [13], which resides on a server at and is maintained by the University of Duisburg-Essen. The collection was indexed with Apache Solr 1.3, which is based on Apache Lucene. Lucene applies a vector space retrieval model. The system is also partially based on the *ezDL* (`http://www.is.inf.uni-due.de/projects/ezdl/`). The basis of the search system is the same as have been used for previous iTracks, but the interface has been modified extensively to accommodate the new data set, and a set of new functionalities have been developed.

Figure 1 shows the interface of the system. The main features available to the user are

- When a search term is entered, the searcher can choose to search on "content", "reviews", or both together. "Content" searches all the "formalized" text connected to each book – title, keywords, publisher's description etc. "Reviews" allows search in the text of any user reviews of the book. In both cases the search index bases result rankings on term occurrence. In addition, there is field-based search available on author, title or publication year.

- The system can order the search results according to "relevance" (which books the system considers to be most relevant to your search terms), "year" (publication year of the book), or "average rating" (in the cases where people have rated the quality of the books).

- The system will show results twenty titles at a time, with features to assist in moving further forwards or backwards in the result list.

- A double click on an item in the result list will show the book details in the "Details" window. If the book has been reviewed, the reviews can be seen by clicking the "Reviews" tab at the bottom of this window.

- The relevance of any which is examined should be determined, as "Relevant", "Partially relevant" or "Not relevant", by clicking markers at the bottom of the screen. Any book decided to constitute part of the answer to the search task should be moved to a result basket by clicking the "Add to basket" button next to the relevance buttons.

- When the first search term has been entered, the system will use the task window to suggest search terms which might be relevant to the task. A double click on a term in this list will move it to the search term window.

- A "Query history" button in the middle of the screen displays the search terms used so far in the search session.

- A line of yellow dots above an item in the result list is used to indicate the system's estimate of how closely related to the query the item is considered to be.

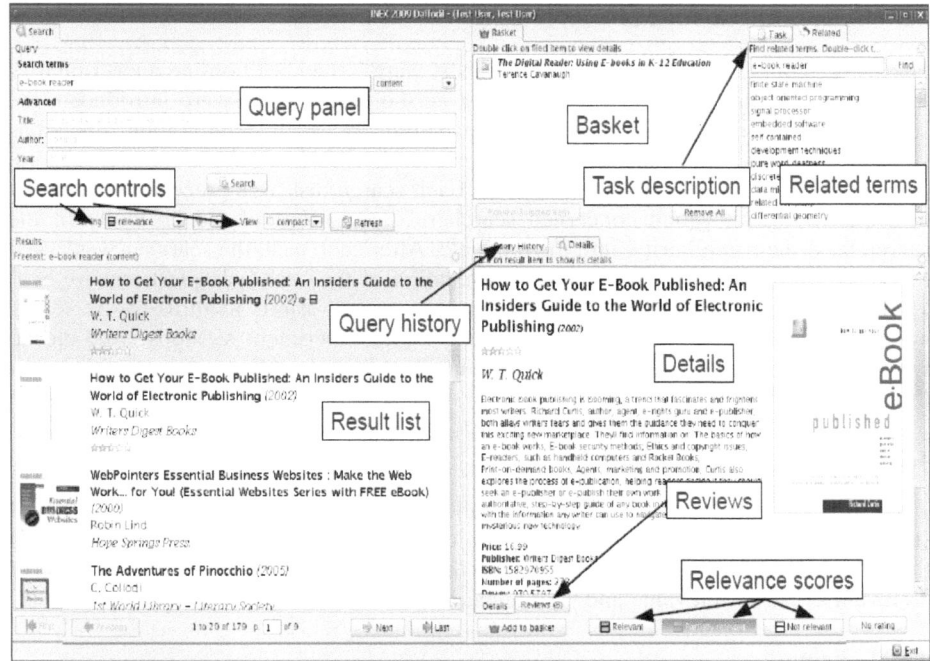

Fig. 1. Daffodil interface

4.2 Document Corpus

The collection contains metadata of 2 780 300 English-language books. The data has been crawled from the online bookstore of *Amazon* and the social cataloging web site *LibraryThing* in February/March 2009 by the University of Duisburg-Essen. The MySQL database containing the crawled data has size of about 190 GB. Cover images are available for over one million books (100 GB of the database). Several millions of customer reviews were crawled.

The XML-coded records present book descriptions on a number of levels: formalized author, title and other bibliographic data; controlled subject descriptions and user-provided content-descriptive tags; book cover images; full text reviews and publisher-supplied content descriptions. The following listing shows what data was crawled from either Amazon or LibraryThing:

Amazon
isbn, title, binding, label, list price, number of pages, publisher, dimensions, reading level, release date, publication date, edition, Dewey classification, title page images, creators, similar products, height, width, length, weight, reviews (rating, author id, total votes, helpful votes, date, summary, content) editorial reviews (source, content)

LibraryThing
tags (including occurrence frequency), blurbs, dedications, epigraphs, first words, last words, quotations, series, awards, browse nodes, characters, places, subjects.

4.3 Online Questionnaires

During the course of the experiment, searchers were issued brief online questionnaires to support the analysis of the log data. Before the search tasks were introduced, the searchers were given a pre-experiment questionnaire, with demographic questions such as searchers' age, education and experience in information searching in general and in searching and buying books online. Each search task was preceded with a pre-task questionnaire, which concerned searchers' perceptions of the difficulty of the search task, their familiarity with the topic etc. After each task, the searcher was asked to fill out a post-task questionnaire. The intention of the post-task questionnaire is to learn about the searchers' use of and their opinion on various features of the search system, in relation to the just completed task. The experiment sessions were closed with a post-experiment questionnaire, which elicited the searchers' general opinion of the search system.

4.4 Relevance Assessments

The users' task was partly to indicate the relevance of any item in the result list found sufficiently interesting for them to view in detail, partly to collect a result set which they considered to constitute an answer to their task. A three-part relevance scale of "relevant", "partly relevant" and "not relevant" was used.

4.5 Logging

All search sessions were logged and saved to a database. The logs register and time stamp the events in the session and the actions performed by the searcher, as well as the responses from the system. In addition to system logs, some participating institutions have been logging additional data through eye-tracking, screen image capture etc.

4.6 System Comparison

A modified version of the search system (the B version) was developed at the University of Duisburg-Essen. This special version was less interactive and powerful due to missing reviews, tools (related terms, query history) and search options (content & review, review).

12 of the 24 participants in Duisburg used the B version, while the other 12 used the A version (the standard version employed by all other participants in this track). Additionally, the experiments were also recorded by an eyetracking system. It is expected that users behave differently with a more traditional, less interactive search system.

5 Experimental Procedure

Each experiment has been performed following the standard procedure outlined below. Steps 7 to 10 were repeated for each of the three tasks performed by the searcher.

1. Experimenter briefs the searcher, and explains format of study. The searcher reads and signs the Consent Form.
2. The experimenter logs the searchers into the experimental system. Tutorial of the system is given with a training task provided by the system. The experimenter hands out and explains the system features document.
3. Any questions answered by the experimenter.
4. The experimenter administers the pre-experiment questionnaire.
5. Topic descriptions for the first task category administered, and a topic selected.
6. Pre-task questionnaire administered.
7. Task begins by clicking the link to the search system. Maximum duration for a search is 15 minutes, at which point the system issues a "timeout" warning. Task ended by clicking the "Finish task" button.
8. Post-task questionnaire administered.
9. Steps 5-8 repeated for the second and third task.
10. Post-experiment questionnaire administered.

6 Data Analysis

As the experiment phase was delayed, only a preliminary analysis of the questionnaire data is available at the deadline for this report. Log analysis, combined with further questionnaire analysis, will continue and will be reported elsewhere.

The questionnaires included open-question invitations for comments by the participants on both the system and the search experience. The positive comments include the following items:

+ well arranged interface
+ everything fits on the screen, no scrolling
+ reviews are very useful

Some users experienced technical problems. Also, missing highlighting and filtering as well as too many books without enough metadata were points of negative criticism:

- technical problems (search, query syntax, drag and drop)
- "related terms" are not always useful
- no highlighting of query terms in results
- some books do not have enough details
- no filtering

From the quantitative questionnaire data we have attempted to analyze the effect of the different types of search task on searchers' use of the various types of metadata available.

The searchers were asked to indicate on a five point scale how useful (5 for very useful) different types of metadata were for solving their search tasks. From Table 1 we see that document titles, publishers' book descriptions and reviews (by users) were

Table 1. The influence of task category on searchers' preferences of metadata field

	Task category 1	Task category 2	Task category 3	Overall
Title	3.79	3.81	3.96	3.85
Author	1.62	1.57	2.17	1.78
Year	2.55	1.91	2.83	2.43
Publisher's name	1.75	1.51	1.83	1.70
Keywords	3.28	3.29	2.82	3.13
User tags	2.65	2.75	2.58	2.66
Reviews	3.23	3.34	3.38	3.32
Publisher's description	3.45	3.64	3.42	3.50
Image	2.36	2.85	2.45	2.55
Relevance score	2.98	2.98	2.98	2.98

the three most popular metadata fields. It is also worth noting that the searchers found keywords (from Amazon) to be more useful than the user-created tags. It seems that searchers put more trust in authoritative sources that use a controlled vocabulary than users' idiosyncratic tagging.

We see that the variation between the different categories of task only differs significantly with respect to the usefulness of "year". We believe the reason that searchers find year to be more important for the textbook tasks (category 1 and 3) is the sheer number of relevant documents generated by these queries. The Category 2 tasks are more specific and the relevant documents are probably easier to select from the result list without reference to additional distinguishing factors such as publication year.

We have also looked at the searchers' familiarity with the topics and seen how this correlates with the usefulness of the metadata components. Our finding is that there is no systematic correlation between topic familiarity and metadata preference.

References

[1] Malik, S., Trotman, A., Lalmas, M., Fuhr, N.: Overview of INEX 2006. In: Fuhr, N., Lalmas, M., Trotman, A. (eds.) INEX 2006. LNCS, vol. 4518, pp. 1–11. Springer, Heidelberg (2007)
[2] Ruthven, I.: Interactive Information Retrieval. Annual Review of Information Science and Technology 42, 43–91 (2008)
[3] Tombros, A., Larsen, B., Malik, S.: The Interactive Track at INEX 2004. In: Fuhr, N., Lalmas, M., Malik, S., Szlávik, Z. (eds.) INEX 2004. LNCS, vol. 3493, pp. 410–423. Springer, Heidelberg (2005)

[4] Larsen, B., Malik, S., Tombros, A.: The interactive track at INEX 2005. In: Fuhr, N., Lalmas, M., Malik, S., Kazai, G. (eds.) INEX 2005. LNCS, vol. 3977, pp. 398–410. Springer, Heidelberg (2006)

[5] Larsen, B., Malik, S., Tombros, A.: The Interactive track at INEX 2006. In: Fuhr, N., Lalmas, M., Trotman, A. (eds.) INEX 2006. LNCS, vol. 4518, pp. 387–399. Springer, Heidelberg (2007)

[6] Pharo, N., Nordlie, R.: Context Matters: An Analysis of Assessments of XML Documents. In: Crestani, F., Ruthven, I. (eds.) CoLIS 2005. LNCS, vol. 3507, pp. 238–248. Springer, Heidelberg (2005)

[7] Hammer-Aebi, B., Christensen, K.W., Lund, H., Larsen, B.: Users, structured documents and overlap: interactive searching of elements and the influence of context on search behaviour. In: Ruthven, I., et al. (eds.) Information Interaction in Context: International Symposium on Information Interaction in Context: IIiX 2006, Proceedings, October 18-20, pp. 80–94. Royal School of Library and Information Science, Copenhagen (2006)

[8] Malik, S., Klas, C.-P., Fuhr, N., Larsen, B., Tombros, A.: Designing a user interface for interactive retrieval of structured documents: lessons learned from the INEX interactive track? In: Gonzalo, J., Thanos, C., Verdejo, M.F., Carrasco, R.C. (eds.) ECDL 2006. LNCS, vol. 4172, pp. 291–302. Springer, Heidelberg (2006)

[9] Kim, H., Son, H.: Users Interaction with the Hierarchically Structured Presentation in XML Document Retrieval. In: Fuhr, N., Lalmas, M., Malik, S., Kazai, G. (eds.) INEX 2005. LNCS, vol. 3977, pp. 422–431. Springer, Heidelberg (2006)

[10] Kazai, G., Trotman, A.: Users' perspectives on the Usefulness of Structure for XML Information Retrieval. In: Dominich, S., Kiss, F. (eds.) Proceedings of the 1st International Conference on the Theory of Information Retrieval, pp. 247–260. Foundation for Information Society, Budapest (2007)

[11] Larsen, B., Malik, S., Tombros, A.: A Comparison of Interactive and Ad-Hoc Relevance Assessments. In: Fuhr, N., Kamps, J., Lalmas, M., Trotman, A. (eds.) INEX 2007. LNCS, vol. 4862, pp. 348–358. Springer, Heidelberg (2008)

[12] Pharo, N.: The effect of granularity and order in XML element retrieval. Information Processing and Management 44(5), 1732–1740 (2008)

[13] Fuhr, N., Klas, C.P., Schaefer, A., Mutschke, P.: Daffodil: An integrated desktop for supporting high-level search activities in federated digital libraries. In: Agosti, M., Thanos, C. (eds.) ECDL 2002. LNCS, vol. 2458, pp. 597–612. Springer, Heidelberg (2002)

[14] Pharo, N., Nordlie, R., Fachry, K.N.: Overview of the INEX 2008 Interactive Track. In: Geva, S., Kamps, J., Trotman, A. (eds.) INEX 2008. LNCS, vol. 5631, pp. 300–313. Springer, Heidelberg (2009)

[15] Pehcevski, J.: Relevance in XML retrieval: the user perspective. In: Trotman, A., Geva, S. (eds.) Proceedings of the SIGIR 2006 Workshop on XML Element Retrieval Methodology, Held in Seattle, Washington, USA, August 10, pp. 35–42. Department of Computer Science, University of Otago, Dunedin, New Zealand (2006)

Overview of the INEX 2009 Link the Wiki Track

Wei Che (Darren) Huang[1], Shlomo Geva[1], and Andrew Trotman[2]

[1] Faculty of Science and Technology, Queensland University of Technology, Brisbane
w2.huang@student.qut.edu.au, s.geva@qut.edu.au
[2] Australia Department of Computer Science, University of Otago, Dunedin, New Zealand
andrew@cs.otago.ac.nz

Abstract. In the third year of the Link the Wiki track, the focus has been shifted to anchor-to-bep link discovery. The participants were encouraged to utilize different technologies to resolve the issue of focused link discovery. Apart from the 2009 Wikipedia collection, the Te Ara collection was introduced for the first time in INEX. For the link the wiki tasks, 5000 file-to-file topics were randomly selected and 33 anchor-to-bep topics were nominated by the participants. The Te Ara collection does not contain hyperlinks and the task was to cross link the entire collection. A GUI tool for self-verification of the linking results was distributed. This helps participants verify the location of the anchor and bep. The assessment tool and the evaluation tool were revised to improve efficiency. Submission runs were evaluated against Wikipedia ground-truth and manual result set respectively. Focus-based evaluation was undertaken using a new metric. Evaluation results are presented and link discovery approaches are described.

Keywords: Wikipedia, Focused Link Discovery, Anchor-to-BEP, Assessment, Evaluation.

1 Introduction

The Link the Wiki track was run for the first time in 2007 [1, 2]. It aims to offer an independent evaluation forum for researchers to work together to solve the problem of anchor-to-bep link discovery. The participants are encouraged to utilize different technologies, such as data mining, natural language processing, machine learning, information retrieval, etc., to discover relevant anchors in a new article and link the anchor to best entry points in other documents.

In 2007, the file-to-file (i.e. F2F) runs were evaluated against the Wikipedia ground truth whilst the anchor-to-bep (i.e. A2B) task was introduced in 2008 [3]. High fidelity file-to-file link discovery within the Wikipedia has been achieved as an outcome in 2008, as measured in comparisons with the ground truth. The focus has now been shifted to anchor-to-bep link discovery. Several improvements, including the submission specification, the tools, evaluation methods and metrics, have been made to conduct a better experiment in focused link discovery. Apart from the Wikipedia collection, the Te Ara encyclopedia was introduced and the tasks, *Link Te Ara* and *Link Te Ara to Wiki*, were set up for the first time. Despite its small size,

S. Geva, J. Kamps, and A. Trotman (Eds.): INEX 2009, LNCS 6203, pp. 312–323, 2010.
© Springer-Verlag Berlin Heidelberg 2010

there is a real challenge offered by the Te Ara collection. Since it is not extensively linked, and since page names are not necessarily as informative as Wikipedia page names, both link mining and page-name matching - the methods that work particularly well with the Wikipedia - are ineffective with the Te Ara.

Six groups from different organizations participated in the 2009 track. 16 runs were received for the file-to-file task while 13 runs for the anchor-to-bep task and 8 runs for the F2F on A2B task were submitted. Two groups were also involved in the Te Ara tasks with 7 runs contributed. All link the wiki runs were evaluated against the Wikipedia ground truth. All anchor-to-bep runs were additionally evaluated in different ways such as anchor-to-file and anchor-to-bep. The qrels are obtained through manual assessment. A set of evaluation results is depicted and a brief discussion is presented in this paper.

2 Document Collection

Two collections, the Wikipedia and the Te Ara, were used in the Link the Wiki track in 2009. The Wikipedia corpus consists of 2,666,190 articles with roughly 50GB in size. This collection is much larger than the one used in 2008. For file-to-file link discovery, 5000 articles were randomly selected, but filtered by certain criteria such as the document size and the number of anchors (i.e. links) to control the quality of the documents used in the task. For anchor-to-bep link discovery, the participants nominated 33 topics and submissions were manually assessed by the nominator who is expected to be fully acquainted with the topic content.

The Te Ara Encyclopedia was also used in the Link the Wiki track for 2 designated Te Ara tasks. At the time of writing, the collection contains 3179 articles with around 50MB in size without images. Currently there is no link in the collection and some of documents are still small. New approaches were expected to carry out focused link discovery without taking any advantage of link mining and page name match. The linking was required for the whole collection.

3 Task Specification

3.1 Tasks

The task was specified as twofold: the identification of links from the orphan into the document collection; and the identification of links from the collection into the orphan at both file-to-file and anchor-to-bep levels. Anchor-to-bep link discovery: This task represents the main goal of the Link the Wiki track. Researchers are encouraged to develop focused link discover algorithm, produce reliable assessments and participate in the forum to discuss solutions to focused link discovery. Only 50 anchors and up to 5 beps per anchor were allowed for each topic. At most, 250 incoming links could be specified in the case of the Link the Wiki task. Each incoming link must be from a different document. Only outgoing links were needed for the Te Ara tasks because all documents were used and so all incoming links were discovered anyway. The *Link-Te-Ara* task is to discover anchor texts and link them to best entry points within the collection. *Link-Te-Ara-to-Wiki* is designated to link the anchor text from a Te Ara topic to best entry points in the Wikipedia documents. File-to-file link discovery for the Wikipedia collection: As a special case of the anchor-to-bep task, this task has lower complexity and offers an entry level for newcomers. 5000 documents were selected for file-to-file link discovery. Up to 250 outgoing links and up to 250 incoming links were to be specified per topic. Missing topics were regarded as having a score of zero for the purpose of computing system performance.

3.2 Submission

Each submission run must specify the task (i.e. *LTW_F2F*, *LTW_A2B*, *LTW_F2FonA2B*, *LTAra_A2B* and *LTAraTW_A2B*) performed. The *description* section in the submission format is used to state different link discovery approaches. A sample format in the case of the link the wiki task is presented below.

```
<outgoing>
 <anchor name="Luminiferous aether" offset="1688" length="19">
  <tobep offset="2038">123456</tobep>
  <tobep offset="971">359</tobep>
  ...
 </anchor>
 ...
</outgoing>
<incoming>
 <bep offset="2038">
  < fromanchor offset="799" length="9" file="654321">radiation</fromanchor>
  < fromanchor offset="1019" length="10" file="3162088">medication</fromanchor>
  ...
 </bep>
 ...
</incoming>
```

Fig. 1. Sample link the wiki Submission Format

An anchor text was specified in three parts; the start position of the anchor (i.e. Offset), the Length of the text term and the anchor text itself. The position and length were indicated in characters. The offset specified the anchor starting position within the corresponding text-only document. The anchor text itself was used to verify the specification of the offset-length. The document name could be a unique number in the Wikipedia, or a unique name in the Te Ara collection. A destination link could be specified in two parts: a unique document name and a best entry point. It is the best starting point of the content where the relevant content section starts from.

3.3 Restriction of Linking

An anchor, indicated by a combination of *Offset* and *Length*, must appear only once in a topic - although it may have multiple distinct best entry points. An anchor-text in one document can be linked to several destinations (beps) in other distinct documents. It means that the same set of *Offset* and *Length* should not appear more than once and hence there is no duplicated anchor set for a given topic. For the evaluation purpose, the first 50 anchor sets are extracted and only the first 5 links within the instances of the same anchor offset-length are taken. Document title can also be an anchor, but like any other anchor it can be linked to at most 5 destinations.

3.4 Assistant Program

In order to facilitate the identification of the offset and length for each anchor and bep, several tools have been developed and distributed to participants. A Java program, *XML2FOL*, was created to produce a list of offset-length for all the element nodes in a given XML document. Another Java program, *XML2TXT*, was used to convert the XML document into the text-only content. Apart from the tools, a text-only version of the collection was also available so the offset could be computed by counting the characters from the beginning of the document. These two programs could be embedded into the participant's link discovery system as a parser to identify offset-length for the anchor texts and to produce text only document.

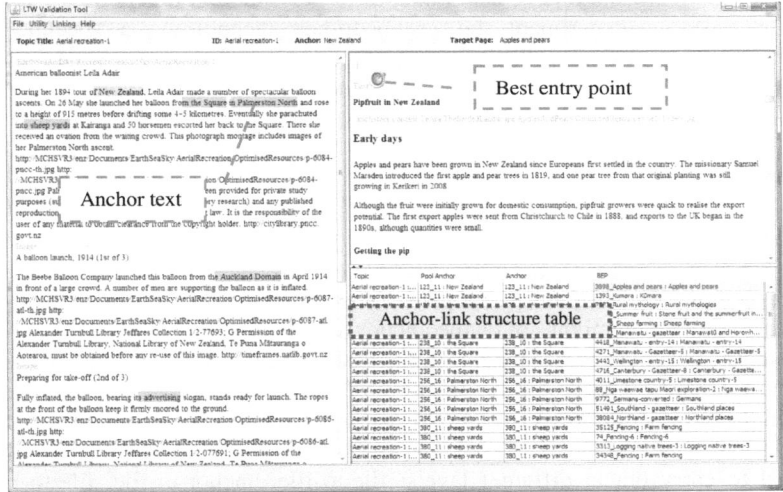

Fig. 2. The Validation Tool

The validation tool was introduced in 2009 and delivered to the participants so as to self-verify their link discovery submissions (see Figure 2). The anchors are highlighted in the left screen while the right screen shows the link content with a best entry point on it and a table recording the hierarchical structure of anchor-links for the given topic. The participants can click on a link in the table to check the particular anchor-link result. This tool intends to bring up what the link discovery application should look like and facilitate to revise the linking results. It can also be seen as a pre-assessment process.

4   Preparation of Qrels

There are two types of qrels used for the evaluation of the link discovery results. One is the Wikipedia ground truth and the other is generated from the manual assessment set. The Wikipedia ground truth is derived from the existing links in the Wikipedia collection. This is a simple way to achieve the automatic evaluation. However, the experiments undertaken in the past 2 years have shown that the comparative evaluation using automatic qrels is unsound in terms of the users' point of view. Some Wikipedia links are topically-obsolete or redundantly assigned. Many of anchors are linked to the documents with the same name. The relevant portions of the document content have not been further discovered. All relevant contents that are not in the Wikipedia are also considered non-relevant for the evaluation. As a consequence, the evaluation result might appear either optimistic or pessimistic. However, evaluation based on the Wikipedia ground-truth does measure performance relative to what is present, and so it is reasonable to use it in comparisons.

Apart from the file-to-file ground truth, Wikipedia can also produce the anchor-to-file ground truth. The offset value is set to the very beginning of the document. Although the Wikipedia does contain anchor-to-bep links, in practice they are rarely used. In order to experiment the anchor-to-bep technology, a special pooling procedure was applied to collect all anchors and links from participants' runs and Wikipedia. The pool for each topic was generated by the following three parts: anchor-to-bep (*A2B*), the file-to-file link discovery on A2B (*F2FonA2B*), and anchor-to-file Wikipedia ground truth. Since not all the offset-length sets were specified preciously and anchor texts could be indicated by different ways, overlapped anchor texts (i.e. offset-length) were merged as *a pool anchor* or **anchor representative**. For example, *quantum theory of atomic motion in solids* is an anchor in the article of *Albert Einstein*. However, *quantum theory*, *atomic* and *atomic motion* could be anchors returned by different participants. Therefore, the anchor texts shown to the assessor on the screen might not be the anchor returned by the system; instead it could be a combined anchor representative. In the case of *F2FonA2B*, the anchor was set as the topic title and linked to the beginning of the target document. The anchor-to-file set from the Wikipedia presents a one-to-one relation and the bep was set at the very beginning of the document. The pool was assessed to completion. The evaluation was expected to carry out at different levels: file-to-file, anchor-to-file, file-to-bep and anchor-to-bep.

5   Assessment and Evaluation

5.1   Manual Assessment

As the assessment is laborious and time consuming we have designed the assessment tool to maximize assessor efficiency. The assessment tool can be seen in Figure 3. Either the anchor representative or the bep link could be identified relevant (or non-relevant). Once the anchor representative was assessed as non-relevant, all anchors and associated links inside this anchor representative became non-relevant. The relevance status could be simply assigned by mouse right or left click. If the target document of the outgoing link was assessed as relevant, the best entry point was indicated by mouse left double-click. Incoming links in the submission were not properly explored in 2009. Most of them were specified in the file-to-file manner, i.e. incoming document title to the beginning of the topic article. Assessing incoming links was achieved for the first time in 2009.

According to the survey carried out after the assessment, a lack of related anchor texts highlighted in the incoming document could be a major obstacle to efficiency. Sometimes it is difficult to identify whether the incoming document is relevant to the topic content or not. Indicating the best entry point in the target document is also a difficult task to achieve without any supplemental information (e.g. system's discovered bep). Highlighting anchor texts or related phrases on the document seems necessary. For instance, a sub-title or a paragraph paired with the linking anchor text (or related phrases) could be a best start point for reading from. Each topic contains around 1000 anchor links and 900 incoming links. A log was created to record all the activities during the assessment. Then time to completion of a topic was estimated at around 4 hours.

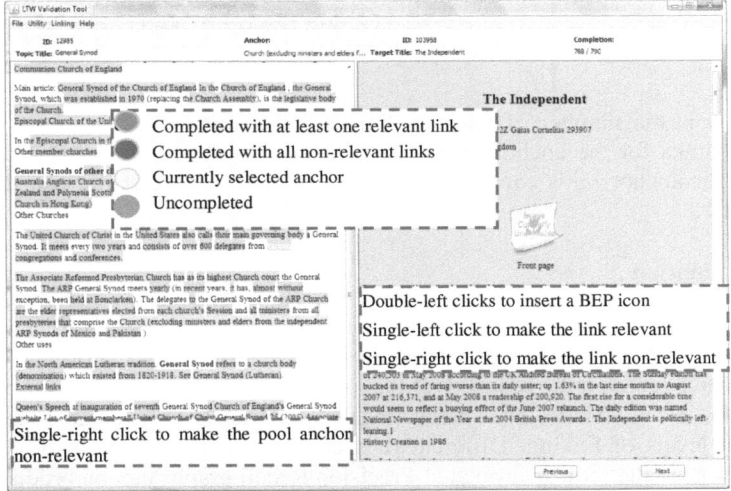

Fig. 3. The Assessment Tool

5.2 Metrics

As with all metrics, it is important to first define the use-case of the application. The assumption at INEX is that link-discovery is a recommendation tasks. The system produces a ranked list of anchors and for each a set of recommended target/bep pairs. The list should also be comprehensive because it is not clear that the document author can know a priori which links will be relevant to a reader of the document. That is, link discovery is a recall oriented task. The Mean Average Precision based metrics are very good at taking rank into account and are recall oriented. A good metric for link discovery should, consequently, be based on MAP. The difficulty is computing the relevance of a single result in the results list. For evaluation purposes it is assumed that if the target is relevant and the anchor overlaps a relevant anchor then the anchor is relevant; $f_{anchor}(i) = 1$.

The assessor might have assessed any number of documents as relevant to the given anchor. If the target of the anchor is in the list of relevant document then it is considered relevant; $f_{doc}(i) = 1$. The contribution of the links' bep is a function of distance from the assessor's bep [4]:

$$f_{bep}(j) = \begin{cases} \dfrac{n - 0.9 \times d(x,b)}{n} & if\ 0 \le d(x,b) \le n \\ 0.1 & if\ d(x,b) > n \end{cases}$$

Where $d(x,b)$ is the distance between submission bep and result bep in character. Therefore, the score of $f_{bep}(j)$ varies between 0.1 (i.e. d is greater than n) and 1 (i.e. the submission and result beps are exactly matched). The score of 0.1 is reserved for the right target document with an indicated bep not in range of n. n typically is set up as 1000 (characters). The score of a result in the results is then:

$$P = \left[\left(f_{anchor}(i) \right) \times \frac{\left(\sum_{j=1}^{m} \left(f_{doc}^{i}(j) \times f_{bep}^{i}(j) \right) \right)}{m_i} \right]$$

Where m is the number of returned links for the anchor and m_i is the number of relevant links for the anchor in the assessments. As the result list is restricted to 5 targets per anchor m_i is capped at 5 for evaluation. A perfect run can thus score a MAP of 1.

5.3   Evaluation

Based on the portable evaluation tool, *ltwEval*, used in 2008, new functionality has been added to achieve a better interaction of the graphs and additional evaluation setup, which increase the usability of the tool. Numerous evaluation metrics including precision, recall, MAP, and precision@R were used to evaluate submission at different levels of linking. Different runs can be evaluated and easily compared to each other via the tool. Interpolated-Precision/Recall graphs can be produced for sets of run.

For the file-to-file evaluation (i.e. F2F and F2FonA2B), the number of outgoing and incoming links have been restricted by 250. Links beyond this number were truncated. The total number of relevant links is based on the ground truth, but at last 250 to make sure the measurement of Recall is meaningful. For the anchor-to-bep evaluation against ground-truth, the first 50 anchors for each topic were taken and the first link from each anchor was collected. As a result, there were 50 outgoing links per topic, used for evaluation. By contrast, first 250 incoming links were taken to do the evaluation since the discovery of bep in the topic document is not that obvious. Most incoming links belong to the same bep. Therefore, in the INEX use case of link discovery it is important to rank the discovered links for presentation to the page author. This use case was modeled in the manual assessment where assessors did

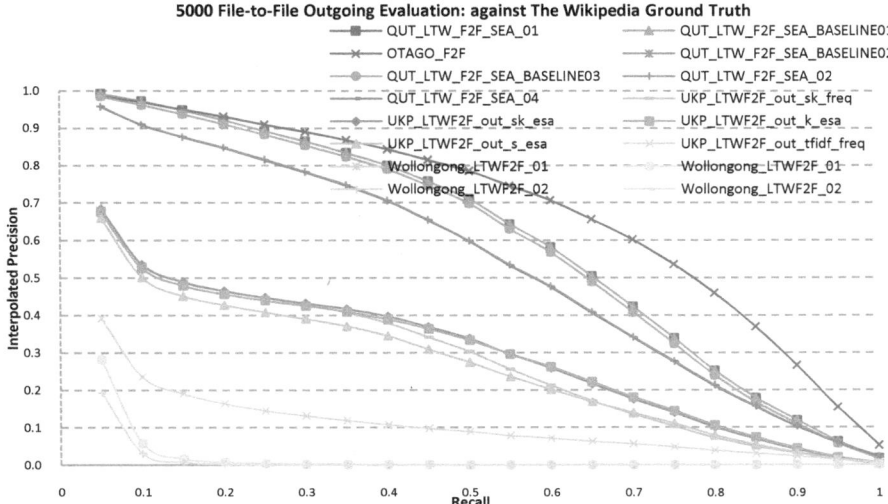

Fig. 4. 5000 F2F Topics Outgoing link discovery evaluated against Wikipedia Ground Truth

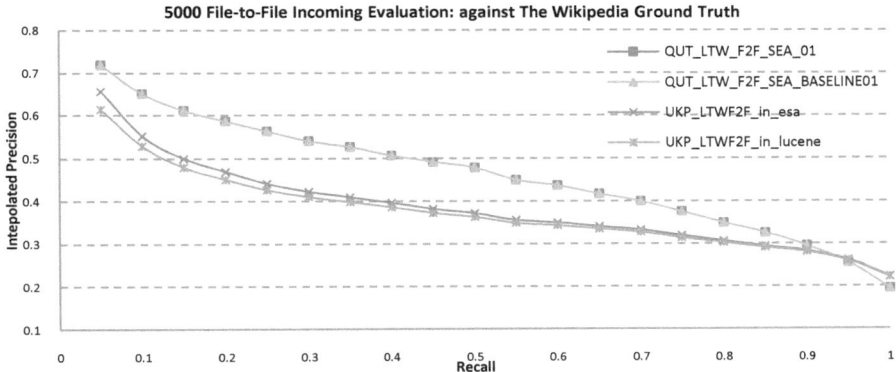

Fig. 5. 5000 F2F Topics Incoming link discovery evaluated against Wikipedia Ground Truth

exactly this. In a realistic link discovery setting the user is unlikely to trudge through hundreds of recommended anchors, so the best anchors should be presented first. The link discovery system must also balance extensive linking against link quality.

6 Results and Discussion

The Queensland University of Technology (i.e. QUT) submitted 6 runs for the file-to-file (F2F) task, 4 runs for the anchor-to-bep (A2B) task and 1 run for the F2FonA2B task. University of Waterloo contributed 2 runs on the A2B task and 5 run for the

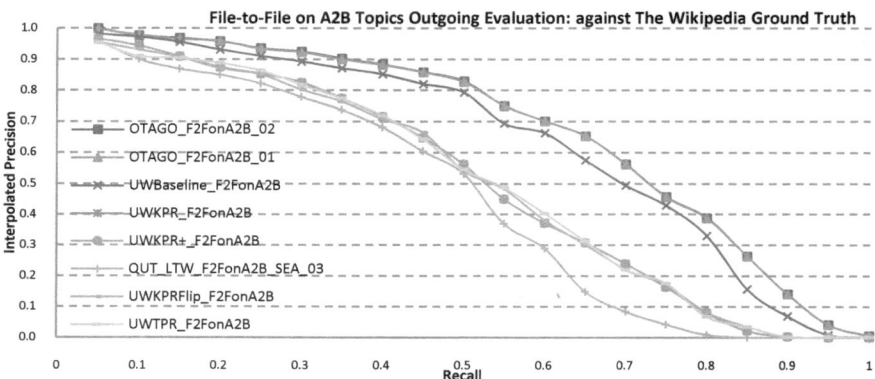

Fig. 6. F2F on A2B Topics Outgoing links evaluated against Wikipedia Ground Truth

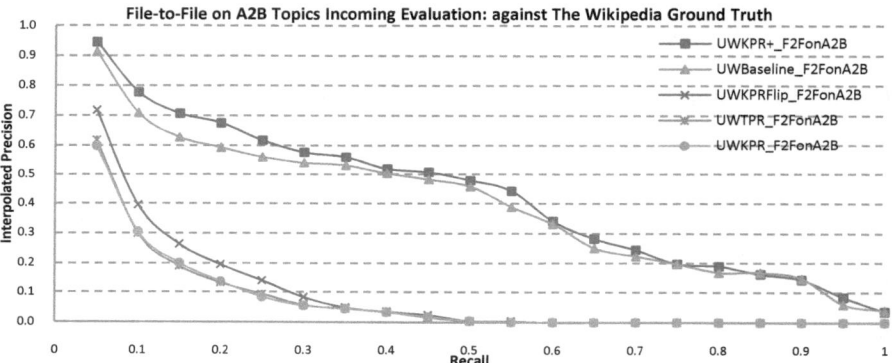

Fig. 7. F2F on A2B Topics Incoming links evaluated against Wikipedia Ground Truth

F2FonA2B task. University of Amsterdam had 5 runs for the A2B task. University of Otago submitted 1 runs for the F2F task, 2 runs for the A2B task and 2 runs for the F2FonA2B task. University of Wollongong submitted 4 runs for the F2F task. Technische Universität Darmstadt contributed 4 runs on the F2F task. Apart from the Link the Wiki tasks, QUT also participated in the Link the Te Ara and Link Te Ara to Wiki tasks by submitting 1 run each. Technische Universität Darmstadt also contributed 5 runs on the Link the Te Ara task. These runs were generated by the anchor-to-bep link discovery technology.

The University of Waterloo (UW) had two approaches, one baseline and the other link-based, to undertake the experiment. For a baseline, UW produced the statistics of the phrase frequency. These phrases were located in the topic files and the most frequent links were returned. For incoming links, we scored the corpus using topic titles as query terms and returned the top documents. The link-based approach computes *PageRank* and *Topical PageRank* values for each file in the corpus for each

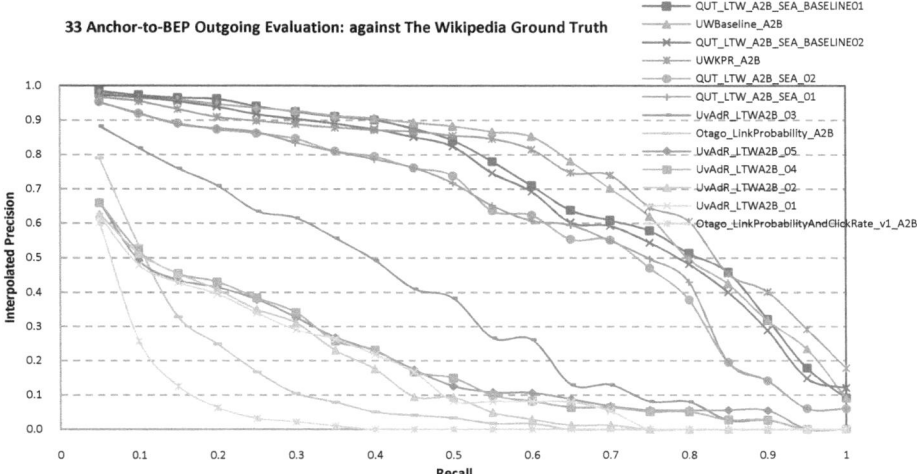

Fig. 8. 33 A2B Topics Outgoing links evaluated against Wikipedia Ground Truth

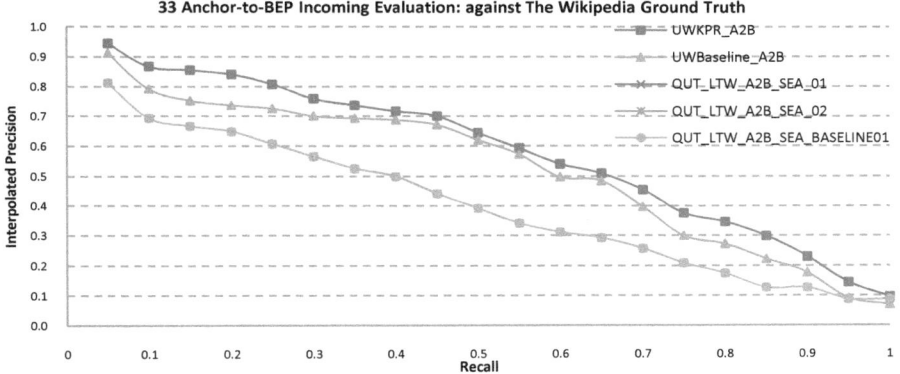

Fig. 9. 33 A2B Topics Incoming links evaluated against Wikipedia Ground Truth

topic, and returned the top scoring pages according to the contribution of *K-L divergence*. For incoming links, UW reversed the graph to get new *PageRank* values and returned the top pages according to the contribution of *K-L divergence* with the new *PageRank* values and the old *Topical PageRank* values.

The Queensland University of Technology (QUT) used the statistical link information of Wikipedia corpus to calculate the probability of anchors and their corresponding target documents for a list of sortable outgoing links. A hybrid approach that combines the results of link analysis method and title matching algorithm for the prediction of potential outgoing links was also undertaken. For the incoming links, the top ranking search results with topic title as the query terms

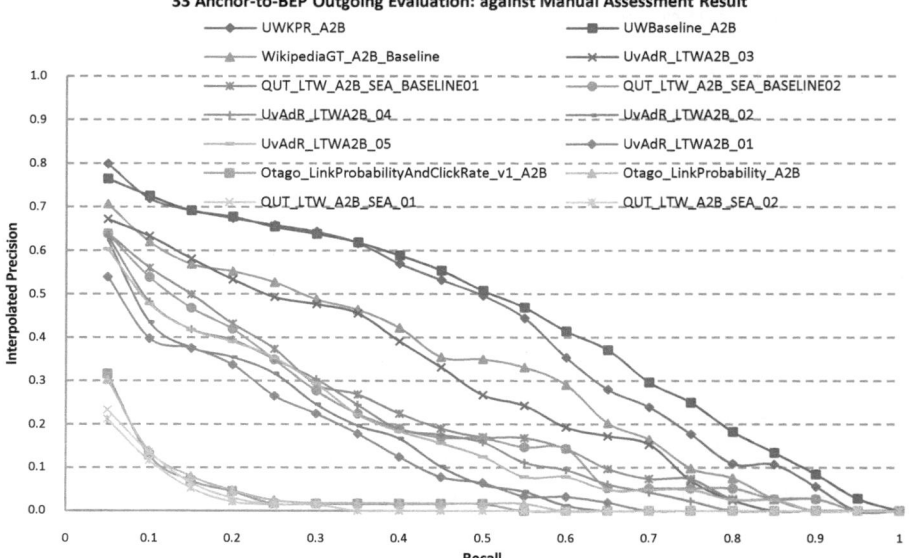

Fig. 10. 33 A2B Topics Outgoing links evaluated against Manual Assessment Set

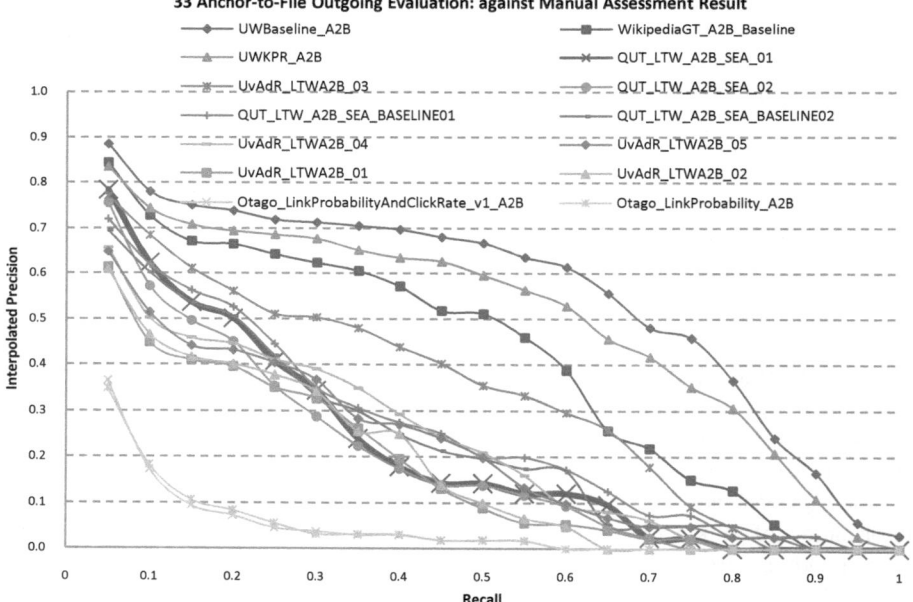

Fig. 11. 33 A2F Topics Outgoing links evaluated against Manual Assessment Set

retrieved from a BM25 ranking search engine were chosen as source documents that can be linked to the topics. In finding the beps for either outgoing or incoming links, QUT tried two different methods: one is that the bep is the position of the phrase in the target document where the terms of the anchor, either the entire words or part of which, appear; the other is that the best entry point is the beginning of a text block which has similar terms features with that of the passage which is extracted from the surrounding text of the anchor in source document.

7 Conclusion and Outlook

This is the third year of the Link-the-Wiki track at INEX. According to the file-to-file experiment, producing Wikipedia links could be achieved by current approaches. In 2009, the focus has been shifted to the anchor-to-bep link discovery and several changes have been made to improve the evaluation procedure. Assistant tools were prepared to self-examine the status of submission. The outcome is twofold: self-verification of the submission to revise the offset-length parser and pre-assessment to improve the link discovery engine. Further experiments were undertaken on the anchor-to-bep runs. The submission was evaluated on anchor-to-file, and anchor-to-bep level to test the usability of approaches provided. This aims to classify the performance of each approach on the contribution of linking for the given topic. The Te Ara collection is introduced for the first time at INEX to bring up the new concept of cross collection link discovery. Through the focus link discovery, the Wikipedia content could be fully explored. Anchors indicated for the given document could be linked to the most relevant content in the collection. Every piece of content discovered in the Wikipedia can be used to provide links for anchors from other document collections. Going through this process, a well defined knowledge network can be constructed. Based on participants' comments and ideas via survey, customization can be made, and the enhancement of evaluation procedure and efficiency is expected. According to the experiment, the contribution of each approach can be classified and future direction of anchor-to-bep link discovery can be possibly pointed out.

References

1. Trotman, A., Geva, S.: Passage Retrieval and other XML-Retrieval Tasks. In: The SIGIR 2006 Workshop on XML Element Retrieval Methodology, pp. 48–50 (2006)
2. Huang, W.C., Xu, Y., Trotman, A., Geva, S.: Overview of INEX 2007 Link the Wiki Track. In: Fuhr, N., Kamps, J., Lalmas, M., Trotman, A. (eds.) INEX 2007. LNCS, vol. 4862, pp. 373–387. Springer, Heidelberg (2008)
3. Huang, W.C., Geva, S., Trotman, A.: Overview of INEX 2008 Link the Wiki Track. In: Fuhr, N., et al. (eds.) INEX 2008. LNCS, vol. 5631, pp. 314–325. Springer, Heidelberg (2009)
4. Huang, W.C., Xu, Y., Trotman, A., Geva, S.: The Methodology of Manual Assessment in the Evaluation of Link Discovery. In: Proceedings of the 14th Australian Document Computing Symposium (2009)

An Exploration of Learning to Link with Wikipedia: Features, Methods and Training Collection

Jiyin He and Maarten de Rijke

ISLA, University of Amsterdam, Science Park 107, 1098 XG Amsterdam, The Netherlands
{j.he,derijke}@uva.nl

Abstract. We describe our participation in the Link-the-Wiki track at INEX 2009. We apply machine learning methods to the anchor-to-best-entry-point task and explore the impact of the following aspects of our approaches: features, learning methods as well as the collection used for training the models. We find that a learning to rank-based approach and a binary classification approach do not differ a lot. The new Wikipedia collection which is of larger size and which has more links than the collection previously used, provides better training material for learning our models. In addition, a heuristic run which combines the two intuitively most useful features outperforms machine learning based runs, which suggests that a further analysis and selection of features is necessary.

1 Introduction

The aim of the LTW track is to automatically identify hyperlinks between documents. We only participated in the task of outgoing link generation within Wikipedia (A2B). In our experiments, we focus on exploring machine learning methods and learning material for link detection. The main purpose of our experiments is two-fold. First, we want to test how our learning methods work on the LTW task, especially how the results learnt from the Wikipedia ground truth will be judged by human assessors. On top of that, since the LTW task is defined as a ranking problem for recommendation purposes, we want to see how a learning to rank approach works as it directly optimizes the rankings instead of assigning binary decisions to candidate links as a classification method would do. Second, we trained our models with different versions of Wikipedia. The two versions used, namely Wikipedia 2008 and Wikipedia 2009, differ in the amount of articles they contain as well as in the amount of links. Presumably, the 2009 version contains more link information but is also more noisy in terms of missing target pages (as some pages are deleted as time passes by). We experiment with both collections so as to see the impact of the training material used. In addition, we explore a set of features for constructing the classifiers/rankers. In order to examine the effectiveness of the features, we also submit a heuristic run that combines the two intuitively most useful features without sophisticated learning methods.

More specifically, we have following research questions:

- When the A2B task is viewed as a ranking problem, is a learning to rank approach more effective than a binary classification approach?

S. Geva, J. Kamps, and A. Trotman (Eds.): INEX 2009, LNCS 6203, pp. 324–330, 2010.
© Springer-Verlag Berlin Heidelberg 2010

Table 1. Features and their corresponding application in different learning stages

Learning Stage	N-gram	N-gram-target	Target	N-gram-topic	Topic-target	1st-stage
Candidate targets ranking		x	x		x	
Candidate links ranking	x	x	x	x	x	x

- Do different versions of the Wikipedia collection (with, potentially, differences in collection size, numbers of links, etc.) result in performance differences when used as training material?
- Are the features used for learning the models effective? Are there single features whose contribution to the linking results is dominant?

The rest of the paper is organized as follows: Section 2 introduces the learning approaches applied to our task, Section 3 describes the features and Section 4 presents the runs we submitted plus an analysis of the evaluation results. We conclude in Section 5.

2   A Two-Stage Learning Procedure

Following [5], we consider the linking task as a two stage procedure, namely *candidate target identification* and *link detection* [2]. First, we extract all n-grams in a topic page and train a target-detector to rank the potential target pages for each n-gram, which we refer to as *candidate target pages*. Then we train a link detector on the (n-gram, target) pairs for the final results.

We experiment with two types of learning methods, viz. classification and learning to rank. For classification, we use SVM to classify the instances in both stages and rank the results by the probability of an instance being positive. For our learning to rank approach, we use RankingSVM [3] to directly optimize the ranking of an instance. In the candidate target identification stage, we train a ranker to rank the target candidates for each n-gram and in the link detection stage a ranker is trained to rank the n-gram target pairs.

For learning both the binary SVM and the RankingSVM, we randomly sample 500 pages from the Wikipedia collection for training and 100 pages for validation. For both SVMs we use the linear kernel and tune the regularization parameter on the validation set. To learn the model for target detection, we use only the real anchor texts and their corresponding candidate target pages as instances, while for training the link detectors we use all n-gram-candidate target pairs as instances.

3   Features

We identify 6 types of feature for learning a preference relation between the candidate links. Table 1 specifies in which stage each type is used and Table 2 lists the features. Here, we discuss the motivations for using them and detail the formulation of some.

N-gram features. The n-gram features suggest how likely a given n-gram would be marked as an anchor text, without any other information such as its context in the topic page, which includes its length, IDF score, the number of candidate targets associated with it, and its ALR (Anchor Likelihood Ratio) scores. IDF is calculated as

$$\log\left(\frac{|D|}{|\{d_i : ng \in d_i\}_{i=1}^N|}\right),$$

where ng is a n-gram, d_i is a page containing this n-gram, and $|D|$ is the total number of pages in the Wikipedia collection. The ALR score can be interpreted as a model selection between two models, the anchor model and the collection model, from which an n-gram is generated. To calculate the probability of an n-gram being generated by either model, the maximum likelihood is used. Specifically, it is calculated as

$$ALR(ng) = \frac{|ng \in A|}{|A|} \cdot \frac{|C|}{|ng \in C|}, \tag{1}$$

where A is the collection of all anchor texts in the Wikipedia collection and C is the collection of all n-grams in the Wikipedia collection, and $|\cdot|$ means the total number of anchor texts/n-grams in the given collection. A large ALR value indicates that the n-gram is more likely to have been generated from the anchor model, i.e., this n-gram is more likely to be an anchor text than a common word sequence from the background collection.

N-gram-target features. The n-gram-target features describe how well an n-gram and its corresponding candidate target page are related. On the assumption that each Wikipedia page is about a specific concept that is usually denoted by its title, the first feature we use is the match between an n-gram and the candidate target page. The second type of feature in this category consists of indicators of how likely a given n-gram ng and a candidate target page $ctar$ are linked, which is expressed by the following two scores: *RatioLink* and *RatioAnchor*. The former is the ratio between the number of times ng and $ctar$ are linked and the number of times $ctar$ is being linked as a target page in the collection. The latter, i.e., *RatioAnchor* is the ratio between the number of times ng and $ctar$ are linked and the number of times ng is used as an anchor text in the collection. Moreover, we adopt retrieval scores between the n-gram and the candidate target pages as features (n-gram as query), which is an obvious description of the relatedness of the two:

$$RatioLink(ng, ctar) = \frac{|link(ng, ctar)|}{|inlink(ctar)|} \tag{2}$$

$$RatioAnchor(ng, ctar) = \frac{|link(ng, ctar)|}{|ng \in A|} \tag{3}$$

Here, $|link(ng, ctar)|$ denotes the number of times that n-gram ng and $ctar$ are linked in Wikipedia, and $|inlink(ctar)|$ denotes the number of times that $ctar$ is used as a target page and linked with some anchor texts in Wikipedia.

Table 2. Features used for learning the preference relation, where ng: n-grams; C: collection; ctar: candidate target pages; topic: topic page

N-gram features	
Length(ng)	Number of words contained in the n-gram
IDF(ng)	The IDF score of the n-gram
ALR(ng)	The ALR score of the n-gram, as detailed in Eq. 1
#Cand(ng)	Number of candidate target pages associated with the n-gram
N-gram - target features	
TitleMatch(ng, ctar)	Three values - 2: exact match; 1: partial match (i.e., either the title contains the n-gram, or the n-gram contains the title); 0: no match
RatioLink(ng, ctar)	Link ratio of the n-gram and the candidate target page, see Eq. 2
RatioAnchor(ng, ctar)	Anchor ratio of the n-gram and the candidate target page, see Eq. 3
Ret_uni(ng, ctar)	Retrieval score with unigram model, i.e., BM25 with default parameter settings
Ret_dep(ng, ctar)	Retrieval scores with dependency model, i.e., Markov Random Field model as described in [4]
Rank_dep(ng, ctar)	The rank of the target page with the dependency retrieval model
Target features	
#Inlinks(ctar)	Number of in-links contained in the candidate target page
#Outlinks(ctar)	Number of out-links contained in the candidate target page
#Categories(ctar)	Number of Wikipedia categories associated with the candidate target page
Gen(ctar)	Generality of the candidate target page as described in [5]
N-gram - topic features	
TFIDF(ng, topic)	The TFIDF score of the n-gram in the topic page
First(ng, topic)	Position of first occurrence of the n-gram in the topic page, normalized by the length of the topic page
Last(ng, topic)	Position of last occurrence of the n-gram in the topic page, normalized by the length of the topic page
Spread(ng, topic)	Distance between first and last occurrence of the n-gram in the topic page normalized by the length of the topic page
Topic-target features	
Sim(ctar, topic)	Cosine similarity between the candidate target page and the topic page
Ret_unigram(ctar, topic)	Retrieval score using the title of the candidate target page as query against the topic page; using BM25 as retrieval model
First stage scores	
score(ng, ctar)	The output of the ranker for the candidate target page given the n-gram
rank(ng, ctar)	The rank of the candidate target page according to the learnt ranker

Target features. The target features are indicators of how likely a candidate target page alone would be linked with some anchor text in the collection. To this end we explore features such as counts of the inlinks and outlinks within the candidate target page, as well as the Wikipedia category information associated with it.

N-gram-topic features. This type of feature describes the importance of the n-gram within its context, i.e., topic page. One would assume that an n-gram being selected as an anchor text should be somewhat important to the understanding of the whole topic

Table 3. Submitted runs

RunID	Description
UvAdR_LTWA2B_01	Binary classification, trained on wiki08
UvAdR_LTWA2B_02	Ranking SVM, trained on wiki08
UvAdR_LTWA2B_03	A heuristic run, combine the ALR and IDF for link ranking, but using rankingSVM for target ranking
UvAdR_LTWA2B_04	Binary classification, trained on wiki09
UvAdR_LTWA2B_05	Ranking SVM, trained on wiki09

page as well as being content-wise related. Here, we use the TFIDF score of the n-gram and its location within the topic page as an indication of the importance of a n-gram within a topic page.

Topic-target features. The topic-target features describe the relatedness between a topic page and a candidate target page. One obvious feature is the similarity between the two pages. In addition, as a candidate target page itself is about a concept, we could measure how important this concept is, or in other words, how well this concept is being expressed in the topic page. We measure it by using the title of the candidate target page as a query and calculating the retrieval score against the topic page.

First stage score. Once the target ranking has been completed (during the first stage), we can get the ranking score for each candidate target, as well as their ranks. In the second stage, we select the top X candidate targets to construct the candidate links with their corresponding n-grams, where the scores and ranks from the first stage are used as features.

4   Five Runs

4.1   Submitted Runs

We submitted 5 runs for the LTW task, as specified in Table 3. We use two Wikipedia collections, wiki08 [1] and wiki09 [6]. Note that we have a heuristic run UvAdR_-LTWA2B_03 that does not a use learning method for link ranking; it only uses Rank-ingSVM for target identification. For link ranking, we filter the candidate links whose ALR score is less than 0.2, and rank the remaining ones with their IDF scores. This run serves as a baseline for other machine learning based approaches. The heuristics used in this run, i.e., the *ALR* and *IDF* scores, however, are the features that are most close to human intuitions, where *ALR* represents how likely an n-gram is involved in a link based on the observation of existing links and *IDF* represents the uncommonness of a n-gram.

For all 5 runs, for each detected link, we set the best entry point to 0, as intuitively, the first paragraph of a Wikipedia page gives a good summary of the concept of the linked anchor text.

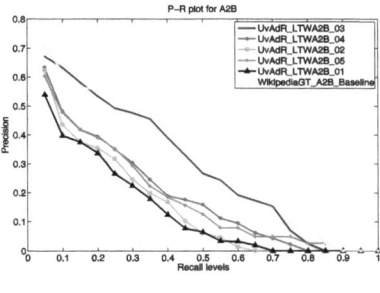
(a) Precision-Recall plot for A2B links

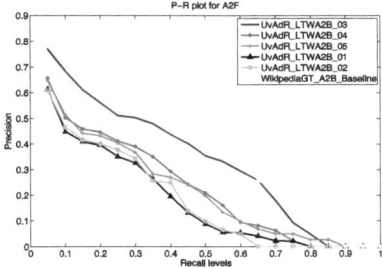
(b) Precision-Recall plot for A2F links

Fig. 1. Precision-Recall plots for the submitted runs

4.2 Results and Analysis

Figure 1 shows the results of our submitted runs using a Precision-Recall plot.

In this year's evaluation, the A2B runs are evaluated at two levels, i.e., at the Anchor-to-BEP level as well as at the Anchor-to-File (A2F) level. For the A2F evaluation, all BEP positions are set to 0. Since we set all BEP positions to 0 in our submitted runs, it is more natural to focus on the results of the A2F evaluation. Table 4 lists the interpolated precision scores at different recall levels. The Wikipedia ground truth is included as a pseudo run and evaluated against the manual assessment.

Table 4. Average precision at different recall levels, evaluated with A2F

RunIDs	R@0.05	R@0.1	R@0.2	R@0.5
UvAdR_LTWA2B_03	**0.77**	**0.68**	**0.56**	**0.35**
UvAdR_LTWA2B_04	0.65	0.50	0.44	0.21
UvAdR_LTWA2B_05	0.64	0.51	0.43	0.20
UvAdR_LTWA2B_01	0.61	0.44	0.39	0.09
UvAdR_LTWA2B_02	0.60	0.46	0.40	0.10
WikipediaGT_A2B_Baseline	0.84	0.73	0.66	0.51

For both levels of evaluation (using A2B and using A2F), the best run is the heuristic run, which outperforms all sophisticated learning methods. This indicates that the two features, ALR and IDF, are very strong features that probably dominate the contribution to the learned models. In addition, this observation suggests that a detailed analysis and selection of features should be conducted.

Next, from the A2F evaluation, we see that runs trained on the Wikipedia 09 collection (i.e., 04 and 05) outperform runs trained on the Wikipedia 08 collection (i.e., 01 and 02). This suggests that a larger collection with more (existing) links provides better training materials. Also, we see that the runs based on binary classification methods (01 and 04) do not differ a lot from the learning to rank based runs (02 and 05). This may be due to the fact that the training examples from Wikipedia do not contain very strong ranking information, i.e., we only have two levels of relevance: relevant (is a link) and non-relevant (not a link).

Finally, none of our runs outperforms the Wikipedia ground truth. This is no surprise, since the models are learned from the Wikipedia ground truth. In order to outperform the Wikipedia ground truth with a learning method, sufficiently many examples with manual labeling should be collected.

5   Conclusions

We have described our approaches and submissions for this year's participation in the INEX Link-the-Wiki track. We submitted 5 runs to the A2B outgoing links detection task. Our main focus was to explore the effectiveness of applying machine learning approaches for the task. Specifically, we experimented with two types of learning approaches, namely classification and learning to rank. We also evaluated the learning material for the task, where we use different sets of training data (based on different versions of Wikipedia). On top of that, we used a heuristic run to exam the impact of the features that are intuitively most effective.

We have found that the learning to rank based approach and the binary classification approach do not differ a lot. The new (2009) Wikipedia collection which is of larger size and has more links than the old (2008) collection, provides better training material for learning the models. In addition, the heuristic run outperforms all machine learning based runs, which suggests that a further analysis and selection of features is necessary.

Acknowledgements. This research was supported by the European Union's ICT Policy Support Programme as part of the Competitiveness and Innovation Framework Programme, CIP ICT-PSP under grant agreement nr 250430, by the DuOMAn project carried out within the STEVIN programme which is funded by the Dutch and Flemish Governments under project nr STE-09-12, and by the Netherlands Organisation for Scientific Research (NWO) under project nrs 612.066.512, 612.061.814, 612.061.815, 640.004.802.

References

[1] Denoyer, L., Gallinari, P.: The Wikipedia XML Corpus. SIGIR Forum (2006)
[2] He, J., de Rijke, M.: A ranking approach to target detection for automatic link generation. In: SIGIR '10: Proceedings of the 28th annual international ACM SIGIR conference on Research and development in information retrieval. ACM, New York (2010)
[3] Herbrich, R., Graepel, T., Obermayer, K.: Large margin rank boundaries for ordinal regression. MIT Press, Cambridge (2000)
[4] Metzler, D., Croft, W.: A markov random field model for term dependencies. In: SIGIR '05: Proceedings of the 28th annual international ACM SIGIR conference on Research and development in information retrieval, pp. 472–479. ACM, New York (2005)
[5] Milne, D., Witten, I.H.: Learning to link with wikipedia. In: CIKM '08: Proceedings of the 17th ACM conference on Information and knowledge management, pp. 509–518. ACM, New York (2008)
[6] Schenkel, R., Suchanek, F., Kasneci, G.: YAWN: A semantically annotated Wikipedia XML corpus. In: BTW 2007 (2007)

University of Waterloo at INEX 2009: Ad Hoc, Book, Entity Ranking, and Link-the-Wiki Tracks

Kelly Y. Itakura and Charles L.A. Clarke

University of Waterloo, Waterloo, ON N2L3G1, Canada
{yitakura,claclark}@cs.uwaterloo.ca

Abstract. This year, University of Waterloo participated in four tracks; Ad Hoc, Book, Entity Ranking, and Link-the-Wiki tracks. In Ad Hoc and Book tracks, we implemented a variation of Okapi BM25F [20,5,18,15] that gave substantial improvements over the baseline BM25 that ranked first in the previous year [12,13], during the training and in the official Ad Hoc-focused results. In Entity ranking track, we used redundancy techniques [4] for question answering to retrieve entities. In Link-the-Wiki track, we employed topic-oriented PageRank with KL divergence in addition to the baseline described in [11].

1 Introduction

This year, University of Waterloo participated in four tracks; Ad Hoc, Book, Entity Ranking, and Link-the-Wiki tracks. For all, except the newly-participating Entity Ranking track, our goal this year is "beat the baseline!"

In INEX 2008, the simple BM25 algorithm scored best for all Ad Hoc tasks, focused, relevant-in-context, and best-in-context [12,13]. For Book track, even though our BM25 approach performed poorly on book retrieval task, the same approach for page-in-context task performed better than the only other participant [12,14]. Though because of the small number of participants, it is too early to say that our simple BM25 approach is optimal for the Book track tasks, for Ad Hoc track, our approach has been consistently performing well for INEX 2004, 2007, and 2008 [3,16,8,13]. Therefore, this year, while keeping our approaches simple, we incorporated BM25F, BM25 with XML field extension [20,5]. As it turned out, our approaches, especially Ad Hoc one, were similar to those of Lu et al. [15] and Robertson et al. [18] that use inherited information. However, our experiments are on a much larger scale on a very different corpus, and the tasks are on focused task, rather than on thorough and relevant-in-context tasks, training are at element level as opposed to document level, we used averaged field length as opposed to a document length, we used length normalization at field level as opposed to at a document level, and we used somewhat different field structure and inheritance. Most importantly, as far as we know, there is no follow-up work on BM25F with inheritance since 2006, but we think there is a room for

S. Geva, J. Kamps, and A. Trotman (Eds.): INEX 2009, LNCS 6203, pp. 331–341, 2010.

improvement and especially application in tasks other than Ad Hoc tasks. As a first step of application, we applied BM25F with inheritance in a book page retrieval setting.

In Entity track, our main question is, "How is it different from a question answering task?" Therefore, we used redundancy technique for question answering [4], combined with category estimation. We also changed a perspective, and instead of trying to retrieve documents that represent entities as many participants did more or less [6, 22], we extracted entities from passages and then located articles corresponding to the entities.

For Link-the-Wiki track, we saw that the simple statistical approach reproduces the Wikipedia as a ground truth well [10, 11, 9] but did not perform well when compared to manual assessments. We think this is due to the fact that the simple statistical approach does not take topicality of anchor phrase into account, and used a modified version of topical PageRank [1, 17].

2   Ad Hoc Track

This year, we only participated in Focused task using CO queries to perform element retrieval. From our previous years' experience, we expect that the performance of our runs in Focused task would behave very similarly in relevant-in-context and in best-in-context tasks.

We submitted two runs (the other two runs had some minor bugs but performed exactly the same), `UWFERBase2` and `UWFERBM25F2`. Our baseline run is `UWFERBase2` that is an element retrieval using BM25 as in the previous years [3, 16, 8, 13]. We used BM25F for another element retrieval run `UWFERBM25F2`.

For both runs, the unit of retrieval is elements. For training set on INEX 2008 collection [7], we used the following elements we used from last year;

> `<p>`, `<section>`, `<normallist>`, `<article>`, `<body>`, `<td>`, `<numberlist>`, `<tr>`, `<table>`, `<definitionlist>`, `<th>` ,`<blockquote>`, `<div>`, `<li>`, `<u>`.

For test set on INEX 2009 collection [21], we used the following elements that correspond to the elements we used for training.

> `<p>`, `<sec>`, `<list>`, `<article>`, `<bdy>`, `<column>`, `<row>`, `<table>`, `<entry>`, `<indent>`,`<ss1>`, `<ss2>`,`<ss3>`,`<ss4>`,`<ss5>`.

For both runs, we processed queries by converting each topic title into a disjunction of query terms without negative query terms. We located positions of all query terms and some XML tags using Wumpus [2]. Note that some query terms appear inside XML tags such as `<image>` tags and we think it is beneficial to consider those terms when scoring. This would be difficult if one used an off-the-shelf XML parser.

2.1   Baseline BM25 Run

For the baseline, `UWFERBase2`, we scored elements using Okapi BM25 [19]. The score of an element E using Okapi BM25 is defined as follows.

$$s(E) \equiv \sum_{t \in Q} W_t \frac{f_{E,t}(k+1)}{f_{E,t} + k(1 - b + b\frac{pl_E}{avgdl})} \quad , \tag{1}$$

where Q is a set of query terms, W_t is an IDF value of the term t in the collection, $f_{E,t}$ is the sum of term frequencies in an element E, pl_E is an element length of E, and $avgdl$ is an average document length in Wikipedia collection to act as a length normalization factor.

We then sorted elements by their scores, and obtained the top elements while removing overlap.

We tuned our parameters to maximize `iP[0.01]` on INEX 2008 training set using `inex_eval.jar` provided from INEX. Our tuning method was repeated fixing of one parameter and changing of another until no further change is observed. The parameters $k = 4$ and $b = 0.6$ gave $iP[0.01] = 0.6607$ on INEX 2008 training set, and we used the same set of parameters on INEX 2009 test set.

2.2   Experimental BM25F Run

For each element we listed in Section 2, we constructed two fields, one for "title" and another for "body". The title field consists of concatenation of an article title and any ancestral and current section titles. The body field contains the rest of the text in the element. Lu et al. [15] and Robertson et al. [18], on the other hand, had separate fields for article title and section title or current title and ancestral titles. Our intuition was that in Wikipedia, section headings do not normally seem to include the article title and therefore, rather vague. We do not know whether our approach is any better or different or not.

For example, in a page titled "Koala" [1], one of the section headings is "Physical description". In our title field for the section, we fit "Koala Physical description". The body field contains the rest of the section but not the current section title of "Physical description".

Unfortunately, because in INEX 2008 collection, we only had one-level section headings, whereas in INEX 2009 collection, we had subsection headings, we only considered the first level of section headings during our test.

Once we collected necessary ingredients, we applied BM25F formula for an element e [5];

$$BM25F(e) = \sum_{t \in q \cap e} \frac{\overline{x}_{e,t}}{K + \overline{x}_{e,t}} W_t \quad ,$$

where q is a query term, $\overline{x}_{e,t}$ is a weighted normalized term frequency that we describe later, and K is a parameter, and W_t is a document-level IDF for a term

---

[1] `http://en.wikipedia.org/wiki/Koala`

t. The weighted normalized term frequency is obtained by first performing length normalization, on a term frequency $x_{e,f,t}$ of a term t of field f in an element e,

$$\overline{x}_{e,f,t} = \frac{x_{e,f,t}}{1 + B_f(\frac{l_{e,f}}{l_f} - 1)} \ ,$$

B_f is a parameter to tune, $l_{e,f}$ is a length of a field f in an element e, l_f is an average field length.

We then multiplied the normalized term frequency $\overline{x}_{e,f,t}$ by field weight W_f,

$$\overline{x}_{e,t} = \sum_f W_f \cdot \overline{x}_{e,f,t} \ .$$

Lu et al. [15] and Robertson et al. [18] use a version of BM25F without individual field length normalization [20].

We fixed $K = 1$ and then trained the rest of the four parameters B_{title}, B_{body},W_{title}, and W_{body}, similarly to our baseline. We created our run with $B_{title} = 0.6$, $B_{body} = 0.6$,$W_{title} = 4$, and $W_{body} = 1.2$ that gave $ip[0.01] = 0.7131$ during training.

2.3   Results

Official results showed that UWFERBM25F2 ranked first with $iP[0.01] = 0.6333$. The score of the second ranked run from University Pierre et Marie Curie is 0.6141, and our other run UWFERBase2 ranked 12th with $iP[0.01] = 0.5940$, so it seems that our experimental BM25F-based run not only improves the baseline BM25 run substantially, but also seems to outperform other participants. We speculate that if we train on INEX 2009 collection with multi-level section headings and test on the same set, the improvement would be more prominent.

Some suggest we use average document length as a training parameter. Our later experiments suggested that using an additional parameter adds little value to the outcome over the existing runs.

3   Book Track

This year, we only participated in Focused Book Search task. As in the Ad Hoc track, our objective is if BM25F-based element retrieval run has any significance over the simple BM25-based element retrieval run that performed well last year. Therefore, we submitted one run, UWBaseline (UWBaseline2 fixed minor bugs) as a baseline BM25 element retrieval run, and another run, UWBM25F as an experimental BM25F element retrieval run. Additionally, we submitted a manual run, UWManual that is the same as UWBM25F run, except some of the query phrases that deemed unimportant are manually deleted. We used the same parameters obtained for UWBM25F on UWManual.

Both UWBaseline and UWBM25F runs are element retrieval run where the unit of element is a page. This is a difference from the last year, where we considered finer element types such as `<regeon>` and `<section>`. Most tags such as `<marker>` and `<regeon>` did not seem useful, and `<line>` tag is below the minimum unit of retrieval in the task specification. Thus it would have been better to work on both `<page>` and `<section>` elements, but since last year's results were converted to a page-level result, we decided to work on page-level element retrieval this year.

For `UWBaseline` run, each page is scored using Equation 1. We accidentally scored and returned our fictious `<title>` elements that contained MARC record used for our other runs, that constitute approximately less than 1% of all the results, so the score adjusted with MARC records removed may be slightly higher.

For `UWBM25F` run, we created two fields for each page; one containing the entire MARC record, and the other containing the actual text in the page. Our original idea was to add the title of the book to each page; however, by examining MARC record, we thought that keywords contained in some of the records may be useful. In the future, we would like to include only the useful information from the records by discarding author name and publishers, and translating LC classification code into phrases. One we obtained fields, we scored pages by Equation 2.2

For both runs, after scoring, we sorted the books by the maximum page scores, and within each book, we sorted pages by the scores. For training, we returned the top 1500 book-page pairs, and used trec_eval.8.1 with qrels from INEX 2008, averaged-relevance/qrel-alltopics.txt.xpath, to maximize MAP. For `UWBaseline` run, we obtained $k = 200$ and $b = 1$ with $MAP = 0.0219$. For `UWBM25F` run, we created our run with $B_{title} = 0.2$, $B_{body} = 0.6$, $W_{title} = 2$, and $W_{body} = 0.4$, with a fixed $K = 1$ that gave $MAP = 0.0356$ during training. For testing, we returned the top 1000 books regardless of the number of pages each book contained.

4   Entity Ranking Track

This year marks our first participation in Entity Ranking track. Our objective is to see if a factoid question answering approach can be effectively applied in an entity ranking task. Therefore, instead of searching for Wikipedia pages that are entities that fits the topic title, we treated titles as questions, searched for answers in passages within Wikipedia pages, and found the Wikipedia pages that correspond to the answers.

We use redundancy technique of Clarke et al. [4] to extract answers. The technique first scores passages using self-information of probability that an extent contain a query term;

$$\sum_{t \in T} \log \frac{N}{f_t} - |T| \log l \ ,$$

where N is the corpus length and f_t is a term frequency in the corpus. The technique then retrieves top k passages and expand the passages to size w around the center. The candidates answers are then ranked according to the number of distinct top k passages that contain the term.

4.1   Entity Ranking Task

We submitted one entity ranking run, 1_Waterloo_ER_TC_qap. The main difference between this run and the question answering technique of Clarke et al. [4]
is in the selection of candidate topics from the top k passages.

We first extracted all titles and the categories listed at the beginning of the
articles in the Wikipedia corpus, call it *wiki_titles* and *wiki_categories* respectively. We also extracted topic titles and the categories from the topic file, call
them *topic_titles* and *topic_categories* respectively.

We collected the highest scoring passages from each article in the corpus using
topic_titles according to Equation 4. We then expanded the passages around the
center to 1000 terms. Call this new set of passages *passage_ER*.

To select the candidate terms among all other terms appearing in *passage_ER*,
we need to extract terms

1. that have corresponding Wikipedia articles
2. whose categories loosely fall under those in *topic_categories*

Therefore, we queried *wikipedia_categories* using *topic_categories* as query terms
with BM25. We collected the top 10000 highest scoring Wikipedia pages that correspond to the top 10000 categories including duplicates, retrieved from
wikipedia_categories. Now the list of the candidate terms are limited to the titles
of these Wikipedia pages.

We ranked the candidate terms by the number of passages in *passage_ER* that
contain the terms. From the list, we removed the topic titles and duplicates to
create the final submission.

4.2   List Completion Task

We submitted one run, 1_Waterloo_LC_TE. The major difference between the
work of Clarke et al. [4] and 1_Waterloo_ER_TC_qap is how to estimate the categories of entities to retrieve.

We extracted all titles and the categories listed at the beginning of the articles
in the Wikipedia corpus, *wiki_titles* and *wiki_categories*. For each topic, we extracted topic titles and examples, and merged each topic title with its examples
to make the final *topic_titles*.

Using the merged *topic_titles*, we collected the highest scoring passages from
each article according to Equation 4. We then expanded the passages around
the center to 1000 terms. Call this new set of passages *passage_LC*.

We estimated categories of retrieval by the union of all categories of examples
according to *wiki_categories*. Call this new set of categories *ex_categories*.

We queried *wikipedia_categories* using *ex_categories* as query terms with BM25.
We collected the top 10000 highest scoring Wikipedia pages that correspond to the
top 10000 categories with duplicates, retrieved from *wikipedia_categories*. The list
of the candidate terms are limited to the titles of these Wikipedia pages.

As in the case of Entity Ranking task, we ranked the candidate terms by the number of passages in *passage_LC* that contain the terms. From the list, we removed
the topic titles, example titles, and duplicates to create the final submission.

4.3   Results

Official results showed that we performed poorly on both tasks. Both entity ranking and list completion run scored the lowest of all the participants scoring 0.0954 and 0.1002 respectively.

In order to analyze the poor performance, we decided to separate the passage retrieval part using qap from the selection of appropriate entities from these passages. We measured the quality of the first step of passage retrieval by assuming that we perform the second step of term selection perfectly. We looked for the set of relevant entities according to qrels in the passages retrieved and made the runs out of them. With the perfect knowledge of term selection, our performance improved to 0.31734 for entity ranking and 0.30768 for list completion tasks. However, these scores are far from perfect, and the best scores among the participants are 0.5173 and 0.5205, both from University of Amsterdam, ISLA group.

We need to improve passage retrieval part by retrieving more passages or expanding from the center of the passages more. In addition, instead of scoring passages by basically how often the query term appears, it may be better to consider how often the entities with the given category appears along with the query terms. For example, two passages may contain the term "EU countries", but one contains more entities categorized as "country" as specified in the topic, than the others. In a sense, this is the main difference between factoid question answering and entity ranking; the number of potential answers. Oftentimes, for entity ranking task, there is a passage that contains the answers for the query, thus the passage scoring function needs to be modified to account the difference. We still think our fundamental approach of extracting solutions from passages, as opposed to retrieving entities-as-pages directly has some promise.

5   Link-the-Wiki Track

This year, our goal is to use link analysis to overcome the problem of the simple statistical approach [11]. We mainly submitted runs for anchor-to-BEP topics, mostly at file level, but also two runs at anchor-to-BEP level. We also submitted one run for Link-Te-Ara-to-the-Wiki task using our baseline.

5.1   Link-the-Wiki Anchor-to-BEP File Level

We submitted the following five runs, UWBaseline_F2FonA2B,UWKPR_F2FonA2B, UWKPRFlip_F2FonA2B, UWTPR_F2FonA2B, and UWKPR+_F2FonA2B.

Baseline: UWBaseline_F2FonA2B. Our implementation of the outgoing link discovery for baseline run is the same as in INEX 2007/2008 [11,12].

After removing topic files from the Wikipedia corpus, we parsed the corpus to obtain the following ratio, γ.

$$\gamma = \frac{\sharp \text{ of files that has a link from anchor } a \text{ to a file } d}{\sharp \text{ of files in which } a \text{ appears at least once}}$$

For each topic file, we located the anchor phrases, and returned the most frequent destination file for each of the top 250 anchor phrases by γ.

For the incoming link discovery, we scored each article in the Wikipedia corpus using topic titles and returned the top 250 articles.

KL-PageRank: `UWKPR_F2FonA2B`. Our baseline approach does not take into account the context of anchor phrases. For example, "Australia" may be frequently linked, but should it be linked from an unrelated topic such as "olive oil"?

We employed PageRank [1] and topical PageRank [17] to balance the frequency of linkage with topicality of the pages to topic files.

For outgoing links, We created a link graph without topic files and then added the links from the topic files according to the outgoing results from the baseline run, `UWBaseline_F2FonA2B`. We obtained PageRank values for each node in the graph. In order to compute topical PageRank, we set the jump vector to be uniform, except for the entry corresponding to the topic file that is set to an arbitrary large number, $1/2$. Once we obtained both PageRank $g(p)$ and topical PageRank values $f(p)$ for each page p, we computed the contribution of pointwise K-L divergence values for each page for each topic as follows:

$$\tau(p) = f(p) \log \frac{f(p)}{g(p)} \ .$$

We returned the top 250 destination files according to τ.

For incoming links, we reversed the link graph obtained from outgoing links and using the same topical PageRank values from the outgoing links, we computed the contribution of point-wise K-L divergence. We returned the top 250 files according to τ.

KPR-Flip: `UWKPRFlip_F2FonA2B`. This run tries to answer the question: Should there be an outgoing link from a page "Marsupial' to "Koala"? But should not there also be an incoming links from "Koala" to "Marsupial"?

Because at file level, it is unclear if there is any difference between the set of incoming links and the set of outgoing links for a topic file, this run returned the set of outgoing links from `UWKPR_F2FonA2B` as incoming links and the set of incoming links from the same run as outgoing links.

Topical PageRank-only: `UWTPR_F2FonA2B`. This run tries to answer the question: How much of the links should come from topically related files?

Because in our `UWKPR_F2FonA2B` run, we added point-wise K-L divergence of Topical PageRank with PageRank values to balance topicality with popularity, in this run, we returned the set of outgoing links and incoming links from `UWKPR_F2FonA2B` run, right after obtaining Topical PageRank, but before applying K-L divergence.

KPR Run with Filtering: `UWKPR+_F2FonA2B`. It is natural to assume that a file that has a link to a topic file contain the topic title within the file more or less. Therefore, in this run, for incoming links, we filtered the result of `UWKPR_F2FonA2B` to include only those files that contain the topic titles.

We may think that this might also hold true for the outgoing links; an anchor phrase must exist for a file to be considered to be a destination from the topic files. We, however, left the outgoing links the same as in UWKPR_F2FonA2B because we were not sure if this holds true for outgoing links at a file level.

5.2   Link-the-Wiki Anchor-to-BEP   Passage Level

We submitted two runs, UWBaseline_A2B and UWKPR_A2B.

The baseline run UWBaseline_A2B implements the file-level version of the baseline run UWBaseline_F2FonA2B at anchor-to-BEP level. Specifically, for outgoing links, we returned the offset and length of anchor phrases in the topic files and the BEP is set to the titles of destination files. The BEP could be set to the beginning of an element with high title term density. We restricted to return only the most frequent destination per anchor in order to compare our results against UWKPR_A2B, which is a KPR-based run that only returns one destination file per anchor phrase. For incoming links, both anchor phrases and destination files are set to the titles of the incoming files and the topic files.

The KPR based run UWKPR_A2B implements UWKPR_F2FonA2B at anchor-to-BEP level. For outgoing links, we performed disjunctive queries using Equation 4 into the topic files, the titles of top destination files according to the values of K-L divergence. Once we obtained the positions of the destination titles in the topic files, we returned those as anchor phrases with BEP being the titles of the destination files. We only returned one destination file per anchor phrase, but we could assign several destination files in order of similarity of the destination file titles to the anchor phrase. For incoming links, we first screened the set of incoming files returned by K-L divergence to only those that contain the topic titles. We then set both the anchor phrase and BEP to be the titles of the incoming and topic files.

5.3   Link-Te-Ara-to-the-Wiki

We only submitted one baseline run at Anchor-to-BEP level, Teara_Baseline_A2B. Due to resource constraints, we only computed γ for those anchor phrases that occurred in more than 100 documents in Wikipedia corpus. Once we have a list of anchor phrases-destination pairs, we queried those phrases in TeAra corpus. As is the case with our other Anchor-to-BEP runs, we only returned the most frequent destination per anchor phrase.

5.4   Results

We report our results based on the graph-based preliminary results [9]. Our main findings is two-fold. First in outgoing link discovery, KPR outperforms the baseline at early precision in manual assessment, but the baseline outperforms KPR at the first half of recall level against Wikipedia ground truth. The second finding was rather unexpected; in the incoming link discovery, the KPR+ run seems to give somewhat better performance than the baseline run that performs better than other participants' runs. This is interesting as most attention in this track was focused on beating our baseline run in the outgoing task. For future work in the incoming link discovery, as comparison, we should filter KPR-Flip run and TPR run by the existence of topic titles in the top documents as our KPR+ run did.

The more important question is how to move from supervised learning to unsupervised learning. Specifically, how can we link all the linkless articles together? This is the Link-the-TeAra task. For outgoing links, our current baseline approach would be useless. On the other hand, our incoming baseline approach still works in this setting. From there, we may be able to apply KPR to improve the baseline incoming links, then flip the graph to apply KPR to get outgoing links.

6   Conclusions

Yes, perhaps, and yes. We think we accomplished this year's goal of beating our baseline runs that performed better than other participants' runs in the previous years. For both Ad Hoc and book tracks, adopting BM25F proved promising, even though we have not received the book track data as of date. For Link-the-Wiki track, adopting KPR not only improved our previous baseline on outgoing links, but also unexpectedly, on incoming links.

The situation in Entity ranking track is rather different; neither us nor others have come up with a simple invincible baseline. We hope we try a simpler approach as well as modifying our question answering approach to entity ranking task, that is essentially the multiple question answering task.

For future work, we would like to integrate our previous work on passage retrieval that we did not do this year, with BM25F to try to come up with a general approach to focused retrieval.

References

1. Brin, S., Page, L.: The anatomy of a large-scale hypertextual web search engine. Computer Networks and ISDN Systems 30(1-7), 107–117 (1998)
2. Büttcher, S.: The Wumpus Search Engine (2007), http://www.wumpus-search.org
3. Clarke, C.L., Tilker, P.L.: Multitext experiments for INEX 2004. In: Fuhr, N., Lalmas, M., Malik, S., Szlávik, Z. (eds.) INEX 2004. LNCS, vol. 3493, pp. 85–87. Springer, Heidelberg (2005)
4. Clarke, C.L.A., Cormack, G.V., Lynam, T.R.: Exploiting redundancy in question answering. In: Proceedings of the 24th Annual International ACM SIGIR Conference on Research and Development in Information Retrieval (SIGIR 2001), pp. 358–365 (2001)
5. Craswell, N., Zaragoza, H., Robertson, S.: Microsoft Cambridge at TREC 14: Enterprise track. In: Proceedings of the 14th Text REtrieval Conference (2005)
6. Demartini, G., Vries, A.P., Iofciu, T., Zhu, J.: Overview of the INEX 2008 entity ranking track. In: Geva, S., Kamps, J., Trotman, A. (eds.) INEX 2008. LNCS, vol. 5631, pp. 243–252. Springer, Heidelberg (2009)

7. Denoyer, L., Gallinari, P.: The Wikipedia XML corpus. SIGIR Forum 40(1), 64–69 (2006)
8. Fuhr, N., Kamps, J., Lalmas, M., Malik, S., Trotman, A.: Overview of the INEX 2007 Ad Hoc Track. In: Fuhr, N., Kamps, J., Lalmas, M., Trotman, A. (eds.) INEX 2007. LNCS, vol. 4862, pp. 1–23. Springer, Heidelberg (2008)
9. Huang, W.C., Geva, S., Trotman, A.: Overview of the INEX 2008 Link the Wiki Track. In: Geva, S., Kamps, J., Trotman, A. (eds.) INEX 2008. LNCS, vol. 5631, pp. 314–325. Springer, Heidelberg (2009)
10. Itakura, K.Y., Clarke, C.L.A.: From passages into elements in XML retrieval. In: Proceedings of the 30th Annual International ACM SIGIR Conference on Research and Development in Information Retrieval (SIGIR 2007) Workshop on Focused Retrieval, pp. 17–27 (2007)
11. Itakura, K.Y., Clarke, C.L.A.: University of Waterloo at INEX 2007: Adhoc and Link-the-Wiki Tracks. In: Fuhr, N., Kamps, J., Lalmas, M., Trotman, A. (eds.) INEX 2007. LNCS, vol. 4862, pp. 417–425. Springer, Heidelberg (2008)
12. Itakura, K.Y., Clarke, C.L.A.: University of Waterloo at INEX 2008: Adhoc, Book, and Link-the-Wiki Tracks. In: Geva, S., Kamps, J., Trotman, A. (eds.) INEX 2008. LNCS, vol. 5631, pp. 132–139. Springer, Heidelberg (2009)
13. Kamps, J., Geva, S., Trotman, A., Woodley, A., Koolen, M.: Overview of the INEX 2008 Ad Hoc Track. In: Geva, S., Kamps, J., Trotman, A. (eds.) INEX 2008. LNCS, vol. 5631, pp. 1–28. Springer, Heidelberg (2009)
14. Kazai, G., Doucet, A., Landoni, M.: Overview of the INEX 2008 Book Track. In: Geva, S., Kamps, J., Trotman, A. (eds.) INEX 2008. LNCS, vol. 5631, pp. 106–123. Springer, Heidelberg (2009)
15. Lu, W., Robertson, S., MacFarlane, A.: Field-weighted XML retrieval based on BM25. In: Fuhr, N., Lalmas, M., Malik, S., Kazai, G. (eds.) INEX 2005. LNCS, vol. 3977, pp. 161–171. Springer, Heidelberg (2006)
16. Malik, S., Lalmas, M., Fuhr, N.: Overview of INEX 2004. In: Fuhr, N., Lalmas, M., Malik, S., Szlávik, Z. (eds.) INEX 2004. LNCS, vol. 3493, pp. 1–15. Springer, Heidelberg (2005)
17. Page, L., Brin, S., Motwani, R., Winograd, T.: The pagerank citation ranking: Bringing order to the web. Technical Report 1999-66, Stanford InfoLab, Previous number = SIDL-WP-1999-0120 (November 1999)
18. Robertson, S., Lu, W., MacFarlane, A.: XML-structured documents: Retrievable units and inheritance. In: Larsen, H.L., Pasi, G., Ortiz-Arroyo, D., Andreasen, T., Christiansen, H. (eds.) FQAS 2006. LNCS (LNAI), vol. 4027, pp. 121–132. Springer, Heidelberg (2006)
19. Robertson, S., Walker, S., Beaulieu, M.: Okapi at TREC-7: Automatic ad hoc, filtering, vlc and interactive track. In: Proceedings of the Seventh Text REtrieval Conference (1998)
20. Robertson, S., Zaragoza, H., Taylor, M.: Simple BM25 extension to multiple weighted fields. In: Proceedings of the 13th ACM Conference on Information and Knowledge Management (CIKM 2007), pp. 42–49 (2004)
21. Schenkel, R., Suchanek, F., Kasneci, G.: YAWN: A semantically annotated Wikipedia XML corpus. In: Kemper, A., Schöning, H., Rose, T., Jarke, M., Seidl, T., Quix, C., Brochhaus, C. (eds.) 12. GI-Fachtagung für Datenbanksysteme in Business, Technologie und Web, Aachen, Germany. Lecture Notes in Informatics, vol. 103, pp. 277–291. Gesellschaft für Informatik (2007)
22. Vries, A.P., Vercoustre, A.-M., Thom, J.A., Craswell, N., Lalmas, M.: Overview of the INEX 2007 entity ranking track. In: Fuhr, N., Kamps, J., Lalmas, M., Trotman, A. (eds.) INEX 2007. LNCS, vol. 4862, pp. 245–251. Springer, Heidelberg (2008)

A Machine Learning Approach to Link Prediction for Interlinked Documents

Milly Kc[1], Rowena Chau[3], Markus Hagenbuchner[1],
Ah Chung Tsoi[2], and Vincent Lee[3]

[1] University of Wollongong, Wollongong, Australia
{millykc,markus}@uow.edu.au
[2] Hong Kong Baptist University, Hong Kong
act@hkbu.edu.hk
[3] Monash University, Melbourne, Australia
{Rowena.Chau,Vincent.lee}@infotech.monash.edu.au

Abstract. This paper provides an explanation to how a recently developed machine learning approach, namely the Probability Measure Graph Self-Organizing Map (PM-GraphSOM) can be used for the generation of links between referenced or otherwise interlinked documents. This new generation of SOM models are capable of projecting generic graph structured data onto a fixed sized display space. Such a mechanism is normally used for dimension reduction, visualization, or clustering purposes. This paper shows that the PM-GraphSOM training algorithm "inadvertently" encodes relations that exist between the atomic elements in a graph. If the nodes in the graph represent documents, and the links in the graph represent the reference (or hyperlink) structure of the documents, then it is possible to obtain a set of links for a test document whose link structure is unknown. A significant finding of this paper is that the described approach is scalable in that links can be extracted in linear time. It will also be shown that the proposed approach is capable of predicting the pages which would be linked to a new document, and is capable of predicting the links to other documents from a given test document. The approach is applied to web pages from Wikipedia, a relatively large XML text database consisting of many referenced documents.

1   Introduction

Self-Organizing Maps (SOMs) are a popular unsupervised machine learning approach for the clustering and projection of high dimensional data vectors [1]. Recent developments extended the algorithms' ability to encode and cluster graph structured data. The methodology has been applied successfully to clustering tasks involving documents retrieved from Wikipedia as part of a participation in the INEX (Initiatives for the Evaluation of XML retrieval) document mining competition [2].

Graph data structures allow the representation of almost any kind of learning problems. A graph consists of a set of nodes and a set of binary relations called links. A link connects any two nodes in a graph if these nodes are related to each other in some way. If the relationship is directed, then this is represented by a directed link. Otherwise it is

S. Geva, J. Kamps, and A. Trotman (Eds.): INEX 2009, LNCS 6203, pp. 342–354, 2010.

undirected. With directed links, the source node is called a *parent* and the destination node is called a *child*. With undirected links, each node connected by a link to a given node is called a *neighbour*. A node with n children and m parents is said to have an out-degree of n and an in-degree of m. A node with k neighbors is said to have a degree of k. A node can be labelled by a numeric vector so as to provide a description of the properties of the associated object. Sequences are a special case of graphs in which the maximum in-degree and maximum out-degree is one. Traditional data vectors are also a special type of graphs for which there exists no link between the nodes. Thus, any approach capable of dealing with graphs can also deal with data sequences and vectors.

This paper addresses a domain which is appropriately represented by a directed graph: hyperlinked web pages. Web pages can contain hyperlinks which point from a given web page to another one. Thus, this defines a directed graph in which the web documents are represented by a node in the graph, and the hyperlinks are the directed links in the graph. A data label associated with each node in the graph can add a description of the content of a document. For example, the well-known Bag of Words approach would summarize the content of a document in vectorial form.

There have been much activities in recent years on approaches that can process graph structured information. Some of the most successful approaches were developed in the area of machine learning. This success stems from the fact that the approaches are scalable to real world data mining tasks. Moreover, some of these approaches have been proven to be capable of solving any given problem involving graphs optimally [3].

This paper describes the latest in the development of the unsupervised machine learning approaches for structured data, and shows that the method can also be used for the purpose of link prediction. Link prediction has particularly important practical applications in the World Wide Web (WWW) domain. For example, a central algorithm in the search engine Google, known as PageRank, relies on link analysis for the purpose of ranking a set of linked documents. PageRank produces a large value for pages with many parents and few children. A known weakness of PageRank is that it neglects pages which have been newly added to the Web. Such new pages do not have any parents by default (as they are not referenced by other web pages due to their newness in the WWW), and hence would be ranked lowly. This in turn causes Google not to rate new pages highly producing an effect known as "the-rich-gets-richer" effect. Another problem with the WWW domain is that it is unregulated. Anyone can add a new document containing any hyperlink and any number of hyperlinks. This is often exploited by companies and individuals who create *link-farms* designed at increasing the rank of a target page. These examples highlight the need for algorithms capable of predicting links between any two documents in the WWW. Such algorithms can then be used to automatically suggest parent nodes which should link to a newly created document, and can be used to verify whether existing links are valid[1].

This paper is organized as follows: Section 2 describes the Probability Measure Graph SOM model, and how it can be used to predict links in a hyperlinked domain. Section 3 applies the proposed approach to a relatively large real world problem on link predictions. Conclusions are drawn in Section 4.

---

[1] Links which exist for the sole purpose of increasing the rank of a given page are said to be *spam* links or *invalid*, otherwise the link is said to be valid.

2   The Approach

The SOM is a well-known algorithm in unsupervised learning of data vectors [1]. The SOM is a topology preserving map capable of mapping data that is close to one another in the high dimensional feature space so that they will remain close in the low dimensional display space. In its most basic form, we consider a two-dimensional display space being discretized into an $N \times M$ grid [1]. Each grid point i, j has an associated m dimensional codebook vector $\mathbf{c}_{i,j}$, where m is also the dimension of the data vectors. The $N \times M$ codebook vectors are initialized randomly. The SOM is trained in two steps:

- **Competition:** A given input vector is compared with each of the m dimensional codebook vectors in the $N \times M$ grid by using the Euclidean distance measure. The best matching codebook $\mathbf{c}_{i_k, j_k}$ is said to be the winner.
- **Parameter adjustment:** The elements of codebook $\mathbf{c}_{i_k, j_k}$ and all its neighbors are pulled closer to the elements of the input vector as follows:

$$\Delta \mathbf{c}_{i,j} = \alpha(t) f(\Delta_{i, \{i_k, j_k\}})(\mathbf{c}_{i,j} - \mathbf{u}) \tag{1}$$

where $\mathbf{u}$ denotes the input vector, α is a learning rate which decreases steadily towards 0, and $\Delta_{i, \{i_k, j_k\}}$ is the neighborhood of the winning vector $\mathbf{c}_{i_k, j_k}$, and $f(\cdot)$ is a nonlinear function, often chosen to be a Gaussian function.

These two steps are repeated for each input vector in a training set, and for a given number of cycles. When the algorithm converges, the elements of the two dimensional grid $N \times M$ encapsulate the set of high dimensional input vectors $\mathbf{u}$.

The SOM assumes that the input vectors are independent. The algorithm needs to be modified if there is a dependency defined on the input vectors. Such dependencies are normally represented as a graph structure. A first approach was made with the introduction of a SOM for Structured Data (SOM-SD) capable of processing trees [4]. A tree is a special class of graphs which is rooted, directed, and acyclic. In this case, the issue is how to encode the tree structure. One way in which this can be obtained is to consider each node in the graph and then to model the node and its offsprings using a SOM. As this is a tree structure consisting of directed links, it makes sense to process the data from the leaf nodes to the root node. The issue is then how to connect the SOM model for each node with the other nodes in the tree (as there is no obvious way in which this can be performed). An approach would be to consider the winning node in the SOM model of each node, and then to pass this information to the parent node[2]. The location of the winning codebook of the SOM model of the node represents the outcome of the SOM model of the node. This information can then be passed to parent nodes through a concatenation of a data label (which may be attached to the parent node) with the information about the mapping of the node's offsprings. In other words, this concatenation passes information about a node's offsprings to the node. Recursively applied, this passes information through all the nodes in a tree structure. If we assume that there are only a fixed number of incoming links from the children nodes, then for nodes which

---

[2] Since we are processing the tree from leaf nodes first towards the root node, the links from a particular node is pointing towards the parents rather than the child.

have lesser number of incoming children links, the input vector can be suitably augmented with zeros. This guarantees that the input is a fixed sized vector, and hence the standard SOM training algorithm can be applied; the only variation is that we will need to weigh the relative importance of the input vector to the node, and the vectors associated with the children nodes. This algorithm was shown to produce good results when processing tree structured data [4].

One way by which this approach can be extended to un-directed graphs and cyclic graphs is by introducing the concept of "state". The *state* is a network's response to a given input. In the case of a SOM, the networks response to a given input is the mapping of the input to a particular codebook. Thus, in the SOM-SD case, a network input was formed through the concatenation of data label and the states of a node's offsprings.

The recently introduced Probability Measure Graph SOM (PM-GraphSOM), the latest version of SOM-SD algorithms, addresses the question on how cyclic dependencies can be encoded by a SOM. To achieve this, the PMGraphSOM generalizes the interpretation of the *state* vector. We associate with each node a *state* which describes the activation of the SOM for this node. In the context of using self organizing map, we may consider the state as the location of the winning node in the $N \times M$ display map. In this case, for the current node, if we assume in the display map, there are additional inputs from the antecedent (parent) nodes, and the descendant (child) nodes, the locations of these additional inputs are the coordinates of the winning nodes in these antecedent nodes and descendant nodes, together with the associated winning vectors [5]. A weakness of this method is that the Euclidean distance measure does not make a distinction as whether any change of a mapping during the update step has been to a nearby location or to a location far away on the map. To counter this behavior it has been proposed to *soft code* the mappings of neighbors to account for the probabilities of any changes in the mapping of nodes. In other words, instead of *hard coding* the mappings of nodes to be either 1 if there is a mapping at a given location, or 0 if there is no mapping at a given location, we encode the likelihood of a mapping in a subsequent iteration with a probability value. We note that due to the effects of the training algorithm it is most likely that the mapping of a node will be unchanged at the next iteration. But since all vectors associated with the grid points in the display map are updated, and since those vectors which are close to a winning entry (as measured by Euclidean distance) are updated more strongly (controlled by the Gaussian function), and, hence, it is more likely that any change of a mapping will be to a nearby location rather than to a location far away from the last update. These likelihoods are directly influenced by the neighborhood function and its spread. Hence, one can incorporate the likelihood of a mapping in subsequent iterations as follows:

$$
M_i = \frac{e^{-\frac{\|\{i_1, j_1\} - \{i_k, j_k\}\|^2}{2\sigma(t)^2}}}{\sqrt{2\pi}\sigma(t)}, \tag{2}
$$

where $\sigma(t)$ decreases with time t towards zero, and $\{i_k, j_k\}$ are the coordinates of the winning vector, while $\{i_1, j_1\}$ are the coordinates of the vector in the display space. The computation is cumulative for all the i-th node's neighbors. Note that the term $\frac{1}{\sqrt{2\pi}\sigma(t)}$ normalizes the states such that $\sum_i M_i \approx 1.0$. It can be observed that this approach accounts for the fact that during the early stages of the training process it is likely that mappings can change significantly, whereas towards the end of the training process, as $\sigma(t) \to 0$, the state vectors become more and more similar to the hard coding method. This approach helps to improve the stability of the GraphSOM significantly, which allows the setting of large learning rates, and reduces the required training time significantly while providing an overall improvement in the clustering performance. This is referred to as the probability mapping GraphSOM (PMGraphSOM) [6].

We note that the PMGraphSOM updates the codebook vectors in the direction of data vectors. We note further that the input vectors contain *state* information which represents the dependencies on other nodes in the graph. Hence, an idea was born to derive the links to a new node from the state component of the the the best matching codebook vector. In other words, given a PMGraphSOM which has been trained on a Web graph, and given a new document whose links are not yet known, we can find the best matching codebook vector for this document and obtain the most likely link structure by "reverse engineering" the part of the codebook which represents the state vector.

As a comparison, and as an alternative approach to obtaining links for a new document, we can furthermore consider the following property of the PMGraphSOM: during training, a best matching codebook has been obtained for each of the nodes in a graph. Since each codebook is activated by a number of nodes, and hence, these codebooks are said to be a *representation* of these nodes. Due to the topology preserving ability of the PMGraphSOM, it can be said that all nodes which activated the same codebook are most closely related to each other in terms of content and hyperlinks. Let us now compute a winning codebook for a new document, it makes much sense that this new document shares greatest similarity with all nodes from the training dataset which activated the same codebook. Hence, we can propose a link structure for this new document based on the links contained in the documents which were mapped at the same location.

3   Experiments

Given a new Wikipedia document, the file-to-file link discovery task is to analyze the text and recommend a set of up to 250 incoming and 250 outgoing links from one document to other documents in the collection.

Documents from the INEX 2009 Wikipedia collection are used for this task. This collection contains $2,666,190$ articles; it is a dump of the Wikipedia taken on 8 October 2008. It is annotated with the 2008-w40-2 version of YAGO. It is 50.7GB in size.

A set of 5000 existing Wikipedia documents, randomly selected from the collection, are used as the test set for the file-to-file link discovery task. Since the topics are not truly orphaned documents (as there are orphaned documents which contain no incoming or outgoing links, a situation similar to the newly introduced documents onto the Web), we must delete these orphaned files from the collection to simulate the situation of genuinely introducing new documents into the Web.

This task has specific restrictions about the use of link-based information. It was recommended that the link information be processed so that all links pointing to and coming from the documents in the test set are discarded. This is in fact conducted as recommended. The 5000 documents in the test set were also removed from the training set, as is the common practice for machine learning tasks. This results in a training set consisting of $2,661,190$ documents.

There are 2 other participating groups that worked on the specific task. The group from Technische Universitat Darmstadt, Germany, used an unsupervised learning approach based on the existing link structure between documents coupled with article relatedness obtained from Wiktionary for the outlinks, whereas Lucene retrieval model and ESA ranking over the top 2000 results returned from a search using Lucene were used for the inlinks. The other group from Queensland University of Technology, Australia, used keyword matching between the content of the test set and the titles of documents in the entire dataset, combined with a likelihood score of a document containing links associated with an anchor text for the outlinks, whereas BM25 search engine with the document title as the query term was used for the inlinks. It is interesting to see that both groups deal with the outlinks and inlinks using significantly different approaches, but both resorted to using a search engine for the discovery of inlinks.

Our group adopted a very different approach to other groups in the sense that a clustering algorithm described in the previous section was used for training. Also, in contrast to both of the other groups, a more general approach is adopted, so there is no need for separate algorithms and procedures for the outlinks and inlinks. In our approach, the whole INEX 2009 Wikipedia collection is represented as a directed graph where each node is a wiki page. Each page is then represented by a state vector encoding both its contextual and link structural information.

Before the data can be trained by PMGraphSOM, decisions need to be taken regarding the selection of features to be used as node labels for each document. It was important to select a feature which could represent each document, but such features do not contain any link information, as otherwise the testing process would not be able to use the same feature as node labels to represent the test data. Some analysis revealed that the category information of each document could be used as a representative feature. However, the category extraction process identified $8,918,924$ categories in total, within which there are $362,251$ unique categories. The maximum number of categories in a document reached $2,022$, but there are also $118,209$ documents with no associated category information at all.

The un-processed category information as node labels would prevent the training process from being completed within a reasonable amount of time; therefore, some dimension reduction is required. Singular Value Decomposition (SVD) was considered for a dimension reduction step. However, SVD requires the building of a $2,661,190$ x $362,251$ matrix which is far too large for the capacity of computers which we had available for this project. Hence, another approach to dimension reduction was taken. This second approach utilizes a well known Multi-Layer Perceptron (MLP) algorithm to assist in dimension reduction. The MLP algorithm is generally applied to neural network architectures which consist of an input layer, followed by one or more hidden layers of neurons, and then an output layer of neurons, where all neurons in a layer are fully connected to all neurons in the next layer. To utilize MLP for dimension reduction we use an architecture known as the "Auto-associative Memory" (AAM) architecture. In an AAM, both the input and output dimension are the number of unique categories $(362,251)$, and the dimension of the single hidden layer will be the dimension which we would like to reduce to. Using this configuration, the number of neurons (dimension) in the hidden layer is less than the dimension of the input layer, so the encoding process

Table 1. Statistics of the training set's link structure

	Max	Min	Standard dev.	Number of documents with no link
Out-degree	5295	0	92.96	36, 978
In-degree	549, 658	0	476.18	304, 518

Table 2. Statistics of the test set's link structure

	Total	Max	Min	Mean	Std dev.	Number of documents with no link
Out-degree	461, 741	245	1	92.35	46.05	0
In-degree	311, 423	3095	0	62.28	131.50	372

in the hidden layer is, in a sense, compressing the information from the input. Then the connection between the single hidden layer neurons and the output layer neurons can be seen as uncompressing the information back to the original dimension. For training purposes, the input data is also used as the target, so that the MLP can learn a mapping which loses the least amount of information through the compression (hidden) layer of an auto-associative memory. It is known that an MLP trained in this fashion results in a dimension reduction which is qualitatively equivalent to those obtained by using SVD algorithm but without the need of having to store a large matrix in memory.

After the node labels are reduced to a more manageable dimension (here we use 16) using the MLP technique as indicated, attention was shifted to the incorporation of link structures within the training set. PMGraphSOM is capable of incorporating link information to assist in the training process, therefore the link structure of the training set after the required pre-processing could be used for training purposes. Some analyses of the links reveal that the number of links from the $2, 661, 190$ training documents total $136, 304, 216$, this equates to a mean of approximately 51.22 links per page. These links are unlikely to be distributed equally, therefore, separate analysis of the out-links and in-links were also carried out. The statistical results can be found in Table 1. The standard deviation indicates that the number of in-links varies much more than the number of out-links. This property is important since it implies that the dataset is unbalanced with respect to the incoming and outgoing links.

Such unbalances in the training dataset are known to potentially cause problems with any machine learning approach.

Although the link structure of the test set will not be used during training, some analyses were carried out to investigate whether the test set is comparable to the training set. This is especially important for machine learning tasks, as a training dataset, which is representative of the testing dataset, will be able to provide more accurate results for the documents the network has not encountered previously, which is the test set in this case. The statistics of the link structure for the testing dataset are included in Table 2.

As can be observed from comparing the statistical information of the link structure, the training dataset and the testing dataset have a number of significant differences.

- The number of documents with no links - It can be observed that a little more than 1% of documents in the training set have no out-links. Based on this ratio, approximately 50 documents in the testing dataset are expected to have no out-links, but the testing dataset contains no such type of documents. The in-link also has a similar problem: with the training dataset containing approximately 11% of documents with no link, in comparison with 7% in the testing dataset.
- A higher number of links in the testing dataset - The average number of out-links and in-links per document in the testing dataset are consistently higher than the 51.22 links per page average in the training set.

These differences suggest that the testing dataset is not a random sampling from the problem domain, but rather a subset selected by some (unknown) criteria. Again, this can impose some added challenges to a machine learning approach.

3.1 Learning for Link Discovery

We learn patterns of the Wikipedia links by feeding the training set to a PMGraphSOM. After training, all Wikipedia pages will be mapped onto a 2-dimensional display map space where pages with similar contextual content and link structure will be mapped onto the same codebook or onto nearby codebooks. Based on this property of the SOM training algorithm, we claim to be able to discover links for the $5,000$ testing pages by inferring from the codebooks of a trained map.

We will use two approaches to infer from the trained map. The first approach maps the test data onto the trained map by comparing the model label of the test documents with that of each of the codebook. The best mapping is the one which has the least Euclidean distance. Then, after the best matching codebook is identified, the state vector component of the winning codebook vector, which comprises of in-link and out-link information, is investigated. This provides information about the likelihood of an in-link or out-link to be mapped at each codebook. Based on this, we were able to identify the codebooks which the child or parent documents are most likely to be mapped, and then identify the documents mapped in the most likely codebook as the proposed links. We refer to this as *codebook-based* link inference.

The second approach also performs a mapping of the test data onto the trained map by finding the best matching codebook. Then, we identify the set of training documents which were mapped onto the same codebook, as these documents are likely to have similar contextual content as that of the test document. The links from all the corresponding training documents are then collected as the proposed links for the task. We refer to this as *content-based* link inference.

Each of these two approaches can produce an arbitrary number of links which is only limited by the size of the map (for the codebook-based approach), or by the existing link structure of the training set (for the content-based approach). These inferred links are then ranked in descending order from the most likely link to the least likely link. Then we will truncate the list of links to the maximum allowable 250 for both, the in-links as well as the out-links. Note that the computed rank values will also play an important role in the evaluation of the results.

Three ranking algorithms were considered for this task. The first is based on the energy flow of a page [7]. This is calculated by accumulating scores when a page receives

Table 3. Statistics of the test set's link structure

Submission ID	Map size	μ_1	μ_2	μ_3	Trained map
01	-	-	-	-	-
02	20x40	0.999	0.0005	0.0005	Figure 1
03	10x30	1.0	0.0	0.0	Figure 2
04	20x40	0.991	0.0045	0.0045	Figure 3
05	20x40	0.991	0.0045	0.0045	Figure 3

in-links, but distributing scores when a page contains out-links. The list of proposed out-links for each of the documents in the test set are ordered according to their associated scores. The reverse of accumulating scores from the out-links, and distributing scores to the in-links, is used to order the list of proposed in-links for each test document.

The second ranking algorithm is based on frequency. For example, if many of the training documents indicate that a link is a likely in-link or out-link, then it has a higher frequency of being proposed, and therefore will be ranked higher.

The third ranking algorithm is based on the Euclidean distance of the test document and the training documents. This ensures that the training documents with more contextual feature similarity are ranked higher.

As mentioned previously, there are two approaches of obtaining an estimation of the local link structure: the First one is by analyzing the state vector of the codebooks in the SOM, and the second one is through association with training patterns. We have submitted results for each of these two approaches. We also investigated the impact of three ranking mechanisms which aim at obtaining the most likely links first.

Unless specified otherwise, the training of the PMGraphSOM in this paper had fixed parameters of 3 iterations, a learning rate of 0.95, a radius of 10 and a seed number of 7. Other parameters such as the size of the map, and the weight μ were varied as indicated later in the paper.

Fixing the afore-mentioned training parameters allows other parameters to be varied and tested. The parameters under investigation here are the map size and the weights μ. We attempted a large number of combinations of map sizes and weights, the resulting trained maps were analyzed based on the test data performance. We selected the five most representative results for the submission to the INEX LinkTheWiki track, and for the visualization purposes in this paper. These submitted training tasks produced results which were amongst the best from any of the approaches attempted.

The first submission does not have any associated information about the training process, because it was not trained, but merely ranked. For the other submissions, training using PMgraphSOM was carried out on a computing cluster. The map size refers to the size of the 2-dimensional map used by PMgraphSOM during training. The three parameters μ_1, μ_2 and μ_3 are weights, and $\sum_{i=1}^{3} \mu_i = 1$. μ_1 is the weight associated with the node labels, μ_2 is the weight associated with the out-links (the vectors associated with the out-links), and μ_3 is the weight associated with the in-links (the vectors associated with the in-links). A large number of variation of weights were used during training, and these are the weights that produced the best performance. It can be observed that the weights associated with the links are significantly less than the weights assigned to node labels; this could be attributed to the differences in the dimensions. For example,

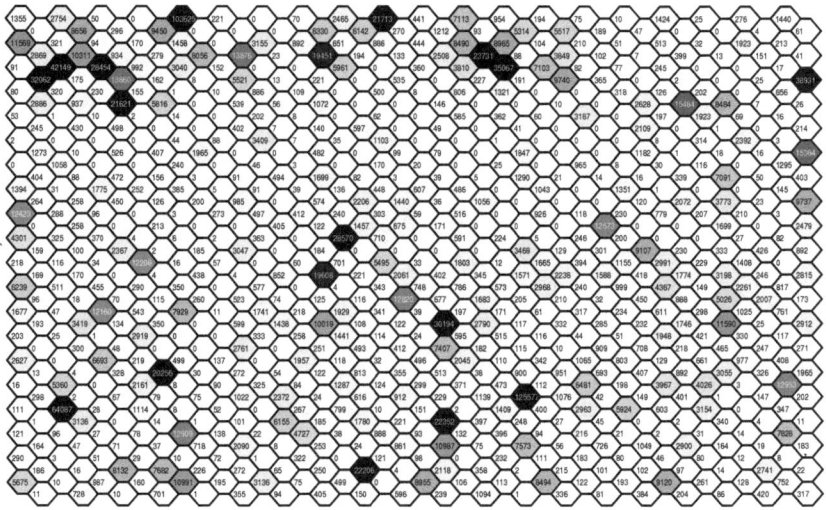

Fig. 1. The trained map of submission 02

in this experimental set up, a 16 dimensional vector is used to represent the node labels, whereas the information about the links (both in-links and out-links) is represented by much larger vectors that are dependent on the map size. It should be noted that submissions 04 and 05 are based on the same trained map, but different ranking algorithms were applied to produce different sets of results.

The frequency of activation of each codebook after the training process (the trained map) can be observed in Figures 1, 2 and 3 respectively.

3.2   Results and Evaluation

The performance measure for the participating challenge includes Precision, Recall, Mean Average Precision (MAP) and interpolated precision (Precision @R). The evaluation mechanism used to compare the results across the participating groups is the interpolated precision. Therefore the evaluation mechanism used for this task is also based on the interpolated precision.

Interpolated precision is precision evaluated on different granularities to observe the performance achievable when varying the number of top ranked links considered. If precision is defined as $\frac{|t \cup r|}{|r|}$, calculating precision for the top n links for each page would be calculated by restricting the size of t to a maximum of n. This is expected to reveal the effectiveness of the ranking algorithms used.

The performance measured using precision and recall for the 5 submissions are included in Table 4. Each of the submissions use a different combination of training configuration and ranking algorithm.

Submission 01 is the best result achievable without using a trained map; a ranking algorithm was simply applied over the entire dataset, and the top few ranked pages were identified. Note that the ranking in this case is only based on the link structure, unlike

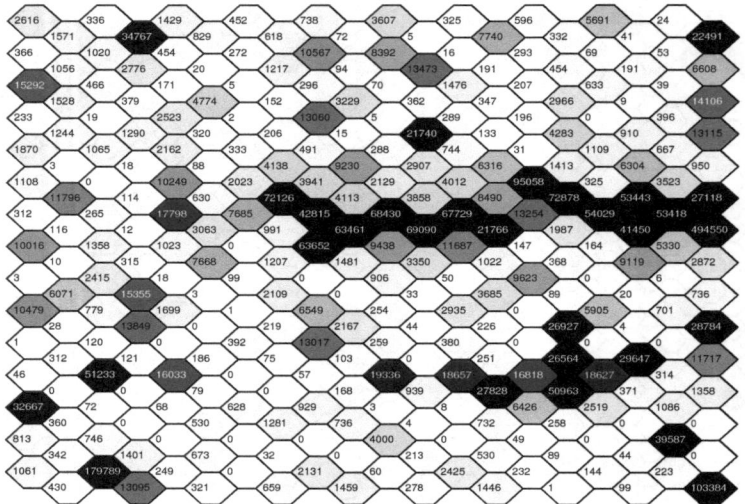

Fig. 2. The trained map of submission 03

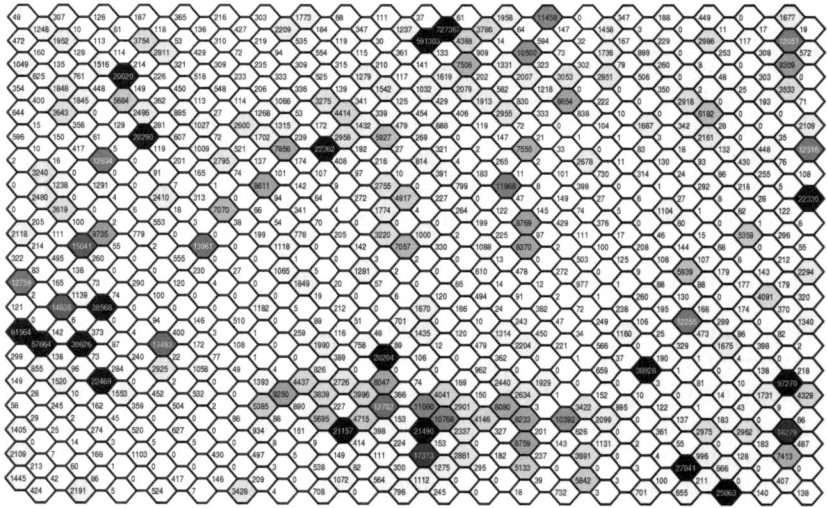

Fig. 3. Visualization of the mappings on the trained map (submissions 04 and 05)

approaches adopted by other participants which involve a combination of link structure and keyword evaluation. As may be observed from Table 4 this simple procedure produces the worst results when compared with those obtained using the training process. The best performance using the trained data, is obtained from submission 03.

Table 4. The performance of proposed links as predicted by our proposed algorithms

Submission ID	01	02	03	04	05
Link inference method	Consider all	Content-based	Content-based	Codebook-based	Content-based
Ranking algorithm	Energy flow	Link freq.	Link freq.	Euclid. distance	Euclid. distance
Recall for out-links	0.00377	0.02942	0.03202	0.00070	0.00448
Prec.@250 out-links	0.00136	0.01057	0.01817	0.00025	0.00161
Prec.@100 out-links	0.00168	0.01795	0.02162	0.00002	0.00147
Prec.@20 out-links	0.00040	0.03941	0.04489	0.00003	0.00268
Recall for in-links	0.00032	0.00050	0.00095	0.00010	0.00029
Prec.@250 in-links	0.00008	0.00012	0.00025	0.00002	0.00007
Prec.@100 in-links	0.00007	0.00008	0.00037	0.00003	0.00008
Precision@20 in-links	0.00007	0.00018	0.00107	0.00001	0.00008

Also seen in Table 4 is that no matter which inference method or ranking algorithm is used, the in-links proposed are less accurate than the out-links proposed.

It is important to observe that the proposed approach performs very significantly better than a random process by 3 to 5 orders of magnitude. This shows that unsupervised approach to learning this dataset is effective. Nevertheless, the precision of the SOM for link prediction tasks is constrained by the discrete nature of the mapping space. In general, by increasing the size of the SOM, this will enhance the precision of the prediction at the cost of additional computational time.

4 Conclusions

Self-Organizing Maps are popularly applied to many data mining tasks for the purpose of clustering, dimension reduction, and visualization. This paper proposes the utilization of a recently developed self-organizing map approach, which is capable of mapping graph structured data, for the purpose of link prediction for "orphaned" documents (the documents which do not have any in-links or out-links yet as they are introduced newly to the Web) in an interlinked domain, Wikipedia. The approach has been applied to a relatively large collection of documents from Wikipedia. It was shown that the approach provides two main alternatives to predict both, incoming and outgoing links for any given "orphaned" document. Some indicative results were obtained by training some relatively small maps. It was shown that the links predicted are substantially better than a random process. Hence, it can be assumed that the accuracy of the prediction will increase with the size of the network. The training of larger networks is left as a future task.

Acknowledgment. This work has received financial support from the Australian Research Council through Discovery Project grant DP0774168 (2007 - 2009).

References

1. Kohonen, T.: Self-Organizing Maps. Springer Series in Information Sciences, vol. 30. Springer, Heidelberg (1995)
2. Hagenbuchner, M., Tsoi, A., Sperduti, A., Kc, M.: Efficient clustering of structured documents using graph self-organizing maps. In: Comparative Evaluation of XML Information Retrieval Systems, pp. 207–221. Springer, Berlin (2008)

3. Scarselli, F., Gori, M., Tsoi, A., Hagenbuchner, M., Monfardini, G.: Computational capabilities of graph neural networks. IEEE Transactions on Neural Networks 20(1), 81–102 (2009)
4. Hagenbuchner, M., Sperduti, A., Tsoi, A.: A self-organizing map for adaptive processing of structured data. IEEE Transactions on Neural Networks 14(3), 491–505 (2003)
5. Hagenbuchner, M., Sperduti, A., Tsoi, A.: Graph self-organizing maps for cyclic and unbounded graphs. Neurocomputing 72(7-9), 1419–1430 (2009)
6. Hagenbuchner, M., Zhang, S., Tsoi, A., Sperduti, A.: Projection of undirected and non-positional graphs using self organizing maps. In: European Symposium on Artificial Neural Networks - Advances in Computational Intelligence and Learning, April 22-24 (2009)
7. Tsoi, A., Hagenbuchner, M., Scarselli, F.: Computing customized page ranks. ACM Transactions on Internet Technology 6(4), 381–414 (2006)

Overview of the 2009 QA Track: Towards a Common Task for QA, Focused IR and Automatic Summarization Systems

Veronique Moriceau[1], Eric SanJuan[2], Xavier Tannier[1], and Patrice Bellot[2]

[1] LIMSI-CNRS, University Paris-Sud 11, France
{moriceau,Xavier.Tannier}@limsi.fr
[2] LIA, Université d'Avignon et des Pays de Vaucluse, France
{patrice.bellot,eric.sanjuan}@univ-avignon.fr

Abstract. QA@INEX aims to evaluate a complex question-answering task. In such a task, the set of questions is composed of factoid, precise questions that expect short answers, as well as more complex questions that can be answered by several sentences or by an aggregation of texts from different documents. Question-answering, XML/passage retrieval and automatic summarization are combined in order to get closer to real information needs. This paper presents the groundwork carried out in 2009 to determine the tasks and a novel evaluation methodology that will be used in 2010.

1 Introduction

Evaluation campaigns for Question-Answering (QA) systems (for example, TREC, CLEF, etc.) aim at evaluating systems that retrieve precise answers rather than documents in response to a question. The question test sets are generally composed of factoid questions (questions which require a single precise answer to be found in the document collection) and sometimes of more complex questions (list, "how" and "why" questions). Complex questions introduced in these campaigns expect a phrase (or a sentence) as an answer for "how" and "why" questions and a set of distinct short items to be found in the whole collection for list questions.

Recent QA evaluation campaigns deal with result aggregation [1,2]. This corresponds to an important user need in several manners:

- Compiling different elements scattered in the collection into a single list of items;
- Finding several valid answers to a single question (*"Who is Nicolas Sarkozy?"* leads to "French president", "former french interior minister", "Carla Bruni's husband", etc.);
- Gathering different answers with different restrictions: temporal (*"Who is the French president?"*: "Jacques Chirac" from 1995 to 2007, "Nicolas Sarkozy" from 2007), spatial or others.

S. Geva, J. Kamps, and A. Trotman (Eds.): INEX 2009, LNCS 6203, pp. 355–365, 2010.
© Springer-Verlag Berlin Heidelberg 2010

This is also the case of more complex questions that have been less studied in details so far: see [3] for an overview, [4,5] for "why" questions and [6] for opinion questions. Questions concerning procedures (in short, "how" questions), reasons ("why") or opinions can hardly find a complete answer in a single part of a document. For example, concerning opinion questions, a QA system should be able to locate opinions in documents and to produce or generate a synthetic "answer" in a suitable way.

The INEX 2009 QA@INEX track aims to compare the performance of QA, XML/passage retrieval and automatic summarization systems on an encyclopedic resource (Wikipedia). The track considers two types of questions: factoid questions and more complex questions whose answers require the aggregation of several passages.

After a brief overview of some other evaluation campaigns (Section 2) and traditional evaluation metrics (Section 3), new methods are proposed in order to compare QA and focused IR systems when a short answer is required (Section 4), as well as QA and summarization systems when aggregated results are expected (Section 5). Finally, in Section 6 we present the tools and resources made available in 2009 and give the timeline for 2010.

2   Overview of Question-Answering Evaluation Campaigns

2.1   Original QA Task in TREC

The first TREC evaluation campaign in questions-answering took place in 1999 [7]. TREC-QA successive tracks evolved over the years to explore new issues. In TREC-8, the participants had to provide 50-bytes or 250-bytes document passages containing answers to some factoid questions. For TREC-9 [8], questions were derived from real user questions selected from Encarta log and from Excite log and a set of variants was proposed in order to study the impact of formulating questions in different ways. For TREC-2001 [9], required passage size of answers was reduced to 50-bytes and questions, extracted from MSNSearch and AskJeeves, were no more guaranteed to have an answer in the collection. More important: list questions were introduced. For this kind of questions, the answers should be mined in several documents and the questions specified the number of instances of items to be retrieved. Context questions were introduced for TREC-2001 as some sets of questions related to each other. For the TREC-2002 question answering track, systems were to return exact answers rather than text passages and were required to provide only one answer to each question. The questions were ranked by the system's confidence [10]. In TREC-2003, definition questions (*e.g.* "*Who is Colin Powell?*") appeared and factoid, list and definition questions were combined in a single task. In TREC-2004, more difficult questions were introduced including temporal constraints, more anaphora and references to previous questions.

In QA@INEX, we mainly consider factoid and definition questions like in TREC-2003. Questions are related to Ad-hoc INEX topics and systems are asked to return passages like in TREC 2001 but each question is guaranteed to have at least one answer. In INEX, list questions are mainly considered in the XML Entity Ranking track (XER). In the QA track, likewise in TREC-2003 we are interested in combining the different types of questions in a single task.

Document collection has fluctuated over the years but it has always been comprised of newspaper and of newswire articles: in the beginning a subset of the TIPSTER corpora, then the AQUAINT Corpus of English News Text[1]. An encyclopedic resource like the Wikipedia was not considered until now.

In parallel of TREC, some other evaluations have taken place. Cross-Language Evaluation Forum (CLEF) introduced its first QA evaluation as a pilot track in 2003 and was centered on European Languages by proposing both monolingual and multilingual tasks. The NTCIR QA campaigns were TREC and CLEF equivalents for Asian languages and Technolangue-EQUER for French [11]. For now, in INEX QA we only consider English Wikipedia , but opening the task to other languages could be easily done.

2.2   Opinion QA Task in TAC

The Question-Answering TREC track last ran in 2007 [12]. However, in recent years, TREC evaluations have led to some new tracks linked to it: Blog Track for exploring seeking behavior in the blogosphere and Entity Track for performing entity-related search on Web data. In the same period, NIST encouraged the use of documents extracted from blogs instead of newspaper articles, opinion mining and links with automatic summarization. This led to some new evaluation campaigns such as TAC (*Text Analysis Conference*)[2]. The aim of the opinion QA task in TAC [13] is to accurately answer questions about opinions expressed in a set of documents (for ex. *"Why do people like Trader Joe's?"*) and to return different aspects of opinion (holder, target and support) in respect of a particular polarity. The answers have to be either named-entities (rigid questions) or complex explanatory answers (squishy questions). The tests were realized on the Blog06 corpus [14] which is composed of 3.2 million texts from more than 100,000 blogs. In 2008, there were 50 series of questions where a serie was composed of 2 to 4 questions and was about a specified target (a person, an organization, a product...). The evaluation of squishy list questions was conducted by employing a pyramidal approach [15].

In QA@INEX we consider both rigid and squishy questions and we link in a single task QA and automatic summarization.

In 2008, 9 teams participated in the opinion QA task in TAC. For the 90 sets of rigid type questions, the best system achieved an F-score of 0.156. In comparison, the scoring of the manual reference was 0.559. The scores of the other 8 teams in 2008 ranged between 0.131 and 0.011. For the series of squishy questions, evaluation of the task was done using the pyramidal method.

---

[1] LDC catalog #LDC93T3A and #LDC2002T31.
[2] http://www.nist.gov/tac/

3   Overview of Traditional Metrics

In current evaluation campaigns, the results of QA systems are presented as a ranked-list of answers (generally between 3 and 5) together with an explanation passage or element involving the answer (supporting text).

Answer assessment generally involves a gradual measure with the following possibilities [16,17]:

- *Incorrect*: the answer is false.
- *Inexact*: the answer is correct but the segmentation is not: either the answer contains more or less than the ideal string.
- *Not supported*: the short answer is correct but the supporting text does not validate it.
- *Locally correct*: the answer is validated by a local passage, but when looking at other documents, it turns out that it is not correct.
- *globally correct*: the answer is correct and validated.

The main global measures are then [18]:

- *accuracy* (rate of good answers at rank 1)
- *MRR* (Mean Reciprocal Rank), the average rank of the first good answer.

Concerning lists, at TREC QA, a system's response to a list question was scored using instance precision (IP) and instance recall (IR) based on the list of known instances of expected items. Let S be the the number of known instances, D be the number of correct, distinct responses returned by the system, and N be the total number of responses returned by the system. Then $IP = D/N$ and $IR = D/S$. Precision and recall were then combined into a F-measure [19].

The evaluation approach of factual questions that we propose in §4 relies on support passages. Locally correct answers are accepted as correct meanwhile not supported answers are rejected. Moreover we consider sets of answers instead of ranked lists.

Complex questions are evaluated in TAC and NTCIR by using the nugget pyramid human-in-the-loop evaluation [20]. A "nugget" is the minimum unit of correct information in answer sentences. It is a number between 0 and 1, depending on how many assessor considered this specific unit was important regarding the question. The value of an answer depends on the number of nuggets it collects and on its length.

Concerning both short and long answers, we propose new evaluation procedures aimed at being common to IR and QA systems and to get closer to the human need.

4   Proposed Evaluation of Short Answers

Short parts of text (one or a very few words) are the most usual way to answer factual questions in so-called question-answering systems. Mostly, these answers are:

- Named Entities (NE: person, date, number):
  When was the storming of the bastille → *14 July 1789.*
- other Entities (objects not included in a NE type):
  Which rocket is handled by Arianespace? → *Ariane*;
  What can be installed on a computer? → *a program, a plant.*
- short nominal phrases, often representing a definition:
  Who was Kurt Cobain? → *the leader of Nirvana*;
  What is Linux? → *an operating system.*

These questions are the classical ("basic") set that state-of-the-art QA systems are expected to answer quite well.

For QA@INEX, participants need to provide:

- A small ordered set (10) of non overlapping XML elements or passages that contains a possible answer to the question.
- For each element or passage, the position of the answer found by the system in the passage.

Answers are evaluated by computing their distance to the real answer. This evaluation methodology differs from traditional QA campaigns, where a short answer must be provided besides the supporting passage. This is a major difference in terms of metrics used to rank the participating systems.

In traditional campaigns, an important technical issue for QA systems is the boundaries of the short answer in the passage. In the quite simple question *"Who is Javier Solana?"*, the following passage would be relevant:

Javier Solana, the Secretary General of NATO, has just announced that the bombing of Yugoslavia may start as soon as the next few hours.

The correct short answer is *the Secretary General of NATO*. This answer is *full*. A system answering only "the Secretary General" (skipping "of NATO") as a short answer would be penalized for its *incomplete* (or *inexact*) answer [17].

However, this metric does not correspond to a real user need. In an end-user QA application, the obvious way to exhibit the answer is to point directly towards it into the supporting text (by highlighting the text for example). In this situation, the user does not need a perfect segmentation of the answer, but rather a good entry point inside the text. He/she is able to estimate the full answer by him/herself, by reading the text surrounding the entry point (as shown by Figure 1).

For this reason, we suggest to assess an answer not through the full/incomplete paradigm, but rather by **the distance between the indicated answer entry point and the real one**.

This new way to evaluate QA systems has an interesting side effect: it allows focused IR systems to participate in this task using the same evaluation, even if they are unable to extract a short answer or if they have very basic techniques to do so. These systems may simply provide the most relevant and short extracted passages they retrieve, and set an entry point wherever they can in this text. For the first time, QA, XML retrieval and other focused IR systems can participate to the same campaign.

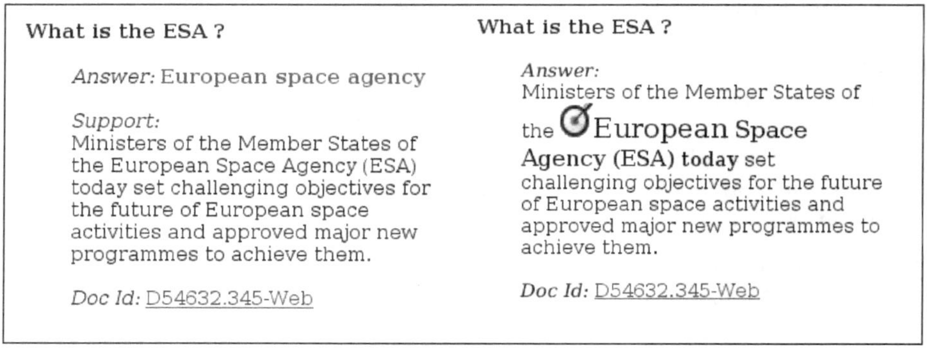

Fig. 1. A traditional QA answer (left), and an INEX-like answer (right)

5   Proposed Evaluation of Long Answers

INEX has a thorough experience in evaluating focused retrieval systems, however the QA "long answer" subtask is new in this context.

Following the first edition of Text Analysis Conference (TAC)[3], that brings together QA and automatic summarization, the idea here is to propose a common task that can be processed by three different kinds of systems: QA systems providing list of answers, automatic summarization systems by extraction and focused IR systems.

In this QA task, answers have to be built by aggregation of several passages from different documents on the Wikipedia. The questions themselves can be the same as in the short answer task. Let us consider again the previous example "Who is Kurt Cobain?". The difference with the short answer task is that here we require a short readable abstract of all the information in the Wikipedia related to this question. In this example, we do not expect only a phrase like "the leader of Nirvana", but also, for example, a short bibliography, important dates or a list of events related to Kurt Cobain.

The maximal length of the abstract being fixed, the systems have to make a selection of the most relevant information. Standard QA systems can produce a list of answers with their support passages. Focused IR systems can return the list of the most relevant XML elements. Note that in this task, IR systems that only retrieve entire documents are strongly handicapped, except if they are combined with automatic summarization systems that build an abstract of the most relevant documents.

Two main qualities of the resulting abstracts need to be evaluated: readability and informative content.

The readability and coherence are evaluated according to "the last point of interest" in the answer which is the counterpart of the "best entry point" in INEX ad-hoc task. It requires a human evaluation where the assessor indicates where

---

[3] http://www.nist.gov/tac/publications/2008/

he misses the point of the answers because of highly incoherent grammatical structures, unsolved anaphora, or redundant passages.

The informative content of the answer has to be evaluated according to the way they overlap with relevant passages that will be assessed by participants as in the INEX ad-hoc task. For that we plan to apply recent results on automatic summary evaluation based on the source text. Given a list of relevant passages, these passages can be whole Wikipedia articles. We intend to compare the word distributions in these passages with the word distribution in the long answer following the experiment in [21] done on TAC 2008 automatic summarization evaluation data. This allows to directly evaluate summaries based on a selection of relevant passages without requiring reference summaries written by experts as in TAC. Indeed, such manual summaries based on such large corpus as the Wikipedia would be very difficult to produce. Therefore, this long answer task is a first tentative of evaluating summarization tools on large data.

Given a set R of relevant passages and a text T, let us denote by $p_X(w)$ the probability of finding a word w from the Wikipedia in $X \in \{R, T\}$. We use standard Dirichlet smoothing with default $\mu = 2500$ to estimate these probabilities over the whole corpus.

– Kullback Leibler divergence:

$$KL(p_T, p_R) = \sum_{w \in R \cup T} p_T(w) \times \log_2 \frac{p_T(w)}{p_R(w)}$$

– Jensen Shannon divergence:

$$JS(p_T, p_R) = \frac{1}{2}(KL(p_T, p_{T \cup R}) + KL(p_R, p_{T \cup R}))$$

Since all answers have to be extracted from the same INEX corpus, we can use smoothing methods that allow to avoid null probabilities. Therefore KL is well founded. JS allows to reduce the impact of smoothing parameters since it is always defined. In [21] this is the metric that obtained the best correlation scores with ROUGE semi automatic evaluations of abstracts used in DUC and TAC. However, since we can compute these probabilities by taking the INEX corpus as referential for the probabilistic space, KL metric should also perform well in this track.

6   General Time Line and Available Resources

2009 has been devoted to fix the tasks and the overall evaluation methodology based on the corpus, topics and qrels from INEX 2009 ad-hoc track. A first list of questions has been released for test. They all deal with 2009 INEX topics, hence answers are part of ad-hoc relevant passages.

We have annotated correct answers among passages for a subset of 100 short type answers and all long type ones. These resources along with the evaluation programs written in Perl are available for active participants. In order to facilitate submissions from Focused IR systems, a Perl program that converts a run in INEX ad-hoc submission FOL format into QA format is also available.

In 2010, we will use the same corpus from ad-hoc INEX 2009 task but without the YAGO annotations. The corpus is available on INEX website. For participants that want to avoid indexing the corpus themselves, we provide two index (with or without stemming) built using Indri engine[4] on LINUX/64 bits. The index sizes are between 11 and 13 Go, they can be both downloaded or used on-line. We also provide the compatible compressed archive of the collection (13 Go). This allows to retrieve passages from the corpus with few disk resources. Active participants will be invited to submit a new set of questions on the Wikipedia. An additional set of questions will be also proposed by organizers. Once these 2010 set of questions released, participants will have a short time period to submit the results of their systems. This period will be set in agreement with participants. Informative content and linguistic quality of answers will be evaluated by participants and organizers based on a short questionnaire.

6.1   Resources for Short Type Question Evaluation

Among the 151 short type questions, a reference set of answers found in the corpus is available for a subset of 100 questions. 31 of them are single entities (names, acronyms, dates or measures), 6 are short definitions and 63 are multiple. Among the multiple answers, 34 are of list type.

Given a question and a reference answer, the Perl evaluation program looks for the closest answer in the participant run. Answers are supposed to be exact short passages extracted from the corpus as they appear without the XML tags.

The scoring measure that we implemented is computed as follows:

$$\sum_{q \in Q} \sum_{a \in A_q} \min_{r \in R_q} pos(a, r)^2$$

where:

- Q is the set of short type questions.
- A_q is the set of reference answers for every $q \in Q$.
- R_q is the set of answers for $q \in Q$ in the participant run.
- $pos(a, r)$ gives the number of words between the offset indicated by the participant and the closest occurrence of a in r. It returns 500 if a is not found.

Participating systems have to minimize this score. Systems that systematically give multiple answers like focused IR systems are not penalized since only the best answers among a maximum of 10 will be considered. On the contrary, this type of ranking favours systems that look for all possible answers.

---

[4] http://www.lemurproject.org/indri/

6.2   Resources for Long Type Questions Evaluation

We use the Perl evaluation package FRESA (a FRamework for Evaluating Summaries Automatically)[22] developed by Juan Manuel Torres-Moreno at LIA, University of Avignon[5] to compute Jensen Shannon divergence on uni-grams and bi-grams. It uses Porter Stemmer and a stop word list. An adapted version is available for QA@INEX participants. This package works with the following datasets:

– For each topic evaluated in ad-hoc 2009 track we have extracted and concatenated the text of all relevant passages indicated in the qrels. This allows us to test the systems taking as questions the ad-hoc topic titles and to compare the resulting scores with measures used in ad-hoc focus, BiC, RiC and thorough sub-tasks.
– For each long type question related to some of the previous topics, we have selected all sub-passages with some relevant answer.

The average size of relevant texts in ad-hoc 2009 qrels is 21060.4 words per topic but with a huge standard deviation of 86117.4. When using this data and software, only topics and long question types with enough relevant text in the corpus (more than 5000 words) should be considered.

7   Conclusion

QA@INEX is offering an evaluation framework combining QA, passage retrieval and automatic summarizing by passage extraction. Its main features are the use of the Wikipedia as referential, its proximity with INEX ad-hoc task and the introduction of new evaluation metrics.

References

1. National Institute of Standards and Technology: Proceedings of the First Text Analysis Conference (TAC 2008). In: Proceedings of the First Text Analysis Conference (TAC 2008), Gaithersburg, Maryland, USA, National Institute of Standards and Technology (November 2008)
2. Mitamura, T., Nyberg, E., Shima, H., Kato, T., Mori, T., Lin, C.Y., Song, R., Lin, C.J., Sakai, T., Ji, D., Kando, N.: Overview of the ntcir-7 aclia tasks: Advanced cross-lingual information access. In: Proceedings of the 7th NTCIR Workshop Meeting on Evaluation of Information Access Technologies, Tokyo, Japan (December 2008)
3. Lee, Y.H., Lee, C.W., Sung, C.L., Tzou, M.T., Wang, C.C., Liu, S.H., Shih, C.W., Yang, P.Y., Hsu, W.L.: Complex Question Answering with ASQA at NTCIR 7 ACLIA. In: Proceeding of the 7th NTCIR Workshop Meeting, Tokyo, Japan, December 2008, pp. 70–76 (2008)

---

[5] juan-manuel.torres@univ-avignon.fr

4. Verberne, S., Boves, L., Oostdijk, N., Coppen, P.A.: Discourse-based answering of why-questions. Traitement Automatique des Langues, Discours et document: traitements automatiques 47(2), 21–41 (2007)
5. Verberne, S., Raaijmakers, S., Theijssen, D., Boves, L.: Learning to Rank Answers to Why-Questions. In: Proceedings of 9th Dutch-Belgian Information Retrieval Workshop (DIR 2009), pp. 34–41 (2009)
6. Yu, H., Hatzivassiloglou, V.: Towards Answering Opinion Questions: Separating Facts from Opinions and Identifying the Polarity of Opinion Sentences. In: Proceedings of 2003 Conference on Empirical Methods in Natural Language Processing (EMNLP), pp. 129–136 (2003)
7. Voorhees Ellen, M., Harman, D.K.: Overview of the eighth text retrieval conference (trec-8). In: The Eighth Text REtrieval Conference (TREC 8), pp. 1–24, NIST Special Publication 500-246 (1999)
8. Voorhees, E.M.: Overview of the trec-9 question answering track. In: The Ninth Text Retrieval Conference (TREC-9), pp. 71–80, NIST Special Publication 500-249 (2000)
9. Voorhees, E.M.: Overview of the trec 2001 question answering track. In: The Tenth Text Retrieval Conference (TREC 2001), pp. 42–50, NIST Special Publication 500-251 (2001)
10. Voorhees, E.M.: Evaluating the evaluation: a case study using the trec 2002 question answering track. In: Proceedings of the 2003 Conference of the North American Chapter of the Association for Computational Linguistics on Human Language Technology, Edmonton, Canada, vol. 1, pp. 181–188. Association for Computational Linguistics (2003), 1073479
11. Ayache, C., Grau, B., Vilnat, A.: Equer: the french evaluation campaign of question answering systems. In: Fifth International Conference on Language Resources and Evaluation (LREC 2006), Genoa, Italy (2006)
12. Dang, H., Kelly, D., Lin, J.: Overview of the TREC 2007 question answering track. In: Proc. of TREC (2007)
13. Dang, H.: Overview of the TAC 2008 Opinion Question Answering and Summarization Tasks. In: Proc. of the First Text Analysis Conference (2008)
14. Macdonald, C., Ounis, I.: The TREC Blogs06 collection: Creating and analysing a blog test collection. Department of Computer Science, University of Glasgow Tech Report TR-2006-224 (2006)
15. Nenkova, A., Passonneau, R.: Evaluating content selection in summarization: The pyramid method. In: Proceedings of HLT-NAACL (2004)
16. Dang, H., Lin, J., Kelly, D.: Overview of the TREC 2006 Question Answering Track. In: Voorhees, E.M., Buckland, L.P. (eds.) Proceedings of the Fifteenth Text REtrieval Conference (TREC), Gaithersburg, MD, National Institute of Standards and Technology, Department of Commerce, National Institute of Standards and Technology (2006)
17. Magnini, B., Vallin, A., Ayache, C., Erbach, G., Peas, A., de Rijke, M., Rocha, P., Simov, K., Sutcliffe, R.: Overview of the CLEF 2004 Multilingual Question Answering Track. In: Peters, C., Clough, P., Gonzalo, J., Jones, G.J.F., Kluck, M., Magnini, B. (eds.) CLEF 2004. LNCS, vol. 3491, pp. 371–391. Springer, Heidelberg (2005)
18. Voorhees, E.: The TREC question answering track. Journal of Natural Language Engineering 7, 361–378 (2001)

19. Voorhees, E.: Overview of the TREC 2004 question answering track. In: Voorhees, E.M., Buckland, L.P. (eds.) Proceedings of the Thirteenth Text REtrieval Conference (TREC), Gaithersburg, MD, USA, National Institute of Standards and Technology, Department of Commerce, National Institute of Standards and Technology. NIST Special Publication 500-261 (2004)
20. Lin, J., Demner-Fushman, D.: Will pyramids built of nuggets topple over? In: Proceedings of the main conference on Human Language Technology Conference of the North American Chapter of the Association of Computational Linguistics (2006)
21. Louis, A., Nenkova, A.: Performance confidence estimation for automatic summarization. In: EACL, The Association for Computer Linguistics, pp. 541–548 (2009)
22. Torres-Moreno, J.M., Saggion, H., da Cunha, I., Velàzquez-Morales, P., SanJuan, E.: Evaluation automatique de résumés avec et sans références. In: TALN, ATALA - Association pour le Traitement Automatique des Langues (to appear, 2010)

Overview of the INEX 2009 XML Mining Track: Clustering and Classification of XML Documents

Richi Nayak[1], Christopher M. De Vries[1], Sangeetha Kutty[1], Shlomo Geva[1], Ludovic Denoyer[2], and Patrick Gallinari[2]

[1] Faculty of Science and Technology
Queensland University of Technology
GPO Box 2434, Brisbane Qld 4001, Australia
{r.nayak,christopher.devries,s.kutty,s.geva}@qut.edu.au
[2] University Pierre et Marie Curie
LIP6 – 104 avenue du président Kennedy
75016 Paris - France
{Ludovic.denoyer,Patrick.gallinari}@lip6.fr

Abstract. This report explains the objectives, datasets and evaluation criteria of both the clustering and classification tasks set in the INEX 2009 XML Mining track. The report also describes the approaches and results obtained by the different participants.

Keywords: XML document mining, INEX, Wikipedia, Structure and content, Clustering, Classification.

1 Introduction

The XML Document Mining track was launched for exploring two main ideas: (1) identifying key problems and new challenges of the emerging field of mining semi-structured documents, and (2) studying and assessing the potential of Machine Learning (ML) techniques for dealing with generic ML tasks in the structured domain i.e. classification and clustering of semi structured documents. This track has run for five editions during INEX 2005, 2006, 2007, 2008 and 2009. The four first editions have been summarized in [2, 3, 4] and we focus here on the 2009 edition.

INEX 2009 included two tasks in the XML Mining track: (1) unsupervised clustering task and (2) semi-supervised classification task where documents are organized in a graph. The clustering task requires the participants to group the documents into clusters without any knowledge of cluster labels using an unsupervised learning algorithm. On the other hand, the classification task requires the participants to label the documents in the dataset into known classes using a supervised learning algorithm and a training set. This report gives the details of clustering and classifications tasks.

S. Geva, J. Kamps, and A. Trotman (Eds.): INEX 2009, LNCS 6203, pp. 366–378, 2010.

2   The Clustering Track

In the last decade, we have observed a proliferation of approaches for clustering XML documents based on their structure and content [9,12]. There have been many approaches developed for diverse application domains. Many applications require data objects to be grouped by similarity of content, tags, paths, structure and semantics. The clustering task in INEX 2009 evaluates clustering approaches in the context of XML information retrieval.

The INEX 2009 clustering task is different from the previous years due to its incorporation of a different evaluation strategy. The clustering task explicitly tests the Jardine and van Rijsbergen cluster hypothesis (1971) [8], which states that documents that are clustered together have a similar relevance to a given query. It uses manual query assessments from the INEX 2009 Ad Hoc track. If the cluster hypothesis holds true, and if suitable clustering can be achieved, then a clustering solution will minimise the number of clusters that need to be searched to satisfy any given query. There are important practical reasons for performing collection selection on a very large corpus. If only a small fraction of clusters (hence documents) need to be searched, then the throughput of an information retrieval system will be greatly improved. INEX 2009 clustering task provides an evaluation forum to measure the performance of clustering methods for collection selection on a huge scale test collection. The collection consists of a set of documents, their labels, a set of information needs (queries), and the answers to those information needs.

2.1   Corpus

The INEX XML Wikipedia collection is used as a dataset in this task. This 60 Gigabyte collection contains 2.7 million English Wikipedia XML documents. The XML mark-up includes explicit tagging of named entities and document structure. In order to enable participation with minimal overheads in data-preparation the collection was pre-processed to provide various representations of the documents such as a bag-of-words representation of terms and frequent phrases in a document, frequencies of various XML structures in the form of trees, links, named entities, etc. These various collection representations made this task a lightweight task that required the participants to submit clustering solutions without worrying about pre-processing this huge data collection.

There are a total of 1,970,515 terms after stemming, stopping, and eliminating terms that occur in a single document for this collection. There are 1,900,075 unique terms that appear more than once enclosed in entity tags. There are 5213 unique entity tags in the collection. There are a total of 110,766,016 unique links in the collection. There are a total of 348,552 categories that contain all documents except for a 118,685 document subset containing no category information. These categories are derived by using the YAGO ontology [16]. The YAGO categories appear to follow a power law distribution as shown in Figure 1. Distribution of documents in the top-10 cluster category is shown in Table 1.

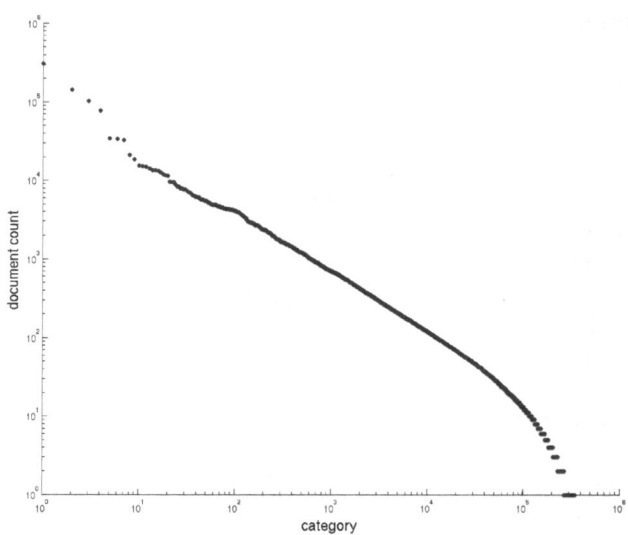

Fig. 1. The YAGO Category Distribution

Table 1. Top-10 Category Distribution

Category	Documents
Living people	307304
All disambiguation pages	143463
Articles with invalid date parameter in template	77659
All orphaned articles	34612
All articles to be expanded	33810
Year of birth missing (living people)	32499
All articles lacking sources	21084
Human name disambiguation pages	18652
United States articles missing geocoordinate data	15363
IUCN Red List least concern species	15241

A subset of the collection, containing about 50,000 documents (of the entire INEX 2009 corpus) was also used in the task to evaluate the categories labels results only, for teams that were unable to process such a large data collection.

2.2   Tasks and Evaluation Measures

The task was to utilize unsupervised classification techniques to group the documents into clusters. Participants were asked to submit multiple clustering solutions containing different numbers of clusters such as 100, 500, 1000, 2500, 5000 and 10000. The clustering solutions are evaluated by two means. Firstly, we utilise the *classes-to-clusters evaluation* which assumes that the classification of the documents in a sample is known (i.e., each document has a class label). Then any clustering of these documents can be evaluated with respect to this predefined classification. It is important to note that the class labels are not used in the process of clustering, but only for the purpose of evaluation of the clustering results.

The standard criterion of purity is used to determine the quality of clusters. These evaluation results were provided online and ongoing, starting from mid-October. Entropy and F-Score were not used in evaluation. The reason behind was that a document in the corpus maps to more than one category. Due to multi labels that a document can have, it was possible to obtain higher value of Entropy and F-Score than the ideal solution. Purity measures the extent to which each cluster contains documents primarily from one class. Each cluster is assigned with the class label of the majority of documents in it. The macro and micro purity of a clustering solution *cs* is obtained as a weighted sum of the individual cluster purity. In general, larger the value of purity, better the clustering solution is.

$$\text{Purity (k)} = \frac{\text{Number of documents with the majority label in cluster k}}{\text{Number of documents in cluster k}}$$

$$\text{Micro-Purity (cs)} = \frac{\sum_{k=0}^{n} \text{Purity(k)} * \text{TotalFoundByClass(k)}}{\sum_{k=0}^{n} \text{TotalFoundByClass(k)}}$$

$$\text{Macro-Purity (cs)} = \frac{\sum_{k=0}^{n} \text{Purity(k)}}{\text{Total Number of Categories}}$$

The clustering solutions are also evaluated to determine the quality of cluster relative to the optimal collection selection goal, given a set of queries. Better clustering solutions in this context will tend to (on average) group together relevant results for (previously unseen) ad-hoc queries. Real Ad-hoc retrieval queries and their manual assessment results are utilised in this evaluation. This novel approach evaluates the clustering solutions relative to a very specific objective - clustering a large document collection in an optimal manner in order to satisfy queries while minimising the search space. The Normalised Cumulative Gain is used to calculate the score of the best possible collection selection according to a given clustering solution of *n* number of clusters. Better the score when the query result set contains more cohesive clusters. The cumulative gain of a cluster (CCG) is calculated by counting the number of relevant documents in a cluster, *c*, for a topic, *t*, where *c* is the set of documents in a cluster and *t* is the set of relevant documents for a topic.

$$CCG(c,t) = |c \cap t|$$

For a clustering solution for a given topic, a (sorted) vector CG is created representing each cluster by its CCG value. Clusters containing no relevant documents are represented by a value of zero. The cumulated gain for the vector CG is calculated and then normalized on the ideal gain vector. Each clustering solution *cs* is scored for how well it has split the relevant set into clusters using *CCG* for the topic *t*.

$$SplitScore(t, cs) = \frac{\sum^{|CG|} \text{cumsum}(CG)}{nr^2}$$

nr = Number of relevant documents in the returned result set for the topic t.

A worst possible split is assumed to place each relevant document in a distinct cluster. Let CG1 be a vector that contains the cumulative gain of every cluster with a document each.

$$MinSplitScore(t, cs) = \frac{\sum^{|CG1|} \text{cumsum}(CG1)}{nr^2}$$

The normalized cluster cumulative gain (nCCG) for a given topic *t* and a clustering solution *cs* is given by,

$$nCCG(t, cs) = \frac{SplitScore(t, cs) - MinSplitScore(t, cs)}{1 - MinSplitScore(t, cs)}$$

The mean and the standard deviation of the nCCG score over all the topics for a clustering solution *cs* are then calculated. n is total number of topics.

$$\text{Mean nCCG(cs)} = \frac{\sum_{t=0}^{n} nCCG(t, cs)}{n}$$

$$\text{Std Dev nCCG(cs)} = \frac{\sum_{t=0}^{n} (nCCG(t, cs) - \text{Mean nCCG(cs)})^2}{n}$$

A total of 68 topics were used to evaluate the quality of clusters generated on the full set of collection of about 2.7 million documents. A total of 52 topics were used to evaluate the quality of clusters generated on the subset of collection of about 50,000 documents. A total number of 4858 documents were found relevant by the manual assessors for the 68 topics. An average number of 71 documents were found relevant for a given topic by manual assessors. The nCCG value varies from 0 to 1.

2.3 Participants, Submissions and Evaluation

A total of six research teams have participated in the INEX 2009 clustering task. Two of them submitted the results for the subset data only. We briefly summarised the approaches employed by the participants.

Exploiting Index Pruning Methods for Clustering XML Collections [1]

[1] used Cover-Coefficient Based Clustering Methodology (C3M) to cluster the XML documents. C3M is a single-pass partitioning type clustering algorithm which measures the probability of selecting a document given a term that has been selected

from another document. As another approach, [1] adapted term-centric and document-centric index pruning techniques to obtain more compact representations of the documents. Documents are clustered with these reduced representations for various pruning levels, again using C3M algorithm. All of the experiments are executed on the subset of INEX 2009 corpus including 50K documents.

Clustering with Random Indexing K-tree and XML Structure [5]
The Random Indexing (RI) K-tree has been used to cluster the entire 2,666,190 XML documents in the INEX 2009 Wikipedia collection. Clusters were created as close as possible to the 100, 500, 1000, 5000 and 10000 clusters required for evaluation. The algorithm produces clusters of many sizes in a single pass. The desired clustering granularity is selected by choosing a particular level in the tree. In the context of document representation, topology preserving dimensionality reduction is preserving document meaning – or at least this is the conjecture which the team tests here Document structure has been represented by using a bag of words and a bag of tags representation derived from the semantic markup in the INEX 2009 collection. The term frequencies were weighted with BM25 where K1 = 2 and b = 0.75. The tag frequencies were not weighted.

Exploiting Semantic Tags in XML Clustering [10]
This technique combines the structure and content of XML documents for clustering. Each XML document in the subset collection is parsed and modeled as a rooted labeled ordered *document tree*. A constrained frequent subtree mining algorithm is then applied to extract the common structural features from these document trees in the corpus. Using the common structural features, the corresponding content features of the XML documents are extracted and represented in a Vector Space Model (VSM). The term frequencies in the VSM model were weighted with both TF-IDF and BM25. There were 100, 500 and 1000 clusters created for evaluation.

Performance of K-Star at the INEX'09 Clustering Task [13]
The employed approach was quite simple and focused on high scalability. The team used a modified version of the Star clustering method which automatically obtains the number of clusters. In each iteration, this clustering method brings together all those items whose similarity value is higher than a given threshold T, which is typically assumed to be the similarity average of the whole document collection and, therefore, the clustering method "discover" the number of clusters by its own. The run submitted to the INEX clustering task split the complete document collection into small subsets which are clustered with the above mentioned clustering method.

Evaluation

Figure 2, Figure 3 and Figure 4 show the performance of various teams in the clustering task. The legends are formatted in the following fashion, [metric] – [institution] (username) [method].

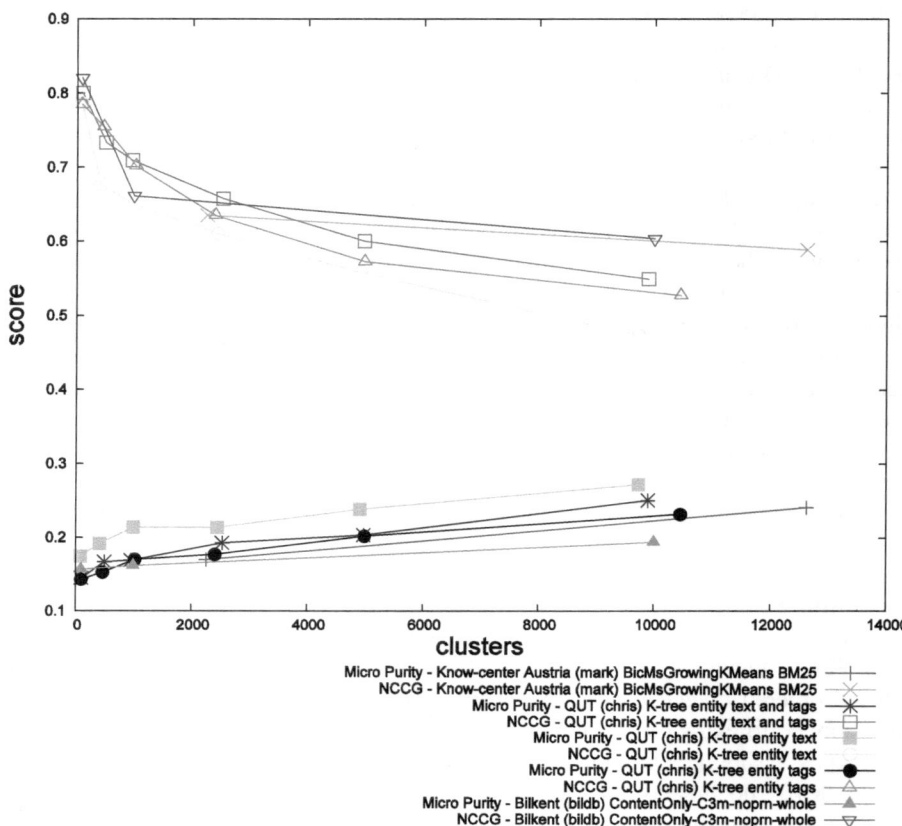

Fig. 2. Purity and NCCG performance of different teams using the entire dataset

3   The Classification Track

Dealing with XML document collections is a particularly challenging task for ML and IR. XML documents are defined by their logical structure and their content (hence the name semi-structured data). Moreover, in a large majority of cases (Web collections for example), XML documents collections are also structured by links between documents (hyperlinks for example). These links can be of different types and correspond to different information: for example, one collection can provide hierarchical links, hyperlinks, citations, ... Most models developed in the field of XML categorization simultaneously use the content information and the internal structure of XML documents (see [2] and [3] for a list of models) but they rarely use the external structure of the collection i.e the links between documents. Some methods using both content and links have been proposed in [4].

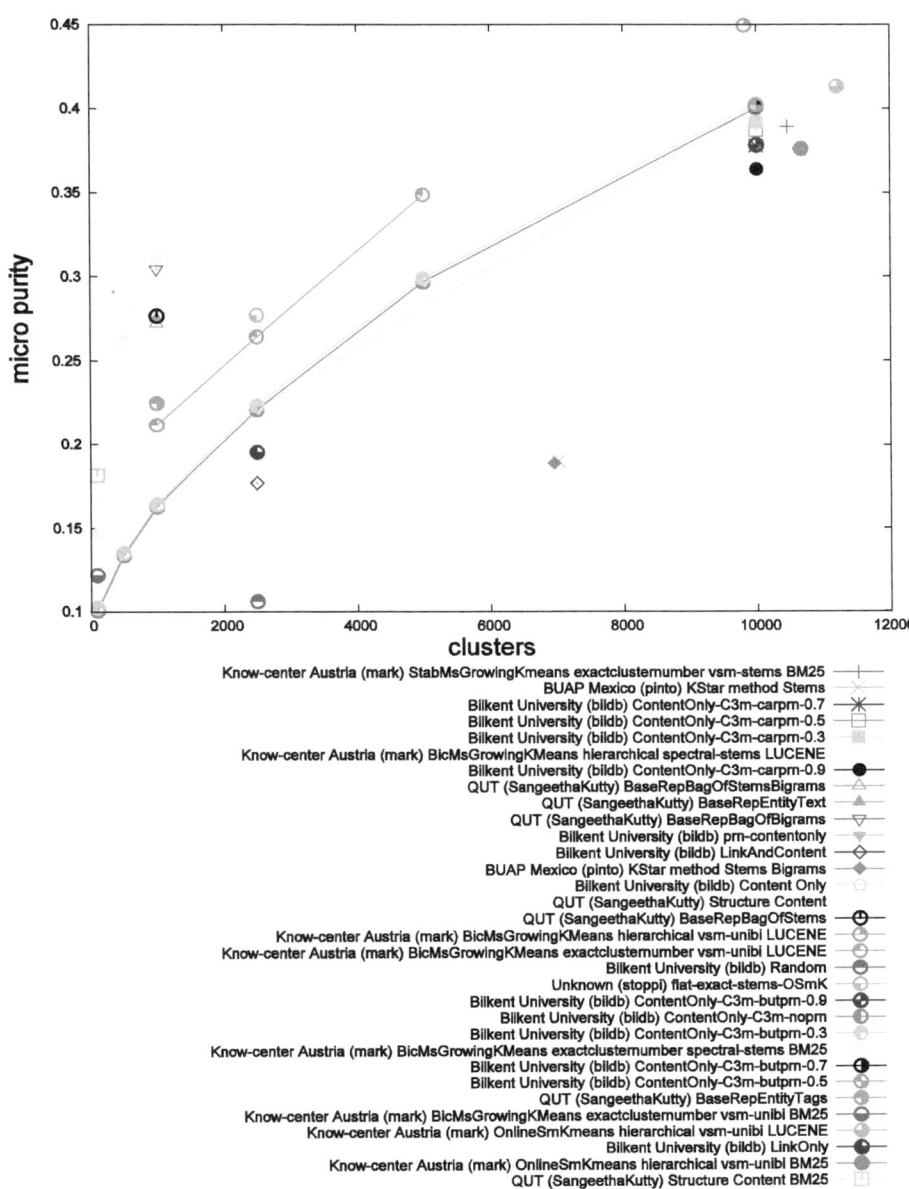

Fig. 3. Purity performance of different teams using the subset data

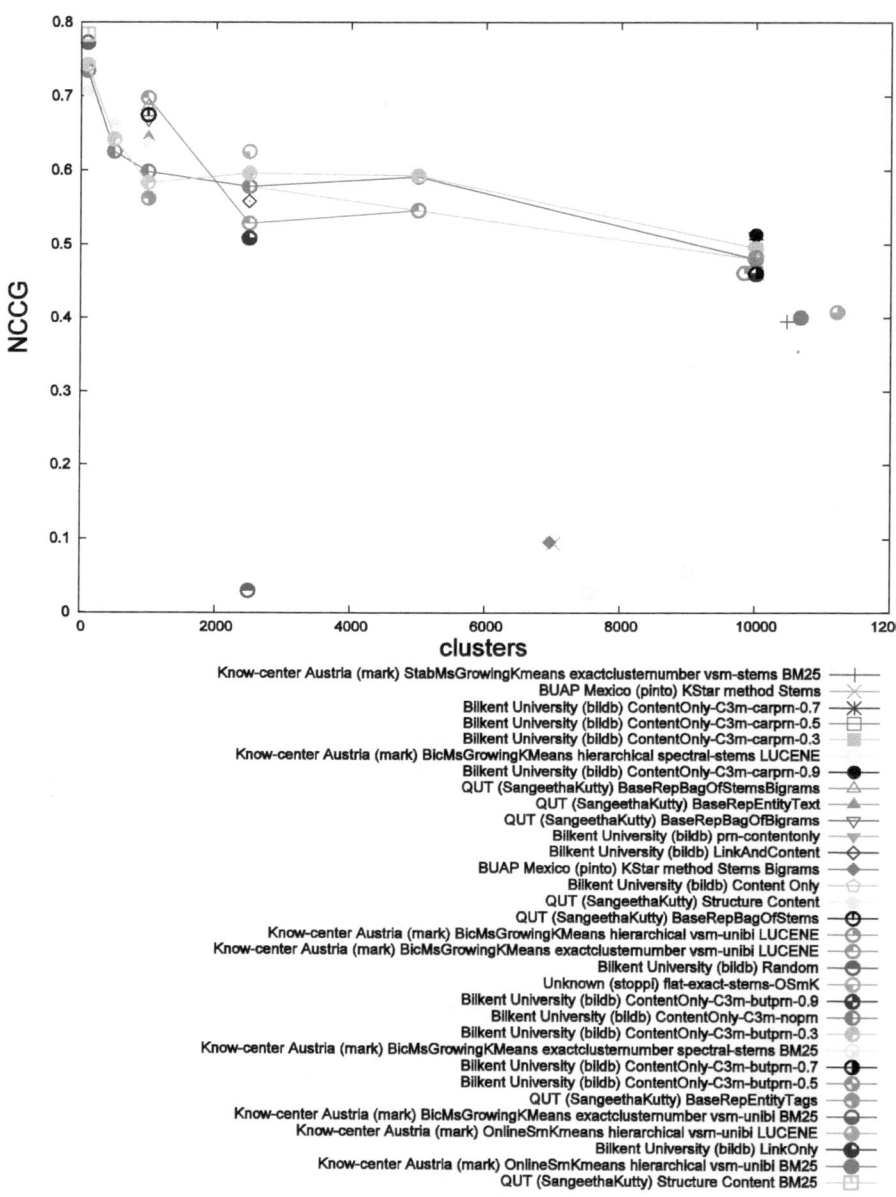

Fig. 4. NCCG performance of different teams using the subset data

The XML Classification Task focuses on the problem of learning to classify documents organized in a graph of documents. Unlike the 2008 track, we consider here the problem of *Multiple labels classification* where a document belongs to one or many different categories. This task considers a transductive context where, during

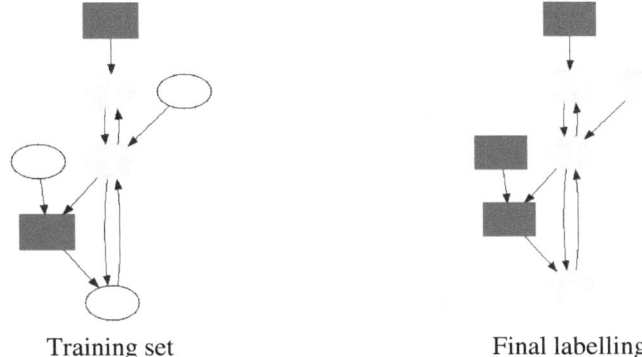

Training set   Final labelling

Fig. 5. The supervised classification task. Colors/Shapes correspond to categories, circle/white nodes are unlabeled nodes. Note that in this track, documents may belong to many categories.

the training phase, the whole graph of documents is known but the labels of only a part of them are given to the participants (Figure 5).

3.1 Corpus

The corpus provided is a subset of the *INEX 2009 Corpus*. We have extracted a set of 54,889 documents and the links between these documents. These links correspond to the links provided by the authors of the Wikipedia articles. The documents have been transformed into TF-IDF vectors by the organizers. The corpus thus corresponds to a set of 54,889 vectors of dimension 186,723. The documents belong to 39 categories that correspond to 39 Wikipedia portals. We have provided the labels of 20 % of the documents. The corpus is composed of 4,554,203 directed links that correspond to hyperlinks between the documents of the corpus. Each document has 84.1 links on average.

Number of documents	54,889
Number of training documents	11,028
Number of test documents	43,861
Number of categories	39
Number of links	4,554,203
Number of distinct words	186,723

3.2 Evaluation Measures

In order to evaluate the submissions of the participants, we have used different measures. The first set of measures are computed over each category and then averaged over the categories (using a micro or a macro average):

- *Accuracy* (ACC) corresponds to the classification error. Note that a system that returns zero irrelevant categories for each document has a quite good accuracy.
- *F1 score* (F1) corresponds to the classical F1 measure and measures the ability of a system to find the relevant categories.

The second set of measures are computed over each document and then averaged over the documents.

- *Average precision* (APR) corresponds to the Average Precision computed over the list of categories returned for each document. It measures the ability of a system to rank correctly the relevant categories. This measure is based on a ranking score of each category for each document.

3.3 Participants and Submissions

Five different teams have participated to the track. They have submitted different runs and we present here only the best results obtained by each team. Note that, due to additional experiments made after the submission deadline, the results presented here and the results presented in the participants' articles can be different.

Team	Micro ACC	Macro ACC	Micro F1	Macro F1	APR
University of Wollongong	92.5	94.6	51.2	47.9	68
University of Peking	94.7	96.2	51.8	48	70.2
XEROX Research Center	96.3	97.4	**60**	**57.1**	67.8
University of Saint Etienne	96.2	97.4	56.4	53	68.5
University of Granada	67.8	75.4	26.2	25.3	**72.9**

3.4 Summary of the Methods

We give here a brief description of the methods submitted by the participants. Please refer to the participants articles for a detailed description of the methods and for the final results obtained by the different teams.

Multi-label Wikipedia classification with textual and graph features [6]

This paper proposes to evaluate different classification methods used on both the textual features of the pages to classify, and also on graph features computed from the structure of the graph. These features include for example the mean centrality, the degree centrality, etc. Different classifiers have been tested to handle the multi-label problem.

Supervised Encoding of Graph-of-Graphs for Classification and Regression Problems [7]

This article proposes a novel method which aims at encoding graphs of graphs structures where data correspond to a graph of elements which are also composed of graphs. The graph to graph structure is described and then used as a classification model based on a back-propagation of the error through the different level of the nested structure.

UJM at INEX 2009 XML Mining Track [11]

The authors use different classification strategies based on a set of content features to handle the classification problem. They mainly compare different features selection methods and thresholding strategies.

Link-Based Text Classification Using Bayesian Networks [14]

The article presents a Bayesian network model that is able to handle both content and links between documents. The proposed model is an extension of the Naïve Bayes model to documents organized in a graph.

Extended VSM for XML Document Classification Using Frequent Subtrees [15]

The last paper proposes the structured link vector model which aims at modeling both the content and the structure of the documents in a vector. Mainly, the authors propose to insert into classical content-based features vectors information about the frequent XML subtrees and the links between documents.

4 Conclusion

The XML Mining track in INEX 2009 brought together researchers from Information Retrieval, Data Mining, Machine Learning and XML fields. The clustering task allowed participants to evaluate clustering methods against a real use case and with significant volumes of data. The task was designed to facilitate participation with minimal effort by providing not only raw data, but also pre-processed data which can be easily used by existing clustering software. The classification task allowed participant to explore algorithmic, theoretical and practical issues regarding the classification of interdependent XML documents.

Acknowledgments

We would like to thank all the participants for their efforts and hard work.

References

1. Altingovde, I., Atilgan, D., Ulusoy, O.: Exploiting Index Pruning Methods for Clustering XML Collections. In: Geva, S., Kamps, J., Trotman, A. (eds.) INEX 2009. LNCS, vol. 6203, pp. 379–386. Springer, Heidelberg (2010)
2. Denoyer, L., Gallinari, P.: Report on the XML Mining Track at Inex 2005 and Inex 2006. Categorization and Clustering of XML Documents 41(1), 79–90 (2007)
3. Denoyer, L., Gallinari, P.: Report on the XML Mining Track at Inex 2007. Categorization and Clustering of XML Documents 42(1), 22–28 (2008)
4. Denoyer, L., Gallinari, P.: Overview of the inex 2008 xml mining track. In: Geva, S., Kamps, J., Trotman, A. (eds.) INEX 2008. LNCS, vol. 5631, pp. 401–411. Springer, Heidelberg (2009)
5. De Vries, C., Geva, S., De Vine, L.: Clustering with Random Indexing K-tree and XML Structure. In: Geva, S., Kamps, J., Trotman, A. (eds.) INEX 2009. LNCS, vol. 6203, pp. 407–415. Springer, Heidelberg (2010)
6. Chidlovskii, B.: Multi-label Wikipedia classification with textual and graph features. In: Geva, S., Kamps, J., Trotman, A. (eds.) INEX 2009. LNCS, vol. 6203, pp. 387–396. Springer, Heidelberg (2010)
7. Hagenbuchner, M., Zhang, S., Scarselli, F., Chung Tsoi, A.: Supervised Encoding of Graph-of-Graphs for Classification and Regression Problems. In: Geva, S., Kamps, J., Trotman, A. (eds.) INEX 2009. LNCS, vol. 6203, pp. 449–461. Springer, Heidelberg (2010)
8. Jardine, N., van Rijsbergen, C.J.: The Use of Hierarchic Clustering in Information Retrieval. Inform. Stor. Retr. 7, 217–240 (1971)
9. Kutty, S., Nayak, R., Li, Y.: HCX: An Efficient Hybrid Clustering Approach for XML Documents. In: Proceedings of the ACM Document Engineering Symposium, Munich, Germany, pp. 94–97 (2009)
10. Kutty, S., Nayak, R., Li, Y.: Clustering XML documents using Multi-feature Model. In: Geva, S., Kamps, J., Trotman, A. (eds.) INEX 2009. LNCS, vol. 6203, pp. 416–425. Springer, Heidelberg (2010)
11. Largeron, C., Moulin, C., Gery, M.: UJM at INEX 2009 XML Mining Track. In: Geva, S., Kamps, J., Trotman, A. (eds.) INEX 2009. LNCS, vol. 6203, pp. 426–433. Springer, Heidelberg (2010)
12. Nayak, R.: XML Data Mining: Process and Applications. In: Song, M., Wu, Y.-F. (eds.) Hand-book of Research on Text and Web Mining Technologies, ch.15, pp. 249–272. Idea Group Inc., USA
13. Pinto, D., Tovar, M., Vilariño, D., Beltran, B., Salazar, H.: BUAP: Performance of K-Star at the INEX 2009 Clustering Task. In: Geva, S., Kamps, J., Trotman, A. (eds.) INEX 2009. LNCS, vol. 6203, pp. 434–440. Springer, Heidelberg (2010)
14. Romero, A.E., de Campos, M.L., Fernandez-Luna, J.M., Huete, J.F., Mase-gosa, A.R.: Link-based text calssification using Bayesian networks. In: Geva, S., Kamps, J., Trotman, A. (eds.) INEX 2009. LNCS, vol. 6203, pp. 397–406. Springer, Heidelberg (2010)
15. Yang, J., Wang, S.: Extended VSM for XML Document Classification using Frequent Subtrees. In: Geva, S., Kamps, J., Trotman, A. (eds.) INEX 2009. LNCS, vol. 6203, pp. 441–448. Springer, Heidelberg (2010)
16. Suchanek, F., Kasneci, G., Weikum, G.: YAGO: A Core of Semantic Knowledge Unifying WordNet and Wikipedia. In: WWW 2007 (2007)

Exploiting Index Pruning Methods for Clustering XML Collections

Ismail Sengor Altingovde, Duygu Atilgan, and Özgür Ulusoy

Department of Computer Engineering, Bilkent University, Ankara, Turkey
{ismaila,atilgan,oulusoy}@cs.bilkent.edu.tr

Abstract. In this paper, we first employ the well known Cover-Coefficient Based Clustering Methodology (C3M) for clustering XML documents. Next, we apply index pruning techniques from the literature to reduce the size of the document vectors. Our experiments show that for certain cases, it is possible to prune up to 70% of the collection (or, more specifically, underlying document vectors) and still generate a clustering structure that yields the same quality with that of the original collection, in terms of a set of evaluation metrics.

Keywords: Cover-coefficient based clustering, index pruning, XML.

1 Introduction

As the number and size of XML collections increase rapidly, there occurs the need to manage these collections efficiently and effectively. While there is still an ongoing research in this area, INEX XML Mining Track fulfills the need for an evaluation platform to compare the performance of several clustering methods on the same set of data. Within the Clustering task of XML Mining Track of INEX campaign, clustering methods are evaluated according to cluster quality measures on a real-world Wikipedia collection.

To this end, in the last few workshops, many different approaches are proposed which use structural, content-based and link-based features of XML documents. In INEX 2008, Kutty et al. [11] use both structure and content to cluster XML documents. They reduce the dimensionality of the content features by using only the content in frequent subtrees of an XML document. In another work, Zhang et al. [13] make use of the hyperlink structure between XML documents through an extension of a machine learning method based on the Self Organizing Maps for graphs. De Vries et al. [9] use K-Trees to cluster XML documents so that they can obtain clusters in good quality with a low complexity method. Lastly, Tran et al. [12] construct a latent semantic kernel to measure the similarity between content of the XML documents. However, before constructing the kernel, they apply a dimension reduction method based on the common structural information of XML documents to make the construction process less expensive. In all of these work mentioned above, not only the quality of the clusters but the efficiency of the clustering process is also taken into account.

S. Geva, J. Kamps, and A. Trotman (Eds.): INEX 2009, LNCS 6203, pp. 379–386, 2010.
© Springer-Verlag Berlin Heidelberg 2010

In this paper, we propose an approach which reduces the dimension of the underlying document vectors without change or with a slight change in the quality of the output clustering structure. More specifically, we use a partitioning type clustering algorithm, so-called Cover-Coefficient Based Clustering Methodology (C^3M) [7], along with some index pruning techniques for clustering XML documents.

2 Approach

2.1 Baseline Clustering with C^3M Algorithm

In this work, we use the well-known Cover-Coefficient Based Clustering Methodology (C^3M) [7] to cluster the XML documents. C^3M is a single-pass partitioning type clustering algorithm which is shown to have good information retrieval performance with flat documents (e.g., see [6]). The algorithm operates on documents represented by vector space model. Using this model, a document collection can be abstracted by a document-term matrix, D; of size m by n whose individual entries, d_{ij} ($1<i<m$; $1<j<n$), indicate the number of occurrences of term j (t_j) in document i (d_i). In C^3M, the document-term matrix[1] D is mapped into an m by m cover-coefficient (C) matrix which captures the relationships among the documents of a database. The diagonal entries of C are used to find the number of clusters, denoted as n_c; and to select the cluster seeds. During the construction of clusters, the relationships between a nonseed document (d_i) and a seed document (d_j) are determined by calculating the c_{ij} entry of C; where c_{ij} indicates the extent to which d_i is covered by d_j.

A major strength of C^3M is that for a given dataset, the algorithm itself can determine the number of clusters, i.e., there is no need for specifying the number of clusters, as in some other algorithms. However, for the purposes of this track, we cluster the XML documents into a given number of clusters (for several values like 1000, 10000, etc.) using C^3M method, as required. In this paper, we simply use the content of XML documents for clustering. Our preliminary experiments that also take the link structure into account did not yield better results than just using the content. Nevertheless; our work in this direction is still under progress.

2.2 Employing Pruning Strategies for Clustering

From the previous works, it is known that static index pruning techniques can reduce the size of an index (and the underlying collection) while providing comparative effectiveness performance with that of the unpruned case [2, 3, 4, 5, 8]. In a more recent study, we show that such pruning techniques can also be adapted for pruning the element-index for an XML collection [1]. Here, with the aim of both improving the quality of clusters and reducing the dataset dimensions for clustering, we apply static pruning techniques on XML documents. We adapt two well-known pruning techniques, namely, term-centric [8] and document-centric pruning [5] from the literature to obtain more compact representations of the documents. Then, we cluster documents with these reduced representations for various pruning levels, again using C^3M algorithm. The pruning strategies we employ in this work can be summarized as follows:

---

[1] Note that, in practice, the document-term matrix only includes non-zero term occurrences for each document.

- *Document-centric pruning (DCP):* This technique is essentially intended to reduce the size of an inverted index by discarding unimportant terms from each document. In the original study, a term's importance for a document is determined by that term's contribution to the document's Kullback-Leibler divergence (KLD) from the entire collection [5]. In a more recent work [2], we show that using the contribution of a term to the retrieval score of a document (by using a function like BM25) also performs quite well. In this paper, we again follow this practice and for each term that appears in a given document, we compute that term's BM25 score for this document. Then, those terms that have the lowest scores are pruned, according to the required pruning level. Once the pruned documents are obtained at a given pruning level, corresponding document vectors are generated to be fed to the C^3M clustering algorithm.

- *Term-centric pruning (TCP)*: This method operates on an inverted index, so we start with creating an index for our collection. Next, we apply term-centric pruning at different pruning levels, and once the pruned index files are obtained, we convert them to the document vectors to be given to the clustering algorithm[2]. In a nutshell, the term-centric pruning strategy works as follows [8]. For each term t, the postings in t's posting list are sorted according to their score with respect to a ranking function, which is BM25 in our case. Next, the kth highest score in the list, z_t, is determined and all postings that have scores less than $z_t * \varepsilon$ are removed, where ε specifies the pruning level. In this paper, we skip this last step, i.e. ε-based tuning, and simply remove the lowest scoring postings of a list for a given pruning percentage.

3 Experiments

In this paper, we essentially use a subset of the INEX 2009 XML Wikipedia collection provided by XML Mining Track. This subset, so-called small collection, contains 54575 documents. On the other hand, the large collection contains around 2.7 million documents and takes 60 GB. It is used only in the baseline experiments for various number of clusters.

As the baseline, we form clusters by applying C^3M algorithm to XML documents represented with the bag of words representation of terms, as provided by the track organizers. For several different number of output clusters, namely 100, 500, 1000, 2500, 5000 and 10000, we obtain the clusters and evaluate them at the online evaluation website of this track. The website reports the standard evaluation criteria for clustering such as micro purity, macro purity, micro entropy, macro entropy, normalized mutual information (NMI), micro F1 score and macro F1 score for a given clustering structure. However, only purity measures are used as the official evaluation

---

[2] It is possible to avoid converting the index to the document vectors by slightly modifying the input requirements of the clustering algorithm. Anyway, we did not spend much effort in this direction as this conversion stage, which is nothing but an inversion of the inverted index, can also be realized in an efficient manner.

Table 1. Micro and macro purity values for the baseline C^3M clustering for different number of clusters using the small collection

No. of clusters	Micro Purity	Macro Purity
100	0.1152	0.1343
500	0.1528	0.1777
1000	0.1861	0.2147
2500	0.2487	0.3031
5000	0.3265	0.4160
10000	0.4004	0.5416

Table 2. Micro and macro purity values for the baseline C^3M clustering for different number of clusters using the large collection

No. of clusters	Micro Purity	Macro Purity
100	0.1566	0.1234
1000	0.1617	0.1669
10000	0.1942	0.2408

criteria for this task. In Tables 1 and 2, we report those results for clustering small and large collection, respectively. For the latter case, due to time limitations, we experimented with three different numbers of clusters such as 100, 1000 and 10000. A quick comparison of the results in Tables 1 and 2 for corresponding cases imply that purity scores are better for the smaller dataset than that of the larger dataset, especially for large number of clusters. We anticipate that better purity scores for the large collection can be obtained by using a higher number of clusters.

Next, we experiment with the clusters produced by the pruning-based approaches. For each pruning technique, namely, TCP and DCP, we obtain the document vectors at four different pruning levels, i.e., 30%, 50%, 70% and 90%. Note that, a document vector includes term id and number of occurrences for each term in a document, stored in the binary format (i.e., as a transpose of an inverted index). In Table 3, we provide results for the small collection and 10000 clusters. Our findings reveal that up to 70% pruning with DCP, quality of the clusters is still comparable to or even superior than the corresponding baseline case, in terms of the evaluation measures.

Regarding the comparison of pruning strategies, clusters obtained with DCP yield better results than those obtained with TCP up to 70% pruning for both micro and macro purity measures. For the pruning levels higher than 70%, DCP and TCP give better results interchangeably for these measures. In [1], we observed a similar behavior regarding the retrieval effectiveness of indexes pruned with TCP and DCP.

From Table 3, we also deduce that DCP-based clustering at 30% pruning level produce the best results for both of the evaluation measures in comparison to the other pruning-based clusters. For this best-performing case, namely DCP at 30% pruning, we also provide performance findings with varying number of clusters (see Table 4).

The comparison of the results in Tables 1 and 4 shows that the DCP-based clusters are inferior to the corresponding baseline clustering up to 10000 clusters, but they provide almost the same performance for the 10000 clusters case.

Table 3. Comparison of the purity scores for clustering structures based on TCP and DCP at various pruning levels using the small collection. Number of clusters is 10000. Prune (%) field denotes the percentage of pruning. Best results for each measure are shown in bold.

Pruning Strategy	Prune (%)	Micro Purity	Macro Purity
No Prune	0%	0.4004	**0.5416**
DCP	30%	**0.4028**	0.5400
TCP	30%	0.3914	0.5229
DCP	50%	0.4019	0.5375
TCP	50%	0.3870	0.5141
DCP	70%	0.4016	0.5302
TCP	70%	0.3776	0.5042
DCP	90%	0.3783	0.4768
TCP	90%	0.3639	0.5073

Table 4. Micro and macro purity values for DCP at 30% pruning for different number of clusters

No. of clusters	Micro Purity	Macro Purity
100	0.1021	0.1265
500	0.1347	0.1539
1000	0.1641	0.1917
2500	0.2234	0.2737
5000	0.2986	0.3854
10000	0.4028	0.5400

In this year's clustering task, other than the standard evaluation criteria, the quality of the clusters relative to the optimal collection selection goal is also investigated. To this end, a set of queries with manual query assessments from the INEX Ad Hoc track are used and each set of clusters obtained is scored according to the result set of each query. According to the clustering hypothesis [10], the documents that cluster together have similar relevance to a given query. Therefore, it is expected that the relevant documents for ad-hoc queries will be in the same cluster in a good clustering solution. In particular, mean Normalised Cluster Cumulative Gain (nCCG) score is used to evaluate the clusters according to the given queries.

In Table 5, we provide the mean and the standard deviation of nCCG values for our baseline C^3M clustering on the small data collection. Regarding the pruning-based approaches, the mean nCCG values obtained from the clusters produced by TCP and DCP for various pruning levels are provided in Table 6. In parallel with the findings obtained by the purity criteria, mean nCCG values of the clusters obtained by DCP are still better than or comparable to the ones obtained by the baseline approach up to 70% pruning level. On the other hand, TCP approach yields better mean nCCG values even at 90% pruning level.

In Table 7, we provide the mean nCCG values obtained from different number of clusters formed by the DCP approach at 30% pruning level. A quick comparison of the results in Table 7 with those in Table 5 reveals that clusters obtained after DCP pruning are more effective than the clusters obtained by the baseline strategy for various number of clusters.

Table 5. Mean and standard deviation of nCCG values for the baseline C^3M clustering for different number of clusters using the small collection

No. of clusters	Mean nCCG	Std. Dev. CCG
100	0.7344	0.2124
500	0.6258	0.2482
1000	0.5986	0.2790
2500	0.5786	0.2352
5000	0.5918	0.2395
10000	0.4799	0.2507

Table 6. Comparison of the mean and standard deviation of nCCG values for clustering structures based on TCP and DCP at various pruning levels using the small collection. Number of clusters is 10000. Prune (%) field denotes the percentage of pruning. Best results for each measure are shown in bold.

Pruning Strategy	Prune (%)	Mean nCCG	Std. Dev. CCG
No Prune	0%	0.4799	0.2507
DCP	30%	0.4950	0.2549
TCP	30%	0.4940	0.2370
DCP	50%	0.4828	0.2467
TCP	50%	0.4618	0.2236
DCP	70%	0.4601	0.2207
TCP	70%	0.5075	0.2176
DCP	90%	0.4613	0.2343
TCP	90%	**0.5132**	0.2804

Table 7. Mean and standard deviation of nCCG values for DCP at 30% pruning for different number of clusters

No. of clusters	Mean nCCG	Std. Dev. CCG
100	0.7426	0.1978
500	0.6424	0.2326
1000	0.5834	0.2804
2500	0.5965	0.2504
5000	0.5929	0.2468
10000	0.4950	0.2549

Finally, in Figure 1 we compare the performance of the C^3M clustering with the other runs submitted to INEX 2009 in terms of the mean nCCG. For each case (i.e., number of clusters), we plot the highest scoring clustering approaches from each group. Note that, for cluster numbers of 100, 500, 2500, and 5000, our strategy (denoted as Altingovde et al.) corresponds to DCP based clustering at 30%; and for the cluster number of 10000 we report the score of TCP based clustering at 90%. For one last case where cluster number is set to 1000, we report the baseline C^3M score, which turns out to be the highest.

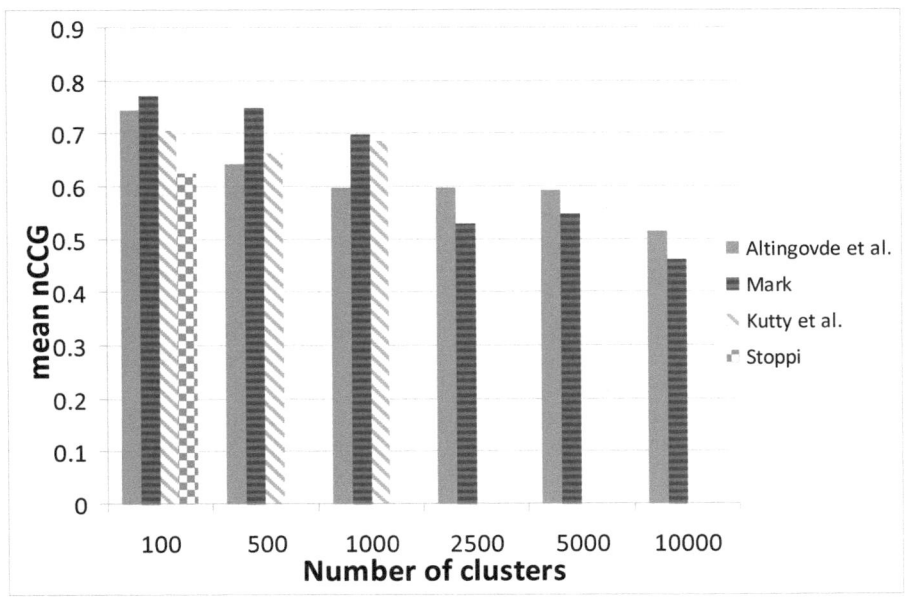

Fig. 1. Comparison of the highest scoring runs submitted to INEX for varying number of clusters on the small collection

4 Conclusion

In this paper, we employ the well-known C^3M algorithm for content based clustering of XML documents. Furthermore, we use index pruning techniques from the literature to reduce the size of the document vectors on which C^3M operates. Our findings reveal that, for a high number of clusters, the quality of the clusters produced by the C^3M algorithm does not degrade when up to 70% of the index (and, equivalently, the document vectors) is pruned.

Our future work involves repeating our experiments with larger datasets and additional evaluation metrics. Furthermore, we plan to extend the pruning strategies to exploit the structure of the XML documents in addition to content.

Acknowledgments. This work is supported by TÜBİTAK under the grant number 108E008.

References

1. Altingovde, I.S., Atilgan, D., Ulusoy, Ö.: XML Retrieval using Pruned Element-Index Files. In: Proc. of ECIR 2010, pp. 306–318 (2010)
2. Altingovde, I.S., Ozcan, R., Ulusoy, Ö.: A practitioner's guide for static index pruning. In: Proc. of ECIR 2009, pp. 675–679 (2009)

3. Altingovde, I.S., Ozcan, R., Ulusoy, Ö.: Exploiting query views for static index pruning in web search engines. In: Proc. of CIKM 2009, pp. 1951–1954 (2009)
4. Blanco, R., Barreiro, A.: Boosting static pruning of inverted files. In: Proc. of SIGIR 2007, The Netherlands, pp. 777–778 (2007)
5. Büttcher, S., Clarke, C.L.: A document-centric approach to static index pruning in text retrieval systems. In: Proc. of CIKM 2006, pp. 182–189 (2006)
6. Can, F., Altingövde, I.S., Demir, E.: Efficiency and effectiveness of query processing in cluster-based retrieval. Information Systems 29(8), 697–717 (2004)
7. Can, F., Ozkarahan, E.A.: Concepts and effectiveness of the cover-coefficient-based clustering methodology for text databases. ACM Transactions on Database Systems 15, 483–517 (1990)
8. Carmel, D., Cohen, D., Fagin, R., Farchi, E., Herscovici, M., Maarek, Y.S., Soffer, A.: Static index pruning for information retrieval systems. In: Proc. of SIGIR 2001, pp. 43–50 (2001)
9. De Vries, C.M., Geva, S.: Document Clustering with K-tree. In: Geva, S., Kamps, J., Trotman, A. (eds.) INEX 2008. LNCS, vol. 5631, pp. 420–431. Springer, Heidelberg (2009)
10. Jardine, N., van Rijsbergen, C.J.: The Use of Hierarchic Clustering in Information Retrieval. Information Storage and Retrieval 7(5), 217–240 (1971)
11. Kutty, S., Tran, T., Nayak, R., Li, Y.: Clustering XML documents using frequent subtrees. In: Geva, S., Kamps, J., Trotman, A. (eds.) INEX 2008. LNCS, vol. 5631, pp. 436–445. Springer, Heidelberg (2009)
12. Tran, T., Kutty, S., Nayak, R.: Utilizing the Structure and Content Information for XML Document Clustering. In: Geva, S., Kamps, J., Trotman, A. (eds.) INEX 2008. LNCS, vol. 5631, pp. 460–468. Springer, Heidelberg (2009)
13. Zhang, S., Hagenbuchner, M., Tsoi, A.C., Sperduti, A.: Self Organizing Maps for the Clustering of Large Sets of Labeled Graphs. In: Geva, S., Kamps, J., Trotman, A. (eds.) INEX 2008. LNCS, vol. 5631, pp. 469–481. Springer, Heidelberg (2009)

Multi-label Wikipedia Classification with Textual and Link Features

Boris Chidlovskii

Xerox Research Centre Europe
6, chemin de Maupertuis, F–38240 Meylan, France

Abstract. We address the problem of categorizing a large set of linked documents with important content and structure aspects, in particular, from the Wikipedia collection proposed at the INEX 2009 XML Mining challenge. We analyze the network of collection pages and turn it into valuable features for the classification. We combine the content-based and link-based features of pages to train an accurate categorizer for unlabelled pages. In the multi-label setting, we revise a number of existing techniques and test some which show a good scalability. We report evaluation results obtained with a variety of learning methods and techniques on the training set of the Wikipedia corpus.

1 Introduction

The objective of the INEX 2009 XML Mining challenge is to develop machine learning methods for structured data mining and to evaluate these methods for XML document mining tasks. The challenge proposes several datasets coming from different XML collections and covering a variety of classification and clustering tasks.

In this work, we address the problem of categorizing a very large set of linked XML documents with important content and link aspects like in Web pages. We cope with the case where there is a small number of labelled pages and a much larger number of unlabelled ones. For example, when categorizing Web pages, some pages have been labelled manually and a huge amount of unlabelled pages is easily retrieved by crawling the Web. Such a semi-supervised approach to categorization is motivated by the high cost of labelling data and the low cost for collecting unlabelled data. Within XML Mining track, the Wikipedia categorization challenge has been set up in the semi-supervised mode. At the training phase, the full data set was available as well as 20% of page labels; the remaining 80%, of labels were kept for the final evaluation.

Wikipedia is a free multilingual encyclopedia project supported by the non-profit Wikipedia foundation[1]. In January 2010, Wikipedia accounted for 3.2 million articles in English and about 21.5 million articles in total (in more than 200 languages). Wikipedia pages are written collaboratively by volunteers around the world, and almost all of its articles can be edited by anyone who can access the Wikipedia website. Launched

---

[1] http://www.wikipedia.org

S. Geva, J. Kamps, and A. Trotman (Eds.): INEX 2009, LNCS 6203, pp. 387–396, 2010.

in 2001, it is currently the largest and most popular general reference work on the Internet. Automated analysis, mining and categorization of Wikipedia pages can serve to improve its internal structure as well as to enable its integration as an external resource in different applications.

Any Wikipedia page is created, revised and maintained according to certain policies and guidelines [10]. Its edition follows certain rules for organizing the content and structuring it in the form of sections, abstract, table of content, citations, links to relevant pages, etc. In the following, we distinguish between different aspects of any Wikipedia page. First, its content is given by the set of words occurred in the page. Second, we are interested in the set of links in the page referring to other pages. In some cases, we might be interested in the page XML structure, given by the set of tags, attributes and their values in the page [3]. These elements control the presentation of the page content to the viewer.

The paper is organized as follows. In Section 2, we present the Wikipedia corpus used in the INEX 2009 XML Mining challenge. Section 3 reports on the page representation and introduces the textual and link features aimed at supporting such a representation. In Section 4, we review the state of art methods for the multi-label classification. Evaluation results on the training set as well as the final track evaluation results are reported in Section 5. Finally Section 6 concludes the paper.

2   INEX 2009 Collection

The training corpus for the INEX 2009 XML Mining track is composed of the following three files[2] :

Category file gives the set of documents considered in this track and the categories of the training documents.
Link file provides the links between the documents.
Content file corresponds to normalized tf-idf vectors computed by the organizers over this collection.

A simple analysis of the collection unveils a small mismatch between information presented in the category, link and content files. The category data includes 54,889 pages, while the tf-idf values and links are available for 54,572 and 54,451 of them, respectively. On the other hand, the graph of 4,554,203 links corresponds to one very large connected component in the Wikipedia corpus.

The training set with the category annotations is composed of 11,028 elements, tf-idf vectors are given for 10,968 and links are given for 10,992 of them.

Figure 1.a shows the link structure of the collection, plotted with the help of the LGL package. Figure 1.b plots the distributions of outcoming and incoming links in pages. One can note that both distributions are close to fitting the power distribution law, which is a frequent case for the Web data.

Unlike the previous editions, INEX 2009 challenge operates in a multi-label setting. The multi-category annotations are available for 20% of data where each category

---

[2] http://www-connex.lip6.fr/~denoyer/inex2009/corpus_train.tar.gz

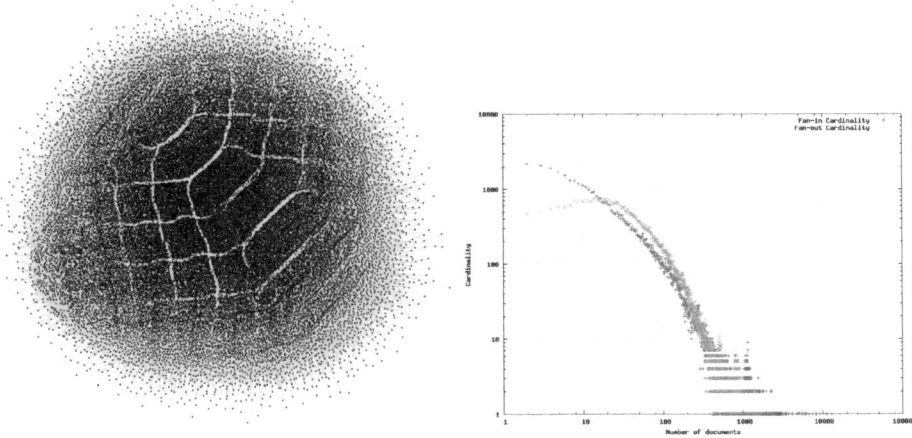

Fig. 1. INEX 2009 collection analysis. a) The link structure as one connected component. b) Incoming and outcoming link distributions.

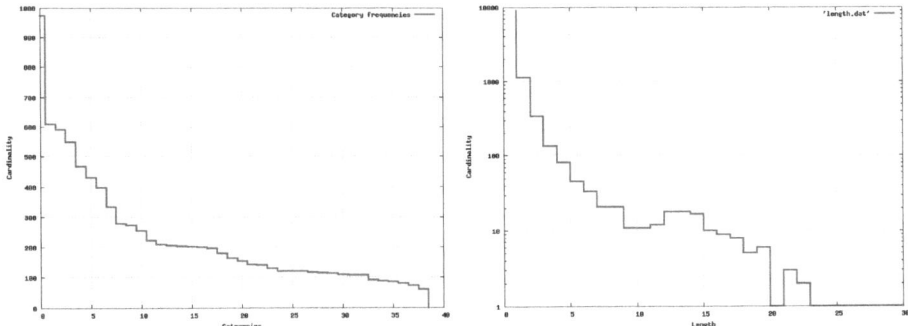

Fig. 2. INEX 2009 collection analysis. a) Category frequencies. b) Length frequencies.

corresponds to one of the 39 Wikipedia portals. Frequencies of categories occurring in the training set are reported in Figure 2.a. In the plot, one can easily recognize the leading category (`Portal:Trains`) as well as 7-8 core categories. Other categories are less frequent, they form a long tail. Due to the multi-label setting, any page can be annotated with one or more categories. Figure 2.b reports the length distribution in the category sets in the training set. Many pages have 1 label only, the longest label set includes 23 labels; on average, one page has 1.46 labels. Figure 3 represents the 39x39 matrix of category co-occurrences in the training set; diagonal elements in the matrix refer to the number of pages annotated with one category only. As one can see the diagonal elements mostly dominate in the rows; this indicates an important level of independence between labels.

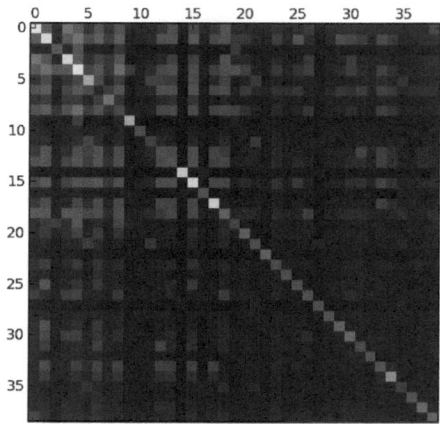

Fig. 3. INEX 2009 collection analysis. Category co-occurrence

3 Page Representation

The most conventional representation of Web pages is given by their textual content. The contents of body and subject fields of a page form a *bag-of-words* representation that has proven to be well suited for classification goals. The idea is to construct a vector where each feature represents a word in the lexicon and the feature values express some weight.

The document frequency df of a word w is the number of documents it occurs in, the term frequency tf is the number of word occurrences in a document d, N is the number of documents. For textual representation of the corpora documents, we consider three different representations: *binary*, *frequency* and *tf-idf* values, defined as follows:

$$\text{binary}(w,d) = \begin{cases} 1 \text{ if } w \text{ occurs in } d \\ 0 \text{ otherwise} \end{cases}$$

$$\text{frequency}(w,d) = tf$$

$$\text{tf-idf}(w,d) = tf \cdot \log(N/df).$$

The feature set based on a bag-of-words representation are high-dimensional (around 180,000) and the feature vectors are very sparse. This makes it particularly suited for the SVM classification with a linear kernel [7], because only a small number of the words actually occur in each respective document.

3.1 Graph Features

Graph-like structures induced by links or communications between different objects have received a lot of attention and have proven to be useful in diverse fields like sociology, biology, engineering, computer science and epidemiology [9]. With the advent of techniques to handle large graphs and the emergence of huge linked structures on the

Internet, the analysis of linked data has become a subject of intensive research. Classifying objects based on the relations/links among them is known as Link Mining [4]. The idea is to base a description of data on the structure it lives in. Typical tasks include the sub-graph discovery, link prediction (i.e. link-related) or classification (i.e. object-related).

The structure of a large linked collection like Wikipedia is not homogeneous. It has been shown that certain properties of nodes in a graph can serve very well to automatically detect their particular role [11]. We adopt and report below four groups of features that represent key properties of nodes in a network [1,11].

We represent the Wikipedia linked structure as a *directed graph* $G = \langle V, E \rangle$, where set V contains N nodes and set E includes M edges. The first feature group represents the immediate characteristics of a node $v \in V$ in the graph:

1. the number of incoming and out-coming links and their sum, $m(v) = in(v) + out(v)$.

We then address two features that have proven to be very valuable and represent the authority that is assigned to nodes by its peers. Nodes with a high number of incoming edges from hubs are considered to be *authorities*, nodes linking to a large number of authorities are *hubs* [8].

2. *hub score* $h(v)$ is given by v-th element of the principal eigenvector of AA^T where A is the adjacency matrix corresponding to graph G;
3. *authority score* $a(v)$ is given by v-th element of the principal eigenvector of $A^T A$.

The next group is a set of different centrality measures that capture the position of a node in the graph. These depend on a *undirected* version $G' = \langle V, E' \rangle$ of graph G. We calculate the shortest paths in G' using a Matlab library for working with large graphs [6]. We obtain the distance d_{st} from node s to t and the number σ_{st} of paths from s to t. The number of paths from s to t via v is denoted as $\sigma_{st}(v)$. Five centrality features for node v are the following :

4. *mean centrality*: $C_M(v) = \frac{N}{\sum_{s \in V} d_{vs}}$;
5. *degree centrality*: $\deg(v) = |s| : (v, s) \in E$;
6. *betweenness centrality*: $C_B(v) = \sum_{s \neq v \neq t} \frac{\sigma_{st}(v)}{\sigma_{st}}$;
7. *closeness centrality*: $C_C(v) = \frac{1}{\max_t d(v,t)}$;
8. *stress centrality*: $C_S(v) = \sum_{s \neq v \neq t} \sigma_{st}(v)$.

One more feature characterizes the connectiveness in the direct neighbourhood of node v:

9. *clustering coefficient*: $CC(v) = \frac{2|(s,t)|}{(\deg(v)(\deg(v)-1))}$, $(v, s), (v, t), (s, t) \in E'$.

The final group of graph features calculates all cliques in the graph using a Matlab implementation of [2]. It includes three following features:

10. the *number* $clq(v)$ *of cliques* the node v is in;

11. a *raw clique score* where each clique in $clq(v)$ of size n is given a weight 2^{n-1},
$CS_R(v) = \sum_{q \in clq(v)} 2^{size(q)-1}$;
12. a *weighted clique score* where each clique is weighted by the sum of activities of its members,

$$CS_w(v) = \sum_{q \in clq(v)} 2^{size(q)-1} \sum_{w \in q} m(w).$$

All scores are scaled to a value in $[0, 1]$ range, where 1 indicates a higher importance. Once all graph features are extracted for all nodes in the graph, we proceed by combining them with the text representation following one of fusion strategies. According to the basic fusion strategy, we concatenate the tf-idf representation of pages with their graph features.

4 Multi-label Classification

Single-label classification is concerned with learning from a set of examples $x_i \in X$ that are associated with a single label y from a set of k disjoint labels Y. Cases $k = 2$ and $k > 2$ are known as *binary* and *multi-class* classification. In *multi-label* classification, examples x_i are associated with a set of labels from Y. We will present these sets as binary vectors $y_i = (y_1, \ldots, y_k)$, where $y_i \in \{0, 1\}$.

Most existing techniques for multi-label classification follow either the *transformation* or *adaptation* approach, see [13] for the review on multi-label learning. Like in the multi-class setting, there is no method performing well in all cases. There exists a number of multi-label techniques [5,12,13]. Below we shortly describe some most important ones.

One-against-all (1AA) is the most common approach to multi-label classification. It transforms the problem in k binary problems and train k independent classifiers h_i, each deciding to label a given x with y_i; then some ranking or thresholding schemes are used. This technique is easy to implement, it is fast but makes a strong assumption of independence among $y \in Y$.

Length-based One-against-all (L-1AA) is a modification of the 1AA guided by the length prediction [12]. It trains additionally a length predictor h_l on the $\{x_i, |y_i|\}$ training set where $|y|$ is the numbers of 1's in y. It assumes a probabilistic setting for all classifiers h_i. h_l first predicts length (size) l of label set for a given x, then x is labelled with those y_i that have top l scores. The L-1AA often improves the 1AA performance and is scalable. However, it is very sensitive to the performance of h_l, it still assumes independence of $y \in Y$.

Unique multi-class (UMC) takes each label set y present in the training set as a unique label. The technique is easy to implement. Yet, it often results in the exponential number of unique labels, with very few examples per label; no generalization is possible to label sets which are not observed in the training set.

Collective multi-label (CML) The previous techniques are only well-suited to problems in which categories are independent. Instead, [5] explores multi-label conditional random field (CRF) classification models that directly parametrize label co-occurrences. Experiments show that the models outperform their single-label counterparts on standard text corpora. Even when multi-labels are sparse, the models improve considerably subset classification error. The method performs well on small datasets; however due to the high complexity of learning and inference, it is not scalable.

Latent variables in Y This method discovers the relevant groups (topics) among y and can replace a group of labels by a topic. The technique requires a priori the number of topics in Y and often incurs important loss when decoding from topics to labels.

Latent variables in $(\mathbf{x}, \mathbf{y})$ is aimed at discovering topics in the joint (X, Y) space [14]. Latent semantic indexing (LSI) is a well-known unsupervised approach for dimensionality reduction in information retrieval. However if the output information (i.e. category labels) is available, it is often beneficial to derive the indexing not only based on the inputs but also on the target values in the training data set. This is of particular importance in applications with multiple labels, in which each document can belong to several categories simultaneously. In [14], the multi-label informed latent semantic indexing (MLSI) algorithm is developed, it preserves the information of inputs and meanwhile captures the correlations between the multiple outputs. The recovered "latent semantics" thus incorporate the human-annotated category information and can be used to greatly improve the prediction accuracy. The main disadvantage of this technique is a limited scalability.

When working with the INEX09 XML Mining Challenge dataset, we tested different multi-label techniques. In order to meet the scalability requirements, our choice was limited to four methods from the list above, namely, 1AA, L-1AA, UMC and Latent-Y.

5   Evaluation

When preparing our submission to the challenge, we run a series of experiments using 5-fold cross validation on the core training set with the pages having both text-based and link-based features. We evaluated the performance of each tested method using the average Micro-F1, Macro-F1 and Exact Match measures. In the experiments we primarily tested three different components of the learning system, as follows:

Feature set: here we considered three possibilities including the tf-idf feature set, the graph feature set and their fusion.

Basic learning method: among different options, we considered the LIBSVM package (C-SVM) with linear, polynomial and sigmoid kernels, and semi-supervised learning (SSL) with the Transductive Support Vector Machines (TSVM) [3].

Multi-label method: four main possibilities include one-against-all (1AA) and its length-based extension (L-1AA), unique multi-class (UMC) and Latent-Y. Additionally, a length-based version of UMC (L-UMC) has been added, by analogy with L-1AA. Among k-top predictions made by UMC method in the probabilistic mode, L-UMC prefers one supported by the length predictor.

Table 1 reports the most important results of the experiments on the training set. Beyond the three components reported above, we also run a number of tests to probe other components and techniques, including the feature selection, classification fusion, etc. However, none of them was able to boost the performance, thus their results are not included in the table.

Table 1. Evaluation results for different methods and feature sets on the training set

Feature set	Method	Multi-label	Avg Exact Match	Avg Micro-F1	Avg Macro-F1
Tf-idf	Linear C-SVM	1AA	56.66	58.43	54.44
	Linear C-SVM	L-1AA	56.69	58.12	54.61
	Linear C-SVM	UMC	56.81	58.54	54.60
	Linear C-SVM	L-UMC	56.73	58.24	54.52
	Linear C-SVM	Y-latent	52.17	58.63	53.32
Graph	Linear C-SVM	1AA	35.34	41.71	39.56
	Linear C-SVM	L-1AA	37.01	40.33	38.98
	Sigmoid C-SVM	UMC	37.41	40.63	38.38
Tf-idf+Graph	Linear C-SVM	1AA	57.34	60.76	56.79
	Linear C-SVM	L-1AA	57.84	61.13	57.13
	Linear C-SVM	UMC	**58.10**	**61.24**	**57.36**
	Linear C-SVM	L-UMC	57.70	61.15	57.19
	SSL-TSVM	L-1AA	51.73	52.25	51.27
	SSL-TSVM	UMC	51.11	52.19	51.51

5.1 Final Submissions and Evaluation

Results of tests on the training set have been analysed in order to prepare our submission to the XML Mining classification track. Our submissions take up three compositions yielding the top performance in Table 1. They all combine the fusion of tf-idf and graph features with the linear C-SVM and deploy the L-1AA, UMC and L-UMC multi-label strategies, respectively.

Moreover, we applied an additional treatment to cope with the partial mismatch between category, link and tf-idf data. The evaluation set of the category data includes hundreds of pages missing either tf-idf or link data. To predict categories for these pages in the evaluation set, we generated the classification models using an available features only, that is, tf-idf features when link data was missing and the graph features when tf-idf vectors were not provided.

The result of final track evaluation are reported in Figure 4. The plot tracks seven measures for all submissions and all research teams participated in the challenge. As the figure shows, our submissions (labelled 'Xerox') came out first by four measures, namely Micro-PRF (F1) and Macro-PRF (F1), Micro-Acc and Macro-Acc. The final performance results are very close and comparable to the results of the preliminary tests. Moreover, the winning combination of main components (the fusion of tf-idf graph features with Linear C-SVM and UMC) coincides with the composition which showed the best performance on the training set.

Two main conclusions can be drawn from the analysis of the final results. First, the major contribution to the performance gain is achieved by extracting the graph features and coupling them with the text features. Instead, different multi-label methods shown similar results with a far modest contribution to performance. Second, more sophisticated methods, including non-linear kernels and latent variables in (X,Y) space are very promising, but the current lack of scalable versions severely compromises their deployment on large datasets.

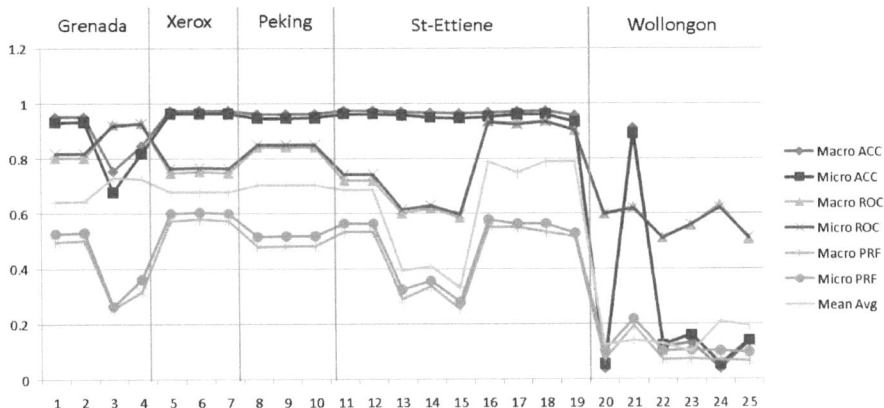

Fig. 4. Track evaluation results

6 Conclusion

We presented our contribution to the INEX 2009 XML Mining track that addresses the multi-label classification of the Wikipedia corpus. We reported a number of complementary techniques which allowed us to make a number of steady improvements over the baseline classification model. A deep analysis of page links allowed us to identify and extract a number of features valuable for the classification goals. Then we tested different fusion methods to combine the content-based and link-based features. We finally implemented a number of alternative multi-label classification techniques in order to determine which one performs best on the evaluation corpus. We reported evaluation results obtained with different combinations on the training set of the Wikipedia corpus. We also included the final evaluation results which confirm the top performance of our submissions according to four core measures, Micro-F1 and Macro-F1, Micro-Acc and Macro-Acc.

Acknowledgement

This work is supported by the Large Scale Integrating Project SHAMAN, co-funded under the EU 7th Framework Programme (http://shaman-ip.eu/shaman/).

References

1. Brandes, U.: A faster algorithm for betweenness centrality. Journal of Mathematical Sociology 25, 163–177 (2001)
2. Bron, C., Kerbosch, J.: Algorithm 457: finding all cliques of an undirected graph. Communications of the ACM 16(9), 575–577 (1973)
3. Chidlovskii, B.: Semi-supervised categorization of wikipedia collection by label expansion. In: Geva, S., Kamps, J., Trotman, A. (eds.) INEX 2008. LNCS, vol. 5631, pp. 412–419. Springer, Heidelberg (2009)

4. Getoor, L., Diehl, C.P.: Link mining: a survey. SIGKDD Explorations 7(2), 3–12 (2005)
5. Ghamrawi, N., McCallum, A.: Collective multi-label classification. In: CIKM 2005: Proceedings of the 14th ACM international conference on Information and knowledge management, pp. 195–200. ACM, New York (2005)
6. Gleich, D.: MatlabBGL: a Matlab Graph Library (2008), http://www.stanford.edu/~dgleich/programs/matlab_bgl
7. Joachims, T.: A statistical learning model of text classification for Support Vector Machines. In: Proc. 24th International ACM SIGIR Conf., pp. 128–136. ACM Press, New York (2001)
8. Kleinberg, J.M.: Authoritative sources in a hyperlinked environment. Journal of the ACM 46(5), 604–632 (1999)
9. Newman, M.E.J.: The structure and function of complex networks. SIAM Review 45, 167–256 (2003)
10. Riehle, D.: How and why Wikipedia works: an interview with Angela Beesley, Elisabeth Bauer, and Kizu Naoko. In: WikiSym 2006: Proceedings of the 2006 international symposium on Wikis, pp. 3–8. ACM, New York (2006)
11. Rowe, R., Creamer, G., Hershkop, S., Stolfo, S.J.: Automated social hierarchy detection through email network analysis. In: Proc. 1st SNA-KDD 2007 Workshop on Web Mining and Social Network Analysis, pp. 109–117. ACM, New York (2007)
12. Tang, L., Rajan, S., Narayanan, V.K.: Large scale multi-label classification via metalabeler. In: WWW 2009: Proceedings of the 18th international conference on World Wide Web, pp. 211–220. ACM, New York (2009)
13. Tsoumakas, G., Katakis, I.: Multi-label classification: An overview. International Journal of Data Warehousing and Mining 3(3), 1–13 (2007)
14. Yu, K., Yu, S., Tresp, V.: Multi-label informed latent semantic indexing. In: SIGIR 2005: Proceedings of the 28th annual international ACM SIGIR conference on Research and development in information retrieval, pp. 258–265. ACM, New York (2005)

Link-Based Text Classification
Using Bayesian Networks

Luis M. de Campos, Juan M. Fernández-Luna, Juan F. Huete,
Andrés R. Masegosa, and Alfonso E. Romero

Departamento de Ciencias de la Computación e Inteligencia Artificial
E.T.S.I. Informática y de Telecomunicación,
CITIC-UGR, Universidad de Granada
18071 – Granada, Spain
{lci,jmfluna,jhg,andrew,aeromero}@decsai.ugr.es

Abstract. In this paper we propose a new methodology for link-based document classification based on probabilistic classifiers and Bayesian networks. We also report the results obtained of its application to the XML Document Mining Track of INEX'09.

1 Introduction

This is the third year that researchers from the University of Granada (specifically from the Uncertainty Treatment in Artificial Intelligence research group) participate on the XML Document Mining Track of the INEX workshop. As in previous editions, we restrict our solutions to the application of probabilistic methods to these problems. To be more precise, we are looking to solve the problem of link-based document classification within the field of Bayesian networks [10] (a special case of probabilistic graphical models).

This year, the proposed problem is rather similar to the one considered in the previous edition of the workshop [6]. A training corpus, composed of labeled XML files is provided, and an unlabeled test corpus is left to the participants, in order to be estimated its labeling. Also, a link file is given, which contains specific relations between pairs of documents (either in the training or the test corpus). Thus, the problem can be seen as a graph labeling problem, where each node has textual (XML) content.

The main difference between this INEX track in 2008 and 2009 is the fact that the corpus is composed of multilabeled documents, that is to say, a document can belong to one or more categories. The rest of the rules are esentially the same, although the document collection and the set of categories are also different.

As we did in the past, we can assume that the XML markup (the "internal structure" of the collection) is not very helpful for categorization. In fact, we did not find it very useful for the task in previous editions [4] (by making several transformations from XML to flat text documents). Moreover, the organizers have provided an indexed file of term vectors representing the documents, where XML marks have been removed.

S. Geva, J. Kamps, and A. Trotman (Eds.): INEX 2009, LNCS 6203, pp. 397–406, 2010.

Like our previous participation [5], we will use explicitly the "external structure" of the collection, i.e. the link file (the graph of documents). There, we provided a "graphical proof" that the category of the documents linked by one tends to be similar to the category of the own document. Several experiments in the same direction showed us the same fact for the 2009 corpus, although we do not reproduce them here. Apart from those experiments, the names of the categories (which are explicitly given in the training set), tend to show categories which are probably coming from a hierarchy (for example Portal:Religion, Portal:Christianity and Portal:Catholicism). The two known facts about the relations are summarized here:

- In this linked corpus, due to its nature, a "hyperlink regularity" is supposed to arise (more precisely an encyclopedia regularity, see [15] for more details).
- There are some categories strongly related a priori, because the probable existence of a (unknown) hierarchy.

Although last year we proposed a method that captures some "fixed" relations among categories, given this different problem setting (multilabel) and its higher dimensionality, this year we pretend to learn those relations automatically from data, leading to a more flexible approach.

2   Base Classifiers

Two base classifiers will be used to label the graph nodes based only on their content. They will serve as the baseline, and next will be combined with the Bayesian network learnt from data with our new methodology. We will briefly describe them, in order to make the paper more self-contained.

Both classifiers are probabilistic, i.e. given a document d_j, they compute the probability values $p(c_i|d_j)$ for each category c_i, and assign them as a degree of confidence in that each c_i is an appropriate label for d_j. The advantage of probability is that it is a very well founded approach, and several different probabilistic approaches can be combined together, because they are dealing with the same measures.

Note that here we deal with a multilabel problem by defining a binary classifier for each label, following the "classical" approach to this task [12].

2.1   Multinomial Naive Bayes

The model is the same used by McCallum et al. [9], adapting it to the case of many binary problems. The naive Bayes, in its multinomial version, is a very fast and well performing method. In this model, we firstly assume that the length of the document is independent of the category. We also assume that the term occurrences are independent on each other, given the category (this is the core of the naive Bayes method).

In the multinomial version of this classifier, we see a document d_j as being drawn from a multinomial distribution of words with as many independent trials as the length $|d_j|$ of d_j.

So, given a category c_i, we express the probability[1] $p_i(c_i|d_j)$ as

$$p_i(c_i|d_j) = \frac{p_i(d_j|c_i)\,p_i(c_i)}{p_i(d_j)}. \tag{1}$$

We can rewrite $p_i(d_j)$ using the law of total probability,

$$p_i(d_j) = p_i(d_j|c_i)\,p_i(c_i) + p_i(d_j|\overline{c}_i)\left(1 - p_i(c_i)\right). \tag{2}$$

The values $p_i(c_i|d_j)$ can be easily computed in terms of the prior probability $p_i(c_i)$ and the probabilities $p_i(d_j|c_i)$ and $p_i(d_j|\overline{c}_i)$.

Besides, prior probabilities are estimated from document counts:

$$\widehat{p}_i(c_i) = \frac{N_{i,doc}}{N_{doc}} \tag{3}$$

where N_{doc} is the number of documents in the training set and $N_{i,doc}$ is the number of documents in the training set which belong to category c_i.

On the other hand, we can estimate $p_i(d_j|c_i)$ and $p_i(d_j|\overline{c}_i)$ as follows (as a multinomial distribution over the words):

$$p_i(d_j|c_i) = p_i(|d_j|)\,\frac{|d_j|!}{\prod\limits_{t_k \in d_j} n_{jk}!} \prod\limits_{t_k \in d_j} p_i(t_k|c_i)^{n_{jk}},$$

and

$$p_i(d_j|\overline{c}_i) = p_i(|d_j|)\,\frac{|d_j|!}{\prod\limits_{t_k \in d_j} n_{jk}!} \prod\limits_{t_k \in d_j} p_i(t_k|\overline{c}_i)^{n_{jk}},$$

where n_{jk} is the frequency of the term t_k in the document d_j.

Substituting and simplifying in equations 1 and 2 we obtain:

$$p_i(c_i|d_j) = \frac{p_i(c_i) \prod\limits_{t_k \in d_j} p_i(t_k|c_i)^{n_{jk}}}{p_i(c_i) \prod\limits_{t_k \in d_j} p_i(t_k|c_i)^{n_{jk}} + \left(1 - p_i(c_i)\right) \prod\limits_{t_k \in d_j} p_i(t_k|\overline{c}_i)^{n_{jk}}}.$$

Finally, individual term probabilities $p_i(t_k|c_i)$ and $p_i(t_k|\overline{c}_i)$ are computed by means of the following formulae (using Laplace smoothing):

$$\widehat{p}_i(t_k|c_i) = \frac{N_{ik} + 1}{N_{i\bullet} + M}, \quad \widehat{p}_i(t_k|\overline{c}_i) = \frac{N_{\bullet k} - N_{ik} + 1}{N - N_{i\bullet} + M}, \tag{4}$$

where N_{ik} is the number of times that the term t_k appears in documents of class c_i, $N_{i\bullet}$ is the total number of words in documents of class c_i ($N_{i\bullet} = \sum_{t_k} N_{ik}$), $N_{\bullet k}$ is the number of times that the term t_k appears in the training documents ($N_{\bullet k} = \sum_{c_i} N_{ik}$), N is the total number of words in the training documents, and M is the size of the vocabulary (the number of distinct words in the documents of the training set).

---

[1] With the notation $p_i(c_i|d_j)$ we are emphasizing that the probability distribution is computed over a binary variable C_i, taking values in $\{c_i, \overline{c}_i\}$. So, we have a different probability distribution over each category.

2.2 Bayesian OR Gate

The Bayesian OR gate classifier was presented in the INEX 2007 workshop by this group [4]. This classifier relies on the assumption that the relationships among the terms and each category follow a so-called *noisy-OR gate* probability distribution. Following the Bayesian networks notation, this model can be graphically represented as a graph having one node for the category C_i (binary variable C_i, ranging in $\{c_i, \bar{c}_i\}$), one node for each term T_k (binary variable T_k, with values in $\{t_k, \bar{t}_k\}$), and arcs going from each term node to the category nodes they appear in (i.e. they form the parent set, $Pa(C_i)$, of the category node C_i).

In the naive Bayes model (a generative one), we are defining $p(d_j|c_i)$, whereas in the Bayesian OR gate (a discriminative model), we are computing directly $p_i(c_i|d_j)$. Instead of using a "general" probability distribution, $p_i(c_i|d_j)$ is modeled by means of a "canonical model" [10], the noisy OR gate, which makes computations and parameter storage feasible tasks.

We can define the probability distribution for this noisy OR gate in the following way:

$$p_i\big(c_i|\,pa(C_i)\big) \;=\; 1 \;-\; \prod_{T_k \in R(pa(C_i))} \big(1 - w(T_k, C_i)\big)$$

$$p_i\big(\bar{c}_i|\,pa(C_i)\big) \;=\; 1 \;-\; p_i\big(c_i|\,pa(C_i)\big),$$

where $R(pa(C_i)) = \{T_k \in Pa(C_i)\,|\,t_k \in pa(C_i)\}$, i.e. $R(pa(C_i))$ is the subset of parents of C_i which are instantiated to its t_k value in the configuration $pa(C_i)$. $w(T_k, C_i)$ is a weight representing the probability that the occurrence of the "cause" T_k alone (T_k being instantiated to t_k and all the other parents T_h instantiated to $\bar{t}_h$) makes the "effect" true (i.e., forces class c_i to occur).

Then, given a certain document d_j, we can compute the posterior probability $p_i(c_i|d_j)$ by instantiating to the value t_k all the terms that appear in the document (i.e. $p_i(t_k|d_j) = 1$), and to the value $\bar{t}_h$ those terms that do not appear in d_j (i.e. $p_i(t_h|d_j) = 0$). The result is [3]:

$$p_i(c_i|d_j) = 1 - \prod_{T_k \in Pa(C_i)} \big(1 - w(T_k, C_i)\,p_i(t_k|d_j)\big)$$

$$= 1 - \prod_{T_k \in Pa(C_i) \cap d_j} \big(1 - w(T_k, C_i)\big).$$

Finally, we have to give a definition for the weights $w(T_k, C_i)$, which is almost the same appearing in [4]:

$$w(T_k, C_i) = \frac{N_{ik}}{nt_i \, N_{\bullet k}} \prod_{h \neq k} \frac{(N_{i\bullet} - N_{ih})N}{(N - N_{\bullet h})N_{i\bullet}}. \tag{5}$$

In this formula, $N_{ik}, N_{\bullet k}, N_{i\bullet}$ and N mean the same than in previous definitions made in the explanation of the multinomial naive Bayes, and nt_i is the number of different terms occurring in documents of the class c_i. The factor nt_i is introduced here to relax the independence assumption among terms, but some other valid definitions for the weights (which do not use this factor) can be found in [3] and [4].

Finally, in order to make the probabilities independent on the length of the document (thus making the scores of different documents comparable), we introduce the following normalization, which is somewhat similar to the $RCut$ thresholding strategy [14], and we return as the final probability $p(c_i|d_j)$:

$$p(c_i|d_j) = \frac{p_i(c_i|d_j)}{\max_{c_k}\{p_k(c_k|d_j)\}}$$

Some experiments [3] have shown that the Bayesian OR gate classifier tends to outperform the multinomial naive Bayes classifier, although the number of parameters needed and the complexity are essentially the same.

3   The Bayesian Network Model

This section describes a new methodology that models a link-based categorization environment using Bayesian networks. We shall build automatically from data a Bayesian network-based model, representing the relationships among the categories of a certain document and the categories present on the related (linked) documents. In this development, we will only use data from incoming links, because we carried several experiments on the corpus, and found them much more informative than outgoing ones. Anyway, information from outgoing links (or even considering undirected links) could also be used in this model.

3.1   Modeling Link Structure between Documents

In this problem, we will consider two binary variables for every category i: one is C_i (with states $\{c_i, \overline{c}_i\}$) which models the probability of a document being (or not) of class c_i, and the variable LC_i (with states $\{lc_i, \overline{lc}_i\}$), which represents if there is a link, or not, from documents of category c_i to the current document[2]. We assume there is a global probability distribution among all these variables, and we will model it with a Bayesian network.

To learn a model from the data, we will use the training documents, each one as an instance whose categories (values for variables C_i) are perfectly known, and the links from other documents. If a document is linked by another training document of category j, we will set $LC_j = lc_j$, setting it to $\overline{lc}_j$ otherwise. Note that a training document could be linked by test documents (whose categories are unknown). In that case, this evidence is ignored, and for the categories j which do not have any document linked to the current document, their variables are set to $\overline{lc}_j$.

---

[2] As we stated before, we also could represent the existence of outgoing links to a document of category c_i, or both types of interactions.

So, we could learn a Bayesian network from training data (see next section) and, for each test document d_j, we could compute $p(c_i|e_j)$, where e_j represents all the evidence given by the information of documents that link this.

Thus, the question is the following: for a certain document d_j, given $p(c_i|d_j)$ and $p(c_i|e_j)$, how could we combine them in an easy way? We want to compute the posterior probability $p(c_i|d_j, e_j)$, the probability of a category given the terms composing the document and the evidence due to link information.

Using Bayes' rule, and assuming that the content and the link information are independent given the category, we get:

$$
\begin{aligned}
p(c_i|d_j, e_j) &= \frac{p(d_j, e_j|c_i)\, p(c_i)}{p(d_j, e_j)} = \frac{p(d_j|c_i)\, p(e_j|c_i)\, p(c_i)}{p(d_j, e_j)} \\
&= \frac{p(c_i|d_j)\, p(d_j)\, p(e_j|c_i)\, p(c_i)}{p(c_i)\, p(d_j, e_j)} = \frac{p(c_i|d_j)\, p(d_j)\, p(c_i|e_j)\, p(e_j)}{p(c_i)\, p(d_j, e_j)} \\
&= \left(\frac{p(d_j)\, p(e_j)}{p(d_j, e_j)} \right) \left(\frac{p(c_i|d_j)\, p(c_i|e_j)}{p(c_i)} \right).
\end{aligned}
$$

The first term of the product is a factor which does not depend on the category. So, we can write the probability as:

$$
p(c_i|d_j, e_j) \propto \frac{p(c_i|d_j)\, p(c_i|e_j)}{p(c_i)}
$$

As $p(c_i|d_j, e_j) + p(\bar{c}_i|d_j, e_j) = 1$, we can easily compute the value of the normalizing factor, and therefore the final expression of $p(c_i|d_j, e_j)$ is:

$$
p(c_i|d_j, e_j) = \frac{p(c_i|d_j)\, p(c_i|e_j)\, /\, p(c_i)}{p(c_i|d_j)\, p(c_i|e_j)\, /\, p(c_i)\ +\ p(\bar{c}_i|d_j)\, p(\bar{c}_i|e_j)\, /\, p(\bar{c}_i)} \tag{6}
$$

We must make some final comments about this equation to make it more clear:

- As we said before, the posterior probability $p(c_i|d_j)$ is the one obtained from a binary probabilistic classifier (one of the two presented before, or any other), which is going to be combined with the information obtained from the link evidence.
- The prior probability used here, $p(c_i)$, is the one computed with propagation over the Bayesian network learnt with link information.
- Because the variables C_i are binary, it is clear that $p(\bar{c}_i|e_j) = 1 - p(c_i|e_j)$, $p(\bar{c}_i) = 1 - p(c_i)$ and $p(\bar{c}_i|d_j) = 1 - p(c_i|d_j)$.

3.2   Learning Link Structure

Given the previous variable setting, from the training documents, their labels and the link file, we can obtain a training set for the Bayesian network learning problem, composed of vectors of binary variables C_i and LC_i (one for each training document).

We have used WEKA package [13] to learn a generic Bayesian network (not a classifier) using a hill climbing algorithm (with the classical operators of addition, deletion and reversal of arcs) [1], with the BDeu metric [8]. In order to reduce the search space, we have limited the number of parents of each node to a maximum of 3.

Once the network has been learnt, we have converted it to the Elvira [7] format. Elvira is a software[3] developed by some Spanish researchers which implements many algorithms for Bayesian networks. In this case, we have used it to carry out the inference procedure. This is done as follows:

1. For each test document d_j, we set in the Bayesian network the LC_i variables to either lc_i or $\overline{lc_i}$, depending whether d_j is linked by at least one document of category i, or not, respectively. This is the evidence coming from the links (represented before as e_j).
2. For each category variable, C_i, we compute the posterior probability $p(c_i|e_j)$. This procedure is what is called *evidence propagation*.

Due to the size of the problem (39 categories, which give rise to a network with 78 variables), instead of exact inference, we have used an approximate inference algorithm [2], firstly to compute the prior probabilities of each category in the network, $p(c_i)$, and secondly to compute the probabilities of each category given the link evidence e_j, for each document d_j in the test set, $p(c_i|e_j)$. The algorithm used is called Importance Sampling algorithm, and is faster than other exact approaches.

4   Results

We have tested our proposal using the INEX'09 XML Document mining corpus, which contains 54572 documents, corresponding to a test/train split of 10968 documents in the training corpus (about 20% of the total), and 43604 in the test set.

The performance measures, selected by the track organizers, are Accuracy (ACC), Area under Roc curve (ROC) and F1 measure (PRF), computed over the categories (micro and macro versions), and Mean average precision by document (MAP), computed over the documents.

4.1   Preliminary Results

Four result files were sent to the organization to participate in the Workshop. Two of them, the baselines (that is to say, no link structure was used to label the documents, only their content), were obtained from the two flat-text classifiers commented in Section 2, Naive Bayes (NB) and the OR Gate (OR). The other two were the Bayesian network model (BN) proposed in Section 3 combined with the two baselines (using equation 6), NB+BN and OR+BN.

---

[3] Available at http://leo.ugr.es/~elvira

Table 1. Preliminary results

	MACC	μACC	MROC	μROC	MPRF	μPRF	MAP
NB	0.95142	0.93284	0.80260	0.81992	0.49613	0.52670	0.64097
NB + BN	0.95235	0.93386	0.80209	0.81974	0.50015	0.53029	0.64235
OR	0.75420	0.67806	0.92526	0.92163	0.25310	0.26268	0.72955
OR + BN	0.84768	0.81891	0.92810	0.92739	0.31611	0.36036	0.72508

The results of the models we sent for this track are displayed in Table 1, where M and μ mean the "macro" and "micro" versions of the performance measures, respectively.

In both cases, the Bayesian network version of the classifier outperforms the "flat" version, though the results on the OR gate are surprisingly poor in ACC and PRF. This fact is due to the nature of the classifier, and to the kind of evaluation. As both ACC and PRF require hard categorization, the evaluation procedure needs a criterion to assign categories to the test documents based on the posterior probabilities. The criterion selected by the organizers was to assign the label c_i to a document d_j if $p(c_i|d_j, e_j) > 0.5$.

But for the OR gate classifier is not known, a priori, what is the appropriate threshold τ_i such that c_i is assigned to d_j if $p(c_i|d_j, e_j) > \tau_i$. This is not a major problem to compute, for example, averaged break-even point [12] or ROC measures, where no hard categorization is needed. In this case, as the threshold 0.5 has been adopted, we need to re-adapt the model to this setting in order to perform better.

In the following section we can see how we estimated a set of thresholds (using only training data) and how we scaled the probability values, in order to match the evaluation criteria, dramatically improving the results.

4.2   Scaled Version of the Bayesian OR Gate Results

We have followed this procedure: using only training data, a classifier has been built (both in its flat and BN versions), and evaluated using cross validation (with five folds). In each fold, for each category, we have searched for the probability threshold that gives the highest $F1$ measure per class and, afterwards, all thresholds have been averaged over the set of cross validation folds.

This is what is called in the literature the *Scut* thresholding strategy [14]. Thus, we obtain, for each category, a threshold τ_i between 0 and 1 (different for each of the two models). We should then transform the original results to a scale where each category threshold is mapped to 0.5.

So, the probabilities of the OR gate model are rescaled using a linear continuous function f_i which verifies $f_i(0) = 0$, $f_i(1) = 1$ and $f_i(\tau_i) = 0.5$. The function is:

$$f_i(x) = \begin{cases} \frac{0.5x}{\tau_i} & \text{if } x \in [0, \tau_i] \\ 1 - \frac{0.5}{1-\tau_i}(1 - x) & \text{if } x \in (\tau_i, 1] \end{cases}$$

Table 2. Results of the OR gate classifier using thresholds

	MACC	μACC	MROC	μROC	MPRF	μPRF	MAP
OR	0.92932	0.92612	0.92526	0.92163	0.45966	0.50407	0.72955
OR + BN	0.96607	0.95588	0.92810	0.92739	0.51729	0.55116	0.72508

Then, the new probability values are computed, using the old values $p(c_i|d_j, e_j)$, as $\hat{p}(c_i|d_j, e_j) = f_i(p(c_i|d_j, e_j))$. Once again, we would like to recall that these new results are only "scaled" versions of the old ones, with thresholds being computed using only the training set. The new results are displayed in Table 2.

Note that, using the scaling procedure, ROC and MAP values remain equal, whereas PRF and ACC, on the contrary, are considerably improved. However, for the Naive Bayes models the results obtained by the scaled and non-scaled versions were almost the same.

5   Conclusions and Future Works

Given the previous results, we can state the two following conclusions:

- The use of the Bayesian network structure for links can moderately improve a basic "flat-text" classifier.
- Our results are fairly well situated in a middle-high point among all participants in this track.

The first statement is clear, particularly in the case of the OR gate classifier, where some measures, like PRF are improved around 10%. Accuracy is improved 3-4%, while ROC stands more or less equal. Only MAP is slightly decreased (less than 1%). The changes on the naive Bayes classifier are more irrelevant, but they are positive too.

The second statement can be easily proved watching at the official table of results. Our best model (OR + BN) performs in a medium position for ACC, slightly better for PRF, fairly well for MAP (where only 4 models beat us) and very well for ROC measures (the third best performing model in each of the two versions of ROC, among all the participants).

The results could probably be improved with the usage of a better probabilistic base classifier. For example, a logistic regression or some probabilistic version of a SVM classifier (like the one proposed by Platt [11]), which are likely to have better results than our base models (although they can be much more inefficient). We expect to carry out more experiments with different basic classifiers in the future.

Acknowledgments. This work has been jointly supported by the Spanish Consejería de Innovación, Ciencia y Empresa de la Junta de Andalucía, Ministerio de Ciencia de Innovación and the research programme Consolider Ingenio 2010, under projects P09-TIC-4526, TIN2008-06566-C04-01 and CSD2007-00018, respectively.

References

1. Buntine, W.L.: A guide to the literature on learning probabilistic networks from data. IEEE Transactions on Knowledge and Data Engineering 8, 195–210 (1996)
2. Cano, A., Moral, S., Salmerón, A.: Algorithms for approximate probability propagation in Bayesian networks. In: Advances in Bayesian Networks, Studies in Fuzziness and Soft Computing, vol. 146, pp. 77–99. Springer, Heidelberg (2004)
3. de Campos, L.M., Fernández-Luna, J.M., Huete, J.F., Romero, A.E.: OR gate Bayesian networks for text classification: a discriminative alternative approach to multinomial naive Bayes. In: XIV Congreso Español sobre Tecnologías y Lógica Fuzzy, pp. 385–390 (2008)
4. de Campos, L.M., Fernández-Luna, J.M., Huete, J.F., Romero, A.E.: Probabilistic methods for structured document classification at INEX'07. In: Fuhr, N., Kamps, J., Lalmas, M., Trotman, A. (eds.) INEX 2007. LNCS, vol. 4862, pp. 195–206. Springer, Heidelberg (2008)
5. de Campos, L.M., Fernández-Luna, J.M., Huete, J.F., Romero, A.E.: Probabilistic methods for link-based classification at INEX'08. In: Geva, S., Kamps, J., Trotman, A. (eds.) INEX 2008. LNCS, vol. 5631, pp. 453–459. Springer, Heidelberg (2009)
6. Denoyer, L., Gallinari, P.: Overview of the INEX 2008 XML Mining Track. In: Geva, S., Kamps, J., Trotman, A. (eds.) INEX 2008. LNCS, vol. 5631, pp. 401–411. Springer, Heidelberg (2009)
7. Elvira Consortium: Elvira: An environment for probabilistic graphical models. In: First European Workshop on Probabilistic Graphical Models, pp. 222–230 (2002)
8. Heckerman, D., Geiger, D., Chickering, D.M.: Learning Bayesian networks: the combination of knowledge and statistical data. Machine Learning 20, 197–243 (1995)
9. McCallum, A., Nigam, K.: A Comparison of event models for Naive Bayes text classification. In: AAAI/ICML Workshop on Learning for Text Categorization, pp. 137–142. AAAI Press, Menlo Park (1998)
10. Pearl, J.: Probabilistic Reasoning in Intelligent Systems: Networks of Plausible Inference. Morgan Kaufmann, San Francisco (1988)
11. Platt, J.: Probabilistic outputs for Support Vector Machines and comparisons to regularized likelihood methods. In: Advances in Large Margin Classifiers, pp. 61–74. MIT Press, Cambridge (1999)
12. Sebastiani, F.: Machine Learning in automated text categorization. ACM Computing Surveys 34, 1–47 (2002)
13. Witten, I.H., Frank, E.: Data Mining: Practical Machine Learning Tools and Techniques, 2nd edn. Morgan Kaufmann, San Francisco (2005)
14. Yang, Y.: A study of thresholding strategies for text categorization. In: 24th Annual International ACM SIGIR Conference on Research and Development in Information Retrieval, pp. 137–145 (2001)
15. Yang, Y., Slattery, S.: A study of approaches to hypertext categorization. Journal of Intelligent Information Systems 18, 219–241 (2002)

Clustering with Random Indexing K-tree and XML Structure

Christopher M. De Vries, Shlomo Geva, and Lance De Vine

Faculty of Science and Technology,
Queensland University of Technology, Brisbane, Australia
chris@de-vries.id.au, {s.geva,l.devine}@qut.edu.au

Abstract. This paper describes the approach taken to the clustering task at INEX 2009 by a group at the Queensland University of Technology. The Random Indexing (RI) K-tree has been used with a representation that is based on the semantic markup available in the INEX 2009 Wikipedia collection. The RI K-tree is a scalable approach to clustering large document collections. This approach has produced quality clustering when evaluated using two different methodologies.

Keywords: INEX, XML, Mining, Documents, Clustering, Structure, K-tree, Random Indexing, Random Projection.

1 Introduction

The cluster hypothesis suggests that documents that cluster together tend to have relevance to similar queries. The clustering task at INEX 2009 aims to evaluate the utility of clustering in collection selection. The goal of clustering is to minimise the spread of relevant results of ad-hoc queries over a clustering solution. The purpose of clustering in this context is to determine the distribution of a collection over multiple machines. We have a dual optimisation problem - it is desirable to maximise the number of clusters while minimising the spread of relevant results of ad-hoc queries over the clusters. Search efficiency can be increased with the distribution of clusters (sub-collections) on more machines. However, since it is not possible to produce clusters that split the collection to perfectly satisfy all conceivable ad-hoc queries, a good clustering solution is expected to optimise the distribution such that for most ad-hoc queries most of the results can be found in a small set of clusters. The goal of collection selection is then to rank the clusters (sub-collections) to identify the order in which they should be searched to satisfy any given query.

We have used K-tree [1,2] to generate clustering solutions. The scalability of K-tree in a document clustering setting has been discussed by De Vries and Geva [3,4]. The original contribution to K-tree in INEX 2009 is the use of Random Indexing (RI) to represent the documents. The K-tree algorithm has also been modified to work with the RI representation. RI facilitates an efficient and economical vector space representation. The RI K-tree provides a scalable approach

S. Geva, J. Kamps, and A. Trotman (Eds.): INEX 2009, LNCS 6203, pp. 407–415, 2010.

to clustering large collections at multiple granularities. The latest Wikipedia collection has included semantic markup that is based on the YAGO ontology. This markup had been used in encoding the documents, and two simple approaches are described in Section 4.

This paper introduces and defines Random Indexing in Section 2 and explains its use with K-tree in Section 3. The representation of semantic markup is discussed in Section 4. The combination of RI K-tree and representation of semantic markup introduced in earlier sections is applied to the INEX clustering task in Sections 5, 6 and 7. The paper ends with a conclusion in Section 8

2   Random Indexing

RI [5] is an efficient, scalable and incremental approach to the implementation of a word space model. Word space models use the distribution of terms in documents to create high dimensional document vectors. The directions of these document vectors represent various semantic meanings and contexts.

Latent Semantic Analysis (LSA) [6] is a popular word space model. LSA creates context vectors from a document term occurrence matrix by performing Singular Value Decomposition (SVD). Dimensionality reduction is achieved through projection of the document term occurrence vectors onto the subspace spanned by the vectors with the largest singular values in the decomposition. This projection is optimal in the sense that it minimises the sum of squares of the difference between the original matrix and the projected matrix components. In contrast, Random Indexing first creates random context vectors of lower dimensionality and then combines them to create a term occurrence matrix in the dimensionally reduced space. Each term in the collection is assigned a random vector and the document term occurrence vector is then a superposition of all the term random vectors. There is no matrix decomposition and hence the process is efficient.

RI is also known as Random Projection and is explained by the Johnson and Linden-Strauss lemma [7]. It states that if points in a high dimensional space are projected into a lower dimensional, randomly selected subspace of sufficient dimensions they will approximately retain the same topology. Any n point set in Euclidean space can be embedded in $O(\log n/\epsilon^2)$ dimensions without distorting the pair-wise distances between points by more than $1 \pm \epsilon$, where $0 < \epsilon < 1$. Dasgupta and Gupta [8] have provided a proof for the Johnson and Linden-Strauss lemma, showing that the proposed bounds of the lemma hold.

The RI mapping is performed by producing r dimensional index vectors for each term in a collection, where r is the desired dimensionality of the reduced space. We have chosen these vectors to be sparse and ternary. Ternary index vectors were introduced by Achlioptas [9] as being friendly for database environments. Bingham and Mannila [10] have found that the sparsity of index vectors does not effect the distortion of the embedding via experimental analysis. Sparse index vectors reduce the time to complete RI as only 10 percent of the dimensions are non-zero. However, other choices exist for index vectors such as binary spatter codes [11] which are randomly selected binary vectors and holographic reduced representations [12] that are dense randomly selected real valued vectors. When indexing the INEX 2009 collection, the index vectors are multiplied by the BM25 weight for each term in each document and added to the RI document vector. The document vector becomes a superposition of the index vectors multiplied by the term weights as determined by BM25.

RI can be viewed as a matrix multiplication of a document by term matrix D and a random projection matrix I resulting in a reduced matrix R. Row vectors of I contain index vectors of r dimensions for each term in D. Moreover, n is the number of documents, t is the number of terms and r is the dimensionality of the reduced spaced. R is the reduced matrix where each row vector represents a document. Equation 1 defines RI as a matrix multiplication.

$$D_{n \times t} I_{t \times r} = R_{n \times r} \tag{1}$$

Note that the RI document vectors themselves are not random. They are composed of a superposition of random term vectors and the superposition result depends on BM25 term weights and document content.

Another way to view RI is to interpret each index vector as a code. These codes are nearly orthogonal to all other codes produced, resulting in minimal interference between terms in the reduced vector space. Orthogonality can be measured by creating a pair-wise distance matrix between index vectors using cosine similarity as a distance measure. If two vectors are orthogonal their cosine similarity will be zero. The closer the vectors are to orthogonal the closer their cosine similarity will be to zero. Therefore, it is expected that the pair-wise distance matrix will contain values close to zero in every position except the main diagonal. Finding truly orthogonal codes is computationally expensive and therefore avoided. Nearly orthogonal codes are found by drawing values in the vector from a normal distribution. Figure 1 shows the addition of index vectors (nearly orthogonal codes) to create a document representation.

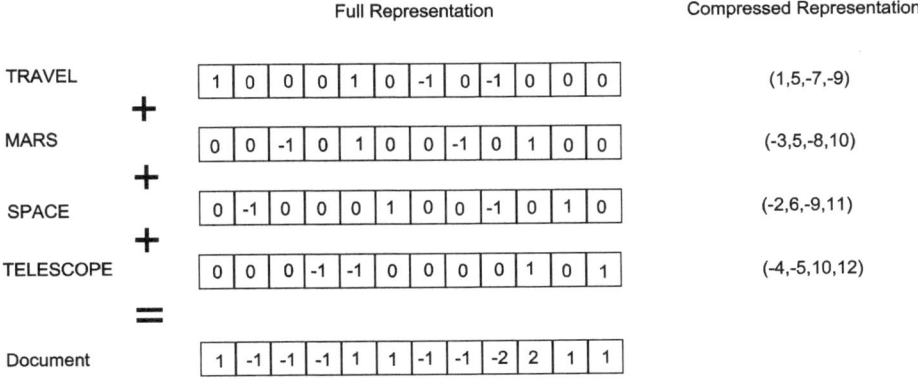

Fig. 1. Random Indexing Example

3   Random Indexing K-tree

The K-tree is an online and dynamic clustering algorithm that scales well by making many local decisions resulting in a hierarchical tree structure. It is a hybrid of the B^+-tree and k-means algorithms where the B^+-tree has been adapted for multi-dimensional data and the k-means algorithm is used to perform splits in the tree. It is built in a bottom-up manner as data arrives. De Vries and Geva [2,3,4] discuss the algorithm and its application to document clustering, including the scalability of the algorithm. K-tree was compared to the popular CLUTO clustering toolkit and found to cluster significantly faster when a large number of clusters are required [4]. The Random Indexing (RI) K-tree [13] combines K-tree with RI to improve the quality of results and run-time performance.

The time complexity of K-tree depends on the length of the document vectors. K-tree insertion incurs two costs, finding the appropriate leaf node for insertion and k-means invocation during node splits. It is therefore desirable to operate with a lower dimensional vector representation.

The combination of RI with K-tree is a good fit. Both algorithms operate in an on-line and incremental mode. This allows it to track the distribution of data as it arrives and changes over time. K-tree insertions and deletions allow flexibility when tracking data in volatile and changing collections. Furthermore, K-tree performs best with dense vectors, such as those produced by RI.

Given the scalable and dynamical properties of the RI K-tree algorithm we propose it is a good solution for clustering large volatile document collections. The logarithmic lookup time of K-tree [13] to find the most similar cluster is also of use in a functioning information retrieval system relying on collection selection. This allows a query broker to direct queries to the most relevant sub-collection in sub-linear time with respect to the size of the collection.

4   Document Representation

Document structure has been represented by using a bag of words and a bag of tags representation derived from the semantic markup in the INEX 2009 collection. Both are vector space representations.

The bag of words is made up of term frequencies contained within any entity tags in the collection. The term frequencies were weighted with BM25 [14] where $K1 = 2$ and $b = 0.75$. We hypothesise that terms contained within entity tags are more likely to indicate the topic of a document. Therefore, documents with the same topic will fall closer together in the vector space representation. This indirectly exploits the XML structure.

The bag of tags representation is made up in an analogous manner of XML entity tag frequencies. The tag frequencies were not weighted. Entity tags consist of concepts such as scientist, location and person. We conjecture that documents with similar tags will belong to the same topic. Future work may compare tag frequency based vectors to set based vectors. In set based vectors each tag would be recorded as existing in a document or not. This way it can be determined if the use of tag frequencies is worthwhile. If a power law distribution exists in tag frequencies the Inverse Document Frequency heuristic may also prove useful as it did with link graphs [3]. The entity tags directly exploit structure by indexing it.

The bag of words and tags representations were combined. This is done by adding the two vector space representations together and then normalising the resulting vector to unit length. As both of the representations are based on RI, the codes between representations will be nearly orthogonal. However, a larger number of dimensions may be required to accommodate the extra information.

5   Run-Time Performance

The performance of RI K-tree has been measured when operating in main memory. The concern is with the performance of the clustering algorithm. Efficiency was not taken into account when indexing or loading the final representation into memory.

All performance figures are for processing all 2,666,190 XML Wikipedia documents. The RI operations took a total of 1860 seconds for the entity text representation. The randomly selected lower dimensional space had 1000 dimensions. The run time of the K-tree algorithm varies between 1200 and 1500 seconds depending on the tree order selected between 15 and 50. This includes the process of re-inserting all vectors to their nearest neighbour leaves upon completion of the tree building process. This produces clustering at many different granularities at once. Table 1 lists the different sized clusters found by trees of order 20, 40, 60, 80 and 100, where m is the tree order.

Table 1. K-tree Clusters

Level	$m = 20$	$m = 40$	$m = 60$	$m = 80$	$m = 100$
1	12	3	8	3	92
2	111	89	356	129	2011
3	542	1260	5610	3090	53174
4	2161	12865	89612	67794	
5	8529	154934			
6	37230				
7	197299				

6   Experimental Setup

The Random Indexing (RI) K-tree has been used to cluster all 2,666,190 XML documents in the INEX 2009 Wikipedia collection. Bag of words and tags representations were used to create different clusters. Both representations were also combined to create a third set of clusters. Clusters were created as close as possible to the 100, 500, 1000, 5000 and 10000 clusters required for evaluation. The

RI K-tree produces clusters in an unsupervised manner where the exact number of clusters can not be precisely controlled. It is determined by the tree order and the randomised seeding process. The algorithm produces clusters of many sizes in a single pass. The desired clustering granularity is selected by choosing a particular level in the tree.

Random Indexing (RI) is an efficient dimensionality reduction technique that projects points in a high dimensional space onto a randomly selected lower dimensional space. It is able to preserve the topology of the points. In the context of document representation, topology preserving dimensionality reduction is preserving document meaning, or at least this is the conjecture which we test here. The RI projection produces dense document vectors that work well with the K-tree algorithm.

Cluster quality has been measured with two metrics this year. Purity is a commonly used metric and it is measured against an external ground truth. In the case of the INEX 2009 collection, the categories were created by YAWN. Purity is the fraction of documents with the majority category in a cluster. Micro purity is the average across all clusters in a solution where each cluster's contribution is weighted by the fraction of documents it contains from the whole collection. Thus, smaller clusters have less influence and larger clusters have more influence on the average. Normalised Cumulative Cluster Gain (NCCG) is a new measure based on relevance judgments from search queries in the ad-hoc track. The ad-hoc track at INEX provides most relevant documents for each topic based on manual human evaluations. Given the relevant results an oracle cluster ranking system can be built, where clusters are sorted in descending order by the number of relevant documents they contain. NCCG measures the spread of relevant documents over the clusters. A score of one is achieved if all relevant documents appear in the first cluster and a score of zero is achieved if relevant documents are evenly spread across all clusters. NCCG rewards placing all relevant documents together. Therefore, it is testing the clustering hypothesis that states that relevant documents for a query tend to cluster together.

7   Experimental Results

Table 2 lists micro purity and NCCG scores for all submissions that clustered the full INEX 2009 collection. The table is split into sections corresponding to the required cluster sizes specified for the track. The RI K-tree, using the entity text representation is clearly the best approach when it comes to finding high purity clusters using an approach that can scale to the full collection at all cluster sizes. The NCCG metric for collection selection favours the combination of entity text and tags over either representation. It changes the ordering of results when compared to the traditional ground truth based approach. The C3m based approach produced higher quality clusters with respect to the NCCG metric on two occasions at 100 and 10,000 clusters.

Guyon et. al. [15] argue that the context of clustering needs to be taken into account during evaluation. The evaluation of this INEX task tests the clustering

Table 2. Clusters

Method	Clusters	Micro Purity	NCCG
RI K-tree Text	88	**0.1744**	0.7859
RI K-tree Tags	99	0.1427	0.7851
RI K-tree Text and Tags	105	0.1450	0.8003
C3m Content Only (bildb)	101	0.1566	**0.8205**
RI K-tree Text	420	**0.1918**	0.6770
RI K-tree Tags	477	0.1526	**0.7546**
RI K-tree Text and Tags	509	0.1668	0.7330
RI K-tree Text	1009	**0.2140**	0.6450
RI K-tree Tags	1026	0.1699	0.7021
RI K-tree Text and Tags	963	0.1690	**0.7092**
C3m Content Only (bildb)	1001	0.1617	0.6614
RI K-tree Text	2450	**0.2136**	0.6100
RI K-tree Tags	2407	0.1769	0.6348
RI K-tree Text and Tags	2536	0.1928	**0.6575**
BM25 BicMsGrowingKMeans (mark)	2263	0.1698	0.6349
RI K-tree Text	4914	**0.2384**	0.5581
RI K-tree Tags	4993	0.2020	0.5729
RI K-tree Text and Tags	4978	0.2038	**0.6003**
RI K-tree Text	9725	**0.2719**	0.4736
RI K-tree Tags	10453	0.2321	0.5274
RI K-tree Text and Tags	9896	0.2509	0.5492
C3m Content Only (bildb)	10001	0.1942	**0.6035**
BM25 BicMsGrowingKMeans (mark)	12636	0.2416	0.5885

hypothesis in the information retrieval specific. Clustering is intended to facilitate document distribution and collection selection for ad-hoc retrieval, and it is tested in that setting. This differs greatly from evaluation where authors assign categories to documents and the categories are then used as the ground truth for the evaluation of clustering. Guyon et. al. [15] argue, and we agree, that ground truth based evaluations are unsound. This is particularly true when it comes to an information retrieval setting where the number of potential topics (clusters) is virtually unconstrained. It is a virtually impossible task to compare alternative clustering possibilities by inspecting large numbers of documents in clusters. In contrast, the evaluation of topics represented as queries in an ad-hoc retrieval system achieves high levels of inter-judge agreement. These relevance judgments have been the backbone of ad-hoc information retrieval system evaluations for many years. They have also been exposed to criticism and review by many of the top researchers in the field. By exploiting this high quality, human generated information, we can have great confidence that we are testing clustering in the context of its use. The context is specifically clustering of documents in an information retrieval setting.

Guyon et. al. go as far to say "In our opinion, this approach [ground truth approach] is dangerous. The underlying assumption is that points with the same class labels form clusters. This might be the case for some data sets but not for others.". If the ground truth reflected the application of clustering in an information retrieval context, then the scores would agree between purity and NCCG. However, they do not. Therefore, we argue that the NCCG scores based on ad-hoc queries are more meaningful in an information retrieval setting.

Relevance of documents to queries can also be derived from click-through data in an operational search engine. This provides a potential mountain of relevance judgments.

8 Conclusion

In conclusion the RI K-tree provided a scalable approach to clustering at multiple granularities in a single pass with quality comparable to other approaches. The hypothesis that combining entity text and tag based representations will improve quality held true for the new ad-hoc based evaluation. Furthermore, the evaluation provided insights into why it is important to take context of use into account when evaluating clustering.

References

1. K-tree project page (2009), http://ktree.sourceforge.net
2. Geva, S.: K-tree: a height balanced tree structured vector quantizer. In: Proceedings of the 2000 IEEE Signal Processing Society Workshop Neural Networks for Signal Processing X, vol. 1, pp. 271–280 (2000)
3. De Vries, C., Geva, S.: Document clustering with k-tree. In: Geva, S., Kamps, J., Trotman, A. (eds.) INEX 2008. LNCS, vol. 5631, pp. 420–431. Springer, Heidelberg (2009)
4. De Vries, C., Geva, S.: K-tree: large scale document clustering. In: SIGIR 2009: Proceedings of the 32nd international ACM SIGIR conference on Research and development in information retrieval, pp. 718–719. ACM, New York (2009)
5. Sahlgren, M.: An introduction to random indexing. In: Methods and Applications of Semantic Indexing Workshop at the 7th International Conference on Terminology and Knowledge Engineering, TKE 2005 (2005)
6. Deerwester, S., Dumais, S., Furnas, G., Landauer, T., Harshman, R.: Indexing by latent semantic analysis. Journal of the American Society for Information Science 41(6), 391–407 (1990)
7. Johnson, W., Lindenstrauss, J.: Extensions of Lipschitz mappings into a Hilbert space. Contemporary mathematics 26(189-206), 1 (1984)
8. Dasgupta, S., Gupta, A.: An elementary proof of the Johnson-Lindenstrauss lemma. Random Structures & Algorithms 22(1), 60–65 (2002)
9. Achlioptas, D.: Database-friendly random projections: Johnson-Lindenstrauss with binary coins. Journal of Computer and System Sciences 66(4), 671–687 (2003)
10. Bingham, E., Mannila, H.: Random projection in dimensionality reduction: applications to image and text data. In: KDD 2001: Proceedings of the seventh ACM SIGKDD international conference on Knowledge discovery and data mining, pp. 245–250. ACM, New York (2001)

11. Kanerva, P.: The spatter code for encoding concepts at many levels. In: ICANN 1994, Proceedings of the International Conference on Artificial Neural Networks (1994)
12. Plate, T.: Distributed representations and nested compositional structure. PhD thesis (1994)
13. De Vries, C., De Vine, L., Geva, S.: Random indexing k-tree. In: ADCS 2009: Australian Document Computing Symposium 2009, Sydney, Australia (2009)
14. Robertson, S., Jones, K.: Simple, proven approaches to text retrieval. Update (1997)
15. Guyon, I., von Luxburg, U., Tubingen, G., Williamson, R., Canberra, A.: Clustering: Science or Art?

Utilising Semantic Tags in XML Clustering

Sangeetha Kutty, Richi Nayak, and Yuefeng Li

Faculty of Science and Technology
Queensland University of Technology
GPO Box 2434, Brisbane Qld 4001, Australia
{s.kutty,r.nayak,y2.li}@qut.edu.au

Abstract. This paper presents an overview of the experiments conducted using Hybrid Clustering of XML documents using Constraints (HCXC) method for the clustering task in the INEX 2009 XML Mining track. This techique utilises frequent subtrees generated from the structure to extract the content for clustering the XML documents. It also presents the experimental study using several data representations such as the structure-only, content-only and using both the structure and the content of XML documents for the purpose of clustering them. Unlike previous years, this year the XML documents were marked up using the Wiki tags and contains categories derived by using the YAGO ontology. This paper also presents the results of studying the effect of these tags on XML clustering using the HCXC method.

1 Introduction

INEX 2009 XML mining track has two tasks namely classification and clustering. The clustering task groups the documents without prior knowledge of the categories. This paper presents an overview of the experiments as a result of our participation in the clustering task.

In this paper, we utilise the Hybrid Clustering of XML documents using Constraints (HCXC) method to cluster the INEX 2009 Wikipedia document collection.HCXC utilises constraints and it is an extension of our previous work, Hybrid Clustering of XML documents(HCX)[7].This method utilises frequent subtrees extracted from the structure of XML documents to obtain the content in order to cluster the documents according to the constrained content.The empirical study reveals that HCXC combines both the structure and the content non-linearly. Also, by using the constrained content the term space for clustering is reduced.

Our overall motivation for participating in INEX 2009 was to investigate the impact of using structure along with content on this new collection of the Wikipedia corpus containing categories derived from YAGO ontology[8]. Hence, we study how these documents were clustered using only the structure or the content of XML documents over the combination of the structure and the content. Also, the presence of semantic tags in the XML documents has motivated us to analyse how does our method perform on this collection with the semantic tags.

S. Geva, J. Kamps, and A. Trotman (Eds.): INEX 2009, LNCS 6203, pp. 416–425, 2010.

The rest of this paper is organized as follows. First, Section 2 provides the overview of our approach. Next, Section 3 covers the details about the pre-processing of structure. Then, in Section 4 and Section 5 the frequent mining algorithms and the clustering process are presented. Finally, in Section 6, we discuss our experimental results and draw some conclusions in Section 7.

2   An Overview

As illustrated in Fig.1, the clustering of XML documents using HCXC is conducted using two phases namely frequent mining of the structures and then utilising these common substructures to extract the content for grouping the documents. The process begins with the pre-processing of XML documents to represent their structure in the form of trees. Next, the frequent mining algorithm, Prefix-based Closed Induced Tree Miner using Constraints (PCITMinerConst) is applied to extract the common structural information in the form of subtrees among the XML document collection. Using these common structural information, the content corresponding to every frequent subtree is then extracted and represented in Vector Space Model(VSM). Finally, a partitional clustering algorithm is applied on this content.

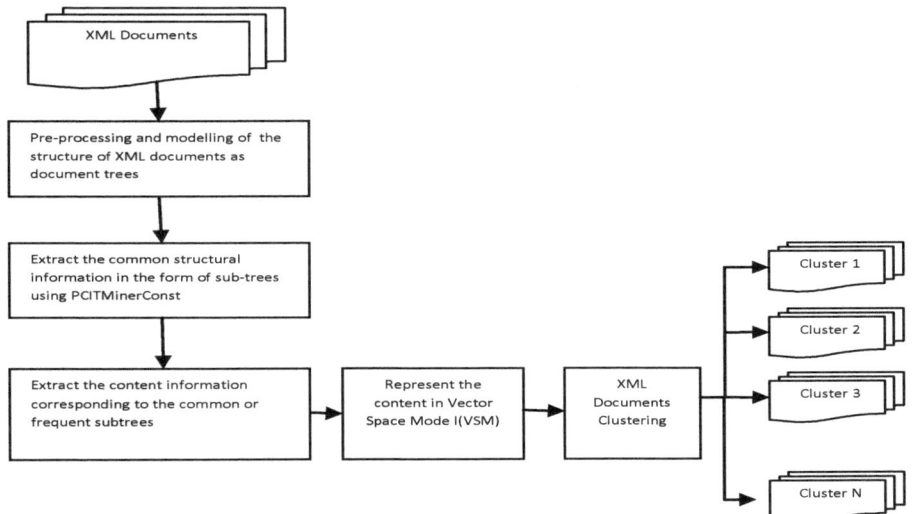

Fig. 1. Overview of HCXC

3   Pre-processing of Structure

The structure of the XML document can have many representations depending on its usability, such as graph, tree and path. We have chosen to use the tree over graph, as the structure of the document did not include any cycles and the problem of mining graphs was known to be NP-hard. On the other hand, paths do not include sibling relationship between the tags hence tree was chosen as the representation. Therefore in the pre-processing phase, each XML document in the collection is modelled as an ordered, labelled and rooted tree. All the documents contain the "article" as the root element hence they are rooted on the "article" element. The tag names are then mapped to unique integers for ease of computation. There are two different types of tags namely formatting tags and descriptive tags which are semantically rich tags. The following table provides the statistics of the tags information.

Table 1. Summary of Entity Tags in INEX 2009 subset

	Subset
Formatting tags	61
Semantically rich tags	9654

Other tag related information such as data types, attributes and constraints is ignored in the representation of structure as the empirical evaluation revealed that this information did not contribute to an improvement in accuracy.

4   Phase 1: Extraction of Common Structural Information

Using the pre-processed structure, PCITMinerConst was applied which generate constrained closed frequent induced subtrees.PCITMinerConst algorithm is an extended version of the PCITMiner [6] algorithm to mine frequent patterns with a constraint. The constraint is applied in the form of the length of the patterns. Let us define the closed frequent induced subtrees.

For a document tree DT_i with node set N and edge set E, a tree dt_i with node set n' and edge set e' is an induced subtree of DT_i iff (1) $n' \in N$; (2) $e' \in E$; (3) the labeling of nodes of n' in dt_i is preserved in DT_i; (4) $(n_1, n_2) \in e'$ where n_1 is the parent node of n_2 in dt_i iff n_1 is the parent node of n_2 in DT_i; and (5) for $n_1, n_2 \in n'$, preorder(n_1)<preorder(n_2) in dt_i iff preorder(n_1)<preorder(n_2) in DT_i. In other words, an induced subtree dt_i preserves the parent-child relationship among the nodes of the document tree, DT_i. Now let us define the frequent induced subtrees and the closed frequent induced subtrees. The frequent induced subtrees in the document tree dataset,DT are the subtrees that have a support that is equal to or more than the user-defined minimum threshold (min_supp). As defined in [6], in DT, there exists two frequent induced subtrees dt_i and dt_j. The frequent induced subtree dt_i is closed of dt_j iff (1) dt_j is an induced subtree of dt_i, supp(dt_i) = supp(dt_j); (2) there exists no supertree for dt_i having the same support as that of dt_i. This property is called as induced closure and dt_i is the closed frequent induced subtree of dt_j.

The PCITMinerConst algorithm computes the length constrained subtrees called Constrained Closed Frequent Induced ($ConstCFI$) subtrees. $ConstCFI$

are generated in a computationally efficient manner as these subtrees have shorter patterns and are concise representation of frequent induced subtrees. As PCITMiner[6] utilises the pattern growth technique and the frequent patterns are in the memory, it is a computationally expensive algorithm for longer structures such as the ones that exists in the INEX 2009 Wikipedia dataset. Also, the empirical analysis showed that computing longer or deeper frequent patterns is not required for this dataset.

5 Phase 2: Extraction of Content and Clustering

Using PCITMinerConst, we generate $ConstCFI$ and use these set of subtrees to extract Structure-Constrained content features from the XML documents . The structure-constrained content features of a given $ConstCFI_i$, $\mathrm{C}(D_i, ConstCFI_i)$ of an XML document D_i, are a collection of node values corresponding to the node labels in $ConstCFI_i$ of D_i. The structure-constrained content features of $ConstCFI$ subtrees, $\mathrm{C}(D_i, ConstCFI)$, for an XML document D_i, are a collection of node values corresponding to node labels in the $ConstCFI$ subtrees of D_i.

The structure-constrained content features of the $ConstCFI$ for every document is extracted.The extracted content is a list of terms which is then pre-processed using the following steps:

1. Stop-word removal - an user-defined stop word list containing 552 words was used to remove stop words from the collection.
2. Stemming using Porters stemming algorithm [2]
3. Integer removal
4. Shorter length words removal- words with length less than or equal to 3 are removed.

The pre-processed content which is a vector of terms for every document in the collection is then represented in VSM with the rows corresponding to the terms in the given document. This VSM is then provided as an input to a clustering algorithm. The k-way clustering algorithm [3] is used in HCXC to group the documents. The k-way clustering solution computes cluster by performing a sequence of k-1 repeated bisections. The input matrix is first clustered into two groups, and then one of these groups is chosen and bisected further. This process of bisection continues until the desired number of clusters is reached.

6 Experiments and Discussion

A number of experiments were conducted on the INEX 2009 Wikipedia clustering task corpus to analyse the impact of using the structure features along with content of XML documents in comparison to using only the structure or the content of these documents. Also, to understand the impact of semantic tags present in this collection for the clustering task.

The INEX 2009 Wikipedia clustering task corpus contained 2.7 million documents and the subset contained 54575 documents.We utilised only the subset for our clustering task submissions. The subset of the collection was extracted using the document ids provided in the classification task excluding 314 files which were missing in the entire document collection.The subset contained 5243 unique entity tags and 1,900,072 unique terms and there were 73,944 categories derived by using the YAGO ontology.

To understand the impact of semantic tags, the experiments are repeated with two sets of collections: 1. *subset with both semantic and formatting tags collection*; and 2.a *semantic subset containing only semantically-rich tags without any formatting tags*. The semantic subset was created by ignoring the formatting tags such as $< p >, < sec >, < i >, < b >$ etc. There were about 61 formatting tags which were detected manually and eliminated.

Firstly, we need to estimate the support threshold(min_supp) and the length constraint($const$) values for the frequent mining algorithm, PCITMinerConst. Hence, we applied support threshold in increasing percentage of 10 on the subset collection.Fig.2 shows the support threshold and the number of subtrees generated using PCITMinerConst.

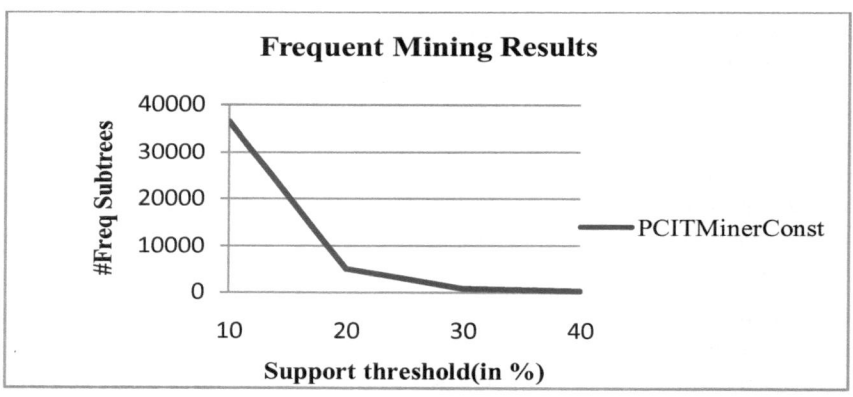

Fig. 2. Frequent mining results using PCITMinerConst

It could be seen that there is an exponential increase in the number of subtrees when the support threshold is reduced below 20% and extracting content for thresholds below 20% might not result in any reduction in the search space. Hence, we chose 20% as the support threshold for generating $ConstCFI$ subtrees. Also, to determine the $const$ value for PCITMinerConst we varied this value from 2 to 14 and determined that a $const$ value of 10 provided $ConstCFI$ subtrees without much information loss.

In order to measure the reduction in the number of terms using HCXC, a Vector Space Model (VSM) was built on all the terms of the documents (Content-only) and clustering was then applied to this representation.Table 2 summarises the dimensionality reduction in both the number of unique terms and the total

number of terms. It can be seen that there is about 54% reductions in the number
of unique terms and the total number of non-zeros by 6% for PCITMinerConst.
There is a drastic reduction in the number of unique terms and the total no. of
non zeroes for the Semantic subset.

Table 2. Summary of Entity Text in INEX 2009 subset

Representation	Unique terms	Total No. Non Zeroes
Content-only	1900072	21480198
Content constrained by $CFIConst$ subtrees for the Original subset	869600	20112292
Content constrained by $CFIConst$ subtrees for the Semantic subset	474426	11105222

We also utilised two document representations namely term frequencies and
BM25 [1]. VSM representation containing term frequencies in TF-IDF were normalized for document length. On the other hand, BM25 works on utilising similar
concepts as that of TF-IDF but has two tuning parameters namely K1 and b. K1
and b influences the effect of term frequency and document length respectively.
The BM25 tuning parameters were set as K1 = 2 and b = 0.75.

6.1   Evaluation Using Purity

The clustering results were evaluated based on two sets of ground truth with
73,944 and 12,804 YAGO categories. The latter ground truth categories have
categories containing at least five documents. Considering each of the YAGO
category as a class, purity measures the extent to which each cluster contains
documents primarily from one class. Each cluster is assigned with the class label
of the majority of documents in it. Then the error is computed as the proportion
of documents with different class and cluster labels. Inversely, the accuracy is
the proportion of documents with the same class and cluster label. Purity is
measured as the ratio of number of documents with the majority label in cluster
to the number of documents in that cluster. The macro and micro purity of
entire clustering solution is obtained as a weighted sum of the individual cluster
purity. In general, larger the value of purity, better the clustering solution is.

$$purity = \frac{Number\ of\ documents\ with\ the\ majority\ label\ in\ cluster\ k}{Number\ of\ document\ in\ cluster\ k} \qquad (1)$$

We present in Fig.3 the results for purity for the subset collection for clustering
solution containing 100 clusters evaluated on the 73,944 YAGO categories as the
ground truth. As there were five submissions with 100 clusters we have chosen
this so that we could compare our method against other approaches without
interpolating the results.

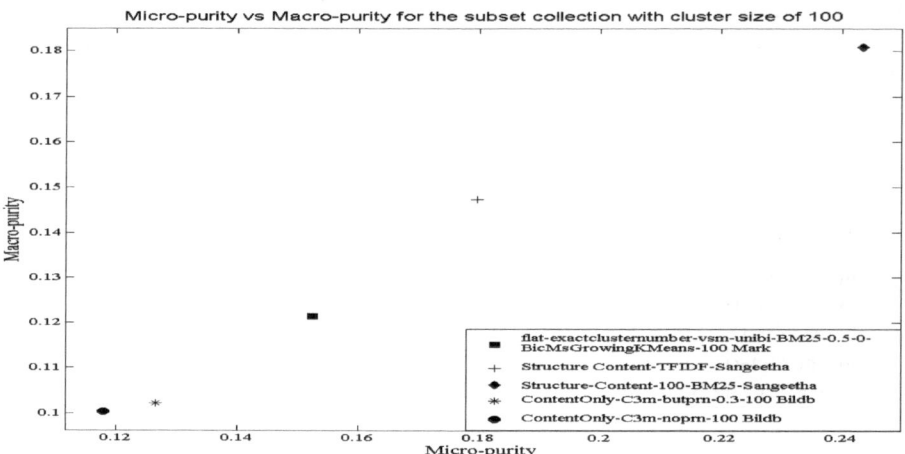

Fig. 3. Micro-purity vs Macro-purity for the subset collection for clustering solution containing 100 clusters

From this figure, we could clearly see that our approach of using structure constrained content performs better than other approaches using content-only. on both micro-purity and macro-purity values The major reason could be that we are effectively using the structure by eliminating non-frequent substructures and the corresponding redundant content is also eliminated. Also, our representation with BM25 outperforms the TF-IDF representation.

6.2   Collection Selection Evaluation Using NCCG Measure

The clustering task in INEX 2009 was also evaluated using a novel evaluation task to determine the quality of clusters relative to the optimal collection selection goal for a given set of queries using manual query assessments from the INEX Ad Hoc track using Normalized Cumulative Cluster gain (NCCG). The details of this metric is provided in the clustering track overview paper [5]. There were 4858 documents for the 69 topics or queries and the subset had 643 documents relevant to 52 topics or queries. The following graph Fig.4 shows the distribution of the topics and its relevant documents in the subset collection. It could be seen that a majority of the topics contained less than 5 relevant documents and there were only 9 topics which contained more than 20 documents. On an average, each topic contained 12 documents.

We also conducted experiments using the Content-only, Structure-only and our structure and content representation to understand how the various representations affect the clustering solution. In the structure-only representation,we used the tags information to group them. It can be seen from Fig.5 that the NCCG values for our representation of using structure and content together is higher than both structure and content-only representation. But the performance gain for our representation is significant in dataset containing both semantic

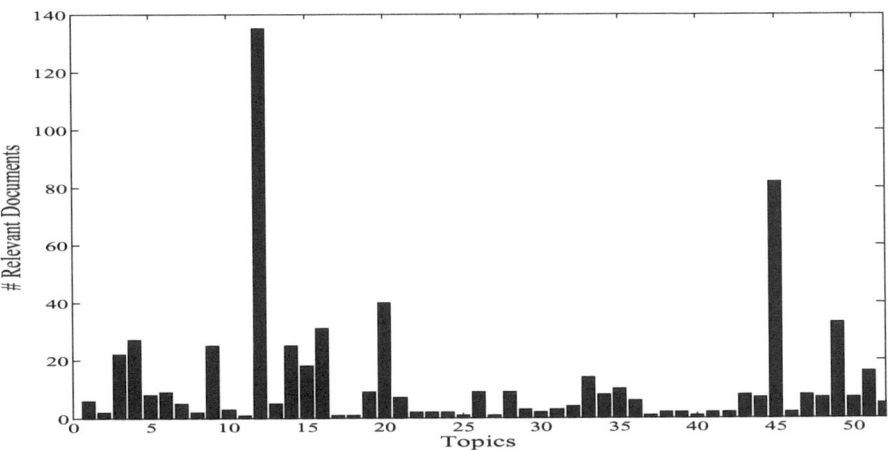

Fig. 4. Distribution of the topics and the relevant documents in the subset collection

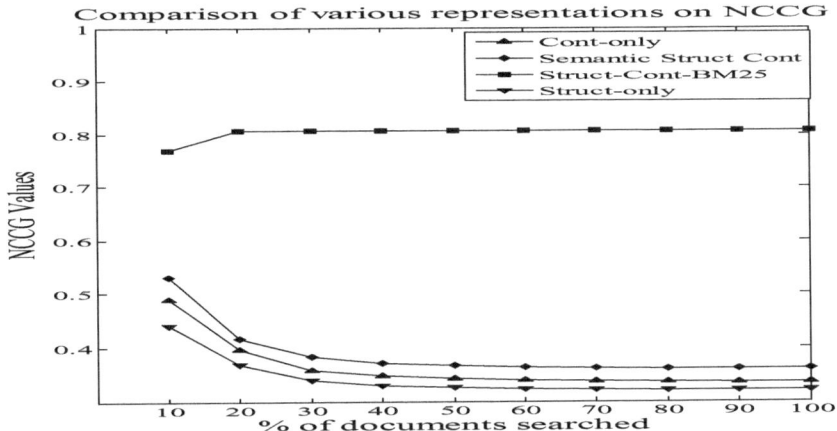

Fig. 5. Comparison of structure-only, content-only and structure-content on semantic subset and original subset for 100 clusters

and formatting tags collection. It also reveals that by searching only 20% of documents our representation could achieve higher NCCG values.

Also, it is an interesting result in comparison to the results for previous years' clustering tasks where clustering only the tags did not provide any useful clustering solution [4]. However, this year due to the use of semantic tags, the clustering solution using just the structure produces better clustering solution in comparison to using content than previous years.

Now we analyse the impact of the number of clusters in the clustering solution on NCCG values in Fig 6. A varied number of clusters ranging from 75 to 1000 were generated for our structure-content BM25 representation.

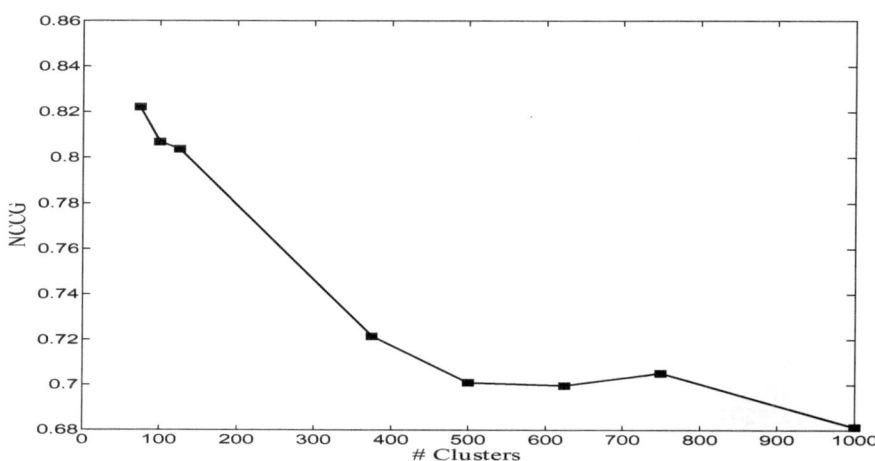

Fig. 6. Comparison on the number of clusters based on NCCG for Structure-Content BM25 representation

From Fig. 6, we could see that with the larger number of clusters, there is a drop in NCCG values. Also, there could be a possibility that we need to be searching a very large number of clusters to find relevant documents. With an effective cluster representation, the number of searches can be reduced with smaller number of clusters. Hence, for our method, smaller number of clusters provide better clustering solution than larger number of clusters.

7   Conclusion

In this paper, we have proposed and presented the results of HCXC clustering approach using constrained frequent subtrees for grouping XML documents using both their structure and content of XML documents in the INEX 2009 Wikipedia dataset. The main aim of this study is to explore the applicability of the proposed method to INEX dataset as well as to understand the importance of the content and structure of the XML documents for the clustering task. We have demonstrated that we were able to apply clustering on both the structure and content of XML documents and also provide more meaningful clusters. By doing so, we could effectively reduce the dimensionality for clustering. On the other hand, our empirical analysis reveals that using only the content enclosed within the semantic tags does not show a significant improvement in the quality of the clustering solution in comparison to using the content enclosed within both the semantic tags and formatting tags. This can be due to the presence of less content in the semantic tags and hence the content corresponding to them does not contain enough information for clustering. As a future work, we are planning to represent the structure and content in higher dimensions and apply clustering on higher dimension data model. Also, the partitional clustering algorithm was not able to scale for clusters more than 1000 and our future work will include developing algorithms to provide clusters more than 1000 and to scale to the entire dataset.

References

1. Robertson, S.E., Jones, K.S.: Simple, Proven Approaches to Text Retrieval. Technical report (1997)
2. Porter, M.F.: An Algorithm for Suffix Stripping. Program 14(3), 130–137 (1980)
3. Karypis, G.: CLUTO - a clustering toolkit. Technical Report #02-017 (November 2003)
4. Denoyer, L., Gallinari, P.: Overview of the INEX 2008 XML Mining Track. In: Geva, S., Kamps, J., Trotman, A. (eds.) INEX 2008. LNCS, vol. 5631, pp. 401–411. Springer, Heidelberg (2009)
5. Nayak, R., De Vries, C., Kutty, S., Geva, S., Denoyer, L., Gallinari, P.: Overview of the INEX 2009 XML Mining Track: Clustering and Classification of XML Documents. In: Geva, S., Kamps, J., Trotman, A. (eds.) INEX 2009. LNCS, vol. 6203, pp. 366–378. Springer, Heidelberg (2010)
6. Kutty, S., Nayak, R., Li, Y.: PCITMiner- Prefix-based Closed Induced Tree Miner for finding closed induced frequent subtrees. In: Christen, P., Kennedy, P.J., Li, J., Kolyshkina, I., Williams, G.J. (eds.) AusDM. CRPIT, vol. 70, pp. 151–160. Australian Computer Society (2007)
7. Kutty, S., Nayak, R., Li, Y.: HCX: an efficient hybrid clustering approach for XML documents. In: Borghoff, U.M., Chidlovskii, B. (eds.) ACM Symposium on Document Engineering, pp. 94–97. ACM, New York (2009)
8. Schenkel, R., Suchanek, F.M., Kasneci, G.: YAWN: A Semantically Annotated Wikipedia XML Corpus. In: Kemper, A., Schöning, H., Rose, T., Jarke, M., Seidl, T., Quix, C., Brochhaus, C. (eds.) BTW, GI. LNI, vol. 103, pp. 277–291 (2007)

UJM at INEX 2009 XML Mining Track*

Christine Largeron, Christophe Moulin, and Mathias Géry

Université de Lyon, F-42023, Saint-Étienne, France
CNRS UMR 5516, Laboratoire Hubert Curien
Université de Saint-Étienne Jean Monnet, F-42023, France
{christine.largeron,christophe.moulin,mathias.gery}@univ-st-etienne.fr

Abstract. This paper reports our experiments carried out for the INEX XML Mining track 2009, consisting in developing categorization methods for multi-labeled XML documents. We represent XML documents as vectors of indexed terms. The purpose of our experiments is twofold: firstly we aim to compare strategies that reduce the index size using an improved feature selection criteria CCD. Secondly, we compare a thresholding strategy ($MCut$) we proposed with common $RCut$, $PCut$ strategies. The index size was reduced in such a way that the results were less good than expected. However, we obtained good improvements with the $MCut$ thresholding strategy.

1 Introduction

This paper describes the participation of Jean Monnet University at the INEX 2009 XML Mining Track. For the categorization task (or classification), given a set of categories, a training set of preclassified documents is provided. Using this training set, the task consists in learning the classes descriptions in order to be able to classify a new document in the categories.

One main difference in the collection of documents provided in INEX 2009 relatively to INEX 2008 lies in the overlapping of the categories and in their dependencies [1]. When each document belongs to one and only one category in INEX 2008, it can belong to several categories in INEX 2009. With the imbalance between the categories, their overlapping poses new challenges and gives opportunities for design machine learning algorithms more suited for XML documents mining.

In this article, we focus on the selection of the set of classes that will label a document for this multi-label text categorization. We explore two approaches. The first one uses a binary classifier which considers one category against the others. The algorithm returns two answers (yes or no) used to decide whether the document belongs or not to this category. In that case, the selection of a set of words characteristic of the category can be essential for improving the performance of the algorithm. Our first contribution to Inex 2009 consists in an

---

* This work has been partly funded by the Web Intelligence project (région Rhône-Alpes: http://www.web-intelligence-rhone-alpes.org)

S. Geva, J. Kamps, and A. Trotman (Eds.): INEX 2009, LNCS 6203, pp. 426–433, 2010.

improvement of the selection criteria that we have introduced in INEX 2008 and which permitted to get the best results of the competition while reducing the index size [2].

The second approach uses a multi-label classifier which considers simultaneously all the categories. Given a document, the classifier returns a score (*i.e.* a numerical value), for each category. In the context of single label classification in which one and only one class must be attributed to each document the decision rule is obvious: it consists to return the class corresponding to the best score. On the contrary, in multi-label categorization, this approach raises the question of the number of classes that must be assigned to each document. In this article, we propose a thresholding strategy for selection of candidate classes and we compare it with the commonly used methods $PCut$ and $RCut$ [7,4].

In the aim of introducing our notations, a brief presentation of the vector space model (VSM [5]), used to represent the documents, is given in section 2, the selection features criteria are defined in the following section. The thresholding strategy for selection of candidate classes is presented in section 4, while the runs and the obtained results are detailed in sections 5 and 6.

2   Document Model for Categorization

Vector space model, introduced by Salton et al. [5], has been widely used for representing text documents as vectors which contain terms weights. Given a collection D of documents, an index $T = \{t_1, t_2, ..., t_{|T|}\}$, where $|T|$ denotes the cardinal of T, gives the list of terms (or features) encountered in the documents of D. A document d_i of D is represented by a vector $\overrightarrow{d_i} = (w_{i,1}, w_{i,2}, ..., w_{i,|T|})$ where $w_{i,j}$ is the weight of the term t_j in the document d_i defined according the TF.IDF formula :

$$w_{i,j} = tf_{i,j} \times idf_j$$

with $tf_{i,j} = \frac{n_{i,j}}{\sum_l n_{i,l}}$ where $n_{i,j}$ is the number of occurrences of t_j in document d_i normalized by the number of occurrences of terms in document d_i and $idf_j = \log \frac{|D|}{|\{d_i:t_j \in d_i\}|}$ where $|D|$ is the total number of documents in the corpus and $|\{d_i : t_j \in d_i\}|$ is the number of documents in which the term t_j occurs at least one time.

3   Criteria for Features Selection

3.1   Category Coverage Criteria (CC)

In the context of text categorization, the number of terms belonging to the index can be exceedingly large and all these terms are not necessarily discriminant features of the categories. It is the reason why, it can be useful to select a subset of T giving a more representative description of the documents belonging to each category. For this purpose, we proposed in a previous work a selection features criteria, called coverage criteria CC and based on the frequency of the documents containing the term [2]. Let $f_j^k = \frac{df_j^k}{|c_k|}$ be the frequency of documents belonging to c_k and including t_j where $df_j^k = |\{d_i \in c_k : t_j \in d_i\}|, k \in \{1, ...r\}$ is the number of documents in the category c_k in which the term t_j appears and $|c_k|$ is the number of documents belonging to c_k. The higher the number of documents of category c_k containing t_j, the higher f_j^k, CC is defined by:

$$CC_j^k = \frac{df_j^k}{|c_k|} * \frac{f_j^k}{\sum_k f_j^k} = \frac{(f_j^k)^2}{\sum_k f_j^k}$$

If the value of CC_j^k is high, then t_j is a characteristic feature of the category c_k.

3.2   Difference Category Coverage Criteria (CCD)

The previous criteria considers only the coverage of the category by one term but it does not take into account the coverage of the other categories. The difference of category coverage permits to overcome this drawback. Thus, the Category Coverage Difference CCD is defined by :

$$CCD_j^k = (CC_j^k - CC_j^{\bar{k}})$$

$$CC_j^{\bar{k}} = \frac{(f_j^{\bar{k}})^2}{f_j^k + f_j^{\bar{k}}}$$

with $f_j^{\bar{k}} = \frac{df_j^{\bar{k}}}{|D| - |c_k|}$ and $df_j^{\bar{k}} = |\{d_i \in D \wedge d_i \notin c_k : t_j \in d_i\}|, k \in \{1, .., r\}$. As previously, if the value of CCD_j^k is high, then t_j is a characteristic feature of the category c_k.

The CC and CCD criteria can be used for multi-label text categorization by binary classifier. For each category and consequently each classifier, they permit to reduce the index to the set of words which are the most characteristic of this category.

4   Thresholding Strategies

When a multi-label algorithm is used in multi-label text categorization, one score $\phi(\overrightarrow{d_i}, c_k)$ is produced by the classifier for each document-category pair (d_i, c_k). Given these scores, the problem consists to determine the set of classes $L(d_i)$ which must be attributed to each document d_i.

To solve this problem, different approaches have been proposed, which consist in applying a threshold to the scores returned by the classifier. In $RCut$ method, given a document d_i, the scores $(\phi(\overrightarrow{d_i}, c_k), k = 1, .., r)$ are ranked and the t top ranked classes are assigned to d_i. The value of the parameter t can be either specified by the user or learned using a training set. In $PCut$ method, given a category c_k, the scores $(\phi(\overrightarrow{d_i}, c_k), i = 1, .., |D|)$ are ranked and the n_k top ranked documents are assigned to the class with:

$$n_k = P(c_k) * x * r$$

where $P(c_k)$ is the prior probability for a document to belong to c_k, r, is the number of categories, and x is a parameter which must be estimated using a training set. A review of these methods can be found in [7,4].

In $PCut$ as well as in $RCut$, the performance of the classifier depends on the value of the parameter (t or x). The main advantage of the thresholding method proposed in this article is that the threshold is automatically fixed.

This method, called $MCut$ (for Maximum Cut) is based on the following principle, explained graphically. Given a document d_i, the scores ($\phi(\overrightarrow{d_i}, c_k), k = 1, .., r$) are ranked in decreasing order. The sorted list obtained is noted $S = (s(l), l = 1, .., r)$ where $s(l) = \phi(\overrightarrow{d_i}, c_k)$ if $\phi(\overrightarrow{d_i}, c_k)$ is the lth highest value in S.

Then, a graph of the scores in their decreasing order is drawned (*i.e.* $s(l), l = 1, .., r$ in function of l). The value t retained as threshold is the middle of the maximum gap for S: $t|(s(t) + s(t+1))/2 = Max\{(s(l) + s(l+1))/2, l = 1, .., r-1\}$

The clusters assigned to d_i are those corresponding to a score $\phi(\overrightarrow{d_i}, c_k)$ higher than t : $L(d_i) = \{c_k \in C/\phi(\overrightarrow{d_i}, c_k) > t\}$.

For instance, the graphs of the scores ($s(l), l = 1, .., r$) versus l obtained for two documents: d_1 (on the left side) and d_2 (on the right one) are presented in figure 1. Using $MCut$, the document d_1 is assigned to one class (on the left) while the document d_2 is assigned to three classes (on the right).

$MCut$ is also compared with the $RCut$ strategy in figure 1. In $RCut_1$ (resp. $RCut_2$), the t parameter is set up to 1 (resp. 2). For the document d_1, the same set of classes is affected by $RCut_1$ and $MCut$, while $RCut_2$ assigns one class more. In the second case, the set of affected classes is different. While d_2 belongs to three classes with $MCut$ strategy, it is associated to one class (resp. two classes) with the $RCut_1$ (resp. $RCut_2$) strategy.

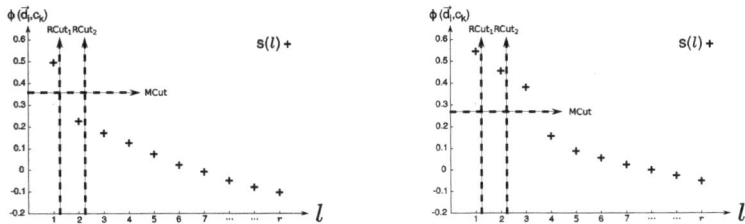

Fig. 1. Illustration comparing RCut and MCut thresholdling strategies

5   Experiments

5.1   Collection INEX XML Mining

The XML Mining collection is composed of about 54 889 XML documents of the Wikipedia XML Corpus. This subset of Wikipedia represents 39 categories, each corresponding to one subject or topic. This year, the collection is multi-label and each document belongs to at least one category. In the XML Mining

Track, the training set is composed of 20% of the collection which corresponds to 11 028 documents. On the training set, the mean of the number of category by document is 1.46 and 9 809 documents belong to only one category.

5.2   Pre-processing and Categorization

The first step of the categorization approach that we propose, consists in a pre-processing of the collection. It begins by the construction of the list all the terms (or features) encountered in the documents of the collection. This index of 1 136 737 terms is built with the LEMUR software[1]. The Porter Algorithm [3] has also been applied in order to reduce different forms of a word to a common form. After this pre-processing, it still remains a large number of irrelevant terms that could degrade the categorization, e.g.: numbers (7277, -1224, 0d254c, etc.), terms with less than three characters, terms that appear less than three times, or terms that appear in almost all the documents of the training set corpus. After their deletion, the index size is reduced to 295 721 terms on all the documents and it will be noted T. Depending on the category c_k, T_k will correspond to the index only composed of terms that appear in documents of the category c_k. In order to reduce the index size, we also define $T_{k_{1000}}$ as the index of the category c_k composed of the most characteristic terms of category c_k according to the CCD criteria introduced in section 3. If CCD^k_{1000} corresponds to the best thousandth score obtained with the CCD criteria for the category c_k, $T_{k_{1000}}$ is composed of all terms t_j for which CCD^k_j is higher than CCD^k_{1000}. All the indexes definitions are summarized in the table 1.

Table 1. Summary of all defined indexes

Index		Definition
T	=	$\{t_j \in d_i \mid d_i \in D\}$
T_k	=	$\{t_j \in d_i \mid d_i \in D \wedge d_i \in c_k\}$
$T_{k_{1000}}$	=	$\{t_j \in T_k \wedge CCD^k_j \geqslant CCD^k_{1000}\}$

The second step is the categorization step itself. The Support Vector Machines (SVM) classifiers are used for the categorization. SVM was introduced by Vapnik for solving two classes pattern recognition problems using Structural Risk Minimization principal[6]. In our experiments, the SVM algorithm available in the Liblinear library[2] has been used.

In the XML Mining Track, the final score ($score(d_i, c_k)$) assigned to a document d_i for the category c_k has to be included in $[0, 1]$ and has to be higher than 0.5 if this document belongs to the category c_k. When the SVM is used as a multi-label classifier (noted $multi - label$), it provides a score $\phi(\vec{d_i}, c_k)$ for each pair document - category (d_i,c_k). In that case, the final score $score(d_i, c_k)$ associated to d_i and c_k corresponds to $\phi(\vec{d_i}, c_k)$ normalized. So, the final result, given d_i, is a set of classes ordered by relevancy. When the SVM is used

---

[1] Lemur is available at the URL http://www.lemurproject.org

[2] http://www.csie.ntu.edu.tw/~cjlin/liblinear/ - L2 loss support vector machine primal.

as a binary classifier (noted $uni - label$), it provides, given a category c_k, two scores $(\phi_k(\vec{d_i}, c_k)$ and $\phi_k(\vec{d_i}, \bar{c_k}))$. In that case, the score $score(d_i, c_k)$ equals 1 if $\phi_k(\vec{d_i}, c_k) > \phi_k(\vec{d_i}, \bar{c_k})$ and equals 0 otherwise. The final result is a set of unordered classes.

5.3 Submitted Runs

In the context of multi-label text categorization, our aim was to evaluate on one hand the influence of the features selection on the performance of the binary classifier (runs $lahc\_2$, $lahc\_3$, $lahc\_4$) and on the other hand the impact of the thresholding strategies on the multi-label classifier (runs $lahc\_5$, $lahc\_6$, $lahc\_7$, $lahc\_8$). We have submitted 8 runs based on different indexes and thresholding strategies, summarized in table 2. Given the SVM score $\phi(\vec{d_i}, c_k)$, the first 4 runs uses the *unordered* method to compute the final score $score(d_i, c_k)$ and the last 4 runs the *ordered* one.

Table 2. Summary of our XML Mining experiments

Run	SVM	Index	Thresholding Strategy	Set of classes
lahc_1_baseline	$multi - label$	T	-	$singleton$
lahc_2_binary	$uni - label$	T	-	$unordered$
lahc_3_binary_Ik	$uni - label$	T_k	-	$unordered$
lahc_4_binary_Ik_1000	$uni - label$	$T_{k_{1000}}$	-	$unordered$
lahc_5_max	$multi - label$	T	$MCut$	$ordered$
lahc_6_pcut	$multi - label$	T	$PCut$	$ordered$
lahc_7_rcut_1	$multi - label$	T	$RCut_1$	$ordered$
lahc_8_rcut_2	$multi - label$	T	$RCut_2$	$ordered$

The first run ($lahc\_1$) corresponds to the baseline. This run only assigns one category for each document. This category corresponds to the highest score provided by the multi-label SVM classifier ($multi - label$).

In order to evaluate the influence of the selection features on the performances, the three next runs consider the SVM as a binary classifier ($uni - label$) employing different indexes. The index T (resp. T_k, $T_{k_{1000}}$) is tested with the second run ($lahc\_2$) (resp. $lahc\_3$, $lahc\_4$). The binary classifier can assign no category to a document. In that case, for $lahc\_3$ and $lahc\_4$ runs, the category c_k provided by the baseline run ($lahc\_1$) is affected to the document.

The last four runs exploit the different thresholding strategies detailed in section 4. The fifth run ($lahc\_5$) uses the $MCut$ strategy.

The run $lahc\_6$ exploits the $PCut$ strategy with x equals to the number of documents in the test set divided by the number of categories. The run $lahc\_7$ (resp. $lahc\_8$) applies the $RCut$ strategy using a parameter t fixed to 1 (resp. 2).

6 Experimental Results

All the results are summarized in table 3. The common criteria (ACC, ROC, PRF and MAP) used for the XML Mining evaluation are presented. In order to rank

Table 3. Summary of all XML Mining results sorted by Mean

Participant	Run	Macro ACC	Micro ACC	Macro ROC	Micro ROC	Macro PRF	Micro PRF	MAP	Mean
lhc	lahc_5_max	0,968	0,952	0,936	0,934	0,549	0,578	0,788	0,820
lhc	lahc_7_rcut_1	0,974	0,962	0,938	0,935	0,531	0,564	0,788	0,817
lhc	lahc_6_pcut	0,973	0,961	0,927	0,925	0,548	0,563	0,748	0,816
lhc	lahc_8_rcut_2	0,959	0,933	0,903	0,906	0,515	0,528	0,788	0,791
xerox	nxQ.3.merge.tfidf	0,975	0,964	0,753	0,767	0,579	0,605	0,678	0,774
xerox	netxQ.4.plus.tfidf	0,974	0,963	0,748	0,765	0,571	0,600	0,679	0,770
xerox	nxQ.4.merge	0,974	0,963	0,748	0,765	0,571	0,600	0,679	0,770
peking	3	0,963	0,948	0,842	0,850	0,480	0,519	0,702	0,767
peking	2	0,963	0,948	0,842	0,850	0,480	0,518	0,702	0,767
peking	1	0,962	0,947	0,842	0,850	0,478	0,516	0,702	0,766
granada	nb_with_links_sub	0,952	0,934	0,802	0,820	0,500	0,530	0,642	0,756
granada	nb_sub	0,951	0,933	0,803	0,820	0,496	0,527	0,641	0,755
lhc	lahc_1_baseline	0,974	0,962	0,721	0,743	0,531	0,564	0,685	0,749
granada	orgate_with_links_sub	0,848	0,819	0,928	0,927	0,316	0,360	0,725	0,700
lhc	lahc_3_binary_Ik	0,967	0,950	0,619	0,629	0,334	0,355	0,407	0,642
granada	orgate_sub	0,754	0,678	0,925	0,922	0,253	0,263	0,730	0,632
lhc	lahc_2_binary	0,971	0,958	0,600	0,613	0,289	0,323	0,393	0,626
lhc	lahc_4_binary_Ik_1000	0,965	0,947	0,585	0,596	0,252	0,279	0,330	0,604
wollongon	bpts2.f1.r3	0,913	0,892	0,625	0,619	0,192	0,218	0,138	0,576
wollongon	bptsext.f1a.r3	0,131	0,160	0,558	0,561	0,072	0,103	0,100	0,264
wollongon	bptsext.f1.r3	0,038	0,055	0,632	0,623	0,071	0,102	0,208	0,253
wollongon	bpts2.f1a.r3	0,038	0,055	0,598	0,599	0,071	0,102	0,125	0,244
wollongon	bptsext.map.r3	0,137	0,141	0,506	0,513	0,065	0,096	0,192	0,243
wollongon	bpts2.map.r3	0,115	0,123	0,511	0,510	0,070	0,101	0,129	0,238

the runs, we introduce the Mean criteria that corresponds to the average obtained for the micro and the macro value for ACC, ROC and PRF. We will firstly discuss results of our baseline. Secondly we will detail the results concerning the selection features criteria and finally those which exploit a thresholding strategy.

Baseline results (run: 1). On table 3, our baseline results are quite good if we compare them to other participant results. As this run limits the number of affectations, it also reduces the number of errors and that is why this is our best run for the ACC criteria. As we only consider a binary score (*unordered*), ROC and MAP criteria are not very good since they take into account the order of returned categories. The PRF criteria, that combines precision and recall, is not very high. That means that we should have a correct precision, but a very low recall because this run considers only one category by document.

Features selection runs (run: 2, 3, 4). Runs 2, 3 and 4 aim to evaluate the influence of the index size using different binary classifiers for each category. As we can see on table 3, all these runs are worse than our baseline for all evaluation criteria. On average on the different evaluation criteria, run 3 is better than runs 2 and 4.

We can conclude that the index reduction is not satisfying and has to be improved. The first idea is to come back to the strategy proposed in INEX 2008 and which consists to define a global index by union of the categories' indexes $\cup_{k \in C} T_{k_{1000}}$. The second idea is to use a different number of terms depending on the category.

Thresholding strategy runs (run: 5, 6, 7, 8). All the runs, that use thresholding strategies, permit globally to improve the baseline results. The accuracy criteria (ACC) is in favour of runs which limit the number of affectations. Indeed, if we use a model that assigns no category to the documents, it will obtain a Macro ACC of 0,963. It is the reason why, our baseline run (run 1) and the $RCut_1$ strategy, which affect only one category, obtain the best accuracy over all our runs. In run 5, 6 and 8 several categories can be assigned to one document, and for this reason,we observe a decrease of the accuracy. The run 8 is globally worse than the others because two classes are systematically affected to each document while the average number of categories by document, estimated on the training set, is around 1.46.

On average, run 5, corresponding to the $MCut$ strategy introduced in this article, is the best of our runs. It is the best for the PRF criteria and it provides the same results as runs 7 and 8 for the MAP criteria since there is only the thresholding strategy that changes. Concerning the ROC criteria, it is slightly worse (Macro: 0.936, Micro: 0.934) than the run 7 (Macro: 0.938, Micro: 0.935).

7   Conclusion

In this article, we focused on the selection of the set of classes that will label a document for the multi-label text categorization. We propose a thresholding strategy, called $MCut$. The results obtained on the Inex XML Mining collection are encouraging. This method is compared to the commonly used approaches $RCut$ and $PCut$. $RCut_1$ and $RCut_2$ give also quite good results but they have the drawback to impose a predefined number of categories by document. Contrary to $RCut$, the number of categories for each document could be different with the $PCut$ strategy. However, this method is not suitable if we want to know the category of a new single document. So, $MCut$ seems to be a good choice because it does not make hypothesis on categories distributions and does not impose the number of category per document.

References

1. Denoyer, L., Gallinari, P.: Overview of the inex 2008 xml mining track. In: Geva, S., Kamps, J., Trotman, A. (eds.) INEX 2008. LNCS, vol. 5631, pp. 401–411. Springer, Heidelberg (2009)
2. Géry, M., Largeron, C., Moulin, C.: Ujm at inex 2008 xml mining track. In: Geva, S., Kamps, J., Trotman, A. (eds.) INEX 2008. LNCS, vol. 5631, pp. 446–452. Springer, Heidelberg (2009)
3. Porter, M.F.: An algorithm for suffix stripping. Readings in information retrieval, 313–316 (1997)
4. Ráez, A.M., López, L.A.U.: Selection strategies for multi-label text categorization. In: Salakoski, T., Ginter, F., Pyysalo, S., Pahikkala, T. (eds.) FinTAL 2006. LNCS (LNAI), vol. 4139, pp. 585–592. Springer, Heidelberg (2006)
5. Salton, G., McGill, M.J.: Introduction to modern information retrieval. McGraw-Hill, New York (1983)
6. Vapnik, V.: The Nature of Statistical Learning Theory. Springer, Heidelberg (1995)
7. Yang, Y.: A study of thresholding strategies for text categorization. In: Proceedings of the ACM SIGIR conference on Research and development in information retrieval, pp. 137–145 (2001)

BUAP: Performance of K-Star at the INEX'09 Clustering Task

David Pinto[1], Mireya Tovar[1], Darnes Vilariño[1], Beatriz Beltrán[1],
Héctor Jiménez-Salazar[2], and Basilia Campos[1]

[1] Faculty of Computer Science
B. Autonomous University of Puebla, Mexico
[2] Department of Information Technologies
Autonomous Metropolitan University, Mexico
{dpinto,mtovar,darnes,bbeltran}@cs.buap.mx, hgimenezs@gmail.com

Abstract. The aim of this paper is to use unsupervised classification techniques in order to group the documents of a given huge collection into clusters. We approached this challenge by using a simple clustering algorithm (K-Star) in a recursive clustering process over subsets of the complete collection.

The presented approach is a scalable algorithm which may automatically discover the number of clusters. The obtained results outperformed different baselines presented in the INEX 2009 clustering task.

1 Introduction

The INEX 2009 clustering task was presented with the purpose of being an evaluation forum for providing a platform to measure the performance of clustering methods over a real-world and high-volume Wikipedia collection.

Clustering analysis refers to the partitioning of a data set into subsets (clusters), so that the data in each subset (ideally) share some common trait, often proximity, according to some defined distance measure [1,2,3].

Clustering methods are usually classified with respect to their underlying algorithmic approaches. Hierarchical, iterative (or partitional) and density-based are some possible categories belonging to this taxonomy. In Figure 1 we can see the taxonomy presented in [4].

Hierarchical algorithms find successive clusters using previously established ones, whereas partitional algorithms determine all clusters at once. Hierarchical algorithms can be agglomerative ("bottom-up") or divisive ("top-down"); agglomerative algorithms begin with each element as a separate cluster and merge the obtained clusters into successively larger clusters. Divisive algorithms begin with the whole set and proceed to divide it into successively smaller clusters. Iterative algorithms start with some initial clusters (their number either being unknown in advance or given a priori) and intend to successively improve the existing cluster set by changing their "representatives" ("centers of gravity" or "centroids"), like in K-Means [3] or by iterative node-exchanging (like in [5]).

S. Geva, J. Kamps, and A. Trotman (Eds.): INEX 2009, LNCS 6203, pp. 434–440, 2010.

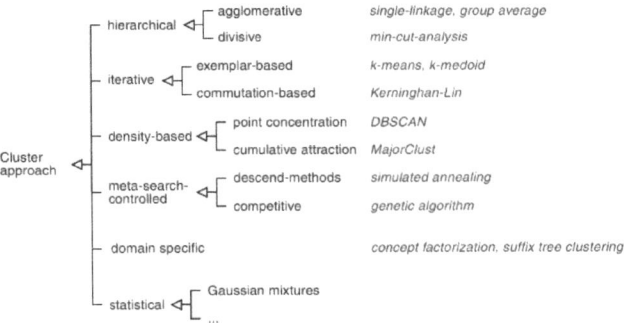

Fig. 1. A taxonomy of clustering methods as presented in [4] (Reproduced with permission of the author)

In this paper we report the obtained results when the K-Star clustering method was applied to the INEX2009_SUBSET collection. Therefore, the complete description of this clustering method is given in the following section.

2   The K-Star Clustering Method

K-Star [6] is a hierarchical agglomerative clustering method which automatically reveals the number of clusters, unknown in advance. As the most of the clustering methods, it requires a similarity matrix of the documents to be clustered (corpus). The K-Star clustering method follows as shown in Algorithm 1.

Algorithm 1. Algorithm of the K-Star clustering method

    Input: A $n \times n$ similarity matrix $\varphi(d_i, d_j)$

    Output: A set of clusters $\{C_1, C_2, \cdots\}$

1 Cluster=1;

2 ClusteredSet $= \emptyset$;

3 **while** $|ClusteredSet| < n$ **do**

4      $C_{Cluster} = C_{Cluster} \bigcup \arg\max_{\{d_i, d_j\}} \varphi(d_i, d_j)$;

5   ClusteredSet = ClusteredSet $\bigcup \{d_i, d_j\}$;

6   **foreach** $d_k \notin ClusteredSet$ **do**

7   **if** $\varphi(d_k, d_i) > \tau$ **then**

8            $C_{Cluster} = C_{Cluster} \bigcup \{d_k\}$;

9   ClusteredSet = ClusteredSet $\bigcup \{d_k\}$;

10   **end**

11   **end**

12   Cluster = Cluster + 1;

13 **end**

14 **return** $C_1, C_2, \cdots$

In this work, we have used a canonic threshold τ defined as the average of the values in the similarity matrix. The rationale of the similarity matrix construction is described in the following section.

3   Construction of the Similarity Matrix

We assume that the complete document clustering task may be carried out by executing at least the following three steps: (1) document representation; (2) calculus of a similarity matrix which represents the similarity degree among all the documents of the collection; and (3) clustering of the documents. In particular, the construction of the similarity matrix was carried out by means of the TF-IDF measure which is described into detail as follows.

The Term Frequency and Inverse Document Frequency (tf-idf) is a statistical measure of weight often used in natural language processing to determine how important a term is in a given corpus, by using a vectorial representation. The importance of each term increases proportionally to the number of times this term appears in the document (frequency), but is offset by the frequency of the term in the corpus. In this document, we will refer to the tf-idf as the complete similarity process of using the tf-idf weight and a special similarity measure proposed by Salton [7] for the Vector Space Model, which is based on the use of the cosine among vectors representing the documents.

The tf component of the formula is calculated by the normalized frequency of the term, whereas the idf is obtained by dividing the number of documents in the corpus by the number of documents which contain the term, and then taking the logarithm of that quotient. Given a corpus D and a document d_j ($d_j \in D$), the tf-idf value for a term t_i in d_j is obtained by the product between the normalized frequency of the term t_i in the document d_j (tf_{ij}) and the inverse document frequency of the term in the corpus ($idf(t_i)$) as follows:

$$tf_{ij} = \frac{tf(t_i, d_j)}{\sum_{k=1}^{|d_j|} tf(t_k, d_j)} \tag{1}$$

$$idf(t_i) = log\left(\frac{|D|}{|d : t_i \in d, d \in D|}\right) \tag{2}$$

$$tf\text{-}idf = tf_{ij} * idf(t_i) \tag{3}$$

Each document can be represented by a vector where each entry corresponds to the tf-idf value obtained by each vocabulary term of the given document. Thus, given two documents in vectorial representation, d_i and d_j, it is possible to calculate the cosine of the angle between these two vectors as follows:

$$Cos_\theta(\vec{d_i}, \vec{d_j}) = \frac{\vec{d_i} \cdot \vec{d_j}}{\left\|\vec{d_i}\right\| \left\|\vec{d_j}\right\|}$$

The similarity matrix is then constructed on the basis of the above formulae, i.e., for each possible pair of documents, we need to calculate how similar they are by using the cosine measure. Once the similarity matrix is calculated, we may proceed with the clustering step. The complete description of the implemented clustering approach is given in the following section.

4   Description of the Approach

We have approached a clustering technique by partitioning the complete document collection. The divide-and-conquer approach described here was motivated by the time and space complexity needed in order to cluster huge volumenes of data. Instead of constructing one similarity matriz of high dimensionality for a document collection D (54,575 × 54,575), we constructed m similarity matrices of low dimensionality ($\frac{|D|}{m} \times \frac{|D|}{m}$). The proposed method actually allow the clustering process to be performed in a considerable lower amount of time in comparison with the traditional approach. The process followed is presented in Algorithm 2, whereas a scheme of the same process is given in Figure 2.

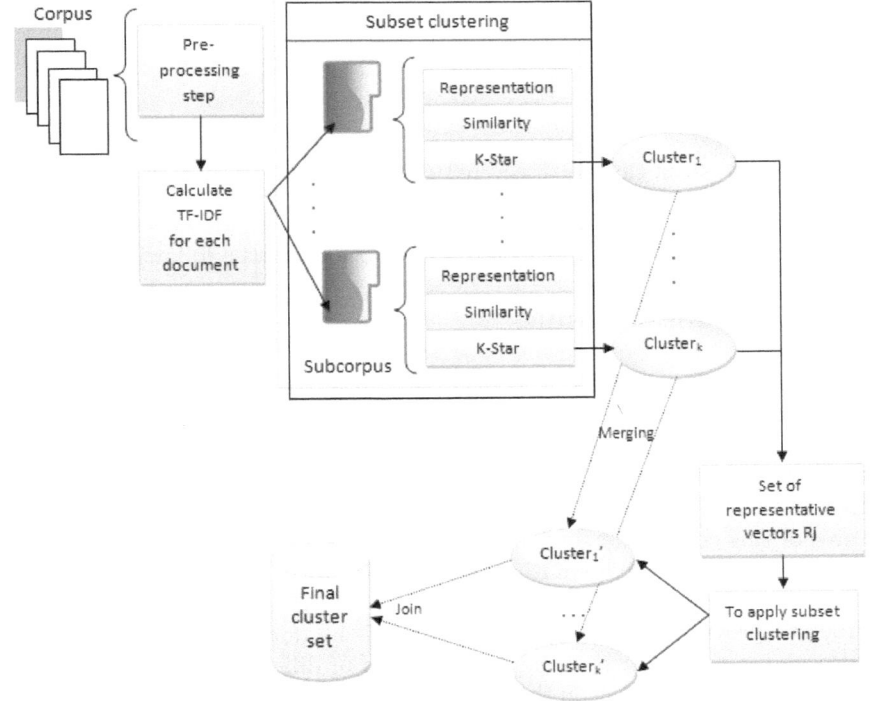

Fig. 2. Two step approach of BUAP Team at the INEX 2009 clustering task

---

Algorithm 2. Algorithm used for clustering the INEX2009_SUBSET with K-Star

---

Input: A document collection D
Output: A set of clusters $C_{i,j}$

1 Compute a global dictionary which contains all unique terms found in a
   collection of documents and their respective frequency of occurrence;
2 Eliminate all those terms whose frequency is lower than β;
3 Represent each document according to TF-IDF;
4 Split D into m subsets D_i made of $\frac{|D|}{m}$ documents;
5 **foreach** $D_i \in D$ *such as* $D_i = \{d_{i,1}, d_{i,2}, d_{i,3}, \cdots d_{i,\frac{|D|}{m}}\}$ **do**
6   $\quad$ Calculate the similarity matrix M_i of D_i using the cosine measure;
7   $\quad$ Apply the K-Star clustering method to M_i in order to discover k clusters
    $\quad$ ($\{C_{i,1}, C_{i,2}, \cdots, C_{i,k}\}$);
8 **end**
9 $Loop = 1$;
10 **while** $Loop \leq MAX\_ITERATIONS$ **do**
11 $\quad$ Select a random representative document $d_{i,j}$ for each cluster $C_{i,j}$ obtained;
12 $\quad$ Let D' be the set of documents $d_{i,j}$, i.e., only those that represent each
    $\quad$ cluster obtained;
13 $\quad$ Calculate the similarity matrix M'_i of D'_i using the cosine measure;
14 $\quad$ Apply the K-Star clustering method to M'_i in order to discover k clusters
    $\quad$ ($\{C'_{i,1}, C'_{i,2}, \cdots, C'_{i,k}\}$);
15 $\quad$ Let $C_{i,j} = C_{i,j} \bigcup C_{i,j'}$, where $d_{i,j} \in C'_{i,r}$ and $d_{i,j'} \in C'_{i,r}$ with $1 \leq r \leq k$ and
    $\quad$ $j <> j'$;
16 $\quad$ $Loop = Loop + 1$;
17 **end**
18 **return** (The set of clusters discovered)$C_{i,j}$

The obtained results are presented and discussed in the following section.

5 Experimental Results

The clustering task of INEX 2009 evaluated unsupervised machine learning solutions against the ground truth categories by using standard evaluation criteria
such as Purity, Entropy and F-score.

Even if the complete description of the dataset used in the clustering task
of INEX 2009 is given in the track overview paper, we may describe general
features of this corpus.

The INEX XML Wikipedia collection used in the experiments is a subset
of the complet corpus. This subset contains 54,575 documents pre-processed
in order to provide various representations of the documents. The aim of this
pre-processing was to enable the participation of different teams with minimal
overheads in data-preparation the collection. It was provided, for instance, a bag-
of-words representation of terms and frequent phrases in a document, frequencies
of various XML structures in the form of tags, trees, links, named entities, etc.

Table 1. Purity of the INEX09_SUBSET clustering evaluation using 73,944 YAGO categories (all YAGO categories)

Run ID	Description	Clusters	Micro Purity	Macro Purity
24	random-54575	54575	1.0	1.0
11	ground truth for 73,944 YAGO categories	73944	0.9999	1.0
67	KStar method Bigrams	35569	0.6812	0.8968
65	KStar method Unigram Stems	7019	0.1894	0.6399
66	KStar method Stems Bigrams	6961	0.1883	0.6350

Table 2. Purity of the INEX09_SUBSET clustering evaluation using 12,803 YAGO categories (containing $\geq$ 5 documents)

Run ID	Description	Clusters	Micro Purity	Macro Purity
12	ground truth for 12,804 YAGO categories	12804	1.0	1.0
24	random-54575	54575	1.0	1.0
67	KStar method Bigrams	35569	0.6964	0.9006
65	KStar method Unigram Stems	7019	0.2076	0.6410
66	KStar method Stems Bigrams	6961	0.2074	0.6407

Table 3. Entropy of the INEX09_SUBSET clustering evaluation using 73,944 YAGO categories (all YAGO categories)

Run ID	Description	Clusters	Micro Entropy	Macro Entropy
66	KStar method Stems Bigrams	6961	0.1883	0.6350
65	KStar method Unigram Stems	7019	0.1894	0.6399
11	ground truth for 73,944 YAGO categories	73944	0.9999	1.0
67	KStar method Bigrams	35569	0.6812	0.8968
24	random-54575	54575	1.0	1.0

Table 4. Entropy of the INEX09_SUBSET clustering evaluation using 12,803 YAGO categories (containing $\geq$ 5 documents)

Run ID	Description	Clusters	Micro Entropy	Macro Entropy
66	KStar method Stems Bigrams	6961	0.2074	0.6407
65	KStar method Unigram Stems	7019	0.2076	0.6410
12	ground truth for 12,804 YAGO categories	12804	1.0	1.0
67	KStar method Bigrams	35569	0.6964	0.9006
24	random-54575	54575	1.0	1.0

In the experiments carried out, we approached three different representations of data: unigram stems, bigrams and bigram stems which were executed in order to observe the K-Star clustering method performance.

In Tables 1 and 2 we may see the obtained results of the three approaches with respect to two baselines, one random assignment (random-54575) and one ground truth over 73,944 and 12,803 YAGO categories, respectively. It may be observed that the bigram representation shows a high Micro and Macro Purity.

Tables 3 and 4 present the K-Star performance in terms of entropy. In general the three aproaches show a low entropy value.

In general, it can be noticed that the two-step approach presented in this paper performs well, given the above mentioned evaluation measures. However, we consider that the performance may be improved by considering a better way of selecting the cluster representative documents that are used in the second step of the algorithm, for instance, the use of cluster centroids.

6   Conclusions

A recursive method based on the K-Star clustering method has been proposed in this paper. The aim of the presented approach was to allow high scalability of the clustering algorithm. Traditional clustering of huge volumes of data requires to calculate a two dimensional similarity matrix. A process which needs quadratic time complexity with respect to the number of documents. The lower the dimensionality of the similarity matrix, the faster the clustering algorithm will be executed. The high scalability is then the contribution of the clustering method presented in this paper. There still however the fact that we are not taking into account the complete information, since we only considered subsets of the complete dataset. However, we observed that the proposed approach is easy of being implemented and obtained good Purity and Entropy results in the INEX 2009 clustering task.

Acknowledgments

This work has been partially supported by CONACYT and PROMEP projects.

References

1. MacKay, D.J.C.: Information Theory, Inference and Learning Algorithms. Cambridge University Press, Cambridge (2003)
2. Mirkin, B.G.: Mathematical Classification and Clustering. Springer, Heidelberg (1996)
3. MacQueen, J.B.: Some methods for classification and analysis of multivariate observations. In: Proc. of the 5th Berkeley Symposium on Mathematical Statistics and Probability, pp. 281–297. University of California Press, Berkeley (1967)
4. Meyer zu Eissen, S.: On information need and categorizing search. PhD dissertation, University of Paderborn, Germany (2007)
5. Kernighan, B.W., Lin, S.: An efficient heuristic procedure for partitioning graphs. Bell Systems Technical Journal 49(2), 291–308 (1970)
6. Shin, K., Han, S.Y.: Fast clustering algorithm for information organization. In: Gelbukh, A. (ed.) CICLing 2003. LNCS, vol. 2588, pp. 619–622. Springer, Heidelberg (2003)
7. Salton, G., Wong, A., Yang, C.: A vector space model for automatic indexing. Communications of the ACM 18(11), 613–620 (1975)

Extended VSM for XML Document Classification Using Frequent Subtrees

Jianwu Yang and Songlin Wang

Institute of Computer Sci. & Tech., Peking University,
Beijing 100871, China
{yangjianwu,wangsonglin}@icst.pku.edu.cn

Abstract. Structured link vector model (SLVM) is a representation proposed for modeling XML documents, which was extended from the conventional vector space model (VSM) by incorporating document structures. In this paper, we describe the classification approach for XML documents based on SLVM in the Document Mining Challenge of INEX 2009, where the closed frequent subtrees as structural units are used for content extraction from the XML document and the Chi-square test is used for feature selection.

Keywords: XML Document, Classification, Vector Space Model (VSM), Structured Link Vector Model (SLVM), Frequent Subtree.

1 Introduction

XML is the W3C recommended markup language for semi-structured data. Its structural flexibility makes it an attractive choice for representing data in application domains. With the rapid growth of XML documents, these arise many issues concerning the management of these documents effectively.

To contrast with ordinary unstructured documents, XML documents represent their syntactic structure via (1) the use of XML elements, each marked by a user-specified tag, and (2) the associated schema specified in either DTD or XML Schema format. In addition, XML documents can be cross-linked by adding IDREF attributes to their elements to indicate the linkage. The conventional document analysis tools developed for unstructured documents [1] fail to take the full advantage of the structural properties of XML documents. Thus, techniques designed for XML document analysis normally take into account the information embedded in both the element tags as well as their associated contents for better performance.

2 Structured Link Vector Model (SLVM): An Overview

Structured Link Vector Model (SLVM), which forms the basis of this paper, was originally proposed in [2] for representing XML documents. It was extended from the conventional vector space model (VSM) [3] by incorporating document structures (represented as term-by-element matrices), referencing links (extracted

S. Geva, J. Kamps, and A. Trotman (Eds.): INEX 2009, LNCS 6203, pp. 441–448, 2010.

based on IDREF attributes), as well as element similarity (represented as an element similarity matrix). The SLVM has been used in the Document Mining Challenge of INEX 2009 [4].

2.1 Basic Representation

Vector Space Model (VSM) [3] has long been used to represent unstructured documents as document feature vectors which contain term occurrence statistics. This bag of terms approach assumes that the term occurrences are *independent* of each other.

Definition 2.1. *Assume that there are n distinct terms in a given set of documents D. Let doc_x denote the xth document and d_x denote the* **document feature vector** *such that*

$$d_x = [d_{x(1)}, d_{x(2)} \cdots\cdots, d_{x(n)}]^T$$
$$d_{x(i)} = TF(w_i, doc_x)\, IDF(w_i)$$

where $TF(w_i, doc_x)$ is the frequency of the term w_i in doc_x, $IDF(w_i) = log(|D|/DF(w_i))$ is the inverse document frequency of w_i for discounting the importance of the frequently appearing terms, $|D|$ is the total number of the documents, and $DF(w_i)$ is the number of documents containing the term w_i.

Applying VSM directly to represent XML documents is not desirable as the document syntactic structure tagged by their XML elements will be ignored. For example, VSM considers two documents with an identical term appearing in, say, their "title" fields to be equivalent to the case with the term appearing in the "title" field of one document and in the "author" field of another. As the "author" field is semantically unrelated to the "title" field, the latter case should be considered as a piece of less supportive evidence for the documents to be similar when compared with the former case. Using merely VSM, these two cases cannot be differentiated.

Structured Link Vector Model (SLVM), proposed in [2], can be considered as an extended version of vector space model for representing XML documents. Intuitively speaking, SLVM represents an XML document as an array of VSMs, each being specific to an XML element (specified by the <element> tag in a DTD).

Definition 2.2. *SLVM represents an XML document doc_x using a* **document feature matrix** $\Delta_x \in R^{n \times m}$ *, given as*

$$\Delta_x = [\Delta_{x(1)}, \Delta_{x(2)}, \cdots\cdots, \Delta_{x(m)}]$$

where m is the number of distinct XML elements, $\Delta_{x(i)} \in R^n$ is the TFIDF feature vector representing the ith XML element (e_i), given as $\Delta_{x(i,j)} = TF(w_j, doc_x.e_i) \cdot IDF(w_j)$ for all j=1 to n, and $TF(w_j, doc_x.e_i)$ is the frequency of the term w_j in the element e_i of doc_x.

Definition 2.3. *The* **normalized document feature matrix** *is defined as*

$$\tilde{\Delta}_{x(i,j)} = \Delta_{x(i,j)} / \sum_{k} \Delta_{x(i,k)}$$

where the factor caused by the varying size of the element content is discounted via normalization.

Example 2.1. Figure 1 shows a simple XML document. Its corresponding document feature vector d_x, document feature matrix Δ_x, and normalized document feature matrix $\tilde{\Delta}_x$ are shown in Figure 2-4 respectively. Here, we assume all the terms share the same *IDF* value equal to one.

```
<article>
    <title>Ontology Enabled Web Search</name>
    <author>John</author>
    <conference>Web Intelligence</conference>
</article>
```

Fig. 1. An XML document

$$d_x = \begin{bmatrix} 1 \\ 1 \\ 2 \\ 1 \\ 1 \\ 1 \end{bmatrix} \begin{array}{l} \text{Ontology} \\ \text{Enabled} \\ \text{Web} \\ \text{Search} \\ \text{John} \\ \text{Intelligence} \end{array}$$

thisDocument

Fig. 2. The document feature vector for the example shown in Figure 1

$$\Delta_x = \begin{bmatrix} 1 & 0 & 0 \\ 1 & 0 & 0 \\ 1 & 0 & 1 \\ 1 & 0 & 0 \\ 0 & 1 & 0 \\ 0 & 0 & 1 \end{bmatrix} \begin{array}{l} \text{Ontology} \\ \text{Enabled} \\ \text{Web} \\ \text{Search} \\ \text{John} \\ \text{Intelligence} \end{array}$$

title   author   Conference

Fig. 3. The document feature matrix for the example in Figure 1

$$\tilde{\Delta}_x = \begin{bmatrix} 0.5 & 0 & 0 \\ 0.5 & 0 & 0 \\ 0.5 & 0 & \sqrt{2}/2 \\ 0.5 & 0 & 0 \\ 0 & 1 & 0 \\ 0 & 0 & \sqrt{2}/2 \end{bmatrix} \begin{array}{l} \text{Ontology} \\ \text{Enabled} \\ \text{Web} \\ \text{Search} \\ \text{John} \\ \text{Intelligence} \end{array}$$

title   author   Conference

Fig. 4. The normalized document feature matrix for the example in Figure 1

2.2 Similarity Measures

Using VSM, similarity between two documents doc_x and doc_y is typically computed as the cosine value between their corresponding document feature vectors, given as

$$sim(doc_x, doc_y) = \frac{d_x d_y}{\| d_x \| \| d_y \|} = \tilde{d}_x \tilde{d}_y^{\ T} = \sum_{i=1}^{k} \tilde{d}_{x(i)} \tilde{d}_{y(i)} \tag{1}$$

where n is the total number of terms and $\tilde{d}_x = d_x / \| d_x \|$ *denotes normalized d_x. So, the similarity measure can also be interpreted as the inner product of the normalized document feature vectors.*

For SLVM, with the objective to model semantic relationships between XML elements, the corresponding document similarity can be defined with an element similarity matrix introduced.

Definition 2.4. The **SLVM-based document similarity** *between two XML documents doc_x and doc_y is defined as*

$$sim(doc_x, doc_y) = \sum_{j=1}^{m} \sum_{i=1}^{m} M_{e(i,j)} \cdot (\tilde{\Delta}_{x(i)}^{\ T} \bullet \tilde{\Delta}_{y(j)}) \tag{2}$$

*where M_e is a matrix of dimension m×m and named as the **element similarity matrix**.*

The matrix M_e in Eq. (2) captures both the similarity between a pair of XML elements as well as the contribution of the pair to the overall document similarity (*i.e.,* the diagonal elements of M_e are not necessarily equal to one). An entry in M_e being small means that the two corresponding XML elements should be unrelated and same words appearing in the two elements of two different documents will not contribute much to the overall similarity of them. If M_e is diagonal, this implies that all the XML elements are not correlated at all with each other, which obviously is not the optimal choice. To obtain an optimal M_e for a specific type of XML data, we proposed in [5] to learn the matrix using pair-wise similar training data in an iterative manner.

2.3 SVM for XML Documents Classification

SVM was introduced by Vapnik in 1995 for solving two-class pattern recognition problems using the Structural Risk Minimization principle [6]. Given a training set containing two kinds of data (one for positive examples, the other for negative examples), which is linearly separable in vector space, this method finds the decision hyper-plane that best separated positive and negative data points in the training set. The problem searching the best decision hyper-plane can be solved using quadratic programming techniques. SVM can also extend its applicability to linearly nonseparable data sets by either adopting soft margin hyper-planes, or by mapping the original data vectors into a higher dimensional space in which the data points are linearly separable. Joachims [7] first applied SVM to text categorization, and compared its performance with other classification methods using the Reuters-21578 corpus. His results show that SVM outperformed all the other methods tested in his experiments.

SVM success in practice is drawn by its solid mathematical foundations which convey the following two salient properties:

- **Margin maximization:** The classification boundary functions of SVM maximize the margin, which in machine learning theory, corresponds to maximizing the *generalization* performance given a set of training data.
- **Nonlinear transformation of the feature space using the kernel trick:** SVM handle a nonlinear classification efficiently using the kernel trick which implicitly transforms the input space into another high dimensional feature space.

The kernel $k(x_i, x_j)$ could be regarded as the similarity function between two data points. For linear boundary, the kernel function is $x_i \cdot x_j$, a scalar product of two data points. The nonlinear transformation of the feature space is performed by replacing $k(x_i, x_j)$ with an advanced kernel, such as polynomial kernel $(x^T x_i + 1)^p$ or RBF kernel $exp(-\frac{1}{2\delta^2} \| x - x_i \|^2)$.

In SLVM, the similarity between two XML documents is defined as definition 2.4, so we consider the kernel $k(x_i, x_j)$ for XML documents classification based on SLVM as:

$$k(x_i, x_j) = sim(doc_x, doc_y) = \sum_{j=1}^{m} \sum_{i=1}^{m} M_{e(i,j)} \cdot (\tilde{\Delta}_{x(i)}^T \bullet \tilde{\Delta}_{y(j)}) \tag{3}$$

3 Frequent Subtree

The form of SLVM studied in [2, 3, 5] is only a simplified one where only the leaf-node elements in the DTD are incorporated without considering their positions in the document DOM tree and their consecutive occurrence patterns.

In this paper, we utilize the frequent substrees as structural units to extract the content information from the XML documents. A series of concepts of the subtree are defined same as in the papers [8, 9].

Let D denote a database where each transaction $s \in D$ is a labeled rooted unordered tree. For a given pattern t, which is a rooted unordered tree, we say t occurs in a transaction s if there exists at least one subtree of s that is isomorphic to t. The *occurrence* $\delta_t(s)$ of t in s is the number of distinct subtrees of s that are isomorphic to t. Let $\sigma_t(s) = 1$ if $\delta_t(s) > 0$, and 0 otherwise.

We say s *supports* pattern t if $\sigma_t(s)$ is 1 and we define the *support* of a pattern t as $supp(t) = \sum_{s \in D} \sigma_t(s)$. A pattern t is called *frequent* if its *support* is greater than or equal to a *minimum support* (*minsup*) specified by a user.

We define a frequent tree t to be *maximal* if none of t's proper supertrees is frequent, and *closed* if none of t's proper supertrees has the same support that t has.

In this paper, we utilize *CMTreeminer* [9] to mining *closed frequent substrees* from XML document collection, and the *closed frequent substrees* as structural units are utilized to extract the content information from the XML documents.

4 Implementation and Evaluation Details

As outlined above, we classify the XML documents based on SLVM in the Document Mining Challenge of INEX 2009, where the closed frequent subtrees as structural units are used to extract content from the XML document.

4.1 Phase 1: Pre-processing of Structure

In the XML document collection of the INEX 2009, a group of "Template Element" is mapped to the tag "<sec>". In order to reduce the complexity of structure, the element "<sec>" is normalized by replacing its sub-element "<st>", whose meaning is "section title", by its attribute.

Each common path is replaced by a node, when any sub-path of it does not appear in the collection.

4.2 Phase 2: Mining Closed Frequent Subtrees

We utilize *CMTreeminer* [9] to mining *closed frequent substrees* from XML document collection.

There are a great deal *closed frequent substrees* for the XML document collection, so we use the Chi-square test [10] to select a part of *substrees* as useful structural units.

4.3 Phase 3: Document Representation

Each document is represented as a matrix based on SLVM, where the selected *closed frequent substrees* are regarded as structural unit.

In order to deal with exception that a document do not include any the selected *closed frequent substrees*, we add the vector of the document based on VSM into the matrix as a column.

In addition, the interconnectivity between the documents based on link should also be considered. We add the vector of all link target document's title based on VSM into the matrix as a column.

4.4 Phase 4: Training, Tuning and Testing

The SVM algorithm in SVMTorch [11] is used for training and testing, with the formula (3) as kernel and the M_e is set as diagonal. For the multi-class document, the document is set as positive example for its each class.

In order to obtain optimal threshold, we split a part of training documents for tuning SVM classification threshold.

5 Experiment Result

In the experiments, all the algorithms were implemented by us in C++, except the SVM algorithm in SVMTorch [11]. All experiments were run on a PC with a 3.0 GHz Intel CPU and 2GB RAM.

In the XML Mining Track, the collection is composed of XML documents of the Wikipedia XML, the training set is composed of 10,969 XML documents, and the test set is composed of 43,606 XML documents.

We utilize *CMTreeminer* [9] to mining *closed frequent substrees* from XML document collection, and select the top 10 *closed frequent substrees* using the Chi-square test [10] for each classification.

The table 1 summarizes the experiment result. The baseline is the approach based on original SLVM [2, 4]. The "No_Link" approach does not use the link information, and the "With_link" approach uses the link information. The "Term_Selection" approach is that the term be selected by the Chi-square test.

Table 1. The experiment result

Approaches	Macro F1	Micro F1	Mean Average precision by document
Baseline	0.241853	0.295418	0.486702
No_Link	0.479748	0.518635	0.7018610
With_Link	0.505916	0.546976	0.712323
Term_Selection	0.483367	0.525132	0.699250

6 Conclusion and Future Works

In this paper, we applied SLVM to XML documents classification, where the closed frequent subtrees as structural units are used for content extraction from the XML document and the Chi-square test is used for feature selection.

For future work, we are interested to study how to use link information, and combine the vector similarity method with graph theory methods.

Acknowledgment

The work reported in this paper was supported by the National Natural Science Foundation of China Grant 60642001 and 60875033.

References

1. Berry, M.: Survey of Text Mining: Clustering, Classification, and Retrieval. Springer, Heidelberg (2003)
2. Yang, J., Chen, X.: A semi-structured document model for text mining. Journal of Computer Science and Technology 17(5), 603–610 (2002)
3. Salton, G., McGill, M.J.: Introduction to Modern information Retrieval. McGraw-Hill, New York (1983)

4. Yang, J., Zhang, F.: XML Document Classification using Extended VSM. In: Fuhr, N., Kamps, J., Lalmas, M., Trotman, A. (eds.) INEX 2007. LNCS, vol. 4862, pp. 234–244. Springer, Heidelberg (2008)

5. Yang, J., Cheung, W.K., Chen, X.O.: Learning Element Similarity Matrix for Semi-structured Document Analysis. Knowledge and Information Systems 19, 53–78 (2009)

6. Vapnic, V.: The Nature of Statistical Learning Theory. Springer, New York (1995)

7. Joachims, T.: Text Categorization with Support Vector Machines: Learning with Many Relevant Features. In: Nédellec, C., Rouveirol, C. (eds.) ECML 1998. LNCS, vol. 1398, pp. 137–142. Springer, Heidelberg (1998)

8. Chi, Y., Nijssen, S., Muntz, R.R., Kok, J.N.: Frequent Subtree Mining -An Overview. Fundamenta Informaticae (2005)

9. Chi, Y., Yang, Y., Xia, Y., Muntz, R.R.: CMTreeMiner: Mining Both Closed and Maximal Frequent Subtrees. In: Dai, H., Srikant, R., Zhang, C. (eds.) PAKDD 2004. LNCS (LNAI), vol. 3056, pp. 63–73. Springer, Heidelberg (2004)

10. Yang, Y., Pedersen, J.O.: A comparative study on feature selection in text categorization. In: Proceedings of the 1998 International Conference on Machine Learning, pp. 412–420 (1997)

11. Collobert, R., Bengio, S.: SVMTorch: support vector machines for large-scale regression problems. Journal of Machine Learning Research 1, 143–160 (2001)

Supervised Encoding of Graph-of-Graphs for Classification and Regression Problems

Shu Jia Zhang[1], Markus Hagenbuchner[1], Franco Scarselli[3], and Ah Chung Tsoi[2]

[1] University of Wollongong, Wollongong, Australia
{sz603,markus}@uow.edu.au
[2] Hong Kong Baptist University, Hong Kong
act@hkbu.edu.hk
[3] University of Siena, Siena, Italy
franco@ing.unisi.it

Abstract. This paper introduces a novel approach for processing a general class of structured information, viz., a graph of graphs structure, in which each node of the graph can be described by another graph, and each node in this graph, in turn, can be described by yet another graph, up to a finite depth. This graph of graphs description may be used as an underlying model to describe a number of naturally and artificially occurring systems, e.g. nested hypertexted documents. The approach taken is a data driven method in that it learns from a set of examples how to classify the nodes in a graph of graphs. To the best of our knowledge, this is the first time that a machine learning approach is enabled to deal with such structured problem domains. Experimental results on a relatively large scale real world problem indicate that the learning is efficient. This paper presents some preliminary results which show that the classification performance is already close to those provided by the state-of-the-art ones.

1 Introduction

The emergence of neural network models capable of encoding graph structured information opened an avenue to solving machine learning problems involving graphs without the need for a pre-processing step in "squashing" the graph structure back into a vectorial form first. These methods are capable of encoding topological information which is available when dealing with structured information, and have been applied with considerable success in a number of real world problems. For example, an MLP based approach known as Recursive Neural Networks along with the training algorithm known as Backpropagation Through Structure (BPTS)[1] was used to solve an image classification problem by processing a set of images represented as a set of directed trees [2]. Similarly, recursive cascade correlation (RCC) [3,4] was used to solve a regression problem by discovering structure-activity relationships of chemical molecules [5]. Another supervised machine learning method is the Graph Neural Network (GNN) which was shown to be capable of solving any practically useful learning

---

[1] The term introduced by [1] is somewhat misleading since BPTS is restricted to the processing of trees. Hence, a more appropriate name would be the Backpropagation Through Trees (BPTT).

S. Geva, J. Kamps, and A. Trotman (Eds.): INEX 2009, LNCS 6203, pp. 449–461, 2010.

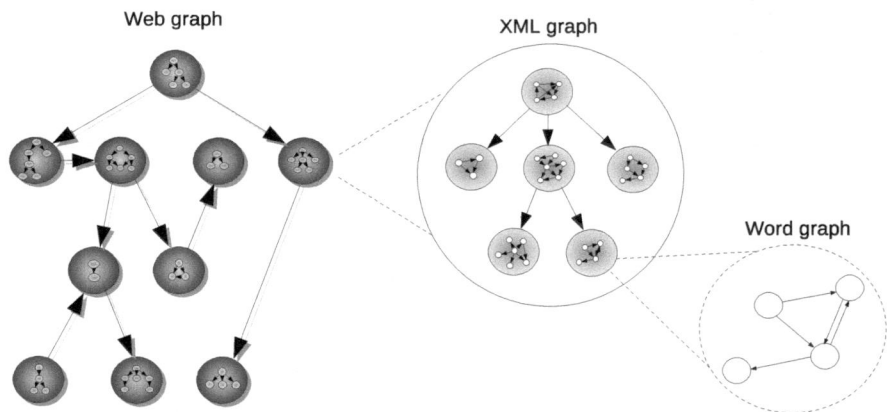

Fig. 1. An example of a graph of graphs structure: Each node in a Web graph represents a document. The document may be represented by an XML graph [7]. The nodes in the XML graph map represent paragraphs where the text in each paragraph may be represented by a word graph [11].

problems involving graphs [6]. There are also unsupervised neural network methods for the encoding of structured information. For example, the Self-Organizing Map for Structured Data (SOM-SD) is capable of clustering tree structured information very efficiently, whereas the Probability Measure Graph SOM (PMGraphSOM) is able to encode much more generic types of graphs such as labeled cyclic graphs which may have feature directed or undirected links. These methods have been very successful in solving practical problems and benchmarking with other algorithms. For example, the SOM-SD and its related algorithms have won several international INEX competitions in text mining (e.g. [7,8]) and has set state-of-the-art performances (e.g. [9,10]).

These existing methods are limited to processing graphs which can be represented by labeled nodes and labeled links. The label attached to nodes and links must be a real valued vector of fixed dimension. However, it is found that there are learning problems for which the label of a node is of a more complex structure, such as a tree or graph. Such learning problems require the encoding of a graph structure the nodes of which can also be graphs. This results in data featuring a *graph of graphs* (GoG) which contains different graphical elements. The situation is best explained using an example (see Figure 1): The World Wide Web consists of a collection of documents which feature a referencing method known as *hyperlinks*. A hyperlink defines a (directed) binary relationship between two documents.

The collective combination of all hyperlinks produces a *Web graph* the nodes of which represent the documents, and the links represent the hyperlinks. The nodes in such a graph can be labeled to describe properties and content of the associated document. In the case of typical Web documents, one may represent the associated document as a graph consisting of nodes representing sections of text encapsulated by HTML formatted elements, and links representing the encapsulation of the HTML formatted elements [7]. The nodes of this document graph may be labeled by a word graph consisting of nodes which may represent the word token, or a set of word tokens, and links among the nodes indicating the relationship between two sets of word tokens [11]. Now, say, in a particular paragraph, there is a hyperlinked document linked to it. Such hyperlinked document in itself can be represented by a graph. This may be represented as a graph within a node of the parent document. This is a graph of graphs (GoG) structure. Obviously there is a possibility that the hyperlinked document also contains hyperlinks to other documents, and these documents in turn, can be represented as graphs within the nodes in the hyperlinked document graph. We note that a GoG is generally of a hybrid nature since the graphs at different levels encode different relationships, and describe different atomic elements. Hence, the GoG is an *embedded hybrid graph* whose components may differ significantly in properties. For example the hybrid graph depicted in Figure 1 features a Web graph which is a directed cyclic graph, XML graphs which are tree structured, and a word graph which may be labeled and undirected.

To the best of our knowledge, there is currently no known approach to machine learning which can directly encode such GoG structures. This paper proposes an approach to modeling such structures, and a supervised learning algorithm is proposed by generalizing an existing supervised machine learning method, viz. backpropagation through structures to encode such structures. The proposed method is recursive, in that it will process one level after the other recursively from the innermost level to the outermost level, and, hence, it will be able to encode GoG structures to any finite depth (e.g. in the case of Figure 1 the depth of the GoG structure is 3).

The structure of this paper is as follows: Section 2 introduces a formal description of the GoG structure, and proposes a data driven approach to encode such structures. Section 3 presents experimental results, and some observations on the limitations of the approach. Finally, Section 4 concludes this paper and provides an outlook for further research in this area.

2   Encoding Graph of Graphs

In this section a formal representation of GoG will be provided. Consider the following situation: we have a node i in a graph with a neighbourhood $\mathcal{N}^0_{[i]}$, which consists of a number of other nodes, where the superscript 0 denotes the topmost level, the parent document level. The nodes in the neighbourhood $\mathcal{N}^0_{[i]}$ all have connections via links to the node i. Each node is described by a set of features, and each link is described by yet another set of features. Each node could have an external input. If we assume that each node is described by an entity called *state* then node i is described by a state $\mathbf{x}^0_i$, an n_0-dimensional vector, The state of node i can be described by Equation 1:

$$\mathbf{x}^0_i = \mathcal{F}^0_i(\mathbf{u}^0_i, \mathcal{C}_0(\mathbf{x}^0_{\mathcal{N}_{[i]}}), \mathbf{x}^1_i) \tag{1}$$

where $\mathbf{u}^0_i$ is the input to node i, $\mathcal{F}^0_i(\cdot, \cdot, \cdot)$ is a nonlinear vector function (which may be considered a hyperbolic tangent function, or a sigmoid function), and $\mathcal{C}_0$ denotes the connections from the neighbourhood $\mathcal{N}^0_{[i]}$ and the vector $\mathbf{x}^1_i$ into the i-th node via connecting links. $\mathbf{x}^1_i$ is an n_1-dimensional vector denoting the *state*[2] of the graph which

---

[2] Here we abuse the notion of state $\mathbf{x}_1$ for simplicity sake. The state here can be the output of the graph, or the concatenation of the individual states of the nodes in the child graph.

is attached to the i-th node. In other words, here we assume an additive model, in which the (state of the) graph attached to a node i is assumed to be an additional input to the node i. The state $\mathbf{x}_i^1$ is described similarly as follows:

$$\mathbf{x}_i^1 = \mathcal{F}_i^1(\mathbf{u}_i^1, \mathcal{C}_1(\mathbf{x}_{N_{[i]}}^1), \mathbf{x}_i^2) \tag{2}$$

where $\mathbf{x}_i^1$ denotes the level 1 state[3], that is the state of the graph (or output of the graph) representing the child document of the i-th node; $\mathbf{u}_i^1$ denotes the input into the level 1 node; $\mathcal{F}_i^1$ denotes a nonlinear vector function; $\mathcal{C}_1$ denotes the connections into the node i in the child graph; and $\mathbf{x}_i^2$ denotes the state of the child graph associated with node i in level 1. From Eq. (1) and Eq. (2) the recursive nature of the approach becomes clear. The recursion stops at the maximum level of encapsulation of graphs. If we assume that there are k levels, then the k-th level will be described by the following equation:

$$\mathbf{x}_i^k = \mathcal{F}_i^k(\mathbf{u}_i^k, \mathcal{C}_k(\mathbf{x}_{N_{[i]}}^k)) \tag{3}$$

At the k-th level by assumption there will not be any inputs from the graph within the node i. Thus k denotes the terminal level; the depth of the GoG structure. Then, a mapping from the state space to the output space takes place as follows:

$$\mathbf{o}_i^0 = \mathcal{G}_i^0(\mathbf{u}_i^0, \mathcal{B}(\mathbf{x}_i^0)), \tag{4}$$

where $\mathcal{B}$ denotes the configuration of the state vector $\mathbf{x}_i^0$ with the output vector $\mathbf{o}_i^0$. The output can then be compared to an associated target value, and a gradient descent method can be applied to update the system with the aim to minimize the squared difference between network output and target values. It is noted that similar equations to Eq. (4) can be written to provide an output to any of the k levels.

Note that Eq. (4) is suitable for node focused applications. With node focused applications, a model is required to produce an output for *any* node in a graph. In contrast, graph focused applications require *one* output for each graph. In the literature, a graph focused behaviour of such systems is achieved by either selecting one node (i.e. the root node) to be representative for the graph as a whole [1,2], or by producing a consolidated mapping from all states [6]. The same principles can be applied here. It is important to note that the model at any level other than level 0 are graph focused ones since the state of the graph (as a whole) which is associated to a given node i is forwarded as an input to node i. In contrast, the model at level 0 can be either a graph focused one or a node focused one depending on the requirements of the underlying learning problem.

It is trivial to observe that the GoG model accepts the common graph model as a special case. Indeed, if $k = 1$, this collapses to the standard graph model considered in [6]. Since the graph model in [6] contains the time series as a special case (a tree), and hence one may observe that the GoG model may be considered as the most general graph model formulated to represent objects so far. An MLP can be used to compute the states in each layer. Hence, the model consists of k MLPs which have forward (and backward) connections to the MLP at the next level.

---

[3] Again we abuse the notion of state here, as it can represent either the output of the child graph associated with node i in level 0, or the concatenation of the individual states of the nodes in the child graph.

Once the GoG model is expressed in recursive form as shown in Eqs. (1) to (3), then it is quite clear how the unknown parameters[4] in the model can be trained using the standard backprop algorithm. The training algorithm can be stated as follows:

Step 0. Initialization. The parameters in the model are initialized randomly.

Step 1. From the deepest level k compute the state, and progressively compute the states in levels $k - 1, k - 2, \ldots, 0$.

Step 2. Compare the outputs at the topmost level 0 with those of the desired ones, and form an error function.

Step 3. Backprop the error from the topmost level through the levels until we reach the innermost level k and update the parameters of the model in the process.

Step 1 through to step 3 are repeated for a limited number of times, or until the sum of squared errors is below a prescribed threshold, then the algorithm stops.

This algorithm while much more complex in terms of notations and concepts, nevertheless is in the same spirit as the backprop through structure algorithm. Hence we will call this algorithm backprop through structures, though in this case, it is the levels that one is concerned with.

3   Experiments

The proposed approach is applied to a real world problem involving hyperlinked documents. The experiments are carried out as part of a participation at the INEX (INitiative for the Evaluation of XML Retrieval) competition on semi-structured document mining for the purpose of classification (the INEX 2009 competition). The dataset provided by INEX is a collection of XML formatted documents from the online encyclopedia, Wikipedia. The documents are interlinked via a xref or hyperlink structure. INEX has provided a subset of Wikipedia for the classification task. The dataset consists of $54, 889$ documents. A target label is available for $11, 028$ of these documents. Hence, these $11, 028$ documents provide a supervising signal to the training algorithm. All remaining $43, 861$ are the test documents. The task is to classify the $43, 861$ documents for which no target is available. The documents are interlinked. However, we found that only $54, 121$ documents contain outgoing links. The maximum out-degree (the maximum number of outgoing links for any one document) for this dataset is $2, 382$, the maximum in-degree is $27, 518$. the dataset contains a total of $4, 554, 203$ links. For simplicity, we removed redundant links (if a document contains several links to the same document, then we count only one such link). The removal of redundant links reduced the maximum out-degree quite significantly to 969, the maximum in-degree to $15, 027$, and the total number of links to $3, 368, 504$. Each document can also be described by the features extracted.

---

[4] Here we assume that the function $\mathcal{F}_i^k$ is characterized by a set of parameters. For example, if the encoding mechanism is a multilayer perceptron, then the unknown parameters will be the strengths of connections from the inputs to hidden layer neurons. In the case if we use the outputs then the set of parameters will include the strengths of connections from the hidden layer neurons to output neurons. For simplicity, we will assume shared weight model, i.e. all $\mathcal{F}_i^k$ in the same level k share the same set of weights.

The result is a graph whose nodes represent the XML documents, and the links represent the hyperlinks. The nodes can be labeled by a combination of the following features:

- **XML tag tree:** Each document is XML formatted. XML is a language which describes the structure of a document, and is naturally represented by a XML (parsing) tree. The tree consists of nodes which represent the XML tags, and links which represent the nesting of the tags. The nodes are labeled by an identifier which uniquely identifies the associated tag. The way to extract the XML tree from XML documents is described in [9].
- **XML tag graph:** Each node in the graph represents a unique tag within the document, and edges represent the relationship between tags. We then apply a common procedure for processing text documents called "rainbow" [12,9] on tags[5] and compute the information gain for each with respect to different categories. Top 100 tags with highest information gain were selected and attached to the nodes in the tag graph as node labels.
- **Concept Link graph:** Concept link graph [13] is a novel text representation scheme which encodes the contextual information of a document using a graph of concepts. Specifically, for a document d, it is represented as a weighted, undirected graph $d = \{N, E\}$ where N is the set of nodes representing the concepts, and E is the set of edges representing the strength of association among concepts. The ConceptLink graph extraction method is described in [13].
- **Term frequency vector:** INEX provided a Term-Frequency vector (presumably obtained by the well-known Bag-of-Words algorithm). The i-th element of the vector lists the number of occurrences of the i-th dictionary word. Hence, the vector encodes the textual content of a given document.
- **Rainbow classification results:** The well-known Rainbow software (which implements the Bag of Word algorithm) was used to classify the test documents over all categories. That is, for each category, we split the training dataset into two classes, one belonging to the category, while the other does not belong to the category. In this manner, Rainbow can produce a probability vector for each test document against all categories.

Thus, we considered to label the nodes in the hyperlink graph by either graphs describing the document of the associated node, by a vector describing the content of the same document, or both.

We extracted the XML structure for each of these documents, then attached the XML structure as a label to the associated node in the hyperlink graph. Thus, there are a total of $54,575$ XML structures. The maximum out-degree of any of the XML structures is $1,006$, and the total number of nodes is $42,668,059$. The latter is equivalent to the total number of XML tags in the dataset. We removed redundant tags by consolidating successively repeated XML tags and obtained a somewhat reduced XML graph with a maximum out-degree of 533. We found that only very few XML trees have an out-degree larger than 60. Hence, we truncated the out-degree to 60 since any algorithm driven by back-propagation discards sparse information anyway.

---

[5] Rainbow is a text classifier based on Naive Bayes approach.

Alternatively, we extract concept link graph from each document in the dataset, and used them as the bottom level graphs. Concept Link Graph is generated in three main steps [13]. First, we discover a set of concepts by clustering related words extracted from the training documents using self-organizing maps. Secondly, given the term clustering result, for each paragraph in a document, we map every existing word to a concept to which it belongs. Then we count the occurrence of every concept paragraph by paragraph. Given the paragraph-based occurrence statistics, we represent each document using a concept-paragraph matrix. Finally, singular value decomposition (SVD) is applied to the concept-paragraph matrix to compute a concept-concept association matrix. Formally speaking, given a matrix A as a concept-paragraph matrix of m concepts and n paragraphs, decomposing A using SVD returns $(U \Sigma V)^T$, where U and V are unitary matrices and Σ is a diagonal matrix with elements arranged in a descending order of magnitudes. Given the SVD result, Σ is interpreted as the "theme" matrix, where each of its diagonal elements represents the strength of its corresponding theme, U is the concept-to-theme relevance matrix and V is paragraph-to-theme relevance matrix. The concept link graph is thus obtained by computing the concept-to-concept association matrix $AA^T = U \Sigma^2 V^T$.

The (given) term frequency vector is of dimension $186,723$ which corresponds to the $186,723$ dictionary words found in the dataset. There is one term frequency vector for each document, and hence, this can also be used to label the nodes in the hyperlink graph. The term frequency vector is very sparse because not every document contains all dictionary words. We were able to reduce the dimensionality of the term frequency vector to 439 by building a matrix of category and feature (the term), then counted the number of documents which belong to a category containing the feature. We retained only those features which exhibited standard deviation of larger than 10. We also used another approach to reduce the term frequency vector: We computed the information gain for each word in the dataset with respect to the different classes, then five words with the highest information gain are selected per class. We ended up with 133 unique words, thus producing a 133 dimensional feature vector.

The classification problem is defined by 39 classes. The classes are known to represent categories to which a document belongs. This produces a 39 dimensional target vector containing binary elements. Note that the target vector may contain several non-zero elements. This indicates that the corresponding document can belong to several categories. We note also that the distribution of the different categories is not balanced (See Table 1). The largest category contains $1,337$ documents, the smallest category contains 191 documents, and the average number of documents per category is 414.

As a result, this dataset is described by a GoG which is similar as was depicted in Figure 1, and is consisting of two levels:

- Level 0: One graph per document, describing the structure, contents or other properties within the document.
- Level 1: One graph where documents are nodes, connected via links representing the hyperlinks between documents.

We trained the proposed machine learning method on the resulting GoG, and varied the labeling mechanisms as is illustrated in Table 2 to identify the impact of these features on the classification performance. We also varied the number of state neurons and

Table 1. The number of documents belonging to the 39 different categories is unbalanced

History	Space	World War I	Weather	Philosophy	Religion	Physics
534	490	262	222	275	438	271
Video games	Saints	Tropical cyclones	Politics	Cricket	Food	Trains
224	299	302	386	241	200	1337
Chess	Chemistry	Medicine	War	Literature	Baseball	Christianity
199	191	236	963	971	286	270
Astronomy	Horror	Pornography	Geography	Nautical	Catholicism	Music
216	280	220	216	232	1240	351
Science	Anarchism	Architecture	Bible	Business	Aviation	Biography
205	349	336	380	207	595	1053
Comics	Formula One	American Civil War	Pharmacy			
522	319	329	518			

Table 2. List of all input data files used for training

ID	Description
1	document-link graph. maxout=2382, maxin=27518, nodelabel=439 (reduced tfidfn vector)
2	document-link graph. maxout=2382, maxin=27518, nodelabel=133 (word counts)
3	document-link graph. maxout=969, maxin=15027, nodelabel=133
4	GoGs. Level 0: tag graphs, maxout=195, nodelabel=1 (tag id); level 1: document-link graph, maxout=969, maxin=15027, nodelabel=encoding of level 0 tag graphs.
5	GoGs. Level 0: tag graphs, maxout=52, nodelabel=39 (tag information gain); level 1: document-link graph, maxout=969, maxin=15027, nodelabel=encoding of level 0 tag graphs.
6	GoGs. Level 0: tag graphs, maxout=52, nodelabel=39 (tag information gain); level 1: document-link graph, maxout=969, maxin=15027, nodelabel=encoding of level 0 tag graphs+133(word counts).
7	GoGs. Level 0: concept link graph, maxout=41, nodelabel=51; level 1: document-link graph, maxout=969, maxin=15027, nodelabel=encoding of level 0 tag graphs+39 dimensional label from classification results of rainbow for each category respectively (NaiveBayes).
8	GoGs. Level 0: concept link graph, maxout=41, nodelabel=51; level 1: document-link graph, maxout=969, maxin=15027, nodelabel=encoding of level 0 tag graphs+39 dimensional label from classification results of rainbow for each category respectively (NaiveBayes with logarithm normalization).

hidden layer neurons to identify the impact of the number of internal network parameters on the classification performance. We also allowed the model to use some mechanisms based on a balanced training dataset in order to improve the results: Balance dataset (See Section 3.1) and label-to-out approach (See Section 3.2).

3.1 Balancing Labeled Data

The labeled data is not balanced among different categories. This is known as a *desirable problem* is machine learning due to noise tolerant abilities of such systems. In order to suitably encode a learning problem which features classes which are much smaller than other classes, then counter measures need to be taken so as to avoid that the network dismisses small classes as *noise*. In order to avoid this, it is useful to balance the distribution of train samples. Since this is a multiple categories task, the traditional methods for balancing data is not applicable. Instead of complementing the number of samples from smaller classes, we modified the error back-propagation algorithm by altering the error which is propagated back during training as follows: For each train sample, the network produced an output vector which has dimension as the number of categories available, and each element in the vector represents the output for corresponding category. Originally for each element i in the vector, the error $\varepsilon_i = o_i - t_i$ is computed. In order to balance train samples from category i, ε_i can be revised by using number of negative samples N_n and number of positive samples N_p for category i, there are two alternative ways which have the same effect on the training algorithm:

1. Method 1: if $t_i = 1$, then $\varepsilon_i = (o_i - t_i) \times N_n$; if $t_i = 0$, then $\varepsilon_i = (o_i - t_i) \times N_p$
2. Method 2: if $t_i = 1$, then $\varepsilon_i = (o_i - t_i) \div N_p$; if $t_i = 0$, then $\varepsilon_i = (o_i - t_i) \div N_n$

3.2 Label-to-Out Approach

We further considered a slight modification of the proposed algorithm, namely the addition of direct links between node label and the output layer[6]. This additional layer of network weights allows the treatment of labels as independent vectors. The effect is a bias input to the output layer, one for each node in the graph. Note that this approach only affects the level 1 graph when the nodes have been produced by rainbow. By using this label-to-out approach, we anticipate that the difficulty of the training task will become simpler to encode by the network. The assumed was confirmed by experimental results. The label-to-out approach and the GoGs of concept link graphs (shown as experiment 7 in Table 3) allowed us to obtain the best results so far: MAP= 0.68.

3.3 Results

Training performance is evaluated according to two measures: the MAP (mean of the average precision) with respect to the document and the Macro F_1 score.

- MAP, the mean of the average precision with respect to each document. This measure is used to evaluate whether the system is capable of retrieving highly relevant categories first. For each document i, we obtain a list of relevant categories by sorting the scores in an ascending order of magnitudes, and then compute:

$$AvgP_i = \frac{\sum_{r=1}^{N}(P(r) \times rel(r))}{N_{relevant}}, \tag{5}$$

---

[6] A direct feedforward input to the output can improve the overall performance of the multilayer perceptron [14]. In the case of a direct state to output connection as suggested here, there is yet a formal proof that this will produce improved results.

Table 3. Results by using different training configuration

Training Configuration	MAP mean	PRF macro	micro	ROC macro	micro	ACC macro	micro
1 graph 6; state=20, hidden=15, outhidden=20, output=10; rprop; seed=45	0.125	0.07	0.10	0.59	0.59	0.03	0.05
2 graph 6; state=20, hidden=15, outhidden=20, output=10; weight control; rprop; seed=45	0.138	0.19	0.21	0.62	0.61	0.91	0.89
3 graph 6; state=15, hidden=10, outhidden=15, output=5; weight control; seed=3	0.129	0.06	0.10	0.51	0.51	0.11	0.12
4 graph 3; state=10, hidden=8, outhidden=6, output=39; balance method 1; seed=37	0.10	0.07	0.10	0.55	0.56	0.13	0.16
5 graph 3; state=30, hidden=20, outhidden=15, output=39; balance method 1; seed=1	0.208	0.07	0.10	0.63	0.62	0.03	0.05
6 graph 3; state=10, hidden=8, outhidden=6, output=39; balance method 2; seed=9	0.192	0.06	0.09	0.50	0.51	0.13	0.14
7 graph 8; state=6, hidden=8, outhidden=10, output=15; jacobian control; seed=91; label-to-out	0.68	0.48	0.51	0.83	0.85	0.95	0.93

where N is the total number of categories, $rel(r)$ is a binary function which indicates whether the results is relevant at position r, and $P(r)$ the precision at a given cut-off rank r. MAP can be in turn computed as:

$$MAP = \frac{\sum_i AvgP_i}{n}, \tag{6}$$

where n is the total number of documents evaluated.

- Macro F-measure score, which is the evenly weighted precision and recall rates. For each category c, we computed:

$$F_c = \frac{2 \times (P_c \times R_c)}{P_c + R_c} \tag{7}$$

then we averaged the F-measure scores for all categories and obtain the Macro F-measure scores.

A range of network architectures and training parameters were tried on this training task. A selection of these are given by Table 3. The main observations are as follows:

- Using label-to-out approach has improved the performance significantly. This indicates that the additional bias is effective in simplifying the given learning task. The simplification arises out of the fact that some features can influence the network output directly without having to travel through the relatively deep network architecture (the unfolded iteration network).
- Larger network is required to produce better results. The more neurons a network features, the more parameters are available to encode a given learning problem. The need for relatively large number of parameters indicates that given learning problem is non-trivial.

Table 4. INEX2009 XML classification task results

Submission	MAP	PRF		ROC		ACC	
	mean	macro	micro	macro	micro	macro	micro
University of Wollongong	0.681	0.479	0.513	0.829	0.849	0.947	0.925
University of Peking	0.702	0.48	0.518	0.842	0.85	0.962	0.948
Xerox Research Center	0.678	0.571	0.6	0.748	0.765	0.974	0.963
University of Saint Etienne	0.788	0.53	0.564	0.937	0.935	0.974	0.962
University of Granada	0.642	0.50	0.53	0.802	0.819	0.952	0.933

Table 5. Best classes

Category	Correct Counts	Total Counts	Percentage
Portal:Pharmacy and Pharmacology	2490	2630	0.9468
Portal:Comics	2343	2583	0.9071
Portal:War	4193	4712	0.8899
Portal:Literature	4051	4688	0.8641
Portal:Trains	5487	6352	0.8638

Table 6. Worst classes

Category	Correct Counts	Total Counts	Percentage
Portal:Pornography	477	1041	0.4582
Portal:Science	524	1101	0.4759
Portal:Nautical	570	1182	0.4822
Portal:Chess	517	1010	0.5119
Portal:Architecture	949	1777	0.534

- Weight control through a Jacobian control mechanism helped to produce better results. This weight control mechanism can aid the training procedure by restricting the movement of weight changes such that the size of weight adjustments remains within a limited range. This aids the stability of the weights during the training, and hence, can result in a general improvement in the quality of the training procedure.

A comparison with other approaches submitted by other groups to this classification problem is given in Table 4. In this comparison, it can be seen that the proposed GoGs approach produced a very reasonable performance. We are in fact very please by these results since these are preliminary results on a system which is still under development.

Table 5 and Table 6 list the categories for which we obtained the best classification result and worst classification result respectively. It can be seen that the better results are generally be obtained for larger classes. Given that we used ways to balance the dataset, and hence, this indicates that the smaller classes in the training set do not sufficiently well cover the problem domain.

4   Conclusions

In this paper, we have deployed a new idea, the graph of graphs model in modeling the connections among the linked documents in the XML dataset. This is quite a novel idea in that it allows us to extend the idea first discussed in graph neural networks [7] which can only be used to describe a single document. A GoG model can be used to model the linked set of documents by considering each document as a graph, and the linkages among the graphs are modeled as links which connecting the documents. Using an extension of the back propagation through structure algorithm, we were able to model the set of documents given in the training dataset as GoG models. This GoG model when combined with two sets of mechanisms: a balanced training dataset, and a label-to-out approach produced very reasonable performance when compared to those obtained by other research groups.

For future work, it is possible to vary some of the parameters in the GoG model, especially the number of state neurons in the model. This state information captures the past information in the training process. A larger number of state neurons will provide a richer model for the information contained in the documents, while a small number of state neurons will compress the information contained in the documents.

Acknowledgment. This work received financial support from the Australian Research Council through Discovery Project grant DP0774168 (2007 - 2009).

References

[1] Gori, M., Küchler, A., Soda, G.: On the implementation of frontier-to-root tree automata in recursive neural networks. IEEE Transactions on Neural Networks 10(6), 1305–1314 (1999)

[2] Frasconi, P., Francesconi, E., Gori, M., Marinai, S., Sheng, J., Soda, G.: Logo recognition by recursive neural networks. In: Kasturi, R., Tombre, K. (eds.) GREC 1997. LNCS, vol. 1389, pp. 104–117. Springer, Heidelberg (1998)

[3] Sperduti, A., Majidi, D., Starita, A.: Extended cascade-correlation for syntactic and structural pattern recognition. In: Perner, P., Rosenfeld, A., Wang, P. (eds.) SSPR 1996. LNCS, vol. 1121, pp. 90–99. Springer, Heidelberg (1996)

[4] Hammer, B., Micheli, A., Sperduti, A.: Universal approximation capability of cascade correlation for structures. Neural Computation 17(5), 1109–1159 (2005)

[5] Bianucci, A., Micheli, A., Sperduti, A., Starita, A.: Quantitative structure-activity relationships of benzodiazepines by recursive cascade correlation. In: IEEE Processing of IJCNN 1998- IEEE World Congress on Computational Intelligence, Anchorage, Alaska, May 1998, pp. 117–122 (1998)

[6] Scarselli, F., Gori, M., Tsoi, A., Hagenbuchner, M., Monfardini, G.: Computational capabilities of graph neural networks. IEEE Transactions on Neural Networks 20(1), 81–102 (2009)

[7] Yong, S., Hagenbuchner, M., Tsoi, A., Scarselli, F., Gori, M.: Document mining using graph neural network. In: Fuhr, N., Lalmas, M., Trotman, A. (eds.) INEX 2006. LNCS, vol. 4518, pp. 458–472. Springer, Heidelberg (2007)

[8] KC, M., Hagenbuchner, M., Tsoi, A., Scarselli, F., Gori, M., Sperduti, S.: Xml document mining using contextual self-organizing maps for structures. LNCS. Springer, Heidelberg (2007)

[9] Hagenbuchner, M., Sperduti, A., Tsoi, A.: A self-organizing map for adaptive processing of structured data. IEEE Transactions on Neural Networks 14(3), 491–505 (2003)

[10] Zhang, S., Hagenbuchner, M., Tsoi, A., Sperduti, A.: Self Organizing Maps for the clustering of large sets of labeled graphs. Springer, Heidelberg (2009)

[11] Mihalcea, R., Tarau, P.: TextRank: Bringing order into texts. In: Proceedings of EMNLP-04 and the 2004 Conference on Empirical Methods in Natural Language Processing (July 2004)

[12] The rainbow software package,
`http://www.cs.umass.edu/~mccallum/bow/rainbow/`

[13] Chau, R., Tsoi, A.C., Hagenbuchner, M., Lee, V.: A conceptlink graph for text structure mining. In: Mans, B. (ed.) Thirty-Second Australasian Computer Science Conference (ACSC 2009), Wellington, New Zealand. CRPIT, vol. 91, pp. 129–137. ACS (2009)

[14] Sontag, E.: Capabilities and training of feedforward nets. In: Neural networks, New Brunswick, NJ, Boston, MA. Academic Press, London (1990)

Erratum: Overview of the INEX 2009 Ad Hoc Track

Shlomo Geva[1], Jaap Kamps[2], Miro Lethonen[3],
Ralf Schenkel[4], James A. Thom[5], and Andrew Trotman[6]

[1] Queensland University of Technology, Brisbane, Australia
s.geva@qut.edu.au
[2] University of Amsterdam, Amsterdam, The Netherlands
kamps@uva.nl
[3] University of Helsinki, Helsinki, Finland
miro.lehtonen@helsinki.fi
[4] Max-Planck-Institut für Informatik, Saarbrücken, Germany
schenkel@mpi-sb.mpg.de
[5] RMIT University, Melbourne, Australia
james.thom@rmit.edu.au
[6] University of Otago, Dunedin, New Zealand
andrew@cs.otago.ac.nz

S. Geva, J. Kamps, and A. Trotman (Eds.): INEX 2009, LNCS 6203, pp. 4–25, 2010.
© Springer-Verlag Berlin Heidelberg 2010

DOI 10.1007/978-3-642-14556-8_46

In the original version, the name of the third author was spelled incorrectly by mistake. It should be Miro Lehtonen.

The original online version for this chapter can be found at
http://dx.doi.org/10.1007/978-3-642-14556-8_4

Author Index